STUDENT SOLUTIONS MANUAL AND STUDY GUIDE

FOR

DISCRETE MATHEMATICS WITH APPLICATIONS

5th EDITION

Susanna S. Epp
DePaul University

with assistance from
Tom Jenkyns
Brock University

Australia · Brazil · Mexico · Singapore · United Kingdom · United States

For product information and technology assistance, contact us at **Cengage Customer & Sales Support, 1-800-354-9706 or support.cengage.com.**

For permission to use material from this text or product, submit all requests online at **www.cengage.com/permissions**.

ISBN: 978-0-357-03520-7

Cengage
20 Channel Center Street
Boston, MA 02210
USA

Cengage is a leading provider of customized learning solutions with employees residing in nearly 40 different countries and sales in more than125 countries around the world. Find your local representative at: **www.cengage.com**.

Cengage products are represented in Canada by Nelson Education, Ltd.

To learn more about Cengage platforms and services, register or access your online learning solution, or purchase materials for your course, visit **www.cengage.com**.

Printed in the United States of America
Print Number: 02   Print Year: 2022

Table of Contents

Note to Students

This *Student Solutions Manual and Study Guide* for the fourth edition of *Discrete Mathematics with Applications* contains complete solutions to every third exercise in the text that is not fully answered in Appendix B. It also contains additional explanation, and review material.

The specific topics developed in the book provide a foundation for virtually every other mathematics or computer science subject you might study in the future. Perhaps more important, however, is the book's recurring focus on the logical principles used in mathematical reasoning and the techniques of mathematical proof.

Why should it matter for you to learn these things? The main reason is that these principles and techniques are the foundation for all kinds of careful analyses, whether of mathematical statements, computer programs, legal documents, or other technical writing. A person who understands and knows how to develop basic mathematical proofs has learned to think in a highly disciplined way, is able to deduce correct consequences from a few basic principles, can build a logically connected chain of statements and appreciates the need for giving a valid reason for each statement in the chain, is able to move flexibly between abstract symbols and concrete objects, and can deal with multiple levels of abstraction. Mastery of these skills opens a host of interesting and rewarding possibilities in a person's life.

In studying the subject matter of this book, you are embarking on an exciting and challenging adventure. I wish you much success!

Acknowledgements

I am very indebted to the work of Tom Jenkyns, whose attention to detail, mathematical knowledge, and understanding of language made a valuable contribution to previous editions of this manual. I am also grateful to my daughter, Caroline Epp, and to my husband, Helmut Epp, for constructing the diagrams, and most especially to my husband, for his support and wise counsel over many years.

Susanna S. Epp

Chapter 1: Speaking Mathematically

The aim of this chapter is to provide some of the basic terminology that is used throughout the book. Section 1.1 introduces special terms that are used to describe aspects of mathematical thinking and contains exercises to help you learn how to express mathematical statements in a variety of ways. To be successful in mathematics, it is important to be able comfortably to translate from formal to informal and from informal to formal modes of expression. Sections 1.2 and 1.3 introduce the basic notions of sets, relations, and functions, and Section 1.4 is designed to acquaint you with the special way the word graph is used in discrete mathematics. The section also includes applications showing how graph models can be used to solve some significant problems.

Section 1.1

1. **a.** $x^2 = -1$ (*Or*: the square of x is -1)   **b.** a real number x

3. **a.** between a and b **b.** distinct real numbers a and b; there is a real number c

5. **a.** r is positive

   b. positive; the reciprocal of r is positive (*Or*: positive; $1/r$ is positive)

   c. is positive; $1/r$ is positive (*Or*: is positive; the reciprocal of r is positive)

6. **a.** s is negative   **b.** negative; the cube root of s is negative (*Or*: $\sqrt[3]{s}$ is negative)

   c. is negative; $\sqrt[3]{s}$ is negative (*Or*: the cube root of s is negative)

7. **a.** There are real numbers whose sum is less than their difference. True. For example, $1 + (-1) = 0, 1 - (-1) = 1 + 1 = 2$, and $0 < 2$.

   c. The square of each positive integer is greater than or equal to the integer.

   True. If n is any positive integer, then $n \geq 1$. Multiplying both sides by the positive number n does not change the direction of the inequality (see Appendix A, T20), and so $n^2 \geq n$.

8. **a.** have four sides   **b.** has four sides   **c.** has four sides   **d.** is a square; has four sides

   e. J has four sides

9. **a.** have at most two real solutions   **b.** has at most two real solutions   **c.** has at most two real solutions   **d.** is a quadratic equation; has at most two real solutions   **e.** E has at most two real solutions

10. **a.** have reciprocals   **b.** a reciprocal   **c.** s is a reciprocal for r

12. **a.** real number; product with every number leaves the number unchanged

   b. a positive square root   **c.** $rs = s$

Section 1.2

1. $A = C$ and $B = D$

2. **a.** The set of all positive real numbers x such that 0 is less than x and x is less than 1

   c. The set of all integers n such that n is a factor of 6

3. **a.** No, $\{4\}$ is a set with one element, namely 4, whereas 4 is just a symbol that represents the number 4

   b. Three: the elements of the set are 3, 4, and 5.

   c. Three: the elements are the symbol 1, the set $\{1\}$, and the set $\{1, \{1\}\}$

5. *Hint:* **R** is the set of all real numbers, **Z** is the set of all integers, and $\mathbf{Z}^+$ is the set of all positive integers.

6. T_2 and T_{-3} each have two elements, and T_0 and T_1 each have one element.

   Justification: $T_2 = \{2, 2^2\} = \{2, 4\}$, $T_{-3} = \{-3, (-3)^2\} = \{-3, 9\}$,
   $T_1 = \{1, 1^2\} = \{1, 1\} = \{1\}$, and $T_0 = \{0, 0^2\} = \{0, 0\} = \{0\}$.

7. **a.** $\{1, -1\}$ **c.** the set has no elements   **d.** **Z** (every integer is in the set)

8. **a.** No, $B \not\subseteq A$ because $j \in B$ and $j \notin A$

   d. Yes, C is a proper subset of A. Both elements of C are in A, but A contains elements (namely c and f) that are not in C.

9. **a.** Yes

   b. No, the number 1 is not a set and so it cannot be a subset.

   c. No: The only elements in $\{1, 2\}$ are 1 and 2, and $\{2\}$ is not equal to either of these.

   d. Yes: $\{3\}$ is one of the elements listed in $\{1, \{2\}, \{3\}\}$.

   e. Yes: $\{1\}$ is the set whose only element is 1.

   f. No, the only element in $\{2\}$ is the number 2 and the number 2 is not one of the three elements in $\{1, \{2\}, \{3\}\}$.

   g. Yes: The only element in $\{1\}$ is 1, and 1 is an element in $\{1, 2\}$.

   h. No: The only elements in $\{\{1\}, 2\}$ are $\{1\}$ and 2, and 1 is not equal to either of these.

   i. Yes, the only element in $\{1\}$ is the number 1, which is an element in $\{1, \{2\}\}$.

   j. Yes: The only element in $\{1\}$ is 1, which is is an element in $\{1\}$. So every element in $\{1\}$ is in $\{1\}$.

10. **a.** No. Observe that $(-2)^2 = (-2)(-2) = 4$, whereas $-2^2 = -(2^2) = -4$. So $((-2)^2, -2^2) = (4, -4)$, whereas $(-2^2, (-2)^2) = (-4, 4)$. And $(4, -4) \neq (-4, 4)$ because $-4 \neq 4$.

    c. Yes. Note that $8 - 9 = -1$ and $\sqrt[3]{-1} = -1$, and so $(8 - 9, \sqrt[3]{-1}) = (-1, -1)$.

11. **a.** $\{(w, a), (w, b), (x, a), (x, b), (y, a), (y, b), (z, a), (z, b)\}$ $A \times B$ has $4 \cdot 2 = 8$ elements.

    b. $\{(a, w), (b, w), (a, x), (b, x), (a, y), (b, y), (a, z), (b, z)\}$ $B \times A$ has $4 \cdot 2 = 8$ elements.

    c. $\{(w, w), (w, x), (w, y), (w, z), (x, w), (x, x), (x, y), (x, z), (y, w), (y, x), (y, y),$
    $(y, z), (z, w), (z, x), (z, y), (z, z)\}$ $A \times A$ has $4 \cdot 4 = 16$ elements.

    d. $\{(a, a), (a, b), (b, a), (b, b)\}$ $B \times B$ has $2 \cdot 2 = 4$ elements.

12. All four sets have nine elements.

    a. $S \times T = \{(2, 1), (2, 3), (2, 5), (4, 1), (4, 3), (4, 5), (6, 1), (6, 3), (6, 5)\}$

    b. $T \times S = \{(1, 2), (3, 2), (5, 2), (1, 4), (3, 4), (5, 4), (1, 6), (3, 6), (5, 6)\}$

    c. $S \times S = \{(2, 2), (2, 4), (2, 6), (4, 2), (4, 4), (4, 6), (6, 2), (6, 4), (6, 6)\}$

    d. $T \times T = \{(1, 1), (1, 3), (1, 5), (3, 1), (3, 3), (3, 5), (5, 1), (5, 3), (5, 5)\}$

13. **a.** $A \times (B \times C) = \{(1, (u, m)), (1, (u, n)), (2, (u, m)), (2, (u, n)), (3, (u, m)), (3, (u, n))\}$

    b. $(A \times B) \times C = \{((1, u), m), ((1, u), n), ((2, u), m), ((2, u), n), ((3, u), m), ((3, u), n)\}$

    c. $A \times B \times C = \{(1, u, m), (1, u, n), (2, u, m), (2, u, n), (3, u, m), (3, u, n)\}$

15. 0000, 0001, 0010, 0100, 1000

Section 1.3

1. **a.** No. Yes. No. Yes.

   b. $R = \{(2,6), (2,8), (2,10), (3,6), (4,8)\}$

   c. Domain of $R = A = \{2,3,4\}$, co-domain of $R = B = \{6,8,10\}$

   d.

   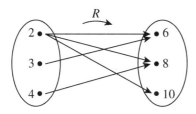

3. **a.** $3\ T\ 0$ because $\frac{3-0}{3} = \frac{3}{3} = 1$, which is an integer.

   $1\ \not{T}\ (-1)$ because $\frac{1-(-1)}{3} = \frac{2}{3}$, which is not an integer.

   $(2,-1) \in T$ because $\frac{2-(-1)}{3} = \frac{3}{3} = 1$, which is an integer.

   $(3,-2) \notin T$ because $\frac{3-(-2)}{3} = \frac{5}{3}$, which is not an integer.

   b. $T = \{(1,-2), (2,-1), (3,0)\}$

   c. Domain of $T = E = \{1,2,3\}$, co-domain of $T = F = \{-2,-1,0\}$

   d.

   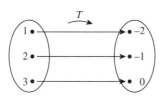

5. **a.** $(2,1) \in S$ because $2 \geq 1$. $(2,2) \in S$ because $2 \geq 2$.
   $2\ \not{S}\ 3$ because $2 \not\geq 3$. $(-1)S(-2)$ because $-1 \geq -2$.

   b.

   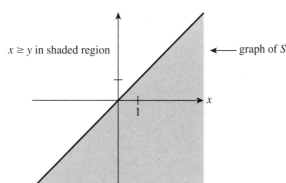

6. **a.** $(2,4) \in R$ because $4 = 2^2$.

   $(4,2) \notin R$ because $2 \neq 4^2$.

   $(-3,9) \in R$ because $9 = (-3)^2$.

   $(9,-3) \notin R$ because $-3 \neq 9^2$.

b.

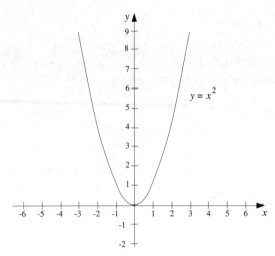

$y = x^2$

7. a.

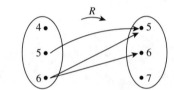

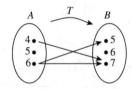

b. R is not a function because it satisfies neither property (1) nor property (2) of the definition. It fails property (1) because $(4, y) \notin R$, for any y in B. It fails property (2) because $(6, 5) \in R$ and $(6, 6) \in R$ and $5 \neq 6$.

S is not a function because $(5, 5) \in S$ and $(5, 7) \in S$ and $5 \neq 7$. So S does not satisfy property (2) of the definition of function.

T is not a function both because $(5, x) \notin T$ for any x in B and because $(6, 5) \in T$ and $(6, 7) \in T$ and $5 \neq 7$. So T does not satisfy either property (1) or property (2) of the definition of function.

9. a. There is only one: $\{(0, 1), (1, 1)\}$

  b. $\{(0, 1)\}, \{(1, 1)\}$

11. $L(0201) = 4, L(12) = 2$

12. $C(x) = yx, C(yyxyx) = yyyxyx$

13. a. Domain $= A = \{-1, 0, 1\}$, co-domain $= B = \{t, u, v, w\}$

  b. $F(-1) = u, F(0) = w, F(1) = u$

15. a. This diagram does not determine a function because 2 is related to both 2 and 6.

  b. This diagram does not determine a function because 5 is in the domain but it is not related to any element in the co-domain.

  c. This diagram does not determine a function because 4 is related to both 1 and 2, which violates property (2) of the definition of function.

  d. This diagram defines a function; both properties (1) and (2) are satisfied.

  e. This diagram does not determine a function because 2 is in the domain but it is not related to any element in the co-domain.

16. $f(-1) = (-1)^2 = 1, f(0) = 0^2 = 0, f\left(\frac{1}{2}\right) = \left(\frac{1}{2}\right)^2 = \frac{1}{4}$.

18. $h\left(-\frac{12}{5}\right) = h\left(\frac{0}{1}\right) = h\left(\frac{9}{17}\right) = 2$

19. For each $x \in \mathbf{R}$, $g(x) = \dfrac{2x^3 + 2x}{x^2 + 1} = \dfrac{2x(x^2 + 1)}{x^2 + 1} = 2x = f(x)$. Therefore, by definition of equality of functions, $f = g$.

Section 1.4

1. $V(G) = \{v_1, v_2, v_3, v_4\}, E(G) = \{e_1, e_2, e_3\}$

   Edge-endpoint function:

Edge	Endpoints
e_1	$\{v_1, v_2\}$
e_2	$\{v_1, v_3\}$
e_3	$\{v_3\}$

3.

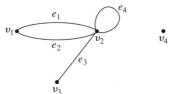

5. Imagine that the edges are strings and the vertices are knots. You can pick up the left-hand figure and lay it down again to form the right-hand figure as shown below.

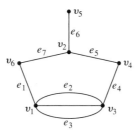

6.

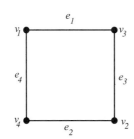

8. (i) e_1, e_2, e_7 are incident on v_1.

   (ii) v_1, v_2, and v_3 are adjacent to v_3.

   (iii) e_2, e_8, e_9, and e_3 are adjacent to e_1.

   (iv) Loops are e_6 and e_7.

   (v) e_8 and e_9 are parallel; e_4 and e_5 are parallel.

   (vi) v_6 is an isolated vertex.

   (vii) degree of $v_3 = 5$

9. (i) e_1, e_2, e_7 are incident on v_1.

(ii) v_1 and v_2 are adjacent to v_3.

(iii) e_2 and e_7 are adjacent to e_1.

(iv) e_1 and e_3 are loops.

(v) e_4 and e_5 are parallel.

(vi) v_4 is an isolated vertex.

(vii) degree of $v_3 = 2$

10. **a.** Yes. According to the graph, *Sports Illustrated* is an instance of a sports magazine, a sports magazine is a periodical, and a periodical contains printed writing.

12. To solve this puzzle using a graph, introduce a notation in which, for example, *wc/fg* means that the wolf and the cabbage are on the left bank of the river and the ferryman and the goat are on the right bank. Then draw those arrangements of wolf, cabbage, goat, and ferryman that can be reached from the initial arrangement (*wgcf/*) and that are not arrangements to be avoided (such as (*wg/fc*)). At each stage ask yourself, "Where can I go from here?" and draw lines or arrows pointing to those arrangements. This method gives the graph shown below.

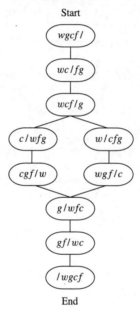

Start

wgcf/

wc/fg

wcf/g

c/wfg   *w/cfg*

cgf/w   *wgf/c*

g/wfc

gf/wc

/wgcf

End

Examining the diagram reveals the solutions

$(wgcf/) \rightarrow (wc/gf) \rightarrow (wcf/g) \rightarrow (w/gcf) \rightarrow (wgf/c) \rightarrow (g/wcf) \rightarrow (gf/wc) \rightarrow (/wgcf)$

and

$(wgcf/) \rightarrow (wc/gf) \rightarrow (wcf/g) \rightarrow (c/wgf) \rightarrow (gcf/w) \rightarrow (g/wcf) \rightarrow (gf/wc) \rightarrow (/wgcf)$

14. *Hint:* The answer is yes. Represent possible amounts of water in jugs A and B by ordered pairs. For instance, the ordered pair $(1, 3)$ would indicate that there is one quart of water in jug A and three quarts in jug B. Starting with $(0, 0)$, draw arrows from one ordered pair to another if it is possible to go from the situation represented by one pair to that represented by the other by either filling a jug, emptying a jug, or transferring water from one jug to another. You need only draw arrows from states that have arrows pointing to them; the other states cannot be reached. Then find a directed path (sequence of directed edges) from the initial state $(0, 0)$ to a final state $(1, 0)$ or $(0, 1)$.

15.

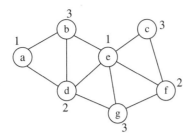

Vertex e has maximal degree, so color it with color #1. Vertex a does not share an edge with e, and so color #1 may also be used for it. From the remaining uncolored vertices, all of d, g, and f have maximal degree. Choose any one of them—say, d—and use color #2 for it. Observe that vertices c and f do not share an edge with d, but they do share an edge with each other, which means that color #2 may be used for one but not the other. Choose to color f with color #2 because the degree of f is greater than the degree of c. The remaining uncolored vertices, b, c, and g, are unconnected, and so color #3 may be used for all three.

16. *Hint:* There are two solutions:

(1) Time 1: hiring, library
Time 2: personnel, undergraduate education, colloquium
Time 3: graduate education

(2) Time 1: hiring, library
Time 2: graduate education, colloquium
Time 3: personnel, undergraduate education

Review Guide: Chapter 1

Variables and Mathematical Statements

- What are the two main ways variables are used?
- What is a universal statement? Give one example.
- What is a conditional statement? Give one example.
- What is an existential statement Give one example.
- Give an example of a universal conditional statement.
- Give an example of a universal existential statement.
- Give an example of an existential universal statement.

Sets

- What does the notation $x \in A$ mean?
- What does the notation $x \notin A$ mean?
- How is the set-roster notation used to define a set?
- What is the axiom of extension?
- What do the symbols $\mathbf{R}$, $\mathbf{Z}$, and $\mathbf{Q}$ stand for?
- What is the set builder notation?
- If S is a set and $P(x)$ is a property that elements may or may not satisfy, how should the following be read out loud: $\{x \in S \mid P(x)\}$?

Subsets

- If A and B are sets, what does it mean for A to be a subset of B? What is the notation that indicates that A is a subset of B?
- What does the notation $A \nsubseteq B$ mean?
- What does it mean for one set to be a proper subset of another?
- How are the symbols $\subseteq$ and $\in$ different from each other?

Cartesian Products

- What does it mean for an ordered pair (a, b) to equal an ordered pair (c, d)?
- What is the Cartesian product of sets $A_1, A_2, A_3, \ldots, A_n$ and what is the notation for it?
- What is the Cartesian plane?
- What is a string of length n over a set A?
- What is the null string?
- What is a bit string?

Relations

- What is a relation from a set A to a set B?
- If R is a relation from A to B, what is the domain of R?
- If R is a relation from A to B, what is the co-domain of R?
- If R is a relation from A to B, what does the notation $x \, R \, y$ mean?
- How should the following notation be read: $x \, \cancel{R} \, y$?
- How is the arrow diagram for a relation drawn?

Functions

- What is a function F from a set A to a set B?
- What are less formal/more formal ways to state the two properties a function F must satisfy?
- Given a function F from a set A to a set B and an element x in A, what is $F(x)$?
- What is the squaring function from $\mathbf{R}$ to $\mathbf{R}$?
- What is the successor function from $\mathbf{Z}$ to $\mathbf{Z}$?
- Give an example of a constant function.
- What is the difference between the notations f and $f(x)$?
- If f and g are functions from A to B, what does it mean for f to equal g?

Graphs

- How are the following terms defined: graph, edge-endpoint function, loop in a graph, parallel edges, adjacent edges, isolated vertex, edge incident on an endpoint, directed graph, degree of a vertex in a graph?
- How can a graph be used to describe a network, represent knowledge, solve a puzzle, color a map, or schedule a meeting?
- What is the four-color theorem?

Fill-in-the-Blank Review Questions

Section 1.1

1. A universal statement asserts that a certain property is ____ for ____.
2. A conditional statement asserts that if one thing is ____ then some other thing ____.
3. Given a property that may or may not be true, an existential statement asserts that ____ for which the property is true.

Section 1.2

1. When the elements of a set are given using the set-roster notation, the order in which they are listed ____.
2. The symbol $\mathbf{R}$ denotes ____.
3. The symbol $\mathbf{Z}$ denotes ____.
4. The symbol $\mathbf{Q}$ denotes ____.
5. The notation $\{x \mid P(x)\}$ is read ____.
6. For a set A to be a subset of a set B means that ____.
7. Given sets A and B, the Cartesian product $A \times B$ is ____.
8. Given sets A, B, and C, the Cartesian product $A \times B \times C$ is ____.
9. A string of length n over a set S is an ordered n-tuple of elements of S, written without ____ or ____.

Section 1.3

1. Given sets A and B, a relation from A to B is ____.
2. A function F from A to B is a relation from A to B that satisfies the following two properties:
   a. for every element x of A, there is ____
   b. for all elements x in A and y and z in B, if ____ then ____.

3. If F is a function from A to B and x is an element of A, then $F(x)$ is _____.

Section 1.4

1. A graph consists of two finite sets: _____ and _____, where each edge is associated with a set consisting of _____.
2. A loop in a graph is _____.
3. Two distinct edges in a graph are parallel if, and only if, _____.
4. Two vertices are called adjacent if, and only if, _____.
5. An edge is incident on _____.
6. Two edges incident on the same endpoint are _____.
7. A vertex on which no edges are incident is _____.
8. In a directed graph, each edge is associated with _____.
9. The degree of a vertex in a graph is _____.

Answers for Fill-in-the-Blank Review Questions

Section 1.1

1. true; all elements of a set
2. is true; also has to be true
3. there is at least one thing

Section 1.2

1. does not matter
2. the set of all real numbers
3. the set of all integers
4. the set of all rational numbers
5. the set of all x such that $P(x)$
6. every element in A is an element in B
7. the set of all ordered pairs (a, b) where a is in A and b is in B
8. the set of all ordered triples of the form (a, b, c) where a is in A, b is in B, and c is in C
9. parentheses; commas

Section 1.3

1. a subset of the Cartesian produce $A \times B$.
2. **a.** an element y of B such that $(x, y) \in F$ (i.e., such that x is related to y by F)
   b. $(x, y) \in F$ and $(x, z) \in F$; $y = z$
3. the unique element of B that is related to x by F

Section 1.4

1. a finite, nonempty set of vertices; a finite set of edges; one or two vertices called its endpoints
2. an edge with a single endpoint
3. they have the same set of endpoints
4. they are connected by an edge
5. each of its endpoints
6. adjacent
7. isolated
8. an ordered pair of vertices called its endpoints
9. the number of edges that are incident on the vertex, with an edge that is a loop counted twice

Chapter 2: The Logic of Compound Statements

The ability to reason using the principles of logic is essential for solving problems in abstract mathematics and computer science and for understanding the reasoning used in mathematical proof and disproof. In Chapters 2 and 3 the various rules used in logical reasoning are developed both symbolically and in the context of their use in everyday language.

Exercise sets for Sections 2.1–2.3 and 3.1–3.4 contain sentences for you to negate, write the contrapositive for, and so forth. These are designed to help you learn to incorporate the rules of logic into your general reasoning processes. Chapters 2 and 3 also present the rudiments of symbolic logic as a foundation for a variety of upper-division courses. Symbolic logic is used in, among others, the study of digital logic circuits, relational databases, artificial intelligence, and program verification.

Section 2.1

1. Common form: If p then q.
    $$p$$
    Therefore, q

    $(a + 2b)(a^2 - b)$ can be written in prefix notation. All algebraic expressions can be written in prefix notation.

3. Common form: $p \lor q$
    $$\sim p$$
    Therefore, q

    My mind is shot. Logic is confusing.

5. **a.** It is a statement because it is a true sentence. 1,024 is a perfect square because $1{,}024 = 32^2$, and the next smaller perfect square is $31^2 = 961$, which has fewer than four digits.

    b. The truth or falsity of this sentence depends on the reference for the pronoun "she." Considered on its own, the sentence cannot be said to be either true or false, and so it is not a statement.

    c. This sentence is false; hence it is a statement.

    d. This is not a statement because its truth or falsity depends on the value of x.

6. **a.** $s \land i$ **b.** $\sim s \land \sim i$

8. **a.** $(h \land w) \land \sim s$ **d.** $(\sim w \land \sim s) \land h$

9. **a.** $p \lor q$ **b.** $r \land p$ **c.** $r \land (p \lor q)$

10. **a.** $p \land q \land r$ **c.** $p \land (\sim q \lor \sim r)$

11. Inclusive or. For instance, a team could win the playoff by winning games 1, 3, and 4 and losing game 2. Such an outcome would satisfy both conditions.

12.

p	q	$\sim p$	$\sim p \land q$
T	T	F	F
T	F	F	F
F	T	T	T
F	F	T	F

14.

p	q	r	q ∧ r	p ∧ (q ∧ r)
T	T	T	T	T
T	T	F	F	F
T	F	T	F	F
T	F	F	F	F
F	T	T	T	F
F	T	F	F	F
F	F	T	F	F
F	F	F	F	F

15.

p	q	r	∼ q	∼ q ∨ r	p ∧ (∼ q ∨ r)
T	T	T	F	T	T
T	T	F	F	F	F
T	F	T	T	T	T
T	F	F	T	T	T
F	T	T	F	T	F
F	T	F	F	F	F
F	F	T	T	T	F
F	F	F	T	T	F

16.

p	q	p ∧ q	p ∨ (p ∧ q)	p
T	T	T	T	T
T	F	F	T	T
F	T	F	F	F
F	F	F	F	F

same truth values

The truth table shows that $p \lor (p \land q)$ and p always have the same truth values. Thus they are logically equivalent. (This proves one of the absorption laws.)

18.

p	t	p ∨ t
T	T	T
F	T	T

same truth values

The truth table shows that $p \lor \mathbf{t}$ and $\mathbf{t}$ always have the same truth values. Thus they are logically equivalent. (This proves one of the universal bound laws.)

21.

p	q	p	q	r	$p \wedge q$	$q \wedge r$	$(p \wedge q) \wedge r$	$p \wedge (q \wedge r)$
T	T		T		T	T	T	T
T	T		F		T	F	F	F
T	F		T		F	F	F	F
T	F		F		F	F	F	F
F	T		T		F	T	F	F
F	T		F		F	F	F	F
F	F		T		F	F	F	F
F	F		F		F	F	F	F

same truth values

The truth table shows that $(p \wedge q) \wedge r$ and $p \wedge (q \wedge r)$ always have the same truth values. Thus they are logically equivalent. (This proves the associative law for $\wedge$.)

23.

p	q	r	$p \wedge q$	$q \vee r$	$(p \wedge q) \vee r$	$p \wedge (q \vee r)$	
T	T	T	T	T	T	T	
T	T	F	T	T	T	T	
T	F	T	F	T	T	T	
T	F	F	F	F	F	F	
F	T	T	F	T	T	F	$\leftarrow$
F	T	F	F	T	F	F	
F	F	T	F	T	T	F	$\leftarrow$
F	F	F	F	F	F	F	

different truth values in rows 5 and 7

The truth table shows that $(p \wedge q) \vee r$ and $p \wedge (q \vee r)$ have different truth values in rows 5 and 7. Thus they are not logically equivalent. (This proves that parentheses are needed with $\wedge$ and $\vee$.)

24.

p	q	r	$p \vee q$	$p \wedge r$	$(p \vee q) \vee (p \wedge r)$	$(p \vee q) \wedge r$	
T	T	T	T	T	T	T	
T	T	F	T	F	T	F	$\leftarrow$
T	F	T	T	T	T	T	$\leftarrow$
T	F	F	T	F	T	F	
F	T	T	T	F	T	T	
F	T	F	T	F	T	F	$\leftarrow$
F	F	T	F	F	F	F	
F	F	F	F	F	F	F	

different truth values in rows 2, 3, and 6

The truth table shows that $(p \vee q) \vee (p \wedge r)$ and $(p \vee q) \wedge r$ have different truth values in rows 2, 3, and 6. Hence they are not logically equivalent.

25. Hal is not a math major or Hal's sister is not a computer science major.

27. The connector is not loose and the machine is not unplugged.

30. The dollar is not at an all-time high or the stock market is not at a record low.

31. **a.** 01, 02, 11, 12

32. $-2 \geq x$ or $x \geq 7$

33. $-10 \geq x$ or $x \geq 2$

34. $2 \leq x \leq 5$

36. $1 \leq x$ or $x < -3$

38. This statement's logical form is $(p \wedge q) \vee r$, so its negation has the form $\sim((p \wedge q) \vee r) \equiv$ $\sim(p \wedge q) \wedge \sim r \equiv (\sim p \vee \sim q) \wedge \sim r$. Thus a negation for the statement is ($num\_orders \leq 100$ or $num\_instock > 500$) and $num\_instock \geq 200$.

39. The statement's logical form is $(p \wedge q) \vee ((r \wedge s) \wedge t)$, so its negation has the form

$$
\begin{aligned}
\sim((p \wedge q) \vee ((r \wedge s) \wedge t)) &\equiv \sim(p \wedge q) \wedge \sim((r \wedge s) \wedge t)) \\
&\equiv (\sim p \vee \sim q) \wedge (\sim(r \wedge s) \vee \sim t)) \\
&\equiv (\sim p \vee \sim q) \wedge ((\sim r \vee \sim s) \vee \sim t)).
\end{aligned}
$$

Thus a negation is ($num\_orders \geq 50$ or $num\_instock \leq 300$) and (($50 > num\_orders$ or $num\_orders \geq 75$) or $num\_instock \leq 500$).

40.

p	q	$\sim p$	$\sim q$	$p \wedge q$	$p \wedge \sim q$	$\sim p \vee (p \wedge \sim q)$	$(p \wedge q) \vee (\sim p \vee (p \wedge \sim q))$
T	T	F	F	T	F	F	T
T	F	F	T	F	T	T	T
F	T	T	F	F	F	T	T
F	F	T	T	F	F	T	T

Since all the truth values of $(p \wedge q) \vee (\sim p \vee (p \wedge \sim q))$ are T, $(p \wedge q) \vee (\sim p \vee (p \wedge \sim q))$ is a tautology.

41.

p	q	$\sim p$	$\sim q$	$p \wedge \sim q$	$\sim p \vee q$	$(p \wedge \sim q)(p \vee q)$
T	T	F	F	F	T	F
T	F	F	T	T	F	F
F	T	T	F	F	T	F
F	F	T	T	F	T	F

Since all the truth values of $(p \wedge \sim q) \wedge (\sim p \vee q)$ are F, $(p \wedge \sim q) \wedge (\sim p \vee q)$ is a contradiction.

42.

p	q	r	$\sim p$	$\sim q$	$\sim p \wedge q$	$q \wedge r$	$((\sim p \wedge q) \wedge (q \wedge r))$	$((\sim p \wedge q) \wedge (q \wedge r)) \wedge \sim q$
T	T	T	F	F	F	T	F	F
T	T	F	F	F	F	F	F	F
T	F	T	F	T	F	F	F	F
T	F	F	F	T	F	F	F	F
F	T	T	T	F	T	T	T	F
F	T	F	T	F	T	F	F	F
F	F	T	T	T	F	F	F	F
F	F	F	T	T	F	F	F	F

all $F's$

Since all the truth values of $((\sim p \wedge q) \wedge (q \wedge r)) \wedge \sim q$ are F, $((\sim p \wedge q) \wedge (q \wedge r)) \wedge \sim q$ is a contradiction.

44. **a.** No real numbers satisfy this inequality

45. Let b be "Bob is a double math and computer science major," m be "Ann is a math major," and a be "Ann is a double math and computer science major." Then the two statements can be symbolized as follows: a. $(b \wedge m) \wedge \sim a$ and   **b.** $\sim (b \wedge a) \wedge (m \wedge b)$. *Note:* The entries in the truth table assume that a person who is a double math and computer science major is also a math major and a computer science major.

b	m	a	$\sim a$	$b \wedge m$	$m \wedge b$	$b \wedge a$	$\sim (b \wedge a)$	$(b \wedge m) \wedge \sim a$	$\sim (b \wedge a) \wedge (m \wedge b)$
T	T	T	F	T	T	T	F	F	F
T	T	F	T	F	T	F	T	T	T
T	F	T	F	T	F	T	F	F	F
T	F	F	T	F	F	F	T	F	F
F	T	T	F	F	F	F	T	F	F
F	T	F	T	F	F	F	T	F	F
F	F	T	F	F	F	F	T	F	F
F	F	F	T	F	F	F	T	F	F

$$\underbrace{\qquad\qquad\qquad\qquad\qquad}_{\text{same truth values}}$$

The truth table shows that $(b \wedge m) \wedge \sim a$ and $\sim (b \wedge a) \wedge (m \wedge b)$ always have the same truth values. Hence they are logically equivalent.

46. **a.** *Solution 1:* Construct a truth table for $p \oplus p$ using the truth values for *exclusive or.*

p	$p \oplus p$
T	F
F	F

because an *exclusive or* statement is false when both components are true and when both components are false, and the two components in $p \oplus p$ are both p.

Since all its truth values are false, $p \oplus p \equiv \mathbf{c}$, a contradiction.

Solution 2: Replace q by p in the logical equivalence $p \oplus q \equiv (p \vee q) \wedge \sim(p \wedge q)$, and simplify the result.

$$
\begin{aligned}
p \oplus p &\equiv (p \vee p) \wedge \sim(p \wedge p) & &\text{by definition of } \oplus \\
&\equiv p \wedge \sim p & &\text{by the identity laws} \\
&\equiv \mathbf{c} & &\text{by the negation law for } \wedge
\end{aligned}
$$

47. There is a famous story about a philosopher who once gave a talk in which he observed that whereas in English and many other languages a double negative is equivalent to a positive, there is no language in which a double positive is equivalent to a negative. To this, another philosopher, Sidney Morgenbesser, responded sarcastically, "Yeah, yeah."

   [Strictly speaking, sarcasm functions like negation. When spoken sarcastically, the words "Yeah, yeah" are not a true double positive; they just mean "no."]

48. **a.** the distributive law   **b.** the commutative law for $\vee$
    c. the negation law for $\vee$ **d.** the identity law for $\wedge$

50. $$
\begin{aligned}
(p \wedge \sim q) \vee p &\equiv p \vee (p \wedge \sim q) & &\text{by the commutative law for } \vee \\
&\equiv p & &\text{by the absorption law (with } \sim q \text{ in place of } q)
\end{aligned}
$$

51. *Solution 1:*
$$
\begin{aligned}
p \wedge (\sim q \vee p) &\equiv p \wedge (p \vee \sim q) & &\text{commutative law for } \vee \\
&\equiv p & &\text{absorption law}
\end{aligned}
$$

   Solution 2:
$$
\begin{aligned}
p \wedge (\sim q \vee p) &\equiv (p \wedge \sim q) \vee (p \wedge p) & &\text{distributive law} \\
&\equiv (p \wedge \sim q) \vee p & &\text{identity law for } \wedge \\
&\equiv p & &\text{by exercise 50.}
\end{aligned}
$$

53. $\sim((\sim p \wedge q) \vee (\sim p \wedge \sim q)) \vee (p \wedge q)$ $\equiv$ $\sim[\sim p \wedge (q \vee \sim q)] \vee (p \wedge q)$ by the distributive law
$\equiv$ $\sim(\sim p \wedge \mathbf{t}) \vee (p \wedge q)$ by the negation law for $\vee$
$\equiv$ $\sim(\sim p) \vee (p \wedge q)$ by the identity law for $\wedge$
$\equiv$ $p \vee (p \wedge q)$ by the double negative law
$\equiv$ p by the absorption law

54. $(p \wedge (\sim (\sim p \vee q))) \vee (p \wedge q)$ $\equiv$ $(p \wedge (\sim (\sim p) \wedge \sim q)) \vee (p \wedge q)$ De Morgan's law
$\equiv$ $(p \wedge (p \wedge \sim q)) \vee (p \wedge q)$ double negative law
$\equiv$ $((p \wedge p) \wedge \sim q)) \vee (p \wedge q)$ associative law for $\wedge$
$\equiv$ $(p \wedge \sim q)) \vee (p \wedge q)$ idempotent law for $\wedge$
$\equiv$ $p \wedge (\sim q \vee q)$ distributive law
$\equiv$ $p \wedge (q \vee \sim q)$ commutative law for $\vee$
$\equiv$ $p \wedge \mathbf{t}$ negation law for $\vee$
$\equiv$ p identity law for $\wedge$

Section 2.2

1. If this loop does not contain a **stop** or a **go to**, then it will repeat exactly N times.

3. If you do not freeze, then I'll shoot.

5.

p	q	$\sim p$	$\sim q$	$\sim p \vee q$	$\sim p \vee q \rightarrow \sim q$
T	T	F	F	T	F
T	F	F	T	F	T
F	T	T	F	T	F
F	F	T	T	T	T

6.

p	q	$\sim p$	$\sim p \wedge q$	$p \vee q$	$(p \vee q) \vee (\sim p \wedge q)$	$(p \vee q) \vee (\sim p \wedge q) \rightarrow q$
T	T	F	F	T	T	T
T	F	F	F	T	T	F
F	T	T	T	T	T	T
F	F	T	F	F	F	T

7.

p	q	r	$\sim q$	$p \wedge \sim q$	$p \wedge \sim q \rightarrow r$
T	T	T	F	F	T
T	T	F	F	F	T
T	F	T	T	T	T
T	F	F	T	T	F
F	T	T	F	F	T
F	T	F	F	F	T
F	F	T	T	F	T
F	F	F	T	F	T

9.

p	q	r	$\sim r$	$p \wedge \sim r$	$q \wedge r$	$p \wedge \sim r \leftrightarrow q \vee r$
T	T	T	F	F	T	F
T	T	F	T	T	T	T
T	F	T	F	F	T	F
T	F	F	T	T	F	F
F	T	T	F	F	T	F
F	T	F	T	F	T	F
F	F	T	F	F	T	F
F	F	F	T	F	F	T

10.

p	q	r	$p \to r$	$q \to r$	$(p \to r) \leftrightarrow (q \to r)$
T	T	T	T	T	T
T	T	F	F	F	T
T	F	T	T	T	T
T	F	F	F	T	F
F	T	T	T	T	T
F	T	F	T	F	F
F	F	T	T	T	T
F	F	F	T	T	T

11.

p	q	r	$q \to r$	$p \to (q \to r)$	$p \wedge q$	$p \wedge q \to r$	$(p \to (q \to r)) \leftrightarrow (p \wedge q \to r)$
T	T	T	T	T	T	T	T
T	T	F	F	F	T	F	T
T	F	T	T	T	F	T	T
T	F	F	T	T	F	T	T
F	T	T	T	T	F	T	T
F	T	F	F	T	F	T	T
F	F	T	T	T	F	T	T
F	F	F	T	T	F	T	T

12. If $x > 2$ then $x^2 > 4$, and if $x < -2$ then $x^2 > 4$.

13. **a.**

p	q	$\sim p$	$p \to q$	$\sim p \wedge q$
T	T	F	T	T
T	F	T	F	F
F	T	F	T	T
F	F	T	T	T

same truth values

The truth table shows that $p \to q$ and $\sim p \vee q$ always have the same truth values. Hence they are logically equivalent.

14. **a.** *Hint:* $p \to q \vee r$ is true in all cases except when p is true and both q and r are false.

15.

p	q	r	$q \to r$	$p \to q$	$p \to (q \to r)$	$(p \to q) \to r$	
T	T	T	T	T	T	T	
T	T	F	F	T	F	F	
T	F	T	T	F	T	T	
T	F	F	T	F	T	T	
F	T	T	T	T	T	T	
F	T	F	F	T	T	F	$\leftarrow$
F	F	T	T	T	T	F	$\leftarrow$
F	F	F	T	T	T	F	$\leftarrow$

different truth values

The truth table shows that $p \to (q \to r)$ and $(p \to q) \to r$ do not always have the same truth values. (They differ for the combinations of truth values for p, q, and r shown in rows 6, 7, and 8.) Therefore they are not logically equivalent.

16. Let p represent "You paid full price" and q represent "You didn't buy it at Crown Books." Thus, "If you paid full price, you didn't buy it at Crown Books" has the form $p \to q$. And "You didn't buy it at Crown Books or you paid full price" has the form $q \vee p$.

p	q	$p \to q$	$q \vee p$	
T	T	T	T	
T	F	F	T	$\leftarrow$
F	T	T	T	$\leftarrow$
F	F	T	F	

different truth values

These two statements are not logically equivalent because their forms have different truth values in rows 2 and 4.

(An alternative representation for the forms of the two statements is $p \to \sim q$ and $\tilde{\sim} q \vee p$. In this case, the truth values differ in rows 1 and 3.)

18. *Part 1*: Let p represent "It walks like a duck," q represent "It talks like a duck," and r represent "It is a duck." The statement "If it walks like a duck and it talks like a duck, then it is a duck" has the form $p \wedge q \to r$. And the statement "Either it does not walk like a duck or it does not talk like a duck or it is a duck" has the form $\sim p \vee \sim q \vee r$.

p	q	r	$\sim p$	$\sim q$	$p \wedge q$	$\sim p \vee \sim q$	$p \wedge q \to r$	$(\sim p \vee \sim q) \vee r$
T	T	T	F	F	T	F	T	T
T	T	F	F	F	T	F	F	F
T	F	T	F	T	F	T	T	T
T	F	F	F	T	F	T	T	T
F	T	T	T	F	F	T	T	T
F	T	F	T	F	F	T	T	T
F	F	T	T	T	F	T	T	T
F	F	F	T	T	F	T	T	T

same truth values

The truth table shows that $p \wedge q \to r$ and $(\sim p \vee \sim q) \vee r$ always have the same truth values. Thus the following statements are logically equivalent: "If it walks like a duck and it talks like a duck, then it is a duck" and "Either it does not walk like a duck or it does not talk like a duck or it is a duck."

Part 2: The statement "If it does not walk like a duck and it does not talk like a duck then it is not a duck" has the form $\sim p \wedge \sim q \to \sim r$.

p	q	r	$\sim p$	$\sim q$	$\sim r$	$p \wedge q$	$\sim p \wedge \sim q$	$p \wedge q \to r$	$(\sim p \wedge \sim q) \to \sim r$	
T	T	T	F	F	F	T	F	T	T	
T	T	F	F	F	T	T	F	F	T	$\leftarrow$
T	F	T	F	T	F	F	F	T	T	
T	F	F	F	T	T	F	F	T	T	
F	T	T	T	F	F	F	F	T	T	
F	T	F	T	F	T	F	F	T	T	
F	F	T	T	T	F	F	T	T	F	$\leftarrow$
F	F	F	T	T	T	F	T	T	T	

different truth values

The truth table shows that $p \wedge q \to r$ and $(\sim p \wedge \sim q) \to \sim r$ do not always have the same truth values. (They differ for the combinations of truth values of p, q, and r shown in rows 2 and 7.) Thus they are not logically equivalent, and so the statement "If it walks like a duck and it talks like a duck, then it is a duck" is not logically equivalent to the statement "If it does not walk like a duck and it does not talk like a duck then it is not a duck." In addition, because of the logical equivalence shown in Part 1, we can also conclude that the following two statements are not logically equivalent: "Either it does not walk like a duck or it does not talk like a duck or it is a duck" and "If it does not walk like a duck and it does not talk like a duck then it is not a duck."

19. False. The negation of an if-then statement is not an if-then statement. It is an *and* statement.

20. **a.** *Negation*: P is a square and P is not a rectangle.

   d. *Negation*: n is prime and both n is not odd and n is not 2. Or: n is prime and n is neither odd nor 2.

   f. *Negation*: Tom is Ann's father and either Jim is not her uncle or Sue is not her aunt.

21. By assumption, $p \to q$ is false. By definition of a conditional statement, the only way this can happen is for the hypothesis, p, to be true and the conclusion, q, to be false.

   a. The only way $\sim p \to q$ can be false is for $\sim p$ to be true and q to be false. But since p is true, $\sim p$ is false. Hence $\sim p \to q$ is not false and so it is true.

   b. Since p is true, then $p \lor q$ is true because if one component of an *and* statement is true, then the statement as a whole is true.

   c. The only way $q \to p$ can be false is for q to be true and p to be false. Thus, since q is false, $q \to p$ is not false and so it is true.

22. **a.** *Contrapositive*: If P is not a rectangle, then P is not a square.

   d. *Contrapositive*: If n is not odd and n is not 2, then n is not prime.

   f. *Contrapositive*: If either Jim is not Ann's uncle or Sue is not her aunt, then Tom is not her father.

23. **a.** *Converse*: If P is a rectangle, then P is a square.
   Inverse: If P is not a square, then P is not a rectangle.

   d. *Converse*: If n is odd or n is 2, then n is prime.
   Inverse: If n is not prime, then n is not odd and n is not 2.

   f. *Converse*: If Jim is Ann's uncle and Sue is her aunt, then Tom is her father.
   Inverse: If Tom is not Ann's father, then Jim is not her uncle or Sue is not her aunt.

24.

p	q	$p \to q$	$q \to p$	
T	T	T	T	
T	F	F	T	←
F	T	T	F	←
F	F	T	T	

different truth values

The truth table shows that $p \to q$ and $q \to p$ have different truth values in the second and third rows. Hence they are not logically equivalent.

26.

p	q	$\sim p$	$\sim q$	$\sim q \to \sim p$	$p \to q$
T	T	F	F	T	T
T	F	F	T	F	F
F	T	T	F	T	T
F	F	T	T	T	T

$\underbrace{\qquad\qquad\qquad}$
same truth values

The truth table shows that $\sim q \to \sim p$ and $p \to q$ always have the same truth values and thus are logically equivalent. It follows that a conditional statement and its contrapositive are logically equivalent to each other.

27.

p	q	$\sim p$	$\sim q$	$q \to p$	$\sim p \to \sim q$
T	T	F	F	T	T
T	F	F	T	T	T
F	T	T	F	F	F
F	F	T	T	T	T

$\underbrace{\qquad\qquad\qquad}$
same truth values

The truth table shows that $q \to p$ and $\sim p \to \sim q$ always have the same truth values and thus are logically equivalent. It follows that the converse and inverse of a conditional statement are logically equivalent to each other.

28. *Hint:* A person who says "I mean what I say" claims to speak sincerely. A person who says "I say what I mean" claims to speak with precision.

29. The corresponding tautology is $(p \to (q \vee r)) \leftrightarrow ((p \wedge \sim q) \to r)$

p	q	r	$\sim q$	$q \vee r$	$p \wedge \sim q$	$p \to (q \vee r)$	$p \wedge \sim q \to r$	$(p \to (q \vee r)) \leftrightarrow ((p \wedge \sim q) \to r)$
T	T	T	F	T	F	T	T	T
T	T	F	F	T	F	T	T	T
T	F	T	T	T	T	T	T	T
T	F	F	T	F	T	F	F	T
F	T	T	F	T	F	T	T	T
F	T	F	F	T	F	T	T	T
F	F	T	T	T	F	T	T	T
F	F	F	T	F	F	T	T	T

The truth table shows that $(p \to (q \vee r)) \leftrightarrow ((p \wedge \sim q) \to r)$ is a tautology because all of its truth values are T.

30. The corresponding tautology is $p \wedge (q \vee r) \leftrightarrow (p \wedge q) \vee (p \wedge r)$

p	q	r	$q \vee r$	$p \wedge q$	$p \wedge r$	$p \wedge (q \vee r)$	$(p \wedge q) \vee (p \wedge r)$	$p \wedge (q \vee r) \leftrightarrow$ $(p \wedge q) \vee (p \wedge r)$
T	T	T	T	T	T	T	T	T
T	T	F	T	T	F	T	T	T
T	F	T	T	F	T	T	T	T
T	F	F	F	F	F	F	F	T
F	T	T	T	F	F	F	F	T
F	T	F	T	F	F	F	F	T
F	F	T	T	F	F	F	F	T
F	F	F	F	F	F	F	F	T

all T's

The truth table shows that $p \wedge (q \vee r) \leftrightarrow (p \wedge q) \vee (p \wedge r)$ is always true. Hence it is a tautology.

32. If this quadratic equation has two distinct real roots, then its discriminant is greater than zero, and if the discriminant of this quadratic equation is greater than zero, then the equation has two real roots.

33. If this integer is even, then it equals twice some integer, and if this integer equals twice some integer, then it is even.

34. If the Cubs do not win tomorrow's game, then they will not win the pennant.
If the Cubs win the pennant, then they will have won tomorrow's game.

36. The Personnel Director did not lie. By using the phrase "only if," the Personnel Director set forth conditions that were necessary but not sufficient for being hired: if you did not satisfy those conditions then you would not be hired. The Personnel Director's statement said nothing about what would happen if you did satisfy those conditions.

37. If a new hearing is not granted, payment will be made on the fifth.

39. If a security code is not entered, then the door will not open.

40. If I catch the 8:05 bus, then I am on time for work.

42. If this number is not divisible by 3, then it is not divisible by 9.
If this number is divisible by 9, then it is divisible by 3.

44. If Jon's team wins the rest of its games, then it will win the championship.

45. If this computer program produces error messages during translation, then it is not correct.

If this computer program is correct, then it does not produce error messages during translation.

46. **a.** This statement is the converse of the given statement, and so it is not necessarily true. For instance, if the actual boiling point of compound X were $200°$C, then the given statement would be true but this statement would be false.

b. This statement must be true. It is the contrapositive of the given statement.

47. **a.** $p \wedge {\sim}q \to r \equiv {\sim}(p \wedge {\sim}q) \vee r$

b. $\begin{aligned} p \wedge {\sim}q \to r \quad &\equiv \quad {\sim}(p \wedge {\sim}q) \vee r && \text{by the identity for } \to \text{ shown in} \\ &&& \text{the directions } \textit{[an acceptable answer]} \\ &\equiv \quad {\sim}[{\sim}({\sim}(p \wedge {\sim}q)) \wedge {\sim}r] && \text{by De Morgan's law } \textit{[another acceptable answer]} \\ &\equiv \quad {\sim}[(p \wedge {\sim}q) \wedge {\sim}r] && \text{by the double negative law} \\ &&& \textit{[another acceptable answer]} \end{aligned}$

Any of the expressions in part (b) would also be acceptable answers for part (a).

48. **a.** $p \vee \sim q \rightarrow r \vee q$ $\equiv$ $\sim (p \vee \sim q) \vee (r \vee q)$ by the identity for $\rightarrow$ shown in the directions *[an acceptable answer]*

$\equiv$ $(\sim p \wedge \sim (\sim q)) \vee (r \vee q)$ by De Morgan's law *[another acceptable answer]*

$\equiv$ $(\sim p \wedge q) \vee (r \vee q)$ by the double negative law *[another acceptable answer]*

b. $p \vee \sim q \rightarrow r \vee q$ $\equiv$ $(\sim p \wedge q) \vee (r \vee q)$ by part (a)

$\equiv$ $\sim (\sim (\sim p \wedge q) \wedge \sim (r \vee q))$ by De Morgan's law

$\equiv$ $\sim (\sim (\sim p \wedge q) \wedge (\sim r \wedge \sim q))$ by De Morgan's law

Any of the expressions in part (b) would also be acceptable answers for part (a).

49. **a.** $(p \rightarrow r) \leftrightarrow (q \rightarrow r)$ $\equiv$ $(\sim p \vee r) \leftrightarrow (\sim q \vee r)$

$\equiv$ $[\sim(\sim p \vee r) \vee (\sim q \vee r)] \wedge [\sim(\sim q \vee r) \vee (\sim p \vee r)]$

by the identity for $\leftrightarrow$ shown in the directions *[an acceptable answer]*

$\equiv$ $[(p \wedge \sim r) \vee (\sim q \vee r)] \wedge [(q \wedge \sim r) \vee (\sim p \vee r)]$

by De Morgan's law *[another acceptable answer]*

b. $(\sim p \vee r) \leftrightarrow (\sim q \vee r)$ $\equiv$ $\sim[\sim(p \wedge \sim r) \wedge \sim(\sim q \vee r)] \wedge \sim[\sim(q \wedge \sim r) \wedge \sim(\sim p \vee r)]$

by De Morgan's law

$\equiv$ $\sim[\sim(p \wedge \sim r) \wedge (q \wedge \sim r)] \wedge \sim[\sim(q \wedge \sim r) \wedge (p \wedge \sim r)]$

by De Morgan's law

Any of the expressions in part (b) would also be acceptable answers for part (a).

51. Yes. As in exercises 47-50, the following logical equivalences can be used to rewrite any statement form in a logically equivalent way using only $\sim$ and $\wedge$:

$$p \rightarrow q \equiv \sim p \vee q \qquad\qquad p \leftrightarrow q \equiv (\sim p \vee q) \wedge (\sim q \vee p)$$
$$p \vee q \equiv \sim (\sim p \wedge \sim q) \qquad\qquad \sim (\sim p) \equiv p$$

The logical equivalence $p \wedge q \equiv \sim (\sim p \vee \sim q)$ can then be used to rewrite any statement form in a logically equivalent way using only $\sim$ and $\vee$.

Section 2.3

1. $\sqrt{2}$ is not rational.

3. Logic is not easy.

6.

		premises		conclusion
p	q	$p \rightarrow q$	$p \rightarrow q$	$p \vee q$
T	T	T	T	T ← critical row
T	F	F	T	
F	T	T	F	
F	T	T	T	F ← critical row

Rows 2 and 4 of the truth table are the critical rows in which all the premises are true, but row 4 shows that it is possible for an argument of this form to have true premises and a false conclusion. Thus this argument form is invalid.

7.

| | | | | premises | | | conclusion |
p	q	r	~q	p	p → q	~q ∨ r	r
T	T	T	F	T	T	T	T ← ——— *critical row*
T	T	F	F	T	T	F	
T	F	T	T	T	F	T	
T	F	F	T	T	F	T	
F	T	T	F	F	T	T	
F	T	F	F	F	T	F	
F	F	T	T	F	T	T	
F	F	F	T	F	T	T	

This row describes the only situation in which all the premises are true. Because the conclusion is also true here, the argument form is valid.

8.

| | | | | premises | | | conclusion |
p	q	r	~q	p ∨ q	p →~ q	p → r	r
T	T	T	F	T	F	T	
T	T	F	F	T	F	F	
T	F	T	T	T	T	T	T ← —*critical row*
T	F	F	T	T	T	F	
F	T	T	F	T	T	T	T ← —*critical row*
F	T	F	F	T	T	T	F ← —*critical row*
F	F	T	T	F	T	T	
F	F	F	T	F	T	T	

This row shows that it is possible for an argument of this form to have true premises and a false conclusion. Thus this argument form is invalid.

9.

| | | | | | | premises | | | conclusion |
p	q	r	~q	~r	p ∧ q	p ∧ q →~ r	p ∨ ~ q	~q → p	~ r
T	T	T	F	F	T	F	T	T	
T	T	F	F	T	T	T	T	T	T ← —*critical row*
T	F	T	T	F	F	T	T	T	F ← —*critical row*
T	F	F	T	T	F	T	T	T	T ← —*critical row*
F	T	T	F	F	F	T	F	T	
F	T	F	F	T	F	T	F	T	
F	F	T	T	F	F	T	T	F	
F	F	F	T	T	F	T	T	F	

Rows 2, 3, and 4 of the truth table are the critical rows in which all the premises are true, but row 3 shows that it is possible for an argument of this form to have true premises and a false conclusion. Hence the argument form is invalid.

12. **a.**

| | | premises | | conclusion |
p	q	p → q	q	p
T	T	T	T	T ← — *critical row*
T	F	F	F	
F	T	T	T	F ← — *critical row*
F	T	T	F	

Rows 1 and 3 of the truth table represent the situations in which all the premises are true, but row 3 shows that it is possible for an argument of this form to have true premises and a false conclusion. Hence the argument form is invalid.

b.

		premises		conclusion
p	q	$p \rightarrow q$	$\sim p$	$\sim q$
T	T	T	F	
T	F	F	T	T ← critical row
F	T	T	F	
F	T	T	T	F ← critical row

Rows 2 and 4 of the truth table represent the situations in which all the premises are true, but row 4 shows that it is possible for an argument of this form to have true premises and a false conclusion. Hence the argument form is invalid.

14.

		premise	conclusion
p	q	p	$p \vee q$
T	T	T	T ← critical row
T	F	F	
F	T	T	T ← critical row
F	F	F	

The truth table shows that in the two situations (represented by rows 1 and 3) in which the premise is true, the conclusion is also true. Therefore, Generalization, version (a), is valid.

15.

		premise	conclusion
p	q	q	$p \vee q$
T	T	T	T ← critical row
T	F	F	
F	T	T	T ← critical row
F	F	F	

The truth table shows that in the two situations (represented by rows 1 and 3) in which the premise is true, the conclusion is also true. Therefore, Generalization, version (b), is valid.

18.

		premises		conclusion
p	q	$p \vee q$	$\sim q$	p
T	T	T	F	
T	F	T	T	T ← critical row
F	T	T	F	
F	T	F	T	

Row 2 represents the only situation in which both premises are true. Because the conclusion is also true here the argument form is valid.

21.

			premises			conclusion
p	q	r	$p \vee q$	$p \rightarrow r$	$q \rightarrow r$	r
T	T	T	T	T	T	T ← critical row
T	T	F	T	F	F	
T	F	T	T	T	T	T ← critical row
T	F	F	T	F	T	
F	T	T	T	T	T	T ← critical row
F	T	F	T	T	F	
F	F	T	F	T	T	
F	F	F	F	T	T	

The truth table shows that in the three situations (represented by rows 1, 3, 5) in which all three premises are true, the conclusion is also true. Therefore, proof by division into cases is valid.

22. Let p represent "Tom is on team A" and q represent "Hua is on team B." Then the argument has the form

$$\sim p \to q$$
$$\sim q \to p$$
$$\therefore \quad \sim p \lor \sim q$$

				premises		conclusion
p	q	$\sim p$	$\sim q$	$\sim p \to q$	$\sim q \to p$	$\sim p \lor \sim q$
T	T	F	F	T	T	F ← critical row
T	F	F	T	T	T	T ← critical row
F	T	T	F	T	T	T ← critical row
F	F	T	T	F	F	

Rows 1, 2, and 3 of the truth table are the critical rows in which all the premises are true, but row 1 shows that it is possible for an argument of this form to have true premises and a false conclusion. Thus this argument form is invalid.

24. form: $p \to q$
    q
    $\therefore \quad p$ invalid: converse error

25. form: $p \lor q$
    $\sim p$
    $\therefore \quad q$ valid: elimination

26. form: $p \to q$
    $q \to r$
    $\therefore \quad p \to r$ valid: transitivity

27. form: $p \to q$
    $\sim p$
    $\therefore \quad \sim q$ invalid: inverse error

30. form: $p \to q$ invalid, converse error
    q
    $\therefore \quad p$

33. A valid argument with a false conclusion must have at least one false premise. In the following example, the second premise is false. (The first premise is true because its hypothesis is false.)

    If the square of every real number is positive, then no real number is negative.

    The square of every real number is positive.

    Therefore, no real number is negative.

36. The program contains an undeclared variable.
    One explanation:
    1. There is not a missing semicolon and there is not a misspelled variable name. *(by (c) and (d) and definition of ∧)*
    2. It is not the case that there is a missing semicolon or a misspelled variable name. *(by (1) and De Morgan's laws)*
    3. There is not a syntax error in the first five lines. *(by (b) and (2) and modus tollens)*
    4. There is an undeclared variable. *(by (a) and (3) and elimination)*

37. The treasure is buried under the flagpole.

One explanation:
1. The treasure is not in the kitchen. *(by (c) and (a) and modus ponens)*
2. The tree in the front yard is not an elm. *(by (b) and (1) and modus tollens)*
3. The treasure is buried under the flagpole. *(by (d) and (2) and elimination)*

38. **a.** *A* is a knave and *B* is a knight.
One explanation:
1. Suppose *A* is a knight.
2. ∴ What *A* says is true. *(by definition of knight)*
3. ∴ *B* is a knight also. *(That's what A said.)*
4. ∴ What *B* says is true. *(by definition of knight)*
5. ∴ *A* is a knave. *(That's what B said.)*
6. ∴ We have a contradiction: *A* is a knight and a knave. *(by (1) and (5))*
7. ∴ The supposition that *A* is a knight is false. *(by the contradiction rule)*
8. ∴ *A* is a knave. *(negation of supposition)*
9. ∴ What *B* says is true. *(B said A was a knave, which we now know to be true.)*
10. ∴ *B* is a knight. *(by definition of knight)*

d. *Hint:* *W* and *Y* are knights; the rest are knaves.

39. The chauffeur killed Lord Hazelton.

One explanation:
1. Suppose the cook was in the kitchen at the time of the murder.
2. ∴ The butler killed Lord Hazelton with strychnine. *(by (c) and (1) and modus ponens)*
3. ∴ We have a contradiction: Lord Hazelton was killed by strychnine and a blow on the head. *(by (2) and (a))*
4. ∴ The supposition that the cook was in the kitchen is false. *(by the contradiction rule)*
5. ∴ The cook was not in the kitchen at the time of the murder. *(negation of supposition)*
6. ∴ Sara was not in the dining room when the murder was committed. *(by (e) and (5) and modus ponens)*
7. ∴ Lady Hazelton was in the dining room when the murder was committed. *(by (b) and (6) and elimination)*
8. ∴ The chauffeur killed Lord Hazelton. *(by (d) and (7) and modus ponens)*

41. (1) $p \rightarrow t$ by premise (d)
     $\sim p$ by premise (c)
     ∴ $\sim p$ by modus tollens

     (2) $\sim p$ by (1)
     ∴ $\sim p \vee q$ by generalization

     (3) $\sim p \vee q \rightarrow r$ by premise (a)
     $\sim p \vee q$ by (2)
     ∴ r by modus ponens

     (4) $\sim p$ by (1)
     r by (3)
     ∴ $\sim p \wedge r$ by conjunction

     (5) $\sim p \wedge r \rightarrow \sim s$ by premise (e)
     $\sim p \wedge r$ by (4)
     ∴ $\sim s$ by modus ponens

     (6) $s \vee \sim q$ by premise (b)
     $\sim s$ by (5)
     ∴ $\sim q$ by elimination

42. (1) $q \to r$ by premise (b)
      $\sim r$ by premise (d)
  $\therefore$ $\sim q$ by modus tollens

   (2) $p \lor q$ by premise (a)
      $\sim q$ by (1)
  $\therefore$ p by elimination

   (3) $\sim q \to u \land s$ by premise (e)
      $\sim q$ by (1)
  $\therefore$ $u \land s$ by modus ponens

   (4) $u \land s$ by (3)
  $\therefore$ s by specialization

   (5) p by (2)
      s by (4)
  $\therefore$ $p \land s$ by conjunction

   (6) $p \land s \to t$ by premise (c)
      $p \land s$ by (5)
  $\therefore$ t by modus ponens

43. (1) $\sim w$ by premise (d)
      $u \lor w$ by premise (e)
  $\therefore$ u by elimination

   (2) $u \to \sim p$ by premise (c)
      u by (1)
  $\therefore$ $\sim p$ by modus ponens

   (3) $\sim p \to r \land \sim s$ by premise (a)
      $\sim p$ by (2)
  $\therefore$ $r \land \sim s$ by modus ponens

   (4) $r \land \sim s$ by (3)
  $\therefore$ $\sim s$ by specialization

   (5) $\sim t \to s$ by premise (b)
      $\sim s$ by (4)
  $\therefore$ $\sim t$ by modus tollens

Section 2.4

1. $R = 1$

3. $S = 1$

5. The input/output table is as follows:

Input		Output
P	Q	R
1	1	1
1	0	1
0	1	0
0	0	1

6. The input/output table is as follows:

Input			Output
P	Q	R	S
1	1	1	1
1	1	0	0
1	0	1	1
1	0	0	1
0	1	1	1
0	1	0	0
0	0	1	1
0	0	0	0

7. The input/output table is as follows:

Input		Output
P	Q	R
1	1	0
1	0	1
0	1	0
0	0	0

9. $P \vee \sim Q$

11. $(P \wedge \sim Q) \vee R$

12. $(P \vee Q) \vee \sim (Q \wedge R)$

13.

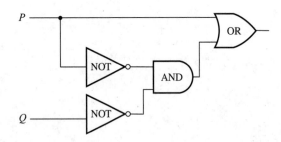

15.

16.

18. **a.** $(P \wedge Q \wedge \sim R) \vee (\sim P \wedge Q \wedge R)$

b.

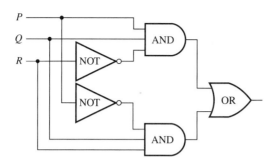

20. **a.** $(P \wedge Q \wedge R) \vee (P \wedge \sim Q \wedge R) \vee (\sim P \wedge \sim Q \wedge \sim R)$

b.

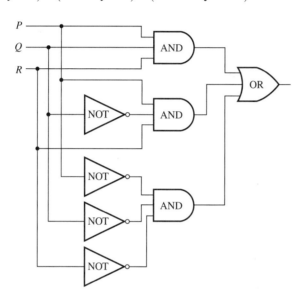

21. **a.** $(P \wedge Q \wedge \sim R) \vee (\sim P \wedge Q \wedge R) \vee (\sim P \wedge Q \wedge \sim R)$

b. One circuit (among many) having the given input/output table is the following:

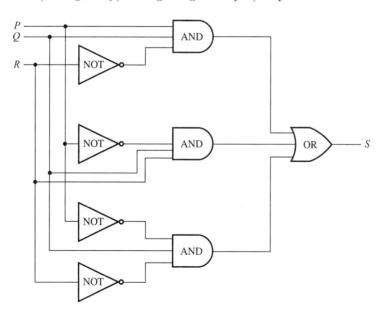

22. The input/output table is

Input			Output
P	Q	R	S
1	1	1	0
1	1	0	1
1	0	1	0
1	0	0	0
0	1	1	0
0	1	0	0
0	0	1	1
0	0	0	0

One circuit (among many) having this input/output table is shown below.

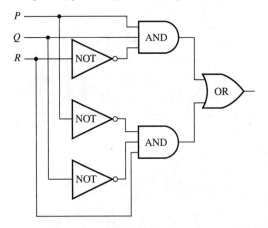

24. Let P and Q represent the positions of the switches in the classroom, with 0 being "down" and 1 being "up." Let R represent the condition of the light, with 0 being "off" and 1 being "on." Initially, $P = Q = 0$ and $R = 0$. If either P or Q (but not both) is changed to 1, the light turns on. So when $P = 1$ and $Q = 0$, then $R = 1$, and when $P = 0$ and $Q = 1$, then $R = 1$. Thus when one switch is up and the other is down the light is on, and hence moving the switch that is down to the up position turns the light off. So when $P = 1$ and $Q = 1$, then $R = 0$. It follows that the input/output table has the following appearance:

Input		Output
P	Q	R
1	1	0
1	0	1
0	1	1
0	0	0

One circuit (among many) having this input/output table is the following:

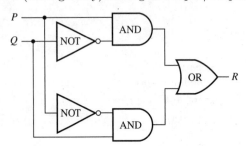

26. The Boolean expression for (a) is $(P \wedge Q) \vee Q$, and for (b) it is $(P \vee Q) \wedge Q$. We must show that if these expressions are regarded as statement forms, then they are logically equivalent. Now

$$
\begin{aligned}
(P \wedge Q) \vee Q &\equiv Q \vee (P \wedge Q) && \text{by the commutative law for } \vee \\
&\equiv (Q \vee P) \wedge (Q \vee Q) && \text{by the distributive law} \\
&\equiv (Q \vee P) \wedge Q && \text{by the idempotent law} \\
&\equiv (P \vee Q) \wedge Q && \text{by the commutative law for } \wedge
\end{aligned}
$$

Alternatively, by the absorption laws, both statement forms are logically equivalent to Q.

27. The Boolean expression for circuit (a) is $\sim P \wedge (\sim (\sim P \wedge Q))$ and for circuit (b) it is $\sim (P \vee Q)$. We must show that if these expressions are regarded as statement forms, then they are logically equivalent. Now

$$
\begin{aligned}
\sim P \wedge (\sim (\sim P \wedge Q)) &\equiv \sim P \wedge (\sim(\sim P) \vee \sim Q) && \text{by De Morgan's law} \\
&\equiv \sim P \wedge (P \vee \sim Q) && \text{by the double negative law} \\
&\equiv (\sim P \wedge P) \vee (\sim P \wedge \sim Q) && \text{by the distributive law} \\
&\equiv (P \wedge \sim P) \vee (\sim P \wedge \sim Q) && \text{by the commutative law for } \wedge \\
&\equiv \mathbf{c} \vee (\sim P \wedge \sim Q) && \text{by the negation law for } \wedge \\
&\equiv (\sim P \wedge \sim Q) \vee \mathbf{c} && \text{by the commutative law for } \vee \\
&\equiv \sim P \wedge \sim Q && \text{by the identity law for } \vee \\
&\equiv \sim (P \vee Q) && \text{by De Morgan's law.}
\end{aligned}
$$

30. $(P \wedge Q) \vee (\sim P \wedge Q) \vee (\sim P \wedge \sim Q) \equiv (P \wedge Q) \vee ((\sim P \wedge Q) \vee (\sim P \wedge \sim Q))$

by inserting parentheses (which is legal by the associative law)

$$
\begin{aligned}
&\equiv (P \wedge Q) \vee (\sim P \wedge (Q \vee \sim Q)) && \text{by the distributive law} \\
&\equiv (P \wedge Q) \vee (\sim P \wedge \mathbf{t}) && \text{by the negation law for } \vee \\
&\equiv (P \wedge Q) \vee \sim P && \text{by the identity law for } \wedge \\
&\equiv \sim P \vee (P \wedge Q) && \text{by the commutative law for } \vee \\
&\equiv (\sim P \vee P) \wedge (\sim P \vee Q) && \text{by the distributive law} \\
&\equiv (P \vee \sim P) \wedge (\sim P \vee Q) && \text{by the commutative law for } \vee \\
&\equiv \mathbf{t} \wedge (\sim P \vee Q) && \text{by the negation law for } \vee \\
&\equiv (\sim P \vee Q) \wedge \mathbf{t} && \text{by the commutative law for } \wedge \\
&\equiv \sim P \vee Q && \text{by the identity law for } \wedge
\end{aligned}
$$

33. **a.**

$$
\begin{aligned}
(P \mid Q) \mid (P \mid Q) &\equiv \sim [(P \mid Q) \wedge (P \mid Q)] && \text{by definition of } \mid \\
&\equiv \sim (P \mid Q) && \text{by the idempotent law for } \wedge \\
&\equiv \sim [\sim (P \wedge Q)] && \text{by definition of } \mid \\
&\equiv P \wedge Q && \text{by the double negative law.}
\end{aligned}
$$

b.

$$
P \wedge (\sim Q \vee R) \equiv (P \mid (\sim Q \vee R)) \mid (P \mid (\sim Q \vee R))
$$

by part (a)

$$
\equiv (P \mid [(\sim Q \mid \sim Q) \mid (R \mid R)]) \mid (P \mid [(\sim Q \mid \sim Q) \mid (R \mid R)])
$$

by Example 2.4.7(b)

$$
\equiv (P \mid [((Q \mid Q) \mid (Q \mid Q)) \mid (R \mid R)]) \mid (P \mid [((Q \mid Q) \mid (Q \mid Q)) \mid (R \mid R)])
$$

by Example 2.4.7(a)

34. **a.**

$$
\begin{aligned}
(P \downarrow Q) \downarrow (P \downarrow Q) &\equiv \sim (P \downarrow Q) && \text{by part (a)} \\
&\equiv \sim [\sim (P \vee Q)] && \text{by definition of } \downarrow \\
&\equiv P \vee Q && \text{by the double negative law}
\end{aligned}
$$

d. *Hint:* Use the results of exercise 13 of Section 2.2 and part (a) and (c) of this exercise.

Section 2.5

1. $19_{10} = 16 + 2 + 1 = 10011_2$

3. $287 = 256 + 16 + 8 + 4 + 2 + 1 = 100011111_2$

4. $458_{10} = 256 + 128 + 64 + 8 + 2 = 111001010_2$

6. $1424 = 1024 + 256 + 128 + 16 = 10110010000_2$

7. $1110_2 = 8 + 4 + 2 = 14_{10}$

9. $110110_2 = 32 + 16 + 4 + 2 = 54_{10}$

10. $1100101_2 = 64 + 32 + 4 + 1 = 101_{10}$

12. $1011011_2 = 64 + 16 + 8 + 2 + 1 = 91_{10}$

13.
```
      1   1   1
      1   0   1   1₂
  +           1   0   1₂
  ─────────────────────
  1   0   0   0   0₂
```

15.
```
      1   1   1       1
      1   0   1   1   0   1₂
  +           1   1   1   0   1₂
  ──────────────────────────
  1   0   0   1   0   1   0₂
```

17.
```
                  1
          1  10  10   1
      1   0   1   0   0₂
  −       1   1   0   1₂
  ─────────────────────
              1   1   1₂
```

18.
```
          10
      0   0  10   0  10
      1   1   0   1   0₂
  −       1   1   0   1₂
  ─────────────────────
          1   1   0   1₂
```

19.
```
                  0  10
      1   0   1   1   0   1₂
  −   1   0   0   1   1₂
  ──────────────────────
      1   1   0   1   0₂
```

21. **a.** $S = 0, T = 1$ *b.* $S = 0, T = 1$ *c.* $S = 0, T = 0$

23. $|-23|_{10} = 23_{10} = (16 + 4 + 2 + 1)_{10} = 00010111_2 \xrightarrow{\text{flip the bits}} 11101000 \xrightarrow{\text{add 1}}$ 11101001. So the answer is 11101001.

24. $|-67|_{10} = 67_{10} = (64 + 2 + 1)_{10} = 01000011_2 \xrightarrow{\text{flip the bits}} 10111100 \xrightarrow{\text{add 1}} 10111101.$ So the 8-bit two's complement is 10111101.

25. $|-4|_{10} = 4_{10} = 00000100_2 \xrightarrow{\text{flip the bits}} 11111011 \xrightarrow{\text{add 1}} 11111100.$ So the answer is 11111100.

27. Because the leading bit is 1, this is the 8-bit two's complement of a negative integer.

    $11010011 \xrightarrow{\text{flip the bits}} 00101100 \xrightarrow{\text{add 1}} 00101101_2 = (32+8+4+1)_{10} = |-45|_{10}$.

    So the answer is -45.

29. Because the leading bit is 1, this is the 8-bit two's complement of a negative integer.

    $11110010 \xrightarrow{\text{flip the bits}} 00001101 \xrightarrow{\text{add 1}} 00001110_2 = (8+4+2)_{10} = |-14|_{10}$.

    So the answer is -14.

30. Because the leading bit is 1, this is the 8-bit two's complement of a negative integer. 10111010
    $\xrightarrow{\text{flip the bits}} 01000101 \xrightarrow{\text{add 1}} 01000110_2 = (64+4+2)_{10} = |-70|_{10}$. So the answer is -70.

31. $57_{10} = (32+16+8+1)_{10} = 111001_2 \rightarrow 00111001$

    $|-118|_{10} = (64+32+16+4+2)_{10} = 01110110_2 \xrightarrow{\text{flip the bits}} 10001001 \xrightarrow{\text{add 1}} 10001010$.

    So the 8-bit two's complements of 57 and -118 are 00111001 and 10001010. Adding the 8-bit two's complements in binary notation gives

$$
\begin{array}{r}
00111001 \\
+\quad 10001010 \\
\hline
11000011
\end{array}
$$

    Since the leading bit of this number is a 1, the answer is negative. Converting back to decimal form gives $11000011 \xrightarrow{\text{flip the bits}} 00111100 \xrightarrow{\text{add 1}} 00111101_2 = (32+16+8+4+1)_{10} = |61|_{10}$.
    So the answer is -61.

32. $62_{10} = (32+16+8+4+2)_{10} = 111110_2 \rightarrow 00111110$

    $|-18|_{10} = (16+2)_{10} = 00010010 \xrightarrow{\text{flip the bits}} 11101101 \xrightarrow{\text{add 1}} 11101110$

    Thus the 8-bit two's complements of 62 and -18 are 00111110 and 10110111. Adding the 8-bit two's complements in binary notation gives

$$
\begin{array}{r}
00111110 \\
+\quad 11101110 \\
\hline
00101100
\end{array}
$$

    Truncating the 1 in the 2^8th position gives 00101100. Since the leading bit of this number is a 0, the answer is positive. Converting back to decimal form gives

$$00101100 \rightarrow 101100_2 = (32+8+4)_{10} = 44_{10}.$$

    So the answer is 44.

33. $|-6|_{10} = (4+2)_{10} = 110_2 \xrightarrow{\text{flip the bits}} 00000110 \rightarrow 11111001 \xrightarrow{\text{add 1}} 11111010$

    $|-73|_{10} = (64+8+1)_{10} = 01001001 \xrightarrow{\text{flip the bits}} 10110110 \xrightarrow{\text{add 1}} 10110111$

    Thus the 8-bit two's complements of -6 and -73 are 11111010 and 10110111. Adding the 8-bit two's complements in binary notation gives

$$
\begin{array}{r}
11111010 \\
+\quad 10110111 \\
\hline
10110001
\end{array}
$$

Truncating the 1 in the 2^8th position gives 10110001. Since the leading bit of this number is a 1, the answer is negative. Converting back to decimal form gives

$$10110001 \xrightarrow{\text{flip the bits}} 01001110 \xrightarrow{\text{add 1}} 01001111_2 = (64 + 8 + 4 + 2 + 1) = 79_{10} = |-79|_{10}$$

So the answer is -79.

36. $123_{10} = (64 + 32 + 16 + 8 + 2 + 1)_{10} = 01111011_2$

$|-94|_{10} = (64 + 16 + 8 + 4 + 2)_{10} = 01011110_2 \xrightarrow{\text{flip the bits}} 10100001 \xrightarrow{\text{add 1}} 10100010$

So the 8-bit two's complements of 123 and -94 are 01111011 and 10100010. Adding the 8-bit two's complements in binary notation gives

$$
\begin{array}{r}
01111011 \\
+ \quad 10100010 \\
\hline
100011101
\end{array}
$$

Truncating the 1 in the 2^8th position gives 00011101. Since the leading bit of this number is a 0, the answer is positive. Converting back to decimal form gives

$$00011101_2 = (16 + 8 + 4 + 1)_{10} = 29_{10}.$$

So the answer is 29.

37. **a.**   The 8-bit two's complement of -128 is computed as follows:

$|-128|_{10} = 128_{10} = (2^7)_{10}$

$= 10000000_2 \xrightarrow{\text{flip the bits}} 01111111 \xrightarrow{\text{add 1}} 10000000.$

So the 8-bit two's complement of -128 is 10000000. If the two's complement procedure is applied to this result, the following is obtained $10000000 \xrightarrow{\text{flip the bits}} 01111111 \xrightarrow{\text{add 1}} 10000000.$ So the 8-bit two's complement of the 8-bit two's complement of -128 is 10000000, which is the 8-bit two's complement of -128.

38. $A2BC_{16} = 10 \cdot 16^3 + 2 \cdot 16^2 + 11 \cdot 16 + 12 = 41{,}660_{10}$

39. $E0D_{16} = 14 \cdot 16^2 + 0 + 13 = 3597_{10}$

41. $0001\ 1100\ 0000\ 1010\ 1011\ 1110_2$

42. $B53DF8_{16} = 1011\ 0101\ 0011\ 1101\ 1111\ 1000_2$

44. $2E_{16}$

45. $1011\ 0111\ 1100\ 0101_2 = B7C5_{16}$

47. **a.** $6 \cdot 8^4 + 1 \cdot 8^3 + 5 \cdot 8^2 + 0 \cdot 8 + 2 \cdot 1 = 25{,}410_{10}$

Review Guide: Chapter 2

Compound Statements

- What is a statement?
- If p and q are statements, how do you symbolize "p but q" and "neither p nor q"?
- What does the notation $a \leq x < b$ mean?
- What is the conjunction of statements p and q?
- What is the disjunction of statements p and q?
- What are the truth table definitions for $\sim p$, $p \wedge q$, $p \vee q$, $p \rightarrow q$, and $p \leftrightarrow q$?
- How do you construct a truth table for a general compound statement form?
- What is exclusive or?
- What is a tautology, and what is a contradiction?
- What is a conditional statement?
- Given a conditional statement, what is its hypothesis (antecedent)? conclusion (consequent)?
- What is a biconditional statement?
- What is the order of operations for the logical operators?

Logical Equivalence

- What does it mean for two statement forms to be logically equivalent?
- How do you test to see whether two statement forms are logically equivalent?
- How do you annotate a truth table to explain how it shows that two statement forms are or are not logically equivalent?
- What is the double negative property?
- What are De Morgan's laws?
- How is Theorem 2.1.1 used to show that two statement forms are logically equivalent?
- What are negations for the following forms of statements?
    a. $p \wedge q$
    b. $p \vee q$
    c. $p \rightarrow q$ (if p then q)

Converse, Inverse, Contrapositive

- What is the contrapositive of a statement of the form "If p then q"?
- What are the converse and inverse of a statement of the form "If p then q"?
- Can you express converses, inverses, and contrapositives of conditional statements in ordinary English?
- If a conditional statement is true, can its converse also be true?
- Given a conditional statement and its contrapositive, converse, and inverse, which of these are logically equivalent and which are not?

Necessary and Sufficient Conditions, Only If

- What does it mean to say that something is true only if something else is true?
- How are statements about only-if statements translated into if-then form.?
- What does it mean to say that something is a necessary condition for something else?
- What does it mean to say that something is a sufficient condition for something else?
- How are statements about necessary and sufficient conditions translated into if-then form?

Validity and Invalidity

- How do you identify the logical form of an argument?
- What does it mean for a form of argument to be valid?
- How do you test to see whether a given form of argument is valid?
- How do you annotate a truth table to explain how it shows that an argument is or is not valid?

- What does it mean for an argument to be sound?
- What are modus ponens and modus tollens?
- Can you give examples for and prove the validity of the following forms of argument?

a.

$$
\begin{array}{c}
p \\
\therefore \quad p \vee q
\end{array}
\quad \text{and} \quad
\begin{array}{c}
q \\
\therefore \quad p \vee q
\end{array}
$$

b.

$$
\begin{array}{c}
p \wedge q \\
\therefore \quad p
\end{array}
\quad \text{and} \quad
\begin{array}{c}
p \wedge q \\
\therefore \quad q
\end{array}
$$

c.

$$
\begin{array}{c}
p \vee q \\
\sim q \\
\therefore \quad p
\end{array}
\quad \text{and} \quad
\begin{array}{c}
p \vee q \\
\sim p \\
\therefore \quad q
\end{array}
$$

d.

$$
\begin{array}{c}
p \rightarrow q \\
q \rightarrow r \\
\therefore \quad p \rightarrow r
\end{array}
$$

e.

$$
\begin{array}{c}
p \vee q \\
p \rightarrow r \\
q \rightarrow r \\
\therefore \quad r
\end{array}
$$

- What are converse error and inverse error?
- Can a valid argument have a false conclusion?
- Can an invalid argument have a true conclusion?
- Which of modus ponens, modus tollens, converse error, and inverse error are valid and which are invalid?
- What is the contradiction rule?
- How do you use valid forms of argument to solve puzzles such as those of Raymond Smullyan about knights and knaves?

Digital Logic Circuits and Boolean Expressions

- Given a digital logic circuit, how do you
  - **a.** find the output for a given set of input signals
  - **b.** find the output for a given set of input signals
  - **c.** find the corresponding Boolean expression?
- What is a recognizer?
- Given a Boolean expression, how do you draw the corresponding digital logic circuit?
- Given an input/output table, how do you draw the corresponding digital logic circuit?
- What is disjunctive normal form?
- What does it mean for two circuits to be equivalent?
- What are NAND and NOR gates?
- What are Sheffer strokes and Peirce arrows?

Binary and Hexadecimal Notation

- How do you transform positive integers from decimal to binary notation and the reverse?
- How do you add and subtract integers using binary notation?
- What is a half-adder?
- What is a full-adder?
- What is the 8-bit two's complement of an integer in binary notation?
- How do you find the 8-bit two's complement of a positive integer that is at most 127?
- What is a convenient way to find the 8-bit two's complement of a negative integer that is greater than or equal to −128?
- How do you find the decimal representation of the integer with a given 8-bit two's complement?

- How do you perform addition and subtraction with integers in two's complement form?
- How do you transform positive integers from hexadecimal to decimal notation?
- How do you transform positive integers from binary to hexadecimal notation and the reverse?
- What is octal notation?

Fill-in-the-Blank Review Questions

Section 2.1

1. An *and* statement is true when, and only when, both components are _____.
2. An *or* statement is false when, and only when, both components are _____.
3. Two statement forms are logically equivalent when, and only when, they always have _____.
4. De Morgan's laws say (1) that the negation of an *and* statement is logically equivalent to the _____ statement in which each component is _____, and (2) that the negation of an *or* statement is logically equivalent to the _____ statement in which each component is _____.
5. A tautology is a statement that is always _____.
6. A contradiction is a statement that is always _____.

Section 2.2

1. An *if-then* statement is false if, and only if, the hypothesis is _____ and the conclusion is _____.
2. The negation of "if p then q" is _____.
3. The converse of "if p then q" is _____.
4. The contrapositive of "if p then q" is _____.
5. The inverse of "if p then q" is _____.
6. A conditional statement and its contrapositive are _____.
7. A conditional statement and its converse are not _____."
8. R is a sufficient condition for S" means "if _____ then _____."
9. "R is a necessary condition for S" means "if _____ then _____."
10. "R only if S" means "if _____ then _____."

Section 2.3

1. For an argument to be valid means that every argument of the same form whose premises _____ has a _____ conclusion.
2. For an argument to be invalid means that there is an argument of the same form whose premises _____ and whose conclusion _____.

3. For an argument to be sound means that its form is _____ and its premises _____. In this case we can be sure that its conclusion _____

Section 2.4

1. The input/output table for a digital logic circuit is a table that shows ____.
2. The Boolean expression that corresponds to a digital logic circuit is ____.
3. A recognizer is a digital logic circuit that ____.
4. Two digital logic circuits are equivalent if, and only if, ____.
5. A NAND-gate is constructed by placing a ____ gate immediately following an ____ gate.
6. A NOR-gate is constructed by placing a ____ gate immediately following an ____ gate.

Section 2.5

1. To represent a nonnegative integer in binary notation means to write it as a sum of products of the form ____, where ____.
2. To add integers in binary notation, you use the facts that $1_2 + 1_2 =$ ____ and $1_2 + 1_2 + 1_2 =$ ____.
3. To subtract integers in binary notation, you use the facts that $10_2 - 1_2 =$ ____ and $11_2 - 1_2 =$ ____.
4. A half-adder is a digital logic circuit that ____, and a full-adder is a digital logic circuit that ____.
5. If a is an integer with $-128 \leq a \leq 127$, the 8-bit two's complement of a is ____ if $a \geq 0$ and is ____ if $a < 0$.
6. To find the 8-bit two's complement of a negative integer a that is at least -128, you ____, ____, and ____.
7. To add two integers in the range -128 through 127 whose sum is also in the range -128 through 127, you ____, ____, ____, and ____.
8. To represent a nonnegative integer in hexadecimal notation means to write it as a sum of products of the form ____, where ____.
9. To convert a nonnegative integer from hexadecimal to binary notation, you ____ and ____.

Answers for Fill-in-the-Blank Review Questions

Section 2.1

1. true   2. false   3. the same truth value   4. *or*; negated; *and*; negated   5. true
6. false

Section 2.2

1. true; false   2. $p \land \sim q$ 3. if q then p 4. if $\sim q$ then $\sim p$ 5. if $\sim p$ then $\sim q$

6. logically equivalent   7. logically equivalent   8. R; S 9. S; R 10. S; R

Section 2.3

1. are all true; true   2. are all true; is false   3. valid; are all true; true

Section 2.4

1. shows the output signal(s) that correspond to all possible combinations of input signals to the circuit

2. a Boolean expression that represents the input signals as variables and indicates the successive actions of the logic gates on the input signals

3. outputs a 1 for exactly one particular combination of input signals and outputs 0's for all other combinations

4. they have the same input/output table

5. NOT; AND

6. NOT; OR

Section 2.5

1. $d \cdot 2^n$; $d = 0$ or $d = 1$, and n is a nonnegative integer

2. 10_2; 11_2

3. 1_2; 10_2

4. outputs the sum of any two binary digits; outputs the sum of any three binary digits

5. the 8-bit binary representation of a; the 8-bit binary representation of $2^8 - a$

6. write the 8-bit binary representation of a; flip the bits; add 1 in binary notation

7. convert both integers to their 8-bit two's complements; add the results using binary notation; truncate any leading 1; convert back to decimal form

8. $d \cdot 16^n$; $d = 0, 1, 2, \ldots 9, A, B, C, D, E, F$, and n is a nonnegative integer

9. write each hexadecimal digit in fixed 4-bit binary notation; juxtapose the results

Chapter 3: The Logic of Quantified Statements

Ability to use the logic of quantified statements correctly is necessary for doing mathematics because mathematics is, in a very broad sense, about quantity. The main purpose of this chapter is to familiarize you with the language of universal and existential statements. The various facts about quantified statements developed in this chapter are used extensively in Chapter 4 and are referred to throughout the rest of the book. Experience with the formalism of quantification is especially useful for students planning to study LISP or Prolog, program verification, or relational databases.

One thing to keep in mind is the tolerance for potential ambiguity in ordinary language, which is typically resolved through context or inflection. For instance, as the "Caution" on page 111 of the text indicates, the sentence "All mathematicians do not wear glasses" is one way to phrase a negation to "All mathematicians wear glasses." (To see this, say it out loud, stressing the word "not.") Some grammarians ask us to avoid such phrasing because of its potentially ambiguity, but the usage is widespread even in formal writing in high-level publications ("All juvenile offenders are not alike," Anthony Lewis, The New York Times, 19 May 1997, Op-Ed page), or in literary works ("All that glisters is not gold," William Shakespeare, The Merchant of Venice, Act 2, Scene 7, 1596-1597).

Even rather complex sentences can be negated in this way. For instance, when asked to write a negation for "The sum of any two irrational numbers is irrational," many people instinctively write "The sum of any two irrational numbers is not irrational." This is an acceptable informal negation (again, say it out loud, stressing the word "not"), but it can lead to genuine mistakes in formal situations. To avoid such mistakes, tell yourself that simply inserting the word "not" is very rarely a good way to express the negation of a mathematical statement.

Section 3.1

1. **a.** False   **b.** True

2. **a.** The statement is true. The integers correspond to certain of the points on a number line, and the real numbers correspond to all the points on the number line.

   b. The statement is false; 0 is neither positive nor negative.

   c. The statement is false. For instance, let $r = -2$. Then $-r = -(-2) = 2$, which is positive.

   d. The statement is false. For instance, the number $\frac{1}{2}$ is a real number, but it is not an integer.

3. **a.** When $m = 25$ and $n = 10$, the statement "m is a factor of n^2" is true because $n^2 = 100$ and $100 = 4 \cdot 25$. But the statement "m is a factor of n" is false because 10 is not a product of 25 times any integer. Thus the hypothesis of $R(m, n)$ is true and the conclusion is false, so the statement as a whole is false.

   b. *Sample answer:* $R(m, n)$ is false when $m = 8$ and $n = 4$ because 8 is a factor of $4^2 = 16$, but 8 is not a factor of 4.

   c. When $m = 5$ and $n = 10$, both statements "m is a factor of n^2" and "m is a factor of n" are true because $n = 10 = 5 \cdot 20 = m \cdot 20$. Thus both the hypothesis and conclusion of $R(m, n)$ are true, and so the statement as a whole is true.

   d. *Two sample answers:*

   (1) Let $m = 2$ and $n = 6$. Then both statements "m is a factor of n^2" and "m is a factor of n" are true because $n = 6 = 2 \cdot 3 = m \cdot 3$ and $n^2 = 36 = 2 \cdot 18 = m \cdot 18$. Thus both the hypothesis and conclusion of $R(m, n)$ are true, and so the statement as a whole is true.

   (2) Let $m = 6$ and $n = 2$. Then both statements "m is a factor of n^2" and "m is a factor of n" are false because $n = 2 \neq 6 \cdot k$, for any integer k, and $n^2 = 4 \neq 6 \cdot j$, for any integer j. Thus

both the hypothesis and conclusion of $R(m, n)$ are false, and so the statement as a whole is true.

4. **a.** $Q(-2, 1)$ is the statement "If $-2 < 1$ then $(-2)^2 < 1^2$." The hypothesis of this statement is $-2 < 1$, which is true. The conclusion is $(-2)^2 < 1^2$, which is false because $(-2)^2 = 4$ and $1^2 = 1$ and $4 \not< 1$. Thus $Q(-2, 1)$ is a conditional statement with a true hypothesis and a false conclusion. So $Q(-2, 1)$ is false.

c. $Q(3, 8)$ is the statement "If $3 < 8$ then $3^2 < 8^2$." The hypothesis of this statement is $3 < 8$, which is true. The conclusion is $3^2 < 8^2$, which is also true because $3^2 = 9$ and $8^2 = 64$ and $9 < 64$. Thus $Q(3, 8)$ is a conditional statement with a true hypothesis and a true conclusion. So $Q(3, 8)$ is true.

5. **a.** The truth set is the set of all integers d such that $6/d$ is an integer, so the truth set is $\{-6, -3, -2, -1, 1, 2, 3, 6\}$.

c. The truth set is the set of all real numbers x with the property that $1 \le x^2 \le 4$, so the truth set is $\{x \in \mathbf{R} \mid -2 \le x \le -1 \text{ or } 1 \le x \le 2\}$. In other words, the truth set is the set of all real numbers between -2 and -1 inclusive together with those between 1 and 2 inclusive.

6. **a.** $\{-9, -8, -7, -6, -5, -4, -3, -2, -1, 0, 1, 2, 3, 4, 5, 6, 7, 8, 9\}$

b. Truth set $= \{1, 2, 3, 4, 5, 6, 7, 8.9\}$

c. Truth set $= \{-8, -6, -4, -2, 0, 2, 4, 6, 8\}$

7. $baa, bab, bac, bba, bbb, bbc, bca, bcb$

9. Counterexample: Let $x = 1/2$. Then $\dfrac{1}{x} = \dfrac{1}{(1/2)} = 2$, and $1/2 \not> 2$. (*This is one counterexample among many.*)

11. Counterexample: Let $m = 1$ and $n = 1$. Then $m \cdot n = 1 \cdot 1 = 1$ and $m + n = 1 + 1 = 2$. But $1 \not> 2$, and so $m \cdot n \not> m + n$. (*This is one counterexample among many.*)

12. Counterexample: Let $x = 1$ and $y = 1$, and note that

$$\sqrt{x + y} = \sqrt{1 + 1} = \sqrt{2}$$

whereas

$$\sqrt{x} + \sqrt{y} = \sqrt{1} + \sqrt{1} = 1 + 1 = 2,$$

and

$$2 \ne \sqrt{2}.$$

(This is one counterexample among many. Any real numbers x and y with $xy \ne 0$ will produce a counterexample.)

13. (a), (e), (f)

14. (b), (c), (e), (f)

15. **a.** *Some acceptable answers*: All rectangles are quadrilaterals. If a figure is a rectangle then that figure is a quadrilateral. Every rectangle is a quadrilateral. All figures that are rectangles are quadrilaterals. Any figure that is a rectangle is a quadrilateral.

b. *Some acceptable answers*: There is a set with sixteen subsets. Some set has sixteen subsets. Some sets have sixteen subsets. At least one set has 16 subsets. There is at least one set that has sixteen subsets.

16. **a.** ∀ dinosaur x, x is extinct.

   c. ∀ irrational number x, x is not an integer.

   e. ∀ integer x, x^2 does not equal $2,147,581,953$.

17. **a.** ∃ an exercise x such that x has an answer.

18. **a.** $\exists s \in D$ such that $E(s)$ and $M(s)$. (*Or:* $\exists s \in D$ such that $E(s) \wedge M(s)$.)

   b. $\forall s \in D$, if $C(s)$ then $E(s)$. (*Or:* $\forall s \in D, C(s) \rightarrow E(s)$.)

   c. $\forall s$, if $C(s)$ then $\sim E(s)$.

   d. $\exists x$ such that $C(s) \wedge M(s)$.

   e. $(\exists s \in D$ such that $C(s) \wedge E(s)) \wedge (\exists s \in D$ such that $C(s) \wedge \sim E(s))$

19. (b), (d), (e)

20. *Partial answer:* The square root of a positive real number is positive.

21. **a.** The total degree of G is even, for any graph G.

   b. The base angles of T are equal, for any isosceles triangle T.

   c. p is even, for some prime number p.

   d. f is not differentiable, for some continuous function f.

22. **a.** $\forall x$, if x is a Java program, then x has at least 5 lines.

23. **a.** $\forall x$ if x is an equilateral triangle, then x is isosceles.

   ∀ equilateral triangles x, x is isosceles.

24. **a.** ∃ a hatter x such that x is mad.

   $\exists x$ such that x is a hatter and x is mad.

25. **a.** ∀ nonzero fraction x, the reciprocal of x is a fraction.

   $\forall x$, if x is a nonzero fraction then the reciprocal of x is a fraction.

   c. ∀ triangle x, the sum of the angles of x is $180°$.

   $\forall x$, if x is a triangle then the sum of the angles of x is $180°$.

   e. ∀ even integers x and y, the sum of x and y is even.

   $\forall x$ and y, if x and y are even integers then the sum of x and y is even.

26. **b.** $\forall x(\text{Int}(x) \rightarrow \text{Ratl}(x)) \wedge \exists x(\text{Ratl}(x) \wedge \sim \text{Int}(x))$

27. **a.** False. Figure b is a circle that is not gray.

   b. True. All the gray figures are circles.

   c. This statement translates as "There is a square that is above d." This is false because the only objects above d are a (a triangle) and b (a circle).

   d. This statement translates as "There is a triangle that has f above it," or, "f is above some triangle." This is true because g is a triangle and f is above g.

28. **b.** If a real number is negative, then when its opposite is computed, the result is a positive real number.

   This statement is true because for each real number x, $-(-|x|) = |x|$ (and any negative real number can be represented as $-|x|$, for some real number x).

   d. There is a real number that is not an integer. This statement is true. For instance, $\frac{1}{2}$ is a real number that is not an integer.

30. **a.** This statement translates as "There is a prime number that is not odd." This is true. The number 2 is prime and it is not odd.

   b. If an integer is prime, then it is not a perfect square.

   This statement is true because a prime number is an integer greater than 1 that is not a product of two smaller positive integers. So a prime number cannot be a perfect square because if it were, it would be a product of two smaller positive integers.

   c. This statement translates as "There is a number that is both an odd number and a perfect square." This is true. For example, the number 9 is odd and it is also a perfect square (because $9 = 3^2$).

31. *Hint:* Your answer should have the appearance shown in the following made-up example:

   Statement: "If a function is differentiable, then it is continuous."

   Formal version: $\forall$ function f, if f is differentiable, then f is continuous.

   Citation: Calculus by D. R. Mathematician, Best Publishing Company, 2019, page 263.

32. **a.** True: Any real number that is greater than 2 is greater than 1.

   c. False: $(-3)^2 > 4$ but $-3 \not> 2$.

33. **a.** True. Whenever both a and b are positive, so is their product.

   b. False. Let $a = -2$ and $b = -3$. Then $ab = 6$, which is not less than zero.

   c. This statement translates as "For all real numbers a and b, if $ab = 0$ then $a = 0$ or $b = 0$," which is true.

   d. This statement translates as "For all real numbers a, b, c, and d, if $a < b$ and $c < d$ then $ac < bd$," which is false.

   Counterexample: Let $a = -2$, $b = 1$, $c = -3$, and $d = 0$. Then $a < b$ because $-2 < 1$ and $c < d$ because $-3 < 0$, but $ac \not< bd$ because $ac = (-2)(-3) = 6$ and $bd = 1 \cdot 0 = 0$ and $6 \not< 0$.

Section 3.2

1. (a) and (e) are negations.

3. **a.** $\exists$ a string s such that s does not have any characters. (*Or:* $\exists$ a string s such that s has no characters.)

   b. $\exists$ a computer c such that c does not have a CPU.

   c. $\forall$ movie m, m is less than or equal to 6 hours long. (*Or:* $\forall$ movie m, m is no more than 6 hours long.)

   d. $\forall$ band b, b has won fewer than 10 Grammy awards.

   In 4-7 there are other correct answers in addition to those shown.

4. **a.** Some dogs are unfriendly. (*Or:* There is at least one unfriendly dog.)

   c. All suspicions were unsubstantiated. (*Or:* No suspicions were substantiated.)

5. **a.** There is a valid argument that does not have a true conclusion. (*Or:* There is at least one valid argument that does not have a true conclusion.)

6. **a.** Sets A and B have at least one point in common.

   Or: There is a point p such that p is in both sets A and B.

   b. There is a road on the map that connects towns P and Q.

   Or: Some road on the map connects towns P and Q.

   Or: Towns P and Q are connected by a road on the map.

   Or: $\exists$ a road r on the map such that r connects towns P and Q.

7. **a.** This vertex is connected to at least one other vertex in the graph.

Or: There is at least one other vertex in the graph to which this vertex is connected.

Or: This vertex is connected to some other vertex in the graph.

9. $\exists$ a real number x such that $x > 3$ and $x^2 \leq 9$.

11. The proposed negation is not correct. The given statement makes a claim about *any* two irrational numbers and means that no matter what two irrational numbers you might choose, the sum of those numbers will be irrational. For this to be false means that there is at least one pair of irrational numbers whose sum is rational. On the other hand, the negation proposed in the exercise ("The sum of any two irrational numbers is rational") means that given *any* two irrational numbers, their sum is rational. This is a much stronger statement than the actual negation: The truth of this statement implies the truth of the negation (assuming that there are at least two irrational numbers), but the negation can be true without having this statement be true.

Correct negation: There are at least two irrational numbers whose sum is rational.

Or: The sum of some two irrational numbers is rational.

12. The proposed negation is not correct. *Correct negation:* There are an irrational number x and a rational number y such that xy is rational. Or: There are an irrational number and a rational number whose product is rational.

13. The proposed negation is not correct. There are two mistakes: The negation of a "for every" statement is not a "for every" statement; and the negation of an if-then statement is not an if-then statement.

Correct negation: There exists an integer n such that n^2 is even and n is not even.

15. **a.** True: All the odd numbers in D are positive.

b. True: The only numbers that are less than 0 are -48, -14, and -8, and all three are even.

c. False: $x = 16, x = 26, x = 32$, and $x = 36$ are all counterexamples. They are even numbers that are not negative.

d. True. (Remember that an *or* statement is true if at least one component is true.)

e. False: $x = 36$ is a counterexample because the ones digit of x is 6 and the tens digit is neither 1 nor 2.

16. $\exists$ a real number x such that $x^2 \geq 1$ and $x \not> 0$. In other words, $\exists$ a real number x such that $x^2 \geq 1$ and $x \leq 0$.

18. $\exists$ a real number x such that $x(x + 1) > 0$ and both $x \leq 0$ and $x \geq -1$.

20. $\exists$ integers a, b, and c such that $a - b$ is even and $b - c$ is even and $a - c$ is not even.

21. $\exists$ an integer n such that n is divisible by 6 and either n is not divisible by 2 or n is not divisible by 3 (or both).

22. There is an integer with the property that the square of the integer is odd but the integer itself is not odd. (*Or:* At least one integer has an odd square but is not itself odd.)

24. **a.** If a person is a child in Tom's family, then the person is female.

If a person is a female in Tom's family, then the person is a child.

The second statement is the converse of the first.

b. If an integer greater than 5 ends in 1, 3, 7, or 9, then the integer is prime.

If an integer greater than 5 is prime, then the integer ends in 1, 3, 7, or 9.

The second statement is the converse of the first.

25. **a.** *Converse*: If $n + 1$ is an even integer, then n is a prime number that is greater than 2.

    Counterexample: Let $n = 15$. Then $n + 1 = 16$, which is even but n is not a prime number that is greater than 2.

26. *Statement*: $\forall$ real number x, if $x^2 \geq 1$ then $x > 0$.

    Contrapositive: $\forall$ real number x, if $x \leq 0$ then $x^2 < 1$.

    Converse:$\forall$ real number x, if $x > 0$ then $x^2 \geq 1$.

    Inverse: $\forall$ real number x, if $x^2 < 1$ then $x \leq 0$.

    The statement and its contrapositive are false. As a counterexample, let $x = -2$. Then $x^2 = (-2)^2 = 4$, and so $x^2 \geq 1$. However $x \not> 0$.

    The converse and the inverse are also false. As a counterexample, let $x = 1/2$. Then $x^2 = 1/4$, and so $x > 0$ but $x^2 \not\geq 1$.

27. *Statement*: $\forall$ integer d, if $6/d$ is an integer then $d = 3$.

    Contrapositive: $\forall$ integer d, if $d \neq 3$ then $6/d$ is not an integer.

    Converse: $\forall$ integer d, if $d = 3$ then $6/d$ is an integer.

    Inverse: $\forall$ integer d, if $6/d$ is not an integer, then $d \neq 3$.

    The converse and inverse of the statement are both true, but both the statement and its contrapositive are false. For example, when $d = 2$, then $d \neq 3$ but $6/d = 3$ is an integer.

28. *Statement:* $\forall x \in \mathbf{R}$, if $x(x + 1) > 0$ then $x > 0$ or $x < -1$.

    Contrapositive: $\forall x \in \mathbf{R}$, if $x \leq 0$ and $x \geq -1$, then $x(x + 1) \leq 0$.

    Converse: $\forall x \in \mathbf{R}$, if $x > 0$ or $x < -1$ then $x(x + 1) > 0$.

    Inverse: $\forall x \in \mathbf{R}$, if $x(x + 1) \leq 0$ then $x \leq 0$ and $x \geq -1$.

    The statement, its contrapositive, its converse, and its inverse are all true.

30. *Statement:* $\forall$ integers a, b, and c, if $a - b$ is even and $b - c$ is even, then $a - c$ is even.

    Contrapositive: $\forall$ integers a, b, and c, if $a - c$ is not even, then $a - b$ is not even or $b - c$ is not even.

    Converse: $\forall$ integers a, b, and c, if $a - c$ is even then $a - b$ is even and $b - c$ is even.

    Inverse: $\forall$ integers a, b, and c, if $a - b$ is not even or $b - c$ is not even, then $a - c$ is not even.

    The statement is true, but its converse and inverse are false. As a counterexample, let $a = 3, b = 2$, and $c = 1$. Then $a - c = 2$, which is even, but $a - b = 1$ and $b - c = 1$, so it is not the case that both $a - b$ and $b - c$ are even.

32. *Statement:* If the square of an integer is odd, then the integer is odd.

    Contrapositive: If an integer is not odd, then the square of the integer is not odd.

    Converse: If an integer is odd, then the square of the integer is odd.

    Inverse: If the square of an integer is not odd, then the integer is not odd.

    The statement, its contrapositive, its converse, and its inverse are all true.

33. *Statement*: If a function is differentiable, then it is continuous.

    Contrapositive: If a function is not continuous, then it is not differentiable.

    Converse: If a function is continuous, then it is differentiable.

    Inverse: If a function is not differentiable, then it is not continuous.

    The statement and its contrapositive are true, but both the converse and inverse are false. For example, take the function f defined by $f(x) = |x|$ for every real number x. This function is continuous for every real number, but it is not differentiable at $x = 0$.

34. **a.** If n is divisible by some prime number between 1 and $\sqrt{n}$ inclusive, then n is not prime.

36. **a.** *One possible answer*: Let $P(x)$ be "$2x \neq 1$." The statement "$\forall x \in \mathbf{Z}, 2x \neq 1$" is true because there is no integer which, when doubled, equals 1. But the statements "$\forall x \in \mathbf{Q}, 2x \neq 1$" and "$\forall x \in \mathbf{R}, 2x \neq 1$" are both false because $x = 1/2$ satisfies the equation $2x = 1$ and $1/2$ is in both $\mathbf{R}$ and $\mathbf{Q}$.

    b. *One possible answer*: Let $P(x)$ be "$x^2 \neq 2$." The statements "$\forall x \in \mathbf{Z}, x^2 \neq 2$" and "$\forall x \in \mathbf{Q}, x^2 \neq 2$" are true because there is not integer or rational number whose square is 2, but the statement "$\forall x \in \mathbf{R}, x^2 \neq 2$" is false because $\left(\sqrt{2}\right) = 2$.

37. The claim is "$\forall x$, if $x = 1$ and x is in the sequence 0204, then x is to the left of all the 0's in the sequence."

    The negation is "$\exists x$ such that $x = 1$ and x is in the sequence 0204, and x is not to the left of all the 0's in the sequence." The negation is false because the sequence does not contain the character 1. So the claim is vacuously true (or true by default)

39. If a person earns a grade of C$^-$ in this course, then the course counts toward graduation.

41. If a person is not on time each day, then the person will not keep this job.

42. If a person does not pass a comprehensive exam, then that person cannot obtain a master's degree. *Or*: If a person obtains a master's degree then that person passed a comprehensive exam.

43. If a number is prime, then it is greater than 1.

45. To say that "Being divisible by 8 is a necessary condition for being divisible by 4" means that, "If a number is not divisible by 8 then that number is not divisible by 4. The negation is, "There is a number that is not divisible by 8 and is divisible by 4."

47. To say that "having a large income is a sufficient condition for being happy" means that "If a person has a large income then that person is happy." The negation is "There is a person who has a large income and is not happy."

48. To say that "Being a polynomial is a sufficient condition for a function to have a real root" means that "If a function is a polynomial, then that function has a real root." The negation is "There is a function that is a polynomial but does not have a real root."

50. No. Interpreted formally, the statement says, "If carriers do not offer the same lowest fare, then you may not select among them."

Section 3.3

1. **a.** True: Tokyo is the capital of Japan.

    b. False: Athens is not the capital of Egypt.

2. **a.** True: $2^2 > 3$ **b.** False: $1^2 \not> 1$

3. **a.** Let $y = \dfrac{1}{2}$. Then $xy = 2\left(\dfrac{1}{2}\right) = 1$.

    b. Let $y = -1$. Then $xy = (-1)(-1) = 1$.

    c. Let $y = \dfrac{4}{3}$. Then $xy = \left(\dfrac{3}{4}\right)\left(\dfrac{4}{3}\right) = 1$.

4. **a.** *One possible answer*: Let $n = 16$. Then $n > x$ because $16 > 15.83$.

5. The statement says that no matter what circle anyone might give you, you can find a square of the same color.

Solution 1: The statement is true because the only circles in the Tarski world are a, b, and c, and given a or c, which are blue, square j is also blue, and given b, which is gray, squares g and h are also gray.

Solution 2: The statement is true. The Tarski world shown in Figure 3.3.1 has exactly three circles: a, b, and c.

Given circle $x =$	Choose square $y =$	Is y the same color as x?
a	j	yes ✓
b	g or h	yes ✓
c	j	yes ✓

6. True. The Tarski world shown in Figure 3.3.1 has exactly four squares: e, g, h, and j.

Given square $x =$	Choose circle $y =$	Is y above x and with a different color from x?
e	a, b, or c	yes ✓
g	a or c	yes ✓
h	a or c	yes ✓
j	b	yes ✓

7. *Solution 1:* The statement is true because the Tarski world has exactly four squares: e, g, h, and j and triangle d is above all of them.

Solution 2: The statement is true. The Tarski world shown in Figure 3.3.1 has exactly four squares: e, g, h, and j.

Choose triangle $x = d$	Choose square $y =$	Is x above y?
	e	yes ✓
	f	yes ✓
	h	yes ✓
	j	yes ✓

9. **a.** There are five elements in D. For each, an element in E must be found so that the sum of the two equals 0. So: for $x = -2$, take $y = 2$; for $x = -1$, take $y = 1$; for $x = 0$, take $y = 0$; for $x = 1$, take $y = -1$; and for $x = 2$, take $y = -2$.

Alternatively, note that for each integer x in D, the integer $-x$ is also in D, including 0 (because $-0 = 0$), and for every integer $x, x + (-x) = 0$.

b. True. *Solution 1*: Let $x = 0$. Then for any real number r, $x + r = r + x = r$ because 0 is an identity for addition of real numbers. Thus, because every element in E is a real number, $\forall y \in E, x + y = y$.

Solution 2: Let $x = 0$. Then $x + y = y$ is true for each individual element y of E:

Choose $x = 0$	Given $y =$	Is $x + y = y$?
	-2	yes: $0 + (-2) = -2$ ✓
	-1	yes: $0 + (-1) = -1$ ✓
	0	yes: $0 + 0 = 0$ ✓
	1	yes: $0 + 1 = 1$ ✓
	2	yes: $0 + 2 = 2$ ✓

10. **a.** True. Every student chose at least one dessert: Uta chose pie, Tim chose both pie and cake, and Yuen chose pie.

c. This statement says that some particular dessert was chosen by every student. This is true: Every student chose pie.

11. **a.** There is a student who has seen *Casablanca*.

    c. Every student has seen at least one movie.

    d. There is a movie that has been seen by every student. (There are many other acceptable ways to state these answers.)

12. **a.** *Negation*: $\exists x$ in D such that $\forall y$ in $E, x + y \neq 1$.

    The negation is true. When $x = -2$, the only number y with the property that $x + y = 1$ is $y = 3$, and 3 is not in E.

    b. *Negation*: $\forall x$ in $D, \exists y$ in E such that $x + y \neq -y$.

    The negation is true because the original statement is false. To see that the original statement is false, take any x in D and choose y to be any number in E with $y \neq -\frac{x}{2}$. Then $2y \neq -x$, and adding x and subtracting y from both sides gives $x + y \neq -y$.

    c. *Negation*: $\exists\, x$ in D such that $\forall\, y$ in $E, xy \not\geq y$. (*Or*: $\exists\, x$ in D such that $\forall\, y$ in $E, xy < y$.)

    The statement is true. For each number x in D, you can find a y in D so that $xy \geq y$. Here is a table showing one way to do this: how all possible choices for x could be matched with a y so that $xy \geq y$.:

Given $x =$	you could take $y =$	Is $xy \geq y$?
-2	-2	$(-2) \cdot (-2) = 4 \geq -2$ ✓
-1	0	$(-1) \cdot 0 = 0 \geq -1$ ✓
0	1	$0 \cdot 1 = 0 \geq 0$ ✓
1	1	$1 \cdot 1 = 1 \geq 1$ ✓
2	2	$2 \cdot 2 = 4 \geq 2$ ✓

    d. *Negation*: $\forall x$ in $D, \exists\, y$ in E such that $x \not\leq y$. (*Or*: $\forall x$ in $D, \exists\, y$ in E such that $x > y$.)

    The statement is true. It says that there is a number in D that is less than or equal to every number in D. In fact, -2 is in D and -2 is less than or equal to every number in D ($-2, -1$, $0, 1$, and 2).

    In 13-19 there are other correct answers in addition to those shown.

13. **a.** *Statement*: For every color, there is an animal of that color.

    Or: There are animals of every color.

    b. *Negation*: $\exists$ a color C such that $\forall$ animal A, A is not colored C.

    Or: For some color, there is no animal of that color.

14. **a.** *Statement*: There is a book that every person has read.

    Or: $\exists$ a book b such that $\forall$ person p, p has read **b**.

    b. *Negation*: There is no book that every person has read.

    Or: $\forall$ book $b, \exists$ a person p such that p has not read b.

15. **a.** *Statement*: For every odd integer n, there is an integer k such that $n = 2k + 1$.

    Given any odd integer, there is another integer for which the given integer equals twice the other integer plus 1. Given any odd integer n, we can find another integer k so that $n = 2k + 1$.

    An odd integer is equal to twice some other integer plus 1.

    Every odd integer has the form $2k + 1$ for some integer k.

    b. *Negation*: $\exists$ an odd integer n such that $\forall$ integer $k, n \neq 2k + 1$.

    There is an odd integer that is not equal to $2k + 1$ for any integer k.

    Some odd integer does not have the form $2k + 1$ for any integer k.

18. **a.** *Statement*: For every real number x, there is a real number y such that $x + y = 0$.

   Given any real number x, there exists a real number y such that $x + y = 0$.

   Given any real number, we can find another real number (possibly the same) such that the sum of the given number plus the other number equals 0.

   Every real number can be added to some other real number (possibly itself) to obtain 0.

   b. *Negation*: $\exists$ a real number x such that $\forall$ real number $y, x + y \neq 0$.

   There is a real number x for which there is no real number y with $x + y = 0$.

   There is a real number x with the property that $x + y \neq 0$ for any real number y.

   Some real number has the property that its sum with any other real number is nonzero.

19. **a.** There is a real number whose sum with any real number equals zero.

   b. *Negation*: Given any real number x, there is a real number y such that $x + y \neq 0$.

   Or: $\forall x \in \mathbf{R}$, $\exists y \in \mathbf{R}$ such that $x + y \neq 0$.

20. **a.** Statement (1) says that no matter what square anyone might give you, you can find a triangle of a different color. This is true because the only squares are e, g, h, and j, and given squares g and h, which are gray, you could take triangle d, which is black; given square e, which is black, you could take either triangle f or i, which are gray; and given square j, which is blue, you could take either triangle f or h, which are gray, or triangle d, which is black. In each case the chosen triangle has a different color from the given square.

21. **a.** (1) The statement "$\forall$ real number x, $\exists$ a real number y such that $2x + y = 7$" is true. Given any real number x, take y to be $7 - 2x$.

   (2) The statement "$\exists$ a real number x such that $\forall$ real number y, $2x + y = 7$" is false. If it were true, the single number x would equal $\frac{7-y}{2}$ for every real number y, and that is impossible.

   b. Both statements (1) "$\forall$ real number x, $\exists$ a real number y such that $x + y = y + x$" and (2) "$\exists$ a real number x such that $\forall$ real number y, $x + y = y + x$" are true.

   c. Statement (1) is true because $x^2 - 2xy + y^2 = (x - y)^2$. Thus given any real number x, take $y = x$, then $x - y = 0$, and so $x^2 - 2xy + y^2 = 0$.

   Statement (2) is false. Given any real number x, choose a real number y with $y \neq x$. Then $x^2 - 2xy + y^2 = (x - y)^2 \neq 0$.

   d. Statement (1) is true because no matter what real number x might be chosen, y can be taken to be 1 so that $(x - 5)(y - 1) = (x - 5) \cdot 0 = 0$.

   Statement (2) is also true. Take $x = 5$. Then for all real numbers y, $(x - 5)(y - 1) = 0 (y - 1) = 0$.

   e. Statements (1) and (2) are both false because all real numbers have nonnegative squares and the sum of any two nonnegative real numbers is nonnegative. Hence for all real numbers x and y, $x^2 + y^2 \neq -1$.

22. **a.** Given any real number, you can find a real number so that the sum of the two is zero. In other words, every real number has an additive inverse. This statement is true.

   b. There is a real number with the following property: No matter what real number is added to it, the sum of the two will be zero. In other words, there is one particular real number whose sum with any real number is zero. This statement is false; no one number will work for all numbers. For instance, if $x + 0 = 0$, then $x = 0$, but in that case $x + 1 = 1 \neq 0$.

24. **a.** $\begin{aligned} \sim(\forall x \in D\ (\forall y \in E\ (P(x, y)))) \ &= \ \exists x \in D\ (\sim(\forall y \in E\ (P(x, y)))) \\ &= \ \exists x \in D\ (\exists y \in E\ (\sim P(x, y))) \end{aligned}$

   b. $\begin{aligned} \sim(\exists x \in D\ (\exists y \in E\ (P(x, y)))) \ &= \ \forall x \in D\ (\sim(\exists y \in E\ (P(x, y)))) \\ &= \ \forall x \in D\ (\forall y \in E\ (\sim P(x, y))) \end{aligned}$

25. The statement says that all of the circles are above all of the squares. This statement is true because the circles are a, b, and c, and the squares are e, g, h, and j, and all of a, b and c lie above all of e, g, h, and j.

    Negation: There is a circle x and a square y such that x is not above y. In other words, at least one of the circles is not above at least one of the squares.

27. The statement says that there are a circle and a square with the property that the circle is above the square and has a different color from the square. This statement is true. For example, circle a lies above square e and is differently colored from e. (Several other examples could also be given.)

29. **a.** *Version with interchanged quantifiers*: $\exists x \in \mathbf{R}$ such that $\forall y \in \mathbf{R}$, $x < y$.

    b. The given statement says that for any real number x, there is a real number y that is greater than x. This is true: For any real number x, let $y = x + 1$. Then $x < y$. The version with interchanged quantifiers says that there is a real number that is less than every other real number (including the negative ones). This is false.

30. **a.** $\forall x \in \mathbf{R}$, $\exists y \in \mathbf{R}^-$ such that $x > y$.

    b. The original statement says that there is a real number that is greater than every negative real number. This is true. For instance, 0 is greater than every negative real number.

    The statement with interchanged quantifiers says that no matter what real number might be given, it is possible to find a negative real number that is smaller. This is also true. If the number x that is given is positive, y could be taken to be -1. Then $x > y$. On the other hand, if the number x that is given is 0 or negative, y could be taken to be $x - 1$. In this case also, $x > y$.

31. $\forall$ person x, $\exists$ a person y such that x is older than y.

32. $\exists$ a person x such that $\forall$ person y, x is older than y.

33. **a.** *Formal version*: $\forall$ person x, $\exists$ a person y such that x loves y.

    b. *Negation*: $\exists$ a person x such that $\forall$ person y, x does not love y. In other words, there is someone who does not love anyone.

34. **a.** *Formal version*: $\exists$ a person x such that $\forall$ person y, x loves y.

    b. *Negation*: $\forall$ person x, $\exists$ a person y such that x does not love y. In other words, everyone has someone whom they do not love.

36. **a.** *Formal version*: $\exists$ a person x such that $\forall$ person y, x trusts y.

    b. *Negation*: $\forall$ person x, $\exists$ a person y such that x does not trust y.

    Or: Nobody trusts everybody.

37. **a.** *Formal version*: $\forall$ even integer n, $\exists$ an integer k such that $n = 2k$.

    b. *Negation*: $\exists$ an even integer n such that $\forall$ integer k, $n \neq 2k$.

    There is some even integer that is not equal to twice any other integer.

39. **a.** *Formal version*: $\exists$ a program P such that $\forall$ question Q posed to P, P gives the correct answer to Q.

    b. *Negation*: $\forall$ program P, there is a question Q that can be posed to P such that P does not give the correct answer to Q.

40. **a.** $\forall$ minutes m, $\exists$ a sucker s such that s was born in minute m.

41. **a.** The statement says that given any positive integer, there is a positive integer such that the first integer is 1 more than the second integer. This is false. Given the positive integer $x = 1$, the only integer with the property that $x = y + 1$ is $y = 0$, and 0 is not a positive integer.

b. The statement says that given any integer, there is an integer such that the first integer is 1 more than the second integer. This is true. Given any integer x, take $y = x - 1$. Then y is an integer, and $y + 1 = (x - 1) + 1 = x$.

e. The statement is true because the real number 0 has the property that $\forall y \in \mathbf{R}, 0 + y = y$.

f. The statement says that the difference of any two positive integers is a positive integer. This is false because, for example, $2 - 3 = -1$.

42. $\exists \varepsilon > 0$ such that $\forall$ integer N, $\exists$ an integer n such that $n > N$ and either $L - \varepsilon \geq a_n$ or $a_n \geq L + \varepsilon$. In other words, there is a positive number ε such that for every integer N, it is possible to find an integer n that is greater than N and has the property that a_n does not lie between $L - \varepsilon$ and $L + \varepsilon$.

44. **a.** The statement is true. The unique real number with the given property is 1. Note that
$1 \cdot y = y$ for all real numbers y,

45. $\exists! x \in D$ such that $P(x) \equiv \exists x \in D$ such that $(P(x) \wedge (\forall y \in D, \text{ if } P(y) \text{ then } y = x))$

Or: There exists a unique x in D such that $P(x)$.

Or: There is one and only one x in D such that $P(x)$.

46. **a.** The statement says that there is a triangle that is above all the squares. It is true because both triangles a and c lie above all the squares.

b. *Formal version*: $\exists x (\text{Triangle}(x) \wedge (\forall y (\text{Square}(y) \rightarrow \text{Above}(x, y))))$

c. *Formal negation*: $\forall x (\sim (\text{Triangle}(x) \wedge (\forall y (\text{Square}(y) \rightarrow \text{Above}(x, y)))))$
$$\equiv \forall x (\sim \text{Triangle}(x) \ \vee \sim (\forall y (\text{Square}(y) \rightarrow \text{Above}(x, y))))$$
$$\equiv \forall x (\sim \text{Triangle}(x) \vee (\exists y (\text{Square}(y) \wedge \sim \text{Above}(x, y))))$$

48. **a.** The statement says that given any circle, there is a square that is to the right of the circle. This is false because there is no square to the right of circle k.

b. *Formal version*: $\forall x (\text{Circle}(x) \rightarrow (\exists y (\text{Square}(y) \wedge \text{RightOf}(y, x))))$

c. *Formal negation*: $\exists x (\text{Circle}(x) \wedge \sim (\exists y (\text{Square}(y) \wedge \text{RightOf}(y, x))))$
$$\equiv \exists x (\text{Circle}(x) \wedge \forall y (\sim \text{Square}(y) \vee \sim \text{RightOf}(y, x)))$$

49. **a.** The statement says that given any circle there is a square with the same color. This is false because, for example, circle d is gray and there is no square that is colored gray.

b. *Formal version*: $\forall x (\text{Circle}(x) \rightarrow \exists y (\text{Square}(y) \wedge \text{SameColor}(y, x)))$

c. *Formal negation*: $\exists x (\sim (\text{Circle}(x) \rightarrow \exists y (\text{Square}(y) \wedge \text{SameColor}(y, x))))$
$$\equiv \exists x (\text{Circle}(x) \wedge \sim (\exists y (\text{Square}(y) \wedge \text{SameColor}(y, x))))$$
$$\equiv \exists x (\text{Circle}(x) \wedge \forall y (\sim (\text{Square}(y) \wedge \text{SameColor}(y, x))))$$
$$\equiv \exists x (\text{Circle}(x) \wedge \forall y (\sim \text{Square}(y)) \vee \sim (\text{SameColor}(y, x)))$$

51. **a.** The statement is true: Square e has the property that every triangle above it has the same color e because e is colored blue and the only triangles above e, namely a and c, are also colored blue.

b. *Formal version*: $\exists x (\text{Square}(x) \wedge (\forall y (\text{Triangle}(y) \wedge \text{Above}(y, x)) \rightarrow \text{SameColor}(y, x)))$

c. *Formal negation*:

$\forall x(\sim(\text{Square}(x) \wedge (\forall y((\text{Triangle}(y) \wedge \text{Above}(y,x)) \rightarrow \text{SameColor}(y,x)))))$

$\equiv \forall x(\sim\text{Square}(x) \vee (\sim(\forall y((\text{Triangle}(y) \wedge \text{Above}(y,x)) \rightarrow \text{SameColor}(y,x)))))$

$\equiv \forall x(\sim\text{Square}(x) \vee \exists y(\sim((\text{Triangle}(y) \wedge \text{Above}(y,x)) \rightarrow \text{SameColor}(y,x))))$

$\equiv \forall x(\sim\text{Square}(x) \vee \exists y((\text{Triangle}(y) \wedge \text{Above}(y,x)) \wedge (\sim\text{SameColor}(y,x))))$

53. **a.** The statement is true: Circle b and squares h and j are all colored black.

  b. *Formal version*: $\exists x(\text{Circle}(x) \wedge \exists y(\text{Square}(y) \wedge \text{SameColor}(x,y)))$

  c. *Formal negation*: $\forall x(\sim\text{Circle}(x) \wedge \exists y(\text{Square}(y) \wedge \text{SameColor}(y,x)))$

$\equiv \forall x(\sim\text{Circle}(x) \vee \sim(\exists y(\text{Square}(y) \wedge \text{SameColor}(y,x))))$

$\equiv \forall x(\sim\text{Circle}(x) \vee \forall y(\text{Square}(y) \vee \sim\text{SameColor}(y,x)))$

54. **a.** The statement is false. It says that there are a circle and a triangle that have the same color, which is false because all the triangles are blue, and no circles are blue.

  b. *Formal version:* $\exists x(\text{Circle}(x) \wedge (\exists y\ (\text{Triangle}(y) \wedge \text{SameColor}(x,y))))$

  c. *Formal negation:* $\forall x(\sim\text{Circle}(x) \vee \sim (\exists y\ (\text{Triangle}(y) \wedge \text{SameColor}(x,y))))$

$\equiv \forall x(\sim\text{Circle}(x) \vee (\forall y\ (\sim\text{Triangle}(y) \vee \sim\text{SameColor}(x,y))))$

55. No matter what the domain D or the predicates $P(x)$ and $Q(x)$ are, the given statements have the same truth value. If the statement "$\forall x$ in D, $(P(x) \wedge Q(x))$" is true, then $P(x) \wedge Q(x)$ is true for every x in D, which implies that both $P(x)$ and $Q(x)$ are true for every x in D. But then $P(x)$ is true for every x in D, and also $Q(x)$ is true for every x in D. So the statement "$(\forall x$ in D, $P(x)) \wedge (\forall x$, in D, $Q(x))$" is true. Conversely, if the statement "$(\forall x$ in D, $P(x)) \wedge (\forall x$ in D, $Q(x))$" is true, then $P(x)$ is true for every x in D, and also $Q(x)$ is true for every x in D. This implies that both $P(x)$ and $Q(x)$ are true for every x in D, and so $P(x) \wedge Q(x)$ is true for every x in D. Hence the statement "$\forall x$ in D, $(P(x) \wedge Q(x))$" is true.

57. These statements do not necessarily have the same truth values. For example, let $D = \mathbf{Z}$, the set of all integers, let $P(x)$ be "x is even," and let $Q(x)$ be "x is odd." Then the statement "$\forall x \in D, (P(x) \vee Q(x))$" can be written "$\forall$ integers x, x is even or x is odd," which is true. On the other hand, "$(\forall x \in D, P(x)) \vee (\forall x \in D,\ Q(x))$" can be written "All integers are even or all integers are odd," which is false.

59. **a.** Yes   **b.** $X = w_1$, $X = w_2$ **c.** $X = b_2$, $X = w_2$

60. **a.** No   **b.** No   **c.** $X = g$

Section 3.4

1. **b.** $(f_i + f_j)^2 = f_i^2 + 2f_if_j + f_j^2$

  c. $(3u + 5v)^2 = (3u)^2 + 2(3u)(5v) + (5v)^2$

  d. $(g(r) + g(s))^2 = (g(r))^2 + 2g(r)g(s) + (g(s))^2$

2. 0 is even.

3. $\dfrac{2}{3} + \dfrac{4}{5} = \dfrac{(2 \cdot 5 + 3 \cdot 4)}{(3 \cdot 5)}\ \left(= \dfrac{22}{15}\right)$

5. $\dfrac{1}{0}$ is not an irrational number.

6. This computer program is not correct.

7. Invalid; converse error

8. Valid by universal modus ponens (or universal instantiation)

9. Invalid; inverse error

10. Valid by universal modus tollens

12. Valid, universal modus tollens

15. Invalid, converse error

16. Invalid; converse error

18. Valid, universal modus tollens

19. $\forall x$, if x is a good car, then x is not cheap.

  a. Valid, universal modus ponens (or universal instantiation)

  b. Invalid, converse error

21. Valid. (A valid argument can have false premises and a true conclusion!)

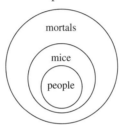

The major premise says the set of people is included in the set of mice. The minor premise says the set of mice is included in the set of mortals. Assuming both of these premises are true, it must follow that the set of people is included in the set of mortals. Since it is impossible for the conclusion to be false if the premises are true, the argument is valid.

23. Valid. The major and minor premises can be diagrammed as follows:

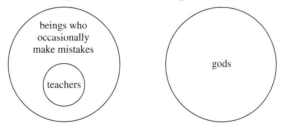

According to the diagram, the set of teachers and the set of gods can have no common elements. Hence, if the premises are true, then the conclusion must also be true, and so the argument is valid.

24. Valid. The only drawing representing the truth of the premises also represents the truth of the conclusion.

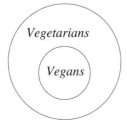

25. Invalid. Let C represent the set of all college cafeteria food, G the set of all good food, and W the set of all wasted food. Then any one of the following diagrams could represent the given premises.

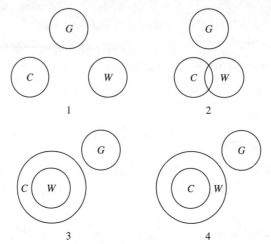

Only in drawing (1) is the conclusion true. Hence it is possible for the premises to be true while the conclusion is false, and so the argument is invalid.

27. Valid. The only drawing representing the truth of the premises also represents the truth of the conclusion.

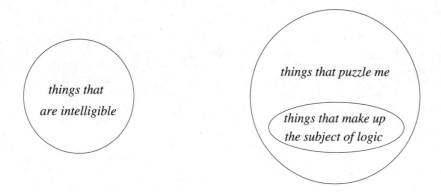

28. 3. (*contrapositive form*): If an object is gray, then it is a circle.

    2. If an object is a circle, then it is to the right of all the blue objects.

    1. If an object is to the right of all the blue objects, then it is above all the triangles.

    ∴ If an object is gray, then it is above all the triangles.

30. 3. If an object is black, then it is a square.

    2. (*contrapositive form*) If an object is a square, then it is above all the gray objects.

    4. If an object is above all the gray objects, then it is above all the triangles.

    1. If an object is above all the triangles, then it is above all the blue objects.

    ∴ If an object is black, then it is above all the blue objects.

31. 4. If an animal is in the yard, then it is mine.

    1. If an animal belongs to me, then I trust it.

    5. If I trust an animal, then I admit it into my study.

3. If I admit an animal into my study, then it will beg when told to do so.

6. If an animal begs when told to do so, then that animal is a dog.

2. If an animal is a dog, then that animal gnaws bones.

∴ If an animal is in the yard, then that animal gnaws bones. (In other words, all the animals in the yard gnaw bones.)

33. 2. If a bird is in this aviary, then it belongs to me.

4. If a bird belongs to me, then it is at least 9 feet high.

1. If a bird is at least 9 feet high, then it is an ostrich.

3. If a bird lives on mince pies, then it is not an ostrich.

(*contrapositive form*): If a bird is an ostrich, then it does not live on mince pies.

∴ If a bird is in this aviary, then it does not live on mince pies. In other words, no bird in this aviary lives on mince pies.

36. The universal form of elimination (part *a*) says that the following form of argument is valid:

$\forall x$ in D, $P(x) \vee Q(x)$.　　　← *major premise*
$\sim Q(c)$ for a particular c in D.　← *minor premise*
∴　$P(c)$

Proof of Validity:

Suppose the major and minor premises of the above argument form are both true.

[We must show that the conclusion $P(c)$ is also true.]

By definition of truth value for a universal statement, "$\forall x$ in D, $P(x) \vee Q(x$" is true if, and only if, the statement "$P(x) \vee Q(x)$" is true for each individual element of D.

So, by universal instantiation, "$P(x) \vee Q(x)$" is true for the particular element c.

Hence "$P(c) \vee Q(c)$" is true.

And since the minor premise says that $\sim Q(c)$, it follows by the elimination rule that $P(c)$ is true.

[This is what was to be shown.]

Review Guide: Chapter 3

Quantified Statements

- What is a predicate?
- What is the truth set of a predicate?
- What is a universal statement, and what is required for such a statement to be true?
- What is required for a universal statement to be false?
- What is the method of exhaustion?
- What is an existential statement, and what is required for such a statement to be true?
- What is required for a existential statement to be false?
- What are some ways to translate quantified statements from formal to informal language?
- What are some ways to translate quantified statements from informal to formal language?
- What is a universal conditional statement?
- What are equivalent ways to write a universal conditional statements?
- What are equivalent ways to write existential statements?
- What is a trailing quantifier?
- What does it mean for a statement to be quantified implicitly?
- What do the notations $\Rightarrow$ and $\Leftrightarrow$ mean?
- What is the relation among $\forall$, $\exists$, $\wedge$, and $\vee$?
- What does it mean for a universal statement to be vacuously true?
- What is the rule for interpreting a statement that contains both a universal and an existential quantifier?
- How are statements expressed in the computer programming language Prolog?

Negations: What are negations for the following forms of statements?

- $\forall x, Q(x)$
- $\exists x$ such that $Q(x)$
- $\forall x$, if $P(x)$ then $Q(x)$
- $\forall x, \exists y$ such that $P(x, y)$
- $\exists x$ such that $\forall y, P(x, y)$

Variants of Conditional Statements

- What are the converse, inverse, and contrapositive of a statement of the form "$\forall x$, if $P(x)$ then $Q(x)$"?
- How are quantified statements involving necessary and sufficient conditions and the phrase only-if translated into if-then form?

Validity and Invalidity

- What is universal instantiation?
- What are the universal versions of modus ponens, modus tollens, converse error, and inverse error, and which of these forms of argument are valid and which are invalid?
- How is universal modus ponens used in a proof?
- How can diagrams be used to test the validity of an argument with quantified statements?

Fill-in-the-Blank Review Questions

Section 3.1

1. If $P(x)$ is a predicate with domain D, the truth set of $P(x)$ is denoted _____. We read these symbols out loud as _____.
2. Some ways to express the symbol $\forall$ in words are _____ or _____ or _____.
3. Some ways to express the symbol $\exists$ in words are _____ or _____ or _____.
4. A statement of the form $\forall x \in D$, $Q(x)$ is true if, and only if, $Q(x)$ is _____ for _____.
5. A statement of the form $\exists x \in D$ such that $Q(x)$ is true if, and only if, $Q(x)$ is _____ for _____.

Section 3.2

1. A negation for "All R have property S" is "There is _____ R that _____.
2. A negation for "Some R have property S" is _____.
3. A negation for "For every x, if x has property P then x has property Q" is _____.
4. The converse of "For every x, if x has property P then x has property Q" is _____.
5. The contrapositive of "For every x, if x has property P then x has property Q" is _____.
6. The inverse of "For every x, if x has property P then x has property Q" is _____.

Section 3.3

1. To establish the truth of a statement of the form "$\forall x$ in D, $\exists y$ in E such that $P(x,y)$," you imagine that someone has given you an element x from D but that you have no control over what that element is. Then you need to find ____ with the property that the x the person gave you together with the_____ you subsequently found satisfy _____.
2. To establish the truth of a statement of the form "$\exists x$ in D such that $\forall y$ in E, $P(x,y)$," you need to find _____ so that no matter what _____ a person might subsequently give you, _____ will be true.
3. Consider the statement "$\forall x$, $\exists y$ such that some property $P(x,y)$ involving x and y is true." A negation for this statement is _____.
4. Consider the statement "$\exists x$ such that $\forall y$, some property $P(x,y)$ involving x and y is true." A negation for this statement is _____.
5. Suppose $P(x,y)$ is some property involving x and y, and suppose the statement "$\forall x$ in D, $\exists y$ in E such that $P(x,y)$" is true. Then the statement "$\exists x$ in D such that $\forall y$ in E, $P(x,y)$"
   a. is true.   b. is false.   c. may be true or may be false.

Section 3.4

1. The rule of universal instantiation says that if some property is true for _____ in a domain, then it is true for _____.
2. If the first two premises of universal modus ponens are written as "If x makes $P(x)$ true, then x makes $Q(x)$ true" and "For a particular value of a _____," then the conclusion can be written as "_____."
3. If the first two premises of universal modus tollens are written as "If x makes $P(x)$ true, then x makes $Q(x)$ true" and "For a particular value of a _____," then the conclusion can be written as "_____."
4. If the first two premises of universal transitivity are written as "Any x that makes $P(x)$ true makes $Q(x)$ true" and "Any x that makes $Q(x)$ true makes $R(x)$ true," then the conclusion can be written as "_____."

5. Diagrams can be helpful in testing an argument for validity. However, if some possible configurations of the premises are not drawn, a person could conclude that an argument was _____ when it was actually _____.

Answers for Fill-in-the-Blank Review Questions

Section 3.1

1. $\{x \in D \mid P(x)\}$; the set of all x in D such that $P(x)$
2. Possible answers: for all, for every, for any, for each, for arbitrary, given any
3. Possible answers: there exists, there exist, there exists at least one, for some, for at least one, we can find a
4. true; every x in D (*Alternative answers:* all x in D, each x in D)
5. true; at least one x in D (*Alternative answer:* some x in D)

Section 3.2

1. some (*Alternative answers:* at least one; an); does not have property S.
2. No R have property S.
3. There is an x such that x has property P and x does not have property Q.
4. For every x, if x has property Q then x has property P.
5. For every x, if x does not have property Q then x does not have property P.
6. For every x, if x does not have property P then x does not have property Q.

Section 3.3

1. an element y in E; y; $P(x, y)$
2. an element x in D; y in E; $P(x, y)$
3. $\exists x$ such that $\forall y$, the property $P(x, y)$ is false.
4. $\forall x$, $\exists y$ such that the property $P(x, y)$ is false.
5. The answer is (c): the truth or falsity of a statement in which the quantifiers are reversed depends on the nature of the property involving x and y.

Section 3.4

1. all elements; any particular element in the domain (*Or:* each individual element of the domain)
2. $P(a)$ is true; $Q(a)$ is true
3. $Q(a)$ is false; $P(a)$ is false
4. "Any x that makes $P(x)$ true makes $R(x)$ true."
5. valid; invalid

Chapter 4: Elementary Number Theory and Methods of Proof

One aim of this chapter is to introduce you to methods for evaluating whether a given mathematical statement is true or false. Throughout the chapter the emphasis is on learning to prove and disprove statements of the form

$$\forall x \text{ in } D, \text{ if } P(x) \text{ then } Q(x).$$

- To prove such a statement directly, you suppose you have a particular but arbitrarily chosen element x in D for which $P(x)$ is true and you show that $Q(x)$ must also be true.

- To disprove such a statement, you show that there is an element x in D (a counterexample) for which $P(x)$ is true and $Q(x)$ is false.

- To prove such a statement by contradiction, you show that no counterexample exists. In other words, you suppose that there is an x in D for which $P(x)$ is true and $Q(x)$ is false and you show that this supposition leads to a contradiction.

Direct proof, disproof by counterexample, and proof by contradiction can, therefore, all be viewed as three aspects of one whole. You arrive at one or the other by a thoughtful examination of the given statement, knowing what it means for a statement of that form to be true or false.

Another aim of the chapter is to help you obtain fundamental knowledge about numbers that is needed in many areas of mathematics and computer science. Be aware that the exercise sets contain problems of varying difficulty, so do not be discouraged if some are difficult for you. And because proofs can be written in a variety of acceptable styles and because the amount of detail in a proof can depend of the mathematical knowledge of the intended reader, you should regard the proofs given in the book as samples. Your instructor may discuss with you the particular range of proof styles that will be considered acceptable in your course.

Section 4.1

1. **a.** Yes: $-17 = 2(-9) + 1$

   b. No. 0 is even because $0 = 0 \cdot 2$.

   c. Yes: $2k - 1 = 2(k - 1) + 1$ and $k - 1$ is an integer because it is a difference of integers.

3. **a.** Yes: $6m + 8n = 2(3m + 4n)$ and $(3m + 4n)$ is an integer because $3, 4, m,$ and n are integers, and products and sums of integers are integers.

   b. Yes: $10mn + 7 = 2(5mn + 3) + 1$ and $5mn + 3$ is an integer because $3, 5, m,$ and n are integers, and products and sums of integers are integers.

   c. Not necessarily. For instance, if $m = 3$ and $n = 2$, then $m^2 - n^2 = 9 - 4 = 5$, which is prime. (However, $m^2 - n^2$ is composite for many values of m and n because of the identity $m^2 - n^2 = (m - n)(m + n)$.)

5. For example, let $m = n = 2$. Then m and n are integers such that $m > 1$ and $n > 1$ and $\dfrac{1}{m} + \dfrac{1}{n} = \dfrac{1}{2} + \dfrac{1}{2} = 1$, which is an integer.

6. For example, let $m = 1$ and $n = -1$. Then

$$\frac{1}{m} + \frac{1}{n} = \frac{1}{1} + \frac{1}{(-1)} = 1 + (-1) = 0.$$

   In fact, if k is any nonzero integer, then $\dfrac{1}{k} + \dfrac{1}{(-k)} = \dfrac{1}{k} + \left(-\dfrac{1}{k}\right) = 0.$

8. For example, let $n = 7$. Then n is an integer such that $n > 5$ and $2^n - 1 = 127$, which is prime.

9. For example, let $x = 60$. Note that to four significant digits

$$2^{60} \cong 1.153 \times 10^{18}$$

and

$$60^{10} \cong 6.047 \times 10^{17},$$

and so

$$2^x \geq x^{10}.$$

Examples can also be found in the approximate range $1 < x < 1.077$. For instance,

$$2^{1.07} \cong 2.099$$

and

$$1.07^{10} \cong 1.967,$$

and so

$$2^{1.07} > 1.07^{10}.$$

10. For example, 25, 9, and 16 are all perfect squares, because $25 = 5^2, 9 = 3^2$, and $16 = 4^2$, and $25 = 9 + 16$. Thus 25 is a perfect square that can be written as a sum of two other perfect squares.

12. **a.** *Negation for the statement*: There exist real numbers a and b such that $a < b$ and $a^2 \not< b^2$.

   b. *Counterexample for the statement*: Let $a = -2$ and $b = -1$. Then $a < b$ because $-2 < -1$, but $a^2 \not< b^2$ because $(-2)^2 = 4$ and $(-1)^2 = 1$ and $4 \not< 1$. *[So the hypothesis of the statement is true and its conclusion is false.]*

14. Counterexample: Let $m = 2$ and $n = 1$. Then

$$2m + n = 2 \cdot 2 + 1 = 5,$$

which is odd. But m is not odd, and so it is false that both m and n are odd. *[This is one counterexample among many.]*

15. Counterexample: Let $p = 2$. Then

$$p^2 - 1 = 3,$$

which is not even.

17. This property is true for some integers and false for other integers. For instance, if $a = 0$ and $b = 1$, the property is true because $(0 + 1)^2 = 0^2 + 1^2$, but if $a = 1$ and $b = 1$, the property is false because $(1 + 1)^2 = 4$ and $1^2 + 1^2 = 2$ and $4 \neq 2$.

18. Counterexample: Let $a = 1$, $b = 2$, $c = 1$, and $d = 2$. Then

$$\frac{a}{b} + \frac{c}{d} = \frac{1}{2} + \frac{1}{2} = 1, \text{ whereas } \frac{a+c}{b+d} = \frac{1+1}{2+2} = \frac{2}{4} = \frac{1}{2}.$$

So

$$\frac{a}{b} + \frac{c}{d} \neq \frac{a+c}{b+d}.$$

19. *Hint:* This property is true for some integers and false for other integers. To justify this answer you need to find examples of both.

21. $2 = 1^2 + 1^2, \quad 4 = 2^2, \quad 6 = 2^2 + 1^2 + 1^2,$
   $8 = 2^2 + 2^2, \quad 10 = 3^2 + 1^2, \quad 12 = 2^2 + 2^2 + 2^2,$
   $14 = 3^2 + 2^2 + 1^2, \quad 16 = 4^2,$
   $18 = 3^2 + 3^2 = 4^2 + 1^2 + 1^2, \quad 20 = 4^2 + 2^2,$
   $22 = 3^2 + 3^2 + 2^2, \quad 24 = 4^2 + 2^2 + 2^2$

23. **a.** If an integer is greater than 1, then its reciprocal is between 0 and 1.

b. *Start of proof:* Suppose m is any integer such that $m > 1$.

Conclusion to be shown: $0 < 1/m < 1$.

24. If a real number is greater than 1, then its square it greater than itself.

Start of Proof: Suppose x is any *[particular but arbitrarily chosen]* real number such that $x > 1$.

Conclusion to be shown: $x^2 > x$.

25. **a.** If the product of two integers is 1, then either both are 1 or both are -1.

b. $x > 1$.*Start of proof:* Suppose m and n are any integers with $mn = 1$.

Conclusion to be shown: $m = n = 1$ or $m = n = -1$.

27. (a) odd integer, (b) $2k + 1$, (c) n, (d) generalizing from the generic particular

28. **a.** $\forall$ integers m and n, if m and n are odd then $m + n$ is odd.

$\forall$ odd integers m and n, $m + n$ is odd.

If m and n are any odd integers, then $m + n$ is odd.

b. (a) definition of odd, (b) substitution, (c) any sum of integers is an integer, (d) definition of even

30. **a.** $\forall$ integers m and n, if m is even and n is odd, then $m + n$ is odd.

$\forall$ even integers m and odd integers $n, m + n$ is odd.

If m is any even integer and n is any odd integer, then $m + n$ is odd.

b. (a) any odd integer, (b) integer r, (c) $2r + (2s + 1)$, (d) $m + n$ is odd

31. **a.** $\forall$ integer n, if n is odd, then $5n^2 + 7$ is even.

Section 4.2

1. Proof:

Suppose n is any *[particular but arbitrarily chosen]* odd integer.

[We must show that $3n + 5$ is even. By definition of even, this means we must show that $3n + 5 = 2 \cdot$ (some integer).]

By definition of odd, $n = 2r + 1$ for some integer r.

Then

$$
\begin{aligned}
3n + 5 &= 3(2r + 1) + 5 && \text{by substitution} \\
&= 6r + 3 + 5 \\
&= 6r + 8 \\
&= 2(3r + 4) && \text{by algebra.}
\end{aligned}
$$

[Idea for the rest of the proof: We want to show that $3n + 5 = 2 \cdot$ (some integer). At this point we know that $3n + 5 = 2(3r + 4)$. So is $3r + 4$ an integer? Yes, because products and sums of integers are integers.]

Let $k = 3r + 4$.

Then $3n + 5 = 2(3r + 4) = 2k$, and k is an integer because products and sums of integers are integers. Hence $3n + 5$ is even by definition of even.

3. Proof:

Suppose n is any *[particular but arbitrarily chosen]* integer.

[We must show that $2n - 1$ is odd.] [By definition of odd, this means we must show that $2n - 1 = 2 \cdot (some\ integer) + 1$.]

Now
$$
\begin{aligned}
2n - 1 &= 2n - 2 + 1 \\
&= 2(n - 1) + 1 \quad \text{by algebra.}
\end{aligned}
$$

So let $t = n - 1$. Then t is an integer because it is a difference of integers, and, by substitution,

$$2n - 1 = 2(n - 1) + 1 = 2t + 1.$$

Hence $2n - 1$ equals 2 times an integer plus 1, and so $2n - 1$ is odd by definition of odd *[as was to be shown]*.

4. *Two versions of a correct proof are given below to illustrate some of the variety that is possible.*

Proof 1: Suppose a is any even integer and b is any odd integer. *[We must show that $a - b$ is odd.]* By definition of even and odd, $a = 2r$ and $b = 2s + 1$ for some integers r and s. By substitution and algebra,

$$a - b = 2r - (2s + 1) = 2r - 2s - 1 = 2(r - s - 1) + 1.$$

Let $t = r - s - 1$. Then t is an integer because differences of integers are integers. Thus, by substitution, $a - b = 2t + 1$, where t is an integer, and so, by definition of odd, $a - b$ is odd *[as was to be shown]*.

Proof 2: Suppose a is any even integer and b is any odd integer. By definition of even and odd, $a = 2r$ and $b = 2s + 1$ for some integers r and s. Then, by substitution,

$$a - b = 2r - (2s + 1) = 2(r - s - 1) + 1.$$

Now $r - s - 1$ is an integer because differences of integers are integers, and so $a - b$ equals twice some integer plus 1. Thus $a - b$ is odd.

6. Proof:

Suppose k is any odd integer and m is any even integer.

[We must show that $k^2 + m^2$ is odd.] [By definition of odd, this means we must show that $2n - 1 = 2 \cdot (some\ integer) + 1$.]

By definition of odd and even, $k = 2a + 1$ and $m = 2b$ for some integers a and b. It follows that
$$
\begin{aligned}
k^2 + m^2 &= (2a + 1)^2 + (2b)^2 \quad \text{by substitution} \\
&= 4a^2 + 4a + 1 + 4b^2 \\
&= 4(a^2 + a + b^2) + 1 \\
&= 2(2a^2 + 2a + 2b^2) + 1 \quad \text{by algebra.}
\end{aligned}
$$

Now $2a^2 + 2a + 2b^2$ is an integer because it is a sum of products of integers.

Thus $k^2 + m^2$ equals twice some integer plus 1, and so $k^2 + m^2$ is odd *[as was to be shown]*.

7. *Hint:* It is convenient to represent two consecutive integers as n and $n + 1$ or as $n - 1$ and n for some integer n.

9. Proof:

Suppose n is any integer such that $n > 4$ and n is a perfect square.

[We must show that $n - 1$ is not prime.]

By definition of perfect square, $n = r^2$ for some positive integer r.

Then

$$
\begin{aligned}
n - 1 &= r^2 - 1 & \text{by substitution} \\
& (r-1)(r+1) & \text{by algebra.}
\end{aligned}
$$

Now because $n = r^2$ and $n > 4$, it follows that $r^2 > 4$, which implies that $r > 2$ since r is positive.

Thus $r - 1 > 1$ and $r + 1 > 1$.

Also both $r - 1$ and $r + 1$ are integers because sums and differences of integers are integers.

Thus $n - 1$ is a product of two positive integers $r - 1$ and $r + 1$, each of which is greater than 1.

Hence $n - 1$ is not a prime number by definition of prime number *[as was to be shown]*.

10. Proof:

Suppose n is any even integer.

[We must show that $(-1)^n = 1$.]

By definition of even, $n = 2k$ for some integer k. Hence

$$(-1)^n = (-1)^{2k} = ((-1)^2)^k = 1^k = 1$$

[by the laws of exponents from algebra].

This is what was to be shown.

12. To prove the given statement is false, we prove that its negation is true.

The negation of the statement is "For every integer $m \geq 3$, $m^2 - 1$ is not prime."

Proof of the negation:

Suppose m is any integer with $m \geq 3$.

By basic algebra, $m^2 - 1 = (m-1)(m+1)$.

Because $m \geq 3$, both $m - 1$ and $m + 1$ are positive integers greater than 1, and each is smaller than $m^2 - 1$.

So $m^2 - 1$ is a product of two smaller positive integers, each greater than 1.

Hence $m^2 - 1$ is not prime.

15. The incorrect proof just shows the theorem to be true in the one case where $k = 2$. A real proof must show that it is true for *every* integer $k > 0$.

16. The mistake in the "proof" is that the same symbol, k, is used to represent two different quantities. By setting $m = 2k$ and $n = 2k + 1$, the proof implies that $n = m + 1$, and thus it deduces the conclusion only for this one situation. When $m = 4$ and $n = 17$, for instance, the computations in the proof indicate that $n - m = 1$, but actually $n - m = 13$. In other words, the proof does not deduce the conclusion for an arbitrarily chosen even integer m and odd integer n, and hence it is invalid.

17. This incorrect proof assumes what is to be proved. The word *since* in the third sentence is completely unjustified. The second sentence tells only what happens *if* $k^2 + 2k + 1$ is composite. But at that point in the proof, it has not been established that $k^2 + 2k + 1$ *is* composite. In fact, that is exactly what is to be proved.

18. This incorrect "proof" assumes what is to be proved. The second sentence states a conclusion that would follow from the assumption that $m \cdot n$ is even, and the next-to-last sentence states this conclusion as if it were known to be true. But it is not known to be true at this point in the proof. In fact, it is the main task of a genuine proof to derive this conclusion, not from the assumption that it is true but from the hypothesis of the theorem.

20. The statement is true.

Proof: Suppose m and n are any odd integers. *[We must show that mn is odd.]*

By definition of odd, $n = 2r + 1$ and $m = 2s + 1$ for some integers r and s. Then

$$\begin{aligned} mn &= (2r+1)(2s+1) & \text{by substitution} \\ &= 4rs + 2r + 2s + 1 \\ &= 2(2rs + r + s) + 1 & \text{by algebra.} \end{aligned}$$

Now $2rs + r + s$ is an integer because products and sums of integers are integers and 2, r, and s are all integers. Hence $mn = 2$ (some integer) $+ 1$, and so, by definition of odd, mn is odd.

21. The statement is true.

Proof:

Suppose n is any odd integer. *[We must show that $-n$ is odd.]*

By definition of odd, $n = 2k + 1$ for some integer k. It follows that

$$\begin{aligned} -n &= -(2k+1) & \text{by substitution} \\ &= -2k - 1 \\ &= -2k - 2 + 1 \\ &= 2(-k-1) + 1 & \text{by algebra.} \end{aligned}$$

Let $t = -k - 1$.

[Note that $-k = 0 - k = (-1) \cdot k$.]

Then t is an integer because products and differences of integers are integers, and $-n = 2t + 1$ by substitution.

Hence $-n$ is odd by definition of odd *[as was to be shown.]*.

22. The statement is false.

Counterexample: Let $a = 1$ and $b = 0$. Then $4a + 5b + 3 = 4 \cdot 1 + 5 \cdot 0 + 3 = 7$, which is odd.

[This is one counterexample among many]

24. The statement is false.

Counterexample: Let $m = 1$ and $n = 3$. Then $m + n = 4$ is even, but neither summand m nor summand n is even.

27. The statement is true.

Proof:

Let m and n be any odd integers.

By definition of odd, $m = 2r + 1$ and $n = 2s + 1$ for some integers r and s.

By substitution,

$$m - n = (2r+1) - (2s+1) = 2(r-s).$$

Since $r - s$ is an integer (being a difference of integers), then $m - n$ equals twice some integer, and so $m - n$ is even by definition of even.

28. *Hint:* The statement is true.

30. The statement is false.

Counterexample: Let $m = 3$. Then $m^2 - 4 = 9 - 4 = 5$, which is not composite.

32. The statement is true.

Proof: Suppose n is any integer. Then

$$4(n^2 + n + 1) - 3n^2 = 4n^2 + 4n + 4 - 3n^2$$
$$= n^2 + 4n + 4 = (n + 2)^2$$

(by algebra). Now $(n + 2)^2$ is a perfect square because $n + 2$ is an integer (being a sum of n and 2). Hence $4(n^2 + n + 1) - 3n^2$ is a perfect square, as was to be shown.

33. The statement is false.

Counterexample: The number 28 cannot be expressed as a sum of three or fewer perfect squares. The only perfect squares that could be used to add up to 28 are those that are smaller than 28: 1, 4, 9, 16, and 25. The method of exhaustion can be used to show that no combination of these numbers add up to 28. (In fact, there are just three ways to express 28 as a sum of four or fewer of these numbers:

$$28 = 25 + 1 + 1 + 1 = 16 + 4 + 4 + 4 = 9 + 9 + 9 + 1,$$

and none of these ways use only three perfect squares.)

34. *Hint:* The statement is true.

36. The statement is true.

Proof: Suppose two consecutive integers are given. Call the smaller one n. Then the larger is $n + 1$.

Let m be the difference of the squares of the numbers. Then, by substitution,

$$m = (n + 1)^2 - n^2 = (n^2 + 2n + 1) - n^2 = 2n + 1.$$

Because n is an integer, $m = 2 \cdot (\text{an integer}) + 1$, and so m is odd by definition of odd.

37. *Hint:* The statement is true.

39. If m and n are perfect squares, then $m = a^2$ and $n = b^2$ for some integers a and b. We may take a and b to be nonnegative because for any real number x, $x^2 = (-x)^2$ and if x is negative then $-x$ is nonnegative. By substitution,

$$\begin{aligned} m + n + 2\sqrt{mn} &= a^2 + b^2 + 2\sqrt{a^2 b^2} \\ &= a^2 + b^2 + 2ab \qquad \text{since } a \text{ and } b \text{ are nonnegative} \\ &= (a + b)^2. \end{aligned}$$

But $a + b$ is an integer (since a and b are integers), and so $m + n + 2\sqrt{mn}$ is a perfect square.

40. *Hint:* The answer is no.

Section 4.3

1. $-\dfrac{35}{6} = \dfrac{-35}{6}$

3. $\dfrac{4}{5} + \dfrac{2}{9} = \dfrac{4 \cdot 9 + 2 \cdot 5}{45} = \dfrac{46}{45}$

4. Let $x = 0.3737373737\ldots$.

Then $100x = 37.37373737\ldots$, and so

$100x - x = 37.37373737\ldots - 0.3737373737\ldots$

Thus $99x = 37$, and hence $x = \dfrac{37}{99}$.

6. Let $x = 320.5492492492\ldots$. Then

$10000x = 3205492.492492492\ldots$, and

$10x = 3205.492492492\ldots$, and so $10000x - 10x = 3205492 - 3205$.

Thus $9990x = 3202287$, and hence $x = \dfrac{3{,}202{,}287}{9{,}990}$.

8. **b.** $\forall$ real numbers x and y, if $x \neq 0$ and $y \neq 0$ then $xy \neq 0$.

9. Given that a and b are integers, both $b - a$ and ab^2 are integers (since differences and products of integers are integers). Also, by the zero product property, $ab^2 \neq 0$ because neither a nor b is zero. Hence $(b - a)/ab^2$ is a quotient of two integers with a nonzero denominator, and so it is rational.

11. Proof: Suppose n is any *[particular but arbitrarily chosen]* integer. Then $n = n \cdot 1$, and so $n = n/1$ by dividing both sides by 1. Now n and 1 are both integers, and $1 \neq 0$. Hence n can be written as a quotient of integers with a nonzero denominator, and so n is rational.

12. (a) any *[particular but arbitrarily chosen]* rational number

    (b) integers a and b

    (c) $(a/b)^2$

    (d) b^2

    (e) zero product property

    (f) r^2 is rational

13. **a.** $\forall$ real number r, if r is rational then $-r$ is rational.

    Or: $\forall r$, if r is a rational number then $-r$ is rational.

    Or: $\forall$ rational number r, $-r$ is rational.

    b. The statement is true.

    Proof:

    Suppose r is a *[particular but arbitrarily chosen]* rational number.

    [We must show that $-r$ is rational.]

    By definition of rational, $r = a/b$ for some integers a and b with b 0. Then

    $$-r = -\frac{a}{b} \quad \text{by substitution}$$
    $$= \frac{-a}{b} \quad \text{by algebra.}$$

    Now since a is an integer, so is $-a$ (being the product of -1 and a). Hence $-r$ is a quotient of integers with a nonzero denominator, and so $-r$ is rational *[as was to be shown]*.

15. Proof: Suppose r and s are rational numbers. *[We must show that rs is rational.]*

    By definition of rational, $r = a/b$ and $s = c/d$ for some integers $a, b, c,$ and d with $b \neq 0$ and $d \neq 0$. Then

    $$rs = \frac{a}{b} \cdot \frac{c}{d} \quad \text{by substitution}$$

    $$= \frac{ac}{bd} \quad \text{by the rules of algebra for multiplying fractions.}$$

    Now ac and bd are both integers (being products of integers) and $bd \neq 0$ (by the zero product property).

    Hence rs is a quotient of integers with a nonzero denominator, and so, by definition of rational, rs is rational *[as was to be shown].*

16. *Hint:* Counterexample: Let r be any rational number and $s = 0$. Then r and s are both rational, but the quotient of r divided by s is not a real number and therefore is not a rational number.

    Revised statement to be proved: For all rational numbers r and s, if $s \neq 0$ then r/s is rational.

17. *Hint:* The conclusion to be shown is that a certain quantity (the difference of two rational numbers) is rational. To show this, you need to show that the quantity can be expressed as a ratio of two integers with a nonzero denominator.

18. The statement is true.

    Proof: Suppose r and s are any two distinct rational numbers. *[We must show that $\dfrac{r+s}{2}$ is rational.]*

    By definition of rational, $r = \dfrac{a}{b}$ and $s = \dfrac{c}{d}$ for some integers $a, b, c,$ and d with $b \neq 0$ and $d \neq 0$.

    By substitution and the laws of algebra,

    $$\frac{r+s}{2} = \frac{\dfrac{a}{b} + \dfrac{c}{d}}{2} = \frac{\dfrac{ad+bc}{bd}}{2} = \frac{ad+bc}{2bd}.$$

    Now $ad + bc$ and $2bd$ are integers because $a, b, c,$ and d are integers and products and sums of integers are integers. And $2bd \neq 0$ by the zero product property.

    Hence $\dfrac{r+s}{2}$ is a quotient of integers with a nonzero denominator, and so $\dfrac{r+s}{2}$ is rational *[as was to be shown].*

19. *Hint:* If $a < b$ then $a + a < a + b$ (by T19 of Appendix A), or equivalently, $2a < a + b$. Thus $a < \frac{a+b}{2}$ (by T20 of Appendix A).

21. The statement is true.

    Proof: Suppose m is any even integer and n is any odd integer. *[We must show that $m^2 + 3n$ is odd.]*

    By properties 1 and 3 of Example 4.3.3, m is even (because $m^2 = m \cdot m$) and $3n$ is odd (because both 3 and n are odd).

    It follows from property 5 *[and the commutative law for addition]* that $m^2 + 3n$ is odd *[as was to be shown].*

24. Proof:

    Suppose r and s are any rational numbers.

    By Theorem 4.3.1, both 2 and 3 are rational, and so, by exercise 15, both $2r$ and $3s$ are rational.

    Hence, by Theorem 4.3.2, $2r + 3s$ is rational.

27. Let x be the right-hand side of the equation. Then

    $$x = \frac{1 - \dfrac{1}{2^{n+1}}}{1 - \dfrac{1}{2}} = \frac{1 - \dfrac{1}{2^{n+1}}}{\dfrac{1}{2}} = \frac{1 - \dfrac{1}{2^{n+1}}}{\dfrac{1}{2}} \cdot \frac{2^{n+1}}{2^{n+1}} = \frac{2^{n+1} - 1}{2^n}.$$

    Now $2^{n+1} - 1$ and 2^n are both integers (since n is a nonnegative integer) and $2^n \neq 0$ by the zero product property. Therefore, x can be expressed as the ratio of the integers $2^{n+1} - 1$ and 2^n, and so x is rational.

30. Proof: Given a quadratic equation

    $$x^2 + bx + c = 0,$$

    where b and c are rational numbers, suppose one solution, r, is rational. Call the other solution s. Then

    $$x^2 + bx + c = (x - r)(x - s) = x^2 - (r + s)x + rs.$$

    By equating the coefficients of x,

    $$b = -(r + s).$$

    Solving for s yields

    $$s = -r - b = -(r + b).$$

    Because s is the negative of a sum of two rational numbers, s also is rational (by Theorem 4.3.2 and exercise 13).

31. Proof: Suppose c is a real number such that

    $$r_3 c^3 + r_2 c^2 + r_1 c + r_0 = 0,$$

    where r_0, r_1, r_2, and r_3 are rational numbers. By definition of rational, $r_0 = a_0/b_0$, $r_1 = a_1/b_1$, $r_2 = a_2/b_2$, and $r_3 = a_3/b_3$ for some integers, a_0, a_1, a_2, a_3, and nonzero integers b_0, b_1, b_2, and b_3. By substitution,

    $$r_3 c^3 + r_2 c^2 + r_1 c + r_0 = \frac{a_3}{b_3} c_3 + \frac{a_2}{b_2} c_2 + \frac{a_1}{b_1} c_1 + \frac{a_0}{b_0}$$

    $$= \frac{b_0 b_1 b_2 a_3}{b_0 b_1 b_2 b_3} c^3 + \frac{b_0 b_1 b_3 a_2}{b_0 b_1 b_2 b_3} c^2 + \frac{b_0 b_2 b_3 a_1}{b_0 b_1 b_2 b_3} c + \frac{b_1 b_2 b_3 a_0}{b_0 b_1 b_2 b_3}$$

    $$= 0.$$

    Multiplying both sides by $b_0 b_1 b_2 b_3$ gives

    $$b_0 b_1 b_2 a_3 \cdot c^3 + b_0 b_1 b_3 a_2 \cdot c^2 + b_0 b_2 b_3 a_1 \cdot c + b_1 b_2 b_3 a_0 = 0.$$

    Let $n_3 = b_0 b_1 b_2 a_3$, $n_2 = b_0 b_1 b_3 a_2$, $n_1 = b_0 b_2 b_3 a_1$, and $n_0 = b_1 b_2 b_3 a_0$. Then n_0, n_1, n_2, and n_3 are all integers (being products of integers). Hence c satisfies the equation

    $$n_3 c^3 + n_2 c^2 + n_1 c + n_0 = 0,$$

    where all of n_0, n_1, n_2, and n_3 are integers. This is what was to be shown.

33. **a.** When $(x - r)(x - s)$ is multiplied out, the result is

$$(x - r)(x - s) = x^2 - (r + s)x + rs.$$

If both r and s are odd integers, then $r + s$ is even and rs is odd (by properties 2 and 3).

If both r and s are even integers, then both $r + s$ and rs are even (by property 1).

If one of r and s is even and the other is odd, then $r + s$ is odd and rs is even (by properties 4 and 5).

Thus either $r + s$ and rs have opposite parity or both are even.

b. It follows from part (a) that $x^2 - 1253x + 255$ cannot be written as a product of the form $(x - r)(x - s)$ because if it could be then $r + s$ would equal 1253 and rs would equal 255, both of which are odd integers, and part (a) showed that this result is impossible.

Note: In Section 4.5, we establish formally that any integer is either even or odd. The type of reasoning used in this solution is called argument by contradiction. It is introduced formally in Section 4.7.

35. This "proof" assumes what is to be proved.

36. This incorrect proof just shows the theorem to be true in the one case where one of the rational numbers is 1/4 and the other is 1/2. It is an example of the mistake of arguing from examples. A correct proof must show the theorem is true for *any* two rational numbers.

37. By setting both r and s equal to a/b, this incorrect proof violates the requirement that r and s be arbitrarily chosen rational numbers. If both r and s equal a/b, then $r = s$.

39. This incorrect proof assumes what is to be proved. The second sentence asserts that a certain conclusion follows when it is assumed that $r + s$ is rational, and the rest of the proof uses that conclusion to deduce that $r + s$ is rational.

Section 4.4

1. Yes, $52 = 13 \cdot 4$

2. Yes, $56 = 7 \cdot 8$

3. Yes, $0 = 0 \cdot 5$.

4. Yes, $(3k + 1)(3k + 2)(3k + 3) = 3[(3k + 1)(3k + 2)(k + 1)]$, and $(3k + 1)(3k + 2)(k + 1)$ is an integer because k is an integer and sums and products of integers are integers.

6. No, $29/3 \cong 9.67$, which is not an integer.

7. Yes, $66 = (-3)(-22)$.

8. Yes, $6a(a + b) = 3a[2(a + b)]$, and $2(a + b)$ is an integer because a and b are integers and sums and products of integers are integers.

9. Yes, $2a \cdot 34b = 4(17ab)$ and $17ab$ is an integer because a and b are integers and products of integers are integers.

10. No, $34/7 \cong 4.86$, which is not an integer.

12. Yes, $n^2 - 1 = (4k + 1)^2 - 1 = (16k^2 + 8k + 1) - 1 = 16k^2 + 8k = 8(2k^2 + k)$, and $2k^2 + k$ is an integer because k is an integer and sums and products of integers are integers.

14. (a) $a|b$

(b) $b = a \cdot r$

(c) $-r$

(d) $a|(-b)$

15. Proof:

Suppose $a, b,$ and c are any integers such that $a|b$ and $a|c$. *[We must show that $a|(b+c)$.]*

By definition of divides, $b = ar$ and $c = as$ for some integers r and s. Then

$$
\begin{aligned}
b + c &= ar + as && \text{by substitution} \\
&= a(r + s) && \text{by algebra (the distributive law).}
\end{aligned}
$$

Let $t = r + s$.

Then t is an integer (being a sum of integers), and thus $b + c = at$ where t is an integer.

Hence, by definition of divides, $a \mid (b + c)$ *[as was to be shown]*.

16. *Hint:* The conclusion to be shown is that a certain quantity is divisible by a. To show this, you need to show that the quantity equals a times some integer.

18. **a.** $\forall$ integer n if n is a multiple of 3 then $-n$ is a multiple of 3.

b. The statement is true.

Proof:

Suppose n is any integer that is a multiple of 3. *[We must show that $-n$ is a multiple of 3.]*

By definition of multiple, $n = 3k$ for some integer k. Then

$$
\begin{aligned}
-n &= -(3k) && \text{by substitution} \\
&= 3(-k) && \text{by algebra.}
\end{aligned}
$$

Now $-k$ is an integer because k is. Hence, by definition of multiple, $-n$ is a multiple of 3 *[as was to be shown]*.

19. Counterexample: Let $a = 2$ and $b = 1$. Then $a + b = 2 + 1 = 3$, and so $3|(a + b)$ because $3 = 3 \cdot 1$.

On the other hand, $a - b = 2 - 1 = 1$, and $3 \nmid 1$ because $1/3$ is not an integer. Thus $3 \nmid (a - b)$. *[So the hypothesis of the statement is true and its conclusion is false. Thus the statement as a whole is false.]*

20. *Hint:* The consecutive integers can be conveniently represented as $n - 1$, n, and $n + 1$ or as $n, n + 1, n + 2$, where n is an integer.

21. The statement is true.

Proof:

Let m and n be any two even integers. *[We must show that mn is a multiple of 4.]*

By definition of even, $m = 2r$ and $n = 2s$ for some integers r and s. Then

$$
\begin{aligned}
mn &= (2r)(2s) && \text{by substitution} \\
&= 4(rs) && \text{by algebra.}
\end{aligned}
$$

Since rs is an integer (being a product of integers), mn is a multiple of 4 (by definition of divisibility).

22. *Hint:* The given statement can be rewritten formally as "$\forall$ integers n, if n is divisible by 6 then n is divisible by 2." This statement is true.

24. The statement is true.

    Proof:

    Suppose a, b, and c are any integers such that $a|b$ and $a|c$. *[We must show that $a|(2b-3c)$.]*

    By definition of divisibility, we know that $b = am$ and $c = an$ for some integers m and n.

    It follows that
    $$\begin{aligned} 2b - 3c &= 2(am) - 3(an) && \text{by substitution} \\ &= a(2m - 3n) && \text{by basic algebra.} \end{aligned}$$

    Let $t = 2m - 3n$. Then t is an integer because it is a difference of products of integers.

    Hence $2b - 3c = at$, where t is an integer, and so $a|(2b-3c)$ by definition of divisibility *[as was to be shown]*.

25. The statement is false.

    Counterexample:

    Let $a = 2, b = 8$, and $c = 8$.

    Then a is a factor of c because $8 = 2 \cdot 4$ and b is a factor of c because $8 = 1 \cdot 8$.

    But $ab = 16$ and 16 is not a factor of 8 because $8 \neq 16 \cdot k$ for any integer k (since $8/16 = 1/2$).

26. *Hint:* The statement is true.

    Proof:

    Suppose a, b, and c are any integers such that $ab \mid c$. *[We must show that $a \mid b$ and $a \mid c$.]*

    By definition of divisibility, $c = r(ab)$ for some integer r.

    Regrouping shows that $c = (ra)b$ and $c = (rb)a$.

    Now both ra and rb are integers because they are products of integers.

    Thus $c = (\text{an integer}) \cdot b$ and $c = (\text{an integer}) \cdot a$.

    It follows by definition of divisibility that $a \mid c$ and $b \mid c$.

27. The statement is false.

    Counterexample:

    Let $a = 2$, $b = 3$, and $c = 1$.

    Then $a \mid (b + c)$ because $b + c = 4$ and $2 \mid 4$ but $a \nmid b$ because $2 \nmid 3$ and $a \nmid c$ because $2 \nmid 1$.

    [This is one counterexample among many.]

30. The statement is false.

    Counterexample:

    Let $a = 4$ and $n = 6$.

    Then $a \mid n^2$ and $a \leq n$ because $4 \mid 36$ and $4 \leq 6$, but $a \nmid n$ because $4 \nmid 6$.

    [This is one counterexample among many.]

32. No. Each of these numbers is divisible by 3, and so their sum is also divisible by 3. But 100 is not divisible by 3. Thus the sum cannot equal $100.

33. No. The values of nickels, dimes, and quarters are all multiples of 5. It follows from exercise 15 that a sum of numbers each of which is divisible by 5 is also divisible by 5. So since $4.72 is not a multiple of 5, $4.72 cannot be obtained using only nickels, dimes, and quarters.

    Note: The form of reasoning used in this answer is called argument by contradiction. It is discussed formally in Section 4.7.

36. **a.** The sum of the digits is 54, which is divisible by 9. Therefore, 637,425,403,705,125 is divisible by 9 and hence also divisible by 3 (by transitivity of divisibility). Because the rightmost digit is 5, then 637,425,403,705,125 is divisible by 5. And because the two rightmost digits are 25, which is not divisible by 4, then 637,425,403,705,125 is not divisible by 4.

   b. Let $N = 12,858,306,120,312$.

   The sum of the digits of N is 42, which is divisible by 3 but not by 9. Therefore, N is divisible by 3 but not by 9.

   The right-most digit of N is neither 5 nor 0, and so N is not divisible by 5.

   The two right-most digits of N are 12, which is divisible by 4. Therefore, N is divisible by 4.

   c. Let $N = 517,924,440,926,512$.

   The sum of the digits of N is 61, which is not divisible by 3 (and hence not by 9 either). Therefore, N is not divisible by either 3 or 9.

   The right-most digit of N is neither 5 nor 0, and so N is not divisible by 5.

   The two right-most digits of N are 12, which is divisible by 4. Therefore, N is divisible by 4.

   d. Let $N = 14,328,083,360,232$.

   The sum of the digits of N is 45, which is divisible by 9 and hence also by 3. Therefore, N is divisible by 9 and by 3.

   The right-most digit of N is neither 5 nor 0, and so N is not divisible by 5.

   The two right-most digits of N are 32, which is divisible by 4. Therefore, N is divisible by 4.

37. **a.** $1{,}176 = 2^3 \cdot 3 \cdot 7^2$

38. **a.** $8{,}424 = 2^3 \cdot 3^4 \cdot 13$

   c. *Hint:* The answer is no. Note that each factor of 10 is comprised of a factor of 2 and a factor of 5.

   d. *Hint:* The answer is 26. Note that in order for $8{,}424 \cdot m$ to be a perfect square, each prime factor must be raised to an even power.

39. **a.** $a^3 = p_1^{3e_1} \cdot p_2^{3e_2} \cdots p_k^{3e_k}$

   b. To solve the problem, find the smallest powers of the factors 2, 3, 7, and 11 by which to multiply the given number so that the exponent of each factor becomes a multiple of 3. The value of k that works is $k = 2^2 \cdot 3 \cdot 7^2 \cdot 11$:

   $$2^4 \cdot 3^5 \cdot 7 \cdot 11^2 \cdot k \ = \ (2^4 \cdot 3^5 \cdot 7 \cdot 11^2) \cdot (2^2 \cdot 3 \cdot 7^2 \cdot 11)$$
   $$= \ 2^6 \cdot 3^6 \cdot 7^3 \cdot 11^3$$

   Then

   $$2^4 \cdot 3^5 \cdot 7 \cdot 11^2 \cdot k = 2^6 \cdot 3^6 \cdot 7^3 \cdot 11^3 = (2^2 \cdot 3^2 \cdot 7 \cdot 11)^3 = 2772^3.$$

40. **a.** Because $12a = 25b$, the unique factorization theorem guarantees that the standard factored forms of $12a$ and $25b$ must be the same. Thus $25b$ contains the factors $2^2 \cdot 3 (=12)$. But since neither 2 nor 3 divides 25, the factors $2^2 \cdot 3$ must all occur in b, and hence $12|b$. Similarly, $12a$ contains the factors $5^2 = 25$, and since 5 is not a factor of 12, the factors 5^2 must occur in a. So $25 \mid a$.

41. *Hint:* $45^8 \cdot 88^5 = (3^2 \cdot 5)^8 \cdot (2^3 \cdot 11)^5 = 3^{16} \cdot 5^8 \cdot 2^{15} \cdot 11^5$. How many factors of 10 does this number contain?

42. **a.** $6! = 6 \cdot 5 \cdot 4 \cdot 3 \cdot 2 \cdot 1 = 2 \cdot 3 \cdot 5 \cdot 2 \cdot 2 \cdot 3 \cdot 2 = 2^4 \cdot 3^2 \cdot 5$

b. $20!$
$$\begin{aligned}
&= 20 \cdot 19 \cdot 18 \cdot 17 \cdot 16 \cdot 15 \cdot 14 \cdot 13 \cdot 12 \cdot 11 \cdot 10 \cdot 9 \cdot 8 \cdot 7 \cdot 6 \cdot 5 \cdot 4 \cdot 3 \cdot 2 \cdot 1 \\
&= 2^2 \cdot 5 \cdot 19 \cdot 2 \cdot 3^2 \cdot 17 \cdot 2^4 \cdot 3 \cdot 5 \cdot 2 \cdot 7 \cdot 13 \cdot 2^2 \cdot 3 \cdot 11 \cdot 2 \cdot 5 \cdot 3^2 \cdot 2^3 \cdot 7 \cdot 2 \cdot 3 \cdot 5 \cdot 2^2 \cdot 3 \cdot 2 \\
&= 2^{18} \cdot 3^8 \cdot 5^4 \cdot 7^2 \cdot 11 \cdot 13 \cdot 17 \cdot 19
\end{aligned}$$

c. Squaring the result of part (b) gives

$$\begin{aligned}
(20!)^2 &= (2^{18} \cdot 3^8 \cdot 5^4 \cdot 7^2 \cdot 11 \cdot 13 \cdot 17 \cdot 19)^2 \\
&= 2^{36} \cdot 3^{16} \cdot 5^8 \cdot 7^4 \cdot 11^2 \cdot 13^2 \cdot 17^2 \cdot 19^2
\end{aligned}$$

When $(20!)^2$ is written in ordinary decimal form, there are as many zeros at the end of it as there are factors of the form $2 \cdot 5 \ (= 10)$ in its prime factorization.

Thus, since the prime factorization of $(20!)^2$ contains eight 5's and more than eight 2's, $(20!)^2$ contains eight factors of 10 and hence eight zeros.

43. *Hint:* There are 108 mathematics students and 120 computer science students at the university.

44. Proof:

Suppose n is a nonnegative integer whose decimal representation ends in 0.

Then $n = 10m + 0 = 10m$ for some integer m.

Factoring out a 5 yields $n = 10m = 5(2m)$, and $2m$ is an integer since m is an integer.

Hence $10m$ is divisible by 5, which is what was to be shown.

45. Proof:

Suppose n is a nonnegative integer whose decimal representation ends in 5.

Then $n = 10m + 5$ for some integer m.

By factoring out a 5, $n = 10m + 5 = 5(2m + 1)$, and $2m + 1$ is an integer since m is an integer.

Hence n is divisible by 5 by definition of divisibility.

47. *Hint:* You may take it as a fact that for any positive integer k, $10^k = \underbrace{99\ldots9}_{k \text{ of these}} + 1$.

That is: $10^k = 9 \cdot 10^{k-1} + 9 \cdot 10^{k-2} + \cdots + 9 \cdot 10^1 + 9 \cdot 10^0 + 1$.

48. Proof:

Suppose n is any nonnegative integer for which the sum of the digits of n is divisible by 3.

By definition of decimal representation, n can be written in the form

$$n = d_k 10^k + d_{k-1} 10^{k-1} + \cdots + d_2 10^2 + d_1 10 + d_0$$

where k is a nonnegative integer and all the d_i are integers from 0 to 9 inclusive. Then

$$\begin{aligned}
n &= d_k (\underbrace{99\ldots9}_{k\ 9\text{'s}} + 1) + d_{k-1}(\underbrace{99\ldots9}_{(k-1)\ 9\text{'s}} + 1) + \cdots + d_2(99 + 1) + d_1(9 + 1) + d_0 \\
&= d_k \cdot \underbrace{99\ldots9}_{k\ 9\text{'s}} + d_{k-1} \cdot \underbrace{99\ldots9}_{(k-1)\ 9\text{'s}} + \cdots + d_2 \cdot 99 + d_1 \cdot 9 + (d_k + d_{k-1} + \cdots + d_2 + d_1 + d_0) \\
&= 9(d_k \cdot \underbrace{11\ldots1}_{k\ 1\text{'s}} + d_{k-1} \cdot \underbrace{11\ldots1}_{(k-1)\ 1\text{'s}} + \cdots + d_2 \cdot 11 + d_1) + (d_k + d_{k-1} + \cdots + d_2 + d_1 + d_0) \\
&= 3[3(d_k \cdot \underbrace{11\ldots1}_{k\ 1\text{'s}} + d_{k-1} \cdot \underbrace{11\ldots1}_{(k-1)\ 1\text{'s}} + \cdots + d_2 \cdot 11 + d_1)] + (d_k + d_{k-1} + \cdots + d_2 + d_1 + d_0) \\
&= (\text{an integer divisible by 3}) + (\text{the sum of the digits of } n).
\end{aligned}$$

Since the sum of the digits of n is divisible by 3, n can be written as a sum of two integers each of which is divisible by 3.

It follows from exercise 15 that n is divisible by 3.

Section 4.5

1. $q = 7, \quad r = 7$

3. $q = 0, \quad r = 36$

5. $q = -5, \quad r = 10$

6. $q = -4, \quad r = 5$

7. **a.** 4   **b.** 7

9. **a.** 5   **b.** 3

11. **a.** When today is Saturday, 15 days from today is two weeks (which is Saturday) plus one day (which is Sunday). Hence $DayN$ should be 0. According to the formula, when today is Saturday, $DayT = 6$, and so when $N = 15$,

$$
\begin{aligned}
DayN &= (DayT + N) \bmod 7 \\
&= (6 + 15) \bmod 7 \\
&= 21 \bmod 7 = 0, \qquad \text{which agrees.}
\end{aligned}
$$

12. Let the days of the week be numbered from 0 (Sunday) through 6 (Saturday) and let $DayT$ and $DayN$ be variables representing the day of the week today and the day of the week N days from today. By the quotient-remainder theorem, there exist unique integers q and r such that

$$DayT + N = 7q + r \text{ and } 0 \le r < 7.$$

Now $DayT + N$ counts the number of days to the day N days from today starting last Sunday (where "last Sunday" is interpreted to mean today if today is a Sunday). Thus $DayN$ is the day of the week that is $DayT + N$ days from last Sunday. Because each week has seven days,

$DayN$ is the same as the day of the week $DayT + N - 7q$ days from last Sunday.

But

$$DayT + N - 7q = r \quad \text{and} \quad 0 \le r < 7.$$

Therefore,

$$DayN = r = (DayT + N) \bmod 7.$$

13. *Solution 1:* $30 = 4 \cdot 7 + 2$. Hence the answer is two days after Monday, which is Wednesday.

    Solution 2: By the formula, the answer is $(1 + 30) \bmod 7 = 31 \bmod 7 = 3$, which is Wednesday.

14. *Hint:* There are two ways to solve this problem. One is to find that $1{,}000 = 7 \cdot 142 + 6$ and note that if today is Tuesday, then 1,000 days from today is 142 weeks plus 6 days from today. The other way is to use the formula $DayN = (DayT + N) \bmod 7$, with $DayT = 2$ (Tuesday) and $N = 1{,}000$.

15. There are 13 leap year days between January 1, 2000 and January 1, 2050 (once every four years in 2000, 2004, 2008, 2012, . . . , 2048). So 13 of the years have 366 days and the remaining 38 years have 365 days. This gives a total of

$$13 \cdot 366 + 37 \cdot 365 = 18{,}263$$

days between the two dates. Using the formula $DayN = (DayT + N) \bmod 7$, and letting $DayT = 6$ (Saturday) and $N = 18{,}263$ gives

$$DayN = (6 + 18263) \bmod 7 = 18269 \bmod 7 = 6,$$

which is also a Saturday.

16. Because $d \mid n$, $n = dq + 0$ for some integer q. Thus the remainder is 0.

18. **a.** <u>Proof</u>:

Consider any two consecutive integers. Call the smaller one n. By the quotient-remainder theorem with $d = 2$, either n is even or n is odd.

Case 1 (n is even): In this case, $n = 2k$ for some integer k. Then $n(n+1) = 2k(2k+1) = 2[k(2k+1)]$. But $k(2k+1)$ is an integer (because products and sums of integers are integers), and so $n(n+1)$ is even.

Case 2 (n is odd): In this case $n = 2k+1$ for some integer k. Then

$$
\begin{aligned}
n(n+1) &= (2k+1)[(2k+1)+1] \quad \text{by substitution} \\
&= (2k+1)(2k+2) \\
&= 2[(2k+1)(k+1)] \quad \text{by algebra.}
\end{aligned}
$$

But $(2k+1)(k+1)$ is an integer (because products and sums of integers are integers), and so $n(n+1)$ is even.

Conclusion: In both of the two possible cases the product $n(n+1)$ is even *[as was to be shown]*.

b. <u>Proof</u>:

Suppose n is any odd integer. *[We must show that $n^2 = 8m + 1$ for some integer m.]*

By definition of odd, $n = 2q + 1$ for some integer q. Then

$$
\begin{aligned}
n^2 &= (2q+1)^2 \quad \text{by substitution} \\
&= 4q^2 + 4q + 1 \\
&= 4(q^2 + q) + 1 \\
&= 4q(q+1) + 1 \quad \text{by algebra.}
\end{aligned}
$$

By the result of part (a), the product $q(q+1)$ is even, so $q(q+1) = 2m$ for some integer m. Thus, by substitution, $n^2 = 4 \cdot 2m + 1 = 8m + 1$ where m is an integer *[as was to be shown]*.

20. Because $a \bmod 7 = 4$, the remainder obtained when a is divided by 7 is 4, and so $a = 7q + 4$ for some integer q. Multiplying this equation through by 5 gives that $5a = 35q + 20 = 35q + 14 + 6 = 7(5q + 2) + 6$. Because q is an integer, $5q + 2$ is also an integer, and so $5a = 7 \cdot$ (an integer) $+ 6$. Thus, because $0 \le 6 < 7$, the remainder obtained when $5a$ is divided by 7 is 6, and so $5a \bmod 7 = 6$.

21. Given that b is an integer and $b \bmod 12 = 5$, it follows that $8b \bmod 12 = 4$.

<u>Proof</u>: When b is divided by 12, the remainder is 5. Thus there exists an integer m so that $b = 12m + 5$. Multiplying this equation by 8 gives

$$
\begin{aligned}
8b &= 8(12m + 5) \quad \text{by substitution} \\
&= 96m + 40 \\
&= 96m + 36 + 4 \\
&= 12(8m + 3) + 4 \quad \text{by algebra.}
\end{aligned}
$$

Since $8m + 3$ is an integer and since $0 \le 4 < 12$, the uniqueness part of the quotient-remainder theorem guarantees that the remainder obtained when $8b$ is divided by 12 is 4.

23. <u>Proof</u>:

Suppose n is any *[particular but arbitrarily chosen]* integer such that $n \bmod 5 = 3$.

Then the remainder obtained when n is divided by 5 is 3, and so $n = 5q + 3$ for some integer q. Squaring both sides gives

$$
\begin{aligned}
n^2 &= (5q+3)^2 \quad \text{by substitution} \\
&= 25q^2 + 30q + 9 \\
&= 25q^2 + 30q + 5 + 4 \\
&= 5(5q^2 + 6q + 1) + 4 \quad \text{by algebra.}
\end{aligned}
$$

Because products and sums of integers are integers, $5q^2 + 6q + 1$ is an integer, and hence $n^2 = 5 \cdot (\text{an integer}) + 4$.

Thus, since $0 \leq 4 < 5$, the remainder obtained when n^2 is divided by 5 is 4, and so $n^2 \bmod 5 = 4$.

24. <u>Proof</u>:

Suppose m and n are any *[particular but arbitrarily chosen]* integers such that $m \bmod 5 = 2$ and $n \bmod 5 = 1$.

Then the remainder obtained when m is divided by 5 is 2 and the remainder obtained when n is divided by 5 is 1, and so $m = 5q + 2$ and $n = 5r + 1$ for some integers q and r. Then

$$\begin{aligned}
mn &= (5q + 2)(5r + 1) & \text{by substitution} \\
&= 25qr + 5q + 10r + 2 \\
&= 5(5qr + q + 2r) + 2 & \text{by algebra.}
\end{aligned}$$

Because products and sums of integers are integers, $5qr + q + 2r$ is an integer, and hence $mn = 5 \cdot (\text{an integer}) + 2$.

Thus, since $0 \leq 2 < 5$, the remainder obtained when mn is divided by 5 is 2, and so $mn \bmod 5 = 2$.

26. *Hint:* You need to show that (1) for each integer n and positive integer d, if n is divisible by d then $n \bmod d = 0$; and (2) for each integer n and positive integer d, if $n \bmod d = 0$ then n is divisible by d.

27. <u>Proof</u>: Let n be any integer. *[We must show that $n^2 = 4k$ or $n^2 = 4k + 1$ for some integer k.]* By the quotient-remainder theorem with divisor equal to 2, $n = 2q$ or $n = 2q + 1$ for some integer q.

Case 1 ($n = 2q$ for some integer q): In this case,

$$\begin{aligned}
n^2 &= (2q)^2 & \text{by substitution} \\
&= 4q^2 & \text{by algebra.}
\end{aligned}$$

Let $k = q^2$. Then k is an integer because it is a product of integers. Hence $n^2 = 4k$ for some integer k.

Case 2 ($n = 2q + 1$ for some integer q): In this case,

$$\begin{aligned}
n^2 &= (q + 1)^2 & \text{by substitution} \\
&= 4q^2 + 4q + 1 \\
&= 4(q^2 + q) + 1 & \text{by algebra.}
\end{aligned}$$

Let $k = q^2 + q$. Then k is an integer because it is a sum of products of integers. Hence $n^2 = 4k + 1$ for some integer k.

Conclusion: It follows that in both possible cases, there is an integer k such that $n^2 = 4k$ or $n^2 = 4k + 1$ *[as was to be shown]*.

28. **a.** *Hint:* Start by supposing that n, $n + 1$, and $n + 2$ are any three consecutive integers. Then use the quotient- remainder theorem to divide into three cases:

Case 1 ($n = 3q$ for some integer q): In this case you will show that n is a multiple of 3.

Case 2 ($n = 3q + 1$ for some integer q): In this case you will show that $n+2$ is a multiple of 3.

Case 3 ($n = 3q + 2$ for some integer q): In this case you will show that $n+1$ is a multiple of 3.

Conclude that in all possible cases one of the integers is a multiple of 3.

29. **a.** *Hint:* Given any integer n, begin by using the quotient-remainder theorem to say that n can be written in one of the three forms: $n = 3q$, or $n = 3q + 1$, or $n = 3q + 2$ for some integer q. Then divide into three cases according to these three possibilities. Show that in each case either $n^2 = 3k$ for some integer k, or $n^2 = 3k + 1$ for some integer k. For instance, when $n = 3q + 2$, then

$$
\begin{aligned}
n^2 &= (3q + 2)^2 &&\text{by substitution} \\
&= 9q^2 + 12q + 4 \\
&= 9q^2 + 12q + 3 + 1 \\
&= 3(3q^2 + 4q + 1) + 1 &&\text{by algebra,}
\end{aligned}
$$

and $3q^2 + 4q + 1$ is an integer because it is a sum of products of integers.

30. **a.** <u>Proof</u>: Suppose n and $n + 1$ are any two consecutive integers. By the quotient-remainder theorem with $d = 3$, we know that $n = 3q$, or $n = 3q + 1$, or $n = 3q + 2$ for some integer q.

Case 1 ($n = 3q$ for some integer q): In this case,

$$
\begin{aligned}
n(n + 1) &= 3q(3q + 1) &&\text{by substitution} \\
&= 3[q(3q + 1)] &&\text{by algebra.}
\end{aligned}
$$

Let $k = q(3q + 1)$. Then k is an integer because sums and products of integers are integers. Hence $n(n + 1) = 3k$ for some integer k.

Case 2 ($n = 3q + 1$ for some integer q): In this case,

$$
\begin{aligned}
n(n + 1) &= (3q + 1)(3q + 2) &&\text{by substitution} \\
&= 9q^2 + 9q + 2 \\
&= 3(3q^2 + 3q) + 2 &&\text{by algebra.}
\end{aligned}
$$

Let $k = 3q^2 + 3q$. Then k is an integer because sums and products of integers are integers. Hence $n(n + 1) = 3k + 2$ for some integer k.

Case 3 ($n = 3q$ for some integer q): In this case,

$$
\begin{aligned}
n(n + 1) &= (3q + 2)(3q + 3) &&\text{by substitution} \\
&= 3[(3q + 2)(q + 1)] \\
&= 3[(3q + 2)(q + 1)] &&\text{by algebra.}
\end{aligned}
$$

Let $k = (3q + 2)(q + 1)$. Then k is an integer because sums and products of integers are integers. Hence $n(n + 1) = 3k$ for some integer k.

Conclusion: In all three cases, the product of the two consecutive integers either equals $3k$ or it equals $3k + 2$ for some integer k *[as was to be shown]*.

b. Given any consecutive integers m and n, $mn \bmod 3 = 0$ or $mn \bmod 3 = 1$.

(Or, equivalently, Given any consecutive integers m and n, $mn \bmod 3 \neq 2$.)

31. **b.** If $m^2 - n^2 = 56$, then $56 = (m + n)(m - n)$. Now $56 = 2^3 \cdot 7$, and by the unique factorization theorem, this factorization is unique. Hence the only representation of 56 as a product of two positive integers are $56 = 7 \cdot 8 = 14 \cdot 4 = 28 \cdot 2 = 56 \cdot 1$. By part (a), m and n must both be odd or both be even. Thus the only solutions are either $m + n = 14$ and $m - n = 4$ or $m + n = 28$ and $m - n = 2$. It follows that the only solutions are either $m = 9$ and $n = 5$ or $m = 15$ and $n = 13$.

32. *Answer:* Given any integers a, b, and c, if and both $a - b$ and $b - c$ are even, then $2a - (b + c)$ is even.

<u>Proof</u>: Suppose $a, b,$ and c are any integers such that both $a - b$ and $b - c$ are even. *[We must show that $2a - (b + c)$ is even.]*

Note first that
$$2a - (b + c) = (a - b) + (a - c).$$

Also note that because $(a - b) + (b - c)$ is a sum of two even integers,

$$(a - b) + (b - c) \text{ is even} \quad \text{by property 1 in Example 4.3.3.}$$

In addition,
$$(a - b) + (b - c) = a - c \quad \text{by combining like terms.}$$

Hence $a - c$ is even, and thus $2a - (b + c)$ is a sum of the two even integers $(a - b)$ and $(a - c)$. So $2a - (b + c)$ is even *[as was to be shown]*.

33. *Answer:* Given any integers a, b, and c if $a - b$ is odd and $b - c$ is even, then $a - c$ is odd.

    Proof: Suppose a, b, and c are any integers such that $a - b$ is odd and $b - c$ is even. *[We must show that $a - c$ is odd.]*

    Then $(a - b) + (b - c)$ is a sum of an odd integer and an even integer and hence is odd (by property 5 in Example 4.3.3).

    But $(a - b) + (b - c) = a - c$, and thus $a - c$ is odd *[as was to be shown]*.

34. *Hint:* Express n using the quotient-remainder theorem with $d = 3$.

36. Proof:

    Suppose n is any integer. *[We must show that $8 \mid n(n + 1)(n + 2)(n + 3)$.]*

    By the quotient-remainder theorem with divisor equal to 4, there is an integer k such that

    $$n = 4k \quad \text{or} \quad n = 4k + 1 \quad \text{or} \quad n = 4k + 2 \quad \text{or} \quad n = 4k + 3.$$

    Case 1 ($n = 4k$ for some integer k): In this case,

    $$\begin{aligned} n(n + 1)(n + 2)(n + 3) &= 4k(4k + 1)(4k + 2)(4k + 3) & \text{by substitution} \\ &= 8[k(4k + 1)(2k + 1)(4k + 3)] & \text{by algebra,} \end{aligned}$$

    and this is divisible by 8 because $k(4k + 1)(2k + 1)(4k + 3)$ is an integer (since k is an integer and sums and products of integers are integers).

    Case 2 ($n = 4k + 1$ for some integer k): In this case,

    $$\begin{aligned} n(n + 1)(n + 2)(n + 3) &= (4k + 1)(4k + 2)(4k + 3)(4k + 4) & \text{by substitution} \\ &= 8[(4k + 1)(2k + 1)(4k + 3)(k + 1)] & \text{by algebra,} \end{aligned}$$

    and this is divisible by 8 because $(4k + 1)(2k + 1)(4k + 3)(k + 1)$ (since k is an integer and sums and products of integers are integers).

    Case 3 ($n = 4k + 2$ for some integer k): In this case,

    $$\begin{aligned} n(n + 1)(n + 2)(n + 3) &= (4k + 2)(4k + 3)(4k + 4)(4k + 5) & \text{by substitution} \\ &= 8[(2k + 1)(4k + 3)(k + 1)(4k + 5)] & \text{by algebra,} \end{aligned}$$

    and this is divisible by 8 because $(2k + 1)(4k + 3)(k + 1)(4k + 5)$ is an integer (since k is an integer and sums and products of integers are integers).

    Case 4 ($n = 4k + 3$ for some integer k): In this case,

    $$\begin{aligned} n(n + 1)(n + 2)(n + 3) &= (4k + 3)(4k + 4)(4k + 5)(4k + 6) & \text{by substitution} \\ &= 8[(4k + 3)(k + 1)(4k + 5)(2k + 3)] & \text{by algebra,} \end{aligned}$$

    and this is divisible by 8 because $4k + 3)(k + 1)(4k + 5)(2k + 3)$ (since k is an integer and sums and products of integers are integers).

Conclusion: In all four possible cases, $8 \mid n(n+1)(n+2)(n+3)$ *[as was to be shown]*.

Alternative proof: One can make use of exercise 18(a) to produce a proof that only requires two cases: n is even and n is odd. Then, since both $n(n+1)$ and $(n+2)(n+3)$ are products of consecutive integers, by the result of exercise 18(a), both products are even and hence contain a factor of 2. Multiplying those two factors shows that there is a factor of 4 in $n(n+1)(n+2)(n+3)$.

37. *Hint:* Given any integer n, consider the two cases where n is even and where n is odd.

39. Proof: Let p be any prime number except 2 or 3. By the quotient-remainder theorem, there is an integer k so that p can be written as

$$p = 6k \quad \text{or} \quad p = 6k+1 \quad \text{or} \quad p = 6k+2 \quad \text{or} \quad p = 6k+3 \quad \text{or} \quad p = 6k+4 \quad \text{or} \quad p = 6k+5.$$

Since p is prime and $p \neq 2$, p is not divisible by 2. Consequently, $p \neq 6k$, $p \neq 6k+2$, and $p \neq 6k+4$ for any integer k *[because all of these numbers are divisible by 2]*.

Furthermore, since p is prime and $p \neq 3$, p is not divisible by 3. Thus $p \neq 6k+3$ *[because this number is divisible by 3]*.

41. *Hint:* There are four cases: Either x and y are both positive, or x is positive and y is negative, or x is negative and y is positive, or both x and y are negative.

42. Proof: Let c be any positive real number and let r be any real number.

Proof Part 1:

Suppose that $-c \leq r \leq c$. (*)

[We must show that $|r| \leq c$.]

By the trichotomy law (see Appendix A, T17), either $r \geq 0$ or $r < 0$.

Case 1 ($r \geq 0$): In this case $|r| = r$, and so by substitution into (*), $-c \leq |r| \leq c$.

In particular, $|r| \leq c$.

Case 2 ($r < 0$): In this case $|r| = -r$, and so $r = -|r|$.

Hence by substitution into (*), $-c \leq -|r| \leq c$.

In particular, $-c \leq -|r|$.

Multiplying both sides by -1 gives $c \geq |r|$, or, equivalently, $|r| \leq c$.

Therefore, regardless of whether $r \geq 0$ or $r < 0$, $|r| \leq c$ *[as was to be shown]*.

Proof Part 2:

Suppose that $|r| \leq c$.(**)

[We must show that $-c \leq r \leq c$.]

By the trichotomy law, either $r \geq 0$ or $r < 0$.

Case 1 ($r \geq 0$): In this case $|r| = r$, and so by substitution into (**), $r \leq c$.

Since $r \geq 0$ and $c \geq r$, then $c \geq 0$ by transitivity of order (Appendix A, T18).

Then, by property T24 of Appendix A, $0 \geq -c$, and, again by transitivity of order, $r \geq -c$.

Hence $-c \leq r \leq c$.

Case 2 ($r < 0$): In this case $|r| = -r$, and so by substitution into (**) $-r \leq c$.

Multiplying both sides of this inequality by -1 gives $r \geq -c$.

Also since $r < 0$ and $0 \leq c$, then $r \leq c$.

Hence $-c \leq r \leq c$.

Conclusion: Regardless of whether $r \geq 0$ or $r < 0$, we have that $-c \leq r \leq c$ *[as was to be shown]*.

43. *Hint:* Apply the triangle inequality with $x = a - b$ and $y = b$ and with $x = b - a$ and $y = a$. Then use the result of exercise 42.

44. **a.** $7,609 + 5 = 7,614$

45. ***Solution 1:*** We are given that M is a matrix with m rows and n columns, stored in row major form at locations $N + k$, where $0 \leq k < mn$. Given a value for k, we want to find indices r and s so that the entry for M in row r and column s, a_{rs}, is stored in location $N + k$. By the quotient-remainder theorem, $k = nQ + R$, where Q and R are integers and $0 \leq R < n$. The first Q rows of M (each of length n) are stored in the first nQ locations: $N + 0, N + 1, \ldots, N + nQ - 1$ with a_{Qn} stored in the last of these. Consider the next row. When $r = Q + 1$,

a_{r1} will be in location $N + nQ$
a_{r2} will be in location $N + nQ + 1$
a_{r3} will be in location $N + nQ + 2$

$\vdots$

a_{rs} will be in location $N + nQ + (s - 1)$

$\vdots$

and a_{rn} will be in location $N + nQ + (n - 1) = N + n(Q + 1) - 1$.

Thus location $N + k$ contains a_{rs} where $r = Q + 1$ and $R = s - 1$. But $Q = k \ div \ n$ and $R = k \ mod \ n$, and hence $r = (k \ div \ n) + 1$ and $s = (k \ mod \ n) + 1$.

Solution 2: *[After the floor notation has been introduced, the following solution can be considered as an alternative.]* To find a formula for r, note that for $1 \leq a \leq m$,

when $(a - 1)n \leq k < an$,   then $r = a$.

Dividing through by n gives that

when $(a - 1) \leq \dfrac{k}{n} < a$,   then $r = a$

or, equivalently,

when $\left\lfloor \dfrac{k}{n} \right\rfloor = a - 1$,   then $r = a$.

But this implies that

when $a = \left\lfloor \dfrac{k}{n} \right\rfloor + 1 = r$,   then $r = a$.

and since

$\left\lfloor \dfrac{k}{n} \right\rfloor = k \ div \ n$,   we have that $r = (k \ div \ n) + 1$.

To find a formula for s, note that

when $k = n \cdot (\text{an integer}) + b$ and $0 \leq b < n$,   then $s = b + 1$.

Thus by the quotient-remainder theorem, $s = (k \ mod \ n) + 1$.

46. *Answer to the first question:* No. Counterexample: Let $m = 1, n = 3$, and $d = 2$. Then $m \ mod \ d = 1$ and $n \ mod \ d = 1$ but $m \neq n$.

Answer to second question: Yes. Proof: Suppose m, n, and d are integers such that $m \ mod \ d = n \ mod \ d$. Let $r = m \ mod \ d = n \ mod \ d$. By definition of mod, $m = dp + r$ and $n = dq + r$ for some integers p and q. Then $m - n = (dp + r) - (dq + r) = d(p - q)$. But $p - q$ is an integer (being a difference of integers), and so $m - n$ is divisible by d by definition of divisibility.

48. *Answer to the first question:* not necessarily

Counterexample: Let $m = n = 3$, $d = 2$, $a = 1$, and $b = 1$. Then

$$m \bmod d = n \bmod d = 3 \bmod 2 = 1 = a = b.$$

But $a + b = 1 + 1 = 2$, whereas $(m + n) \bmod d = 6 \bmod 2 = 0$.

Answer to the second question: yes.

Proof: Suppose m, n, a, b, and d are integers and $m \bmod d = a$ and $n \bmod d = b$. By definition of *mod*, $m = dq_1 + a$ and $n = dq_2 + b$ for some integers q_1 and q_2. Then

$$
\begin{aligned}
m + n &= (dq_1 + a) + (dq_2 + b) && \text{by substitution} \\
&= d(q_1 + q_2) + (a + b)\ (*) && \text{by algebra.}
\end{aligned}
$$

Apply the quotient-remainder theorem to $a + b$ to obtain unique integers q_3 and r such that $a + b = dq_3 + r$ (**) and $0 \le r < d$. By definition of *mod*, $r = (a + b) \bmod d$. Then

$$
\begin{aligned}
m + n &= d(q_1 + q_2) + (a + b) && \text{by (*)} \\
&= d(q_1 + q_2) + (dq_3 + r) && \text{by (**)} \\
&= d(q_1 + q_2 + q_3) + r && \text{by algebra,}
\end{aligned}
$$

where $q_1 + q_2 + q_3$ and r are integers and $0 \le r < d$. Hence by definition of *mod*,

Section 4.6

1. $\lfloor 37.999 \rfloor = 37$, $\quad \lceil 37.999 \rceil = 38$

3. $\lfloor -14.00001 \rfloor = -15$, $\quad \lceil -14.00001 \rceil - 14$

6. If k is an integer, then $\lceil k \rceil = k$ because $k - 1 < k \le k$ and $k - 1$ and k are integers.

8. When the ceiling notation is used, the answer is either $\lceil \frac{n}{7} \rceil - 1$ if $\frac{n}{7}$ is not an integer or $\lceil \frac{n}{7} \rceil$ if $\frac{n}{7}$ is an integer.

9. If the remainder obtained when n is divided by 36 is positive, an additional box beyond those containing exactly 36 units will be needed to hold the extra units.

So, since the ceiling notation rounds each number up to the nearest integer, the number of boxes required will be $\lceil \frac{n}{36} \rceil$.

Also, because the ceiling of an integer is itself, if the number of units is a multiple of 36, the number of boxes required will be $\lceil \frac{n}{36} \rceil$ as well.

Thus the ceiling notation is more appropriate for this problem because the answer is simply $\lceil \frac{n}{36} \rceil$ regardless of the value of n.

If the floor notation is used, the answer is more complicated: if $\frac{n}{36}$ is not an integer, the answer is $\lfloor \frac{n}{36} \rfloor + 1$, but if n is an integer, it is $\lfloor \frac{n}{36} \rfloor$.

10. **a.** (i) $\left(2050 + \lfloor \frac{2049}{4} \rfloor - \lfloor \frac{2049}{100} \rfloor + \lfloor \frac{2049}{400} \rfloor \right) \bmod 7$

$= (2050 + 512 - 20 + 5) \bmod 7 = 2547 \bmod 7$

$= 6$, which corresponds to a Saturday.

b. *Hint:* One day is added every four years, except that each century the day is not added unless the century is a multiple of 400.

12. The equality $\left\lfloor \dfrac{n}{2} \right\rfloor = \dfrac{n-1}{2}$ is what is to be shown. The "proof" assumes what is to be proved. because it substitutes $2k+1$ for n into both sides of the equation, keeps the equal sign, and works with the result as if it were known to be true.

13. <u>Proof:</u>

    Suppose n is any even integer. By definition of even, $n = 2k$ for some integer k. Then

    $$\left\lfloor \frac{n}{2} \right\rfloor = \left\lfloor \frac{2k}{2} \right\rfloor = \lfloor k \rfloor = k$$

    because k is an integer and $k \le k < k - 1$. On the other hand,

    $$k = \frac{n}{2} \qquad \text{because } n = 2k.$$

    So since $\left\lfloor \dfrac{n}{2} \right\rfloor = k$ and $k = \dfrac{n}{2}$, then

    $$\left\lfloor \frac{n}{2} \right\rfloor = \frac{n}{2}$$

    [as was to be shown].

14. <u>Counterexample:</u> Let $x = 2$ and $y = 1.9$. Then

    $$\lfloor x - y \rfloor = \lfloor 2 - 1.9 \rfloor = \lfloor 0.1 \rfloor = 0 \quad \text{whereas} \quad \lfloor x \rfloor - \lfloor y \rfloor = \lfloor 2 \rfloor - \lfloor 1.9 \rfloor = 2 - 1 = 1.$$

    and $0 \neq 1$.

15. <u>Proof:</u> Suppose x is any real number. Let

    $$m = \lfloor x \rfloor.$$

    By definition of floor,
    $$m \le x < m + 1.$$

    Subtracting 1 from all parts of the inequality gives that
    $$m - 1 \le x - 1 < m,$$

    and so, by definition of floor,
    $$\lfloor x - 1 \rfloor = m - 1.$$

    Substituting $\lfloor x \rfloor$ in place of m shows that
    $$\lfloor x - 1 \rfloor = \lfloor x \rfloor - 1.$$

17. *Proof for the case where $n \bmod 3 = 2$:*

    In the case where $n \bmod 3 = 2$, then $n = 3q + 2$ for some integer q by definition of *mod*. Then

    $$\begin{aligned}
    \left\lfloor \frac{n}{3} \right\rfloor &= \left\lfloor \frac{3q+2}{3} \right\rfloor && \text{by substitution} \\[2mm]
    &= \left\lfloor q + \frac{2}{3} \right\rfloor && \text{by algebra} \\[2mm]
    &= q && \text{because } q \text{ is an integer} \\
    & && \text{and } q \le q + 2/3 < q + 1
    \end{aligned}$$

18. <u>Counterexample</u>: Let $x = y = 1.5$. Then

$$\lceil x + y \rceil = \lceil 1.5 + 1.5 \rceil = \lceil 3 \rceil = 3 \quad \text{whereas} \quad \lceil x \rceil + \lceil y \rceil = \lceil 1.5 \rceil + \lceil 1.5 \rceil = 2 + 2 = 4,$$

and $3 \neq 4$.

[This is one counterexample among many.]

19. *Hint:* The statement is true.

21. <u>Proof</u>: Let n be any odd integer. *[We must show that* $\left\lceil \dfrac{n}{2} \right\rceil = \dfrac{n+1}{2}$.*]* By definition of odd, $n = 2k + 1$ for some integer k. The left-hand side of the equation to be proved is

$$\begin{aligned}
\left\lceil \frac{n}{2} \right\rceil &= \left\lceil \frac{2k+1}{2} \right\rceil && \text{by substitution} \\[2mm]
&= \left\lceil k + \frac{1}{2} \right\rceil && \text{by algebra} \\[2mm]
&= k + 1 && \text{by definition of ceiling because } k \text{ is} \\
& && \text{an integer and } k < k + 1/2 \leq k + 1
\end{aligned}$$

On the other hand, the right-hand side of the equation to be proved is

$$\begin{aligned}
\frac{n+1}{2} &= \frac{(2k+1)+1}{2} && \text{by substitution} \\[2mm]
&= \frac{2k+2}{2} && \\[2mm]
&= \frac{2(k+1)}{2} && \\[2mm]
&= k + 1 && \text{by algebra.}
\end{aligned}$$

Thus both the left- and right-hand sides of the equation to be proved equal $k + 1$, and so both are equal to each other. In other words, $\lceil n/2 \rceil = (n+1)/2$ *[as was to be shown].*

23. <u>Proof</u>:

Suppose x is a real number that is not an integer. Let

$$\lfloor x \rfloor = n.$$

Then, by definition of floor and because x is not an integer,

$$n < x < n + 1.$$

Multiplying both sides by -1 gives

$$-n > -x > -n - 1, \quad \text{or equivalently,} \quad -n - 1 < -x < -n.$$

Since $-n - 1$ is an integer, it follows by definition of floor that

$$\lfloor -x \rfloor = -n - 1.$$

Hence

$$\lfloor x \rfloor + \lfloor -x \rfloor = n + (-n - 1) = n - n - 1 = -1,$$

as was to be shown.

24. <u>Proof</u>: Suppose m is any integer and x is any real number that is not an integer.

By definition of floor, $\lfloor x \rfloor = n$ where n is an integer and $n \leq x < n+1$.

Since x is not an integer, $x \neq n$, and so

$$n < x < n+1.$$

Multiply all parts of this inequality by -1 to obtain

$$-n > -x > -n-1.$$

Then add m to all parts to obtain

$$m-n > m-x > m-n-1, \quad \text{or, equivalently,} \quad m-n-1 < m-x < m-n.$$

But $m-n-1$ and $m-n$ are both integers, and so by definition of floor, $\lfloor m-x \rfloor = m-n-1$. By substitution,

$$\lfloor x \rfloor + \lfloor m-x \rfloor = n + (m-n-1) = m-1$$

[as was to be shown].

25. *Hint:* Let $n = \lfloor \frac{x}{2} \rfloor$ and consider the two cases: n is even and n is odd.

26. <u>Proof</u>: Suppose x is any real number such that

$$x - \lfloor x \rfloor < \frac{1}{2}.$$

Multiplying both sides by 2 gives

$$2x - 2\lfloor x \rfloor < 1, \quad \text{or} \quad 2x < 2\lfloor x \rfloor + 1.$$

Now by definition of floor, $\lfloor x \rfloor \leq x$. Hence, $2\lfloor x \rfloor \leq 2x$. Putting the two inequalities involving $2x$ together gives

$$2\lfloor x \rfloor \leq 2x < 2\lfloor x \rfloor + 1.$$

Thus, by definition of floor (and because $2\lfloor x \rfloor$ is an integer), $\lfloor 2x \rfloor = 2\lfloor x \rfloor$. This is what to be shown.

27. <u>Proof</u>: Suppose x is any real number such that

$$x - \lfloor x \rfloor \geq 1/2.$$

Multiply both sides by 2 to obtain

$$2x - 2\lfloor x \rfloor \geq 1 \quad \text{or, equivalently,} \quad 2x \geq 2\lfloor x \rfloor + 1.$$

Now by definition of floor,

$$x < \lfloor x \rfloor + 1 \quad \text{and hence} \quad 2x < 2\lfloor x \rfloor + 2.$$

Put the two inequalities involving x together to obtain

$$2\lfloor x \rfloor + 1 \leq 2x < 2\lfloor x \rfloor + 2.$$

By definition of floor, then, $\lfloor 2x \rfloor = 2\lfloor x \rfloor + 1$.

28. *Hint:* After applying the hypothesis that n is odd, evaluate the two sides of the equation separately and show that the results are equal.

30. <u>Proof</u>: Suppose n is any integer. *[We must show that* $\left\lfloor\dfrac{n}{2}\right\rfloor+\left\lceil\dfrac{n}{2}\right\rceil = n.]$ By the parity principle, n is either even or odd.

   Case 1 (n is even): In this case, $n = 2k$ for some integer k, and so

$$
\begin{aligned}
\left\lfloor\frac{n}{2}\right\rfloor+\left\lceil\frac{n}{2}\right\rceil &= \left\lfloor\frac{2k}{2}\right\rfloor+\left\lceil\frac{2k}{2}\right\rceil && \text{by substitution} \\
&= \lfloor k\rfloor + \lceil k\rceil && \text{by algebra} \\
&= 2k && \text{by definition of floor and ceiling} \\
&= n && \text{by substitution.}
\end{aligned}
$$

   Case 2 (n is odd): In this case, $n = 2k + 1$ for some integer k, and so

$$
\begin{aligned}
\left\lfloor\frac{n}{2}\right\rfloor+\left\lceil\frac{n}{2}\right\rceil &= \left\lfloor\frac{2k+1}{2}\right\rfloor+\left\lceil\frac{2k+1}{2}\right\rceil && \text{by substitution} \\
&= \left\lfloor k+\frac{1}{2}\right\rfloor+\left\lceil k+\frac{1}{2}\right\rceil && \text{by algebra} \\
&= k + k + 1 && \text{by definition of floor and ceiling} \\
& && \text{(Example 4.6.3 \& exercises 6 and 7)} \\
&= 2k + 1 && \text{by algebra} \\
&= n && \text{by substitution.}
\end{aligned}
$$

   Conclusion: In both possible cases, $\left\lfloor\dfrac{n}{2}\right\rfloor+\left\lceil\dfrac{n}{2}\right\rceil = n$ *[as was to be shown]*.

31. *Hint:* Start by dividing the proof into two cases: n is even and n is odd. In case n is odd, use the quotient-remainder theorem with divisor equal to 6 to divide into three cases: $n = 6k + 1, n = 6k + 3$, and $n = 6k + 5$ for some integer k. You will need to consider a total of four cases.

33. **a.** <u>Proof</u>: Suppose n and d are integers such that $d \neq 0$ and $d \mid n$.

   By definition of divisibility, $n = d \cdot k$ for some integer k.

   By substitution and algebra,

$$
\left\lfloor\frac{n}{d}\right\rfloor = \left\lfloor\frac{d\cdot k}{d}\right\rfloor = \lfloor k\rfloor,
$$

   and $\lfloor k\rfloor = k$ because $k \leq k < k + 1$ and both k and $k + 1$ are integers.

   Hence

$$
\left\lfloor\frac{n}{d}\right\rfloor = k = \frac{n}{d},
$$

   and, therefore,

$$
n = d\cdot\left\lfloor\frac{n}{d}\right\rfloor
$$

   [as was to be shown].

   b. <u>Proof</u>: Suppose n and d are integers with $d \neq 0$ and $n = d\cdot\left\lfloor\dfrac{n}{d}\right\rfloor$.

   By definition of floor, $\left\lfloor\dfrac{n}{d}\right\rfloor$ is an integer.

   Hence, $n = d\cdot$ (*some integer*), and so by definition of divisibility, $d \mid n$.

Section 4.7

1. **a.** a contradiction

   b. a positive real number

   c. x

   d. both sides by 2

   e. contradiction

3. Proof:

   Suppose not. That is, suppose there is an integer n such that $3n + 2$ is divisible by 3. *[We must show that this supposition leads to a contradiction.]*

   By definition of divisibility,

   $$3n + 2 = 3k \quad \text{for some integer } k.$$

   Subtracting 2 from both sides gives that

   $$3n = 3k - 2,$$

   and subtracting $3k$ from both sides gives

   $$3n - 3k = -2,$$

   which implies that

   $$3(n - k) = -2,$$

   by factoring out 3. Dividing both sides by 3 gives that

   $$n - k = -\frac{2}{3}.$$

   Now $n - k$ is an integer (because it is a difference of integers) and $-2/3$ is not an integer. Since an integer cannot equal a non-integer, we have reached a contradiction.

   [This contradiction shows that the supposition is false, and so for every integer n, $3n + 2$ is not divisible by 3.]

5. *Negation for the statement:* There is a greatest even integer.

   Proof of the statement (by contradiction): Suppose not. That is, suppose there is a greatest even integer; call it N. Then N is an even integer, and N is greater than every even integer.*[We must show that this supposition leads logically to a contradiction.]*

   Let $M = N + 2$.

   Then M is an even integer since it is a sum of even integers.

   Also M is greater than N since $M = N + 2$.

   This contradicts the supposition that N is greater than *every* even integer. *[Hence the supposition is false and the given statement is true.]*

6. *Negation for the statement*: There is a greatest negative real number.

   Proof of the statement (by contradiction):

   Suppose not. That is, suppose there is a greatest negative real number. Call it a.

   [We must show that this supposition leads logically to a contradiction.]

By supposition,

$$a < 0 \quad \text{and} \quad a \geq x \quad \text{f or every negative real number } x.$$

Consider the number $a/2$. Now $a/2$ is a quotient of two real numbers (with a nonzero denominator), and so $a/2$ is a real number. In addition, because a is negative and 2 is positive, $a/2$ is negative. So, by supposition,

$$a \geq \frac{a}{2}.$$

On the other hand, multiplying all parts of the inequality $1 > \frac{1}{2}$ by the negative number a changes the direction of the inequality and gives that

$$a < \frac{a}{2}.$$

Thus $a/2$ is a negative real number that is greater than a, which contradicts the supposition that a is the greatest negative real number.

[Hence the supposition is false and the given statement is true.]

8. **a.** a rational number

   b. an irrational number

   c. $\dfrac{a}{b}$

   d. $\dfrac{c}{d}$

   e. integers

   (f) $\dfrac{a}{b} - \dfrac{c}{d}$

   g. 3 integers

   h. zero product property

   i. rational

9. **a.** The mistake in this proof occurs in the second sentence where the negation written by the student is incorrect: instead of being existential, it is universal. The problem is that if the student proceeds in a logically correct manner, all that is needed to reach a contradiction is one example of a rational and an irrational number whose difference is irrational. To prove the given statement, however, it is necessary to show that there is *no* rational number and *no* irrational number whose difference is rational.

   b. <u>Proof by contradiction:</u>

   Suppose not. That is, suppose there is an irrational number and a rational number whose difference is rational. In other words, suppose there are real numbers x and y such that x is irrational, y is rational and $x - y$ is rational. *[We must show that this supposition leads logically to a contradiction.]*

   By definition of rational,

   $$y = \frac{a}{b} \quad \text{and} \quad x - y = \frac{c}{d} \quad \text{for some integers } a, b, c, \text{ and } d \text{ with } b \neq 0 \text{ and } d \neq 0.$$

   Then, by substitution,

   $$x - \frac{a}{b} = \frac{c}{d}.$$

   Solve this equation for x to obtain

   $$x = \frac{c}{d} + \frac{a}{b} = \frac{bc}{bd} + \frac{ad}{bd} = \frac{bc + ad}{bd}.$$

Now both $bc + ad$ and bd are integers because products and sums of integers are integers, and $bd \neq 0$ by the zero product property.

Hence x is a ratio of integers with a nonzero denominator, and so x is rational by definition of rational.

This contradicts the supposition that x is irrational. *[Hence the supposition is false and the given statement is true.]*

Note: The fact that order matters in subtraction implies that the truth of the statement in exercise 8 does not automatically imply the truth of the statement in this exercise.

10. The mistake is that the negation for S that was used in the "proof" is incorrect. Thus deducing a contradiction from it fails to prove that S is true. (The actual negation is "There exist positive real numbers r and s such that $\sqrt{r + s} = \sqrt{r} + \sqrt{s}$.")

12. **a.** *Negation for R:* There exists an irrational number whose square root is rational.

    b. Proof of R by contradiction: Suppose not. That is, suppose there exists an irrational number x such that the square root of x is rational. *[We must show that this supposition leads logically to a contradiction.]*

    By definition of rational, $\sqrt{x} = \dfrac{a}{b}$ for some integers a and b with $b \neq 0$. By substitution,

$$(\sqrt{x})^2 = \left(\frac{a}{b}\right)^2,$$

    and so, by algebra,

$$x = \frac{a^2}{b^2}.$$

    But a^2 and b^2 are both products of integers and thus are integers, and b^2 is nonzero by the zero product property.

    Thus $\dfrac{a^2}{b^2}$ is rational, and, since $x = \dfrac{a^2}{b^2}$, it follows that x is both irrational and rational, which is a contradiction. *[This is what was to be shown.]*

13. **a.** *Negation for S:* There exist an irrational number and a nonzero rational number whose product is rational.

14. **b.** *Hint:* Recall that to say $a \bmod 6 = 3$ means that there exists an integer r such that $a = 6r + 3$.

15. *Answer*: There are no odd integers a, b, and c such that $a^2 + b^2 = c^2$.

    Proof 1 by contradiction: (This proof argues directly from the definition of odd.)

    Suppose not. That is, suppose there exist odd integers a, b, and c such that $a^2 + b^2 = c^2$.

    [We must show that this supposition leads logically to a contradiction.]

    By definition of odd, $a = 2k + 1$, $b = 2m + 1$, and $c = 2n + 1$ for some integers k, m, and n, so we can substitute into the equation $c^2 = a^2 + b^2$ to obtain

$$(2n + 1)^2 \quad = \quad (2k + 1)^2 + (2m + 1)^2.$$

    Hence
$$4n^2 + 4n + 1 \;=\; 4k^2 + 4k + 1 + 4m^2 + 4m + 1$$
$$= \; 2(2k^2 + 2k + 2m^2 + 2m + 1) \quad \text{by algebra.}$$

    Subtracting $4n^2 + 4n$ from both sides gives that

$$1 \;=\; 2(2k^2 + 2k + 2m^2 + 2m + 1) - 4n^2 - 4n \quad \text{by subtracting } 4n^2{+}4n \text{ from both sides}$$
$$= \; 2(2k^2 + 2k + 2m^2 + 2m + 1 - 2n^2 - 2n) \quad \text{by algebra.}$$

Thus

$$\frac{1}{2} = 2k^2 + 2k + 2m^2 + 2m + 1 - 2n^2 - 2n.$$

But the left-hand side of this equation is not an integer and the right-hand side is an integer, and so

the equation is false. It follows that the supposition that there exist odd integers a, b, and c such that

a, b, and c are all odd and $a^2 + b^2 = c^2$ is false, and thus the given statement is true.

Proof 2 by contradiction: (This proof uses a result from Example 4.3.2)

Suppose not. That is, suppose there exist odd integers a, b, and c such that $a^2 + b^2 = c^2$.

[We must show that this supposition leads logically to a contradiction.]

By Example 4.3.2(3), a^2, b^2, and c^2, being products of odd integers, are all odd.

Thus there exist integers r, s, and t such that $a^2 = 2r + 1$, $b^2 = 2s + 1$, and $c^2 = 2t + 1$.

So we can substitute into the equation $c^2 = a^2 + b^2$ to obtain

$$\begin{aligned} 2r + 1 &= (2s + 1) + (2t + 1) &&\text{by substitution} \\ &= 2(s + t) + 2 \\ &= 2(s + t + 1) &&\text{by algebra.} \end{aligned}$$

Subtracting $2r$ from both sides of the equation gives that

$$\begin{aligned} 1 &= 2(s + t + 1) - 2r \\ &= 2(s + t + 1 - r) &&\text{by algebra.} \end{aligned}$$

And dividing both sides of this equation by 2 gives that

$$\frac{1}{2} = k + m + 1 - n.$$

But the left-hand side of this equation is not an integer and the right-hand side is an integer, and so the equation is false.

It follows that the supposition that there exist integers a, b, and c such that a, b, and c are all odd

and $a^2 + b^2 = c^2$ is false, and thus the given statement is true.

16. Proof 1 by contradiction: Suppose not. That is, suppose that there exist odd integers a and b such that $b^2 - a^2 = 4$. *[We must show that this supposition leads logically to a contradiction.]* Factoring gives that
$$b^2 - a^2 = (b + a)(b - a) = 4.$$

Now $b > a$ because $b^2 - a^2 = 4 > 0$, and the only way to factor 4 is either $4 = 2 \cdot 2$ or $4 = 4 \cdot 1$. Hence either $b + a = b - a = 2$, or $b + a = 4$ and $b - a = 1$ or $b + a = 1$ and $b - a = 4$.

(a) In case $b + a = b - a = 2$, then $-a = a$ and so $a = 0$, which is not an odd integer.

(b) In case $b + a = 4$ and $b - a = 1$, then $2b = 5$ and so $b = 5/2$, which is not an odd integer.

(c) In case $b + a = 1$ and $b - a = 4$, then $2b = 5$ and so $b = 5/2$, which is not an odd integer.

Thus there are no odd integers a and b such that $b^2 - a^2 = 4$, which contradicts the supposition. *[Hence the supposition is false and the given statement is true.]*

17. *Hint:* Use the fact that $a^2 = c^2 - b^2 = (c - b)(c + b)$ and apply the unique factorization of integers theorem.

18. <u>Proof by contradiction</u>: Suppose not. That is, suppose there exist rational numbers a and b such that $b \neq 0$, r is an irrational number, and $a + br$ is rational.

    [We must show that this supposition leads logically to a contradiction.]

    By definition of rational,

    $$a = \frac{i}{j}, \quad b = \frac{k}{l} \quad \text{and} \quad a + br = \frac{m}{n}$$

    where i, j, k, l, m, and n are integers and $j \neq 0$, $l \neq 0$, and $n \neq 0$.

    Since $b \neq 0$, we also have that $k \neq 0$.

    By substitution

    $$a + br = \frac{i}{j} + \frac{k}{l} \cdot r = \frac{m}{n}, \quad \text{or, equivalently,} \quad \frac{k}{l} \cdot r = \frac{m}{n} - \frac{i}{j}$$

    Solving for r gives

    $$r = \frac{mj - in}{nj} \cdot \frac{l}{k} = \frac{mjl - inl}{njk}.$$

    Now $mjl - inl$ and njk are both integers *[because products and differences of integers are integers]* and $njk \neq 0$ because $n \neq 0, j \neq 0$, and $k \neq 0$.

    It follows, by definition of rational, that r is a rational number, which contradicts the supposition that r is irrational.

    [Hence the supposition is false and the given statement is true.]

19. *Hint:* Suppose $n^2 - 2$ is divisible by 4, and consider the two cases where n is even and n is odd. (An alternative solution uses Proposition 4.7.4.)

20. **a.** $5 \mid n$

    b. $5 \mid n^2$

    c. $5k$

    d. $(5k)^2$

    e. $5 \mid n^2$

21. **a.** <u>Proof by contradiction</u>: Suppose not. That is, suppose there is an integer n such that n^2 is odd and n is even. Show that this supposition leads logically to a contradiction.

    b. <u>Proof by contraposition</u>: Suppose n is any integer such that n is not odd. Show that n^2 is not odd.

23. *Formal version of the statement to be proved:* For every real number x, if x is irrational then $-x$ is irrational.

    a. <u>Proof by contraposition</u>: Suppose x is any real number such that $-x$ is not irrational. By definition of irrational this means that $-x$ is rational. *[We must show that x is not irrational, or, equivalently, that x is rational.]* By definition of rational, $-x = a/b$ for some integers a and b with $b \neq 0$. Then, by algebra,

    $$x = -(-x) = -(a/b) = (-a)/b.$$

    Now $-a$ is an integer because a is an integer and because $-a = (-1)a$. Also b is a nonzero integer. Thus x is a ratio of integers with a nonzero denominator, and hence x is rational *[as was to be shown]*.

b. <u>Proof by contradiction</u>: Suppose there exists a real number x such that x is irrational and $-x$ is not irrational. *[We must show that this supposition leads logically to a contradiction.]* By definition of rational, $-x = a/b$ for some integers a and b with $b \neq 0$. Then, by algebra,

$$x = -(-x) = -(a/b) = (-a)/b.$$

Now $-a$ is an integer because a is an integer and because $-a = (-1)a$. Also b is a nonzero integer. Thus x is a ratio of integers with a nonzero denominator, and hence x is rational. Hence x is both irrational and rational, which is a contradiction *[which shows that the negation is false and therefore that the statement to be proved is true].*

24. *Formal version of the statement to be proved:* For every real number x, if x is irrational then the reciprocal of x is irrational.

a. <u>Proof by contraposition</u>:

Suppose x is any real number such that the reciprocal of x, namely $\dfrac{1}{x}$, is not irrational. *[We must show that x is not irrational.]*

Because $\dfrac{1}{x}$ is not irrational, it is rational, and so there are integers a and b with $b \neq 0$ such that

$$\frac{1}{x} = \frac{a}{b} \ (*).$$

Now

$$\text{since } x \cdot \left(\frac{1}{x}\right) = 1, \quad \text{it follows that} \quad \frac{1}{x} \quad \text{cannot equal zero, and so} \quad a \neq 0.$$

Thus we may solve equation (*) for x to obtain

$$x = \frac{b}{a} \quad \text{where } b \text{ and } a \text{ are integers and } a \neq 0.$$

Hence, by definition of rational, $\dfrac{1}{x}$ is rational and therefore $\dfrac{1}{x}$ is not irrational *[as was to be shown].*

b. <u>Proof by contradiction</u>:

Suppose not. That is, suppose there exists a nonzero irrational number x such that the reciprocal of x, namely $\dfrac{1}{x}$, is not irrational. *[We must show that this supposition leads logically to a contradiction.]*

Because $\dfrac{1}{x}$, is not irrational, it is rational, and so, by definition of rational,

$$\frac{1}{x} = \frac{a}{b} \ (*).$$

where a and b are integers with $b \neq 0$. Now

$$\text{since } x \cdot \left(\frac{1}{x}\right) = 1, \quad \text{it follows that} \quad \frac{1}{x} \quad \text{cannot equal zero, and so } a \neq 0.$$

Thus we may divide both sides of equation (*) to solve for x:

$$x = \frac{b}{a} \quad \text{where } b \text{ and } a \text{ are integers and } a \neq 0.$$

Hence, by definition of rational, $\dfrac{1}{x}$ is rational, which contradicts the supposition that x is irrational. *[Hence the supposition is false and the given statement is true.]*

25. *Hint:* See the answer to exercise 21 and look carefully at the two proofs for Proposition 4.7.4.

26. Proof by contraposition: Suppose a, b, and c are any *[particular but arbitrarily chosen]* integers such that $a \mid b$. *[We must show that $a \mid bc$.]*

    By definition of divides, $b = ak$ for some integer k. By substitution and the associative law,

    $$bc = (ak)c = a(kc).$$

    But kc is an integer (because it is a product of the integers k and c). Hence $a \mid bc$ by definition of divisibility *[as was to be shown]*.

    Proof by contradiction:

    Suppose not. *[We take the negation and suppose it to be true.]*

    Suppose $\exists$ integers a, b, and c such that $a \nmid bc$ and $a \mid b$.

    Since $a \mid b$, there exists an integer k such that $b = ak$ by definition of divides. Then

    $$bc = (ak)c = a(kc)$$

    by substitution and the associative law. But kc is an integer (being a product of integers), and so $a \mid bc$ by definition of divides.

    Thus $a \nmid bc$ and $a \mid bc$, which is a contradiction.

    [This contradiction shows that the supposition is false, and hence the given statement is true.]

27. Proof by contradiction:

    Suppose not. That is, suppose there exist positive real numbers r and s such that

    $$\sqrt{r + s} = \sqrt{r} + \sqrt{s}.$$

    [We must show that this supposition leads logically to a contradiction.]

    Because both r and s are positive, when we square both sides of the equation we obtain

    $$r + s = r + 2\sqrt{r}\sqrt{s} + s.$$

    Subtracting $r + s$ from both sides gives $0 = 2\sqrt{r}\sqrt{s}$.

    But this is impossible because both r and s are positive.

    Note: For this exercise a proof by contradiction is more natural than a proof by contraposition. But a proof by contraposition is possible.

    Proof by contraposition: Suppose r and s are any real numbers such that

    $$\sqrt{r + s} = \sqrt{r} + \sqrt{s}.$$

    We will show that r and s are not both positive. Squaring both sides of the equation gives

    $$r + s = r + 2\sqrt{r}\sqrt{s} + s,$$

    and subtracting $r + s$ from both sides gives

    $$0 = 2\sqrt{r}\sqrt{s}.$$

    By the zero product property, at least one of $\sqrt{r}$ or $\sqrt{s}$ equals 0, and if the square root of a number is zero, then the number is zero. Hence at least one of r or s equals 0, and so at least one of r or s is not positive *[as was to be shown]*.

28. *Hint:* To prove $p \rightarrow q \vee r$, it suffices to prove either $p \wedge \sim q \rightarrow r$ or $p \wedge \sim r \rightarrow q$. See exercise 14 in Section 2.2.

29. *Hints:* (1) The contrapositive is "For all integers m and n, if m and n are not both even and m and n are not both odd, then $m + n$ is not even." *Equivalently:* "For all integers m and n, if one of m and n is even and the other is odd, then $m + n$ is odd."

   (2) The negation of the given statement is the following: $\exists$ integers m and n such that $m + n$ is even, and either m is even and n is odd, or m is odd and n is even.

30. **a.** $\sqrt{7} \simeq 7.28$; list of primes: 2, 3, 5, 7

    It is true that 53 is not divisible by any prime number less than or equal to $\sqrt{53}$ because the only such primes are 2, 3, 5, and 7 and 53 is not divisible by any of these numbers.

    b. The following are negations for S: (i), (iv)

31. **a.** <u>Proof by contraposition:</u>

    Suppose n, r, and s are positive integers and $r > \sqrt{n}$ and $s > \sqrt{n}$. *[We must show that $rs > n$.]*
    By Theorem T27 in Appendix A (with $a = r$, $b = s$, and $c = d = \sqrt{n}$),

    $$rs > \sqrt{n} \cdot \sqrt{n} = n.$$

    Thus the contrapositive of the given statement is true, and so the given statement is also true.

32. **a.** $\sqrt{667} \cong 25.8$, and so the possible prime factors to be checked are 2, 3, 5, 7, 11, 13, 17, 19, and 23. Testing each in turn shows that 667 is not prime because $667 = 23 \cdot 29$.

    b. $\sqrt{557} \cong 23.6$, and so the possible prime factors to be checked are 2, 3, 5, 7, 11, 13, 17, 19, and 23. Testing each in turn shows that none divides 557. Therefore, 557 is prime.

33. After crossing out all multiples of 2, 3, 5, and 7 (the prime numbers less than $\sqrt{100}$), the remaining numbers are prime. They are circled in the following diagram.

34. **a.** $\sqrt{9269} \cong 96.3$, and so the possible prime factors to be checked are all among those you found for exercise 33. Testing each in turn shows that 9,269 is not prime because $9,269 = 13 \cdot 713$.

    b. $\sqrt{9103} \cong 95.4$, and so the possible prime factors to be checked are all among those you found for exercise 33. Testing each in turn shows that none divides 9,103. Therefore, 9,103 is prime.

35. *Hint:* Assuming that n is not composite, show that $n - 4, n - 6$, and $n - 8$ are all prime. Next show that $n - 7$ is divisible by 3 by considering $n - 6, n - 7$, and $n - 8$. Finally, write $n - 4 = (n - 7) + 3$ and show that 3 divides $n - 4$.

36. <u>Proof by contradiction</u>: Suppose not. That is, suppose there exist odd integers a, b, and c and a rational number z such that

$$az^2 + bz + c = 0.$$

[We must show that this supposition leads logically to a contradiction.] By definition of rational, there exist integers m and n such that

$$z = \frac{m}{n} \quad \text{and} \quad n \neq 0.$$

And by canceling common factors if necessary, we may assume that m and n have no common factors. By substitution,

$$a\left(\frac{m}{n}\right)^2 + b\left(\frac{m}{n}\right) + c = 0, \quad \text{and so} \quad am^2 + bmn + cn^2 = 0 \;\; (*)$$

by multiplying both sides by n^2.

Case 1 (m is even): In this case, observe that since $am^2 + bmn = m(am^2 + bn)$, then $am^2 + bmn$ is even *[because it is a product of an even integer, m, and another integer]*. Subtracting $am^2 + bmn$ from both sides of equation (*) gives that

$$cn^2 = 0 - (am^2 + bmn) = 0 - (\text{an even number}),$$

and so cn^2 is a difference of even integers and hence even. Now since c is odd, the only way for cn^2 to be even is for n to be even *[because otherwise cn^2 would be a product of odd integers and hence odd]*. Thus both m and n are even. This means that both m and n have a common factor of 2, which contradicts the assumption that m and n have no common factors.

Case 2 (n is even): In this case, observe that since $bmn + cn^2 = n(bm + cn^2)$, then $bmn + cn^2$ is even *[because it is a product of an even integer, n, and another integer]*. Subtracting $bmn + cn^2$ from both sides of equation (*) gives that

$$am^2 = 0 - (bmn + cn^2) = 0 - (\text{an even number}),$$

and so am^2 is a difference of even integers and hence even. Now since a is odd, the only way for am^2 to be even is for m to be even *[because otherwise am^2 would be a product of odd integers and hence odd]*. Thus both m and n are even. This means that both m and n have a common factor of 2, which contradicts the assumption that m and n have no common factors.

Conclusion: Since in both possible cases a contradiction is reached, the supposition is false and the given statement is true.

Note: Because of the symmetry in the expression $am^2 + bmn + cn^2$, the reasoning for Case 2 is identical except for notation to the reasoning in Case 1.

Section 4.8

1. The value of $\sqrt{2}$ given by a calculator is an approximation. Calculators can give exact values only for numbers that can be represented using at most the number of decimal digits in the calculator display. In particular, every number in a calculator display is rational, but even many rational numbers cannot be represented exactly. For instance, consider the number formed by writing a decimal point and following it with the first million digits of $\sqrt{2}$. By the discussion in Section 4.2, this number is rational, but you could not infer this from the calculator display.

3. Yes. In fact there are infinitely many rational numbers with the same first trillion digits as $\sqrt{2}$. For instance, if you end the number after the first trillion digits, the result is a finite decimal, which is rational. Repeating the first trillion digits of $\sqrt{2}$ forever would create a repeating decimal, which is rational. Or you could follow the first trillion digits of $\sqrt{2}$ by 012343434..., where the digits 34 repeat forever. This is also rational. Try creating other examples.

6. <u>Proof by contradiction:</u>

   Suppose not. That is, suppose $6 - 7\sqrt{2}$ is rational. *[We must prove a contradiction.]* By definition of rational, there exist integers a and b with $b \neq 0$ and
   $$6 - 7\sqrt{2} = \frac{a}{b}.$$
   Then
   $$\begin{aligned} \sqrt{2} &= \frac{1}{-7}\left(\frac{a}{b} - 6\right) & \text{by subtracting 6 from both sides} \\ & & \text{and dividing both sides by } -7 \\ &= \frac{1}{-7}\left(\frac{a}{b} - \frac{6b}{b}\right) \\ &= \frac{a - 6b}{-7b} & \text{by the rules of algebra.} \end{aligned}$$
   Now $a - 6b$ and $-7b$ are both integers (since a and b are integers and products and difference of integers are integers), and $-7b \neq 0$ by the zero product property.

   Hence $\sqrt{2}$ is a ratio of the two integers $a - 6b$ and $-7b$ with $-7b \neq 0$, and so $\sqrt{2}$ is a rational number (by definition of rational).

   This contradicts the fact that $\sqrt{2}$ is irrational, and thus the supposition is false and $6 - 7\sqrt{2}$ is irrational.

8. This is false. $\sqrt{4} = 2 = 2/1$, which is rational.

9. This is false.

   <u>Proof 1 (by using a previous result):</u> $\sqrt{2}/6 = (1/6) \cdot \sqrt{2}$, which is a product of a nonzero rational number and an irrational number. By exercise 13 of Section 4.7, such a product is irrational.

   <u>Proof 2 (by contradiction):</u>

   Suppose not. That is, suppose $\sqrt{2}/6$ is rational. *[We must show that this supposition leads logically to a contradiction.]*

   By definition of rational, there exist integers a and b with
   $$\sqrt{2}/6 = \frac{a}{b} \quad \text{and} \quad b \neq 0.$$
   Solving for $\sqrt{2}$ gives $\sqrt{2} = 6a/b$. But $6a$ is an integer (because products of integers are integers) and b is a nonzero integer.

   Therefore, by definition of rational, $\sqrt{2}$ is rational. This contradicts Theorem 4.8.1 which states that $\sqrt{2}$ is irrational. Hence the supposition is false. In other words $\sqrt{2}/6$ is irrational.

10. <u>Counterexample:</u> Let $x = \sqrt{2}$ and $y = -\sqrt{2}$. Then x and y are irrational, but $x + y = 0 = 0/1$, which is rational.

12. The statement is true.

Formal version of the statement: $\forall$ positive real number r, if r is irrational, then $\sqrt{r}$ is irrational.

<u>Proof by contraposition:</u>

Suppose r is any positive real number such that $\sqrt{r}$ is rational. *[We must show that r is rational.]*

By definition of rational, $\sqrt{r} = \dfrac{a}{b}$ for some integers a and b with $b \neq 0$. So by substitution and algebra,

$$r = (\sqrt{r})^2 = \left(\frac{a}{b}\right)^2 = \frac{a^2}{b^2}.$$

Now both a^2 and b^2 are integers because they are products of integers, and $b^2 \neq 0$ by the zero product property.

Thus r is rational *[as was to be shown].*

<u>Proof by contradiction:</u>

Suppose not. That is, suppose there is a positive real number r such that r is irrational and $\sqrt{r}$ is rational. *[We must show that this supposition leads logically to a contradiction.]*

By definition of rational, $\sqrt{r} = \dfrac{a}{b}$ for some integers a and b with $b \neq 0$. So by substitution and algebra,

$$r = (\sqrt{r})^2 = \left(\frac{a}{b}\right)^2 = \frac{a^2}{b^2}.$$

Now both a^2 and b^2 are integers because they are products of integers, and $b^2 \neq 0$ by the zero product property.

Thus, by definition of rational, r is rational, which contradicts the supposition that r is irrational.

15. <u>Counterexample:</u> $\sqrt{2}$ is irrational. Also $\sqrt{2} \cdot \sqrt{2} = 2$ and 2 is rational because $2 = 2/1$. Thus there exist irrational numbers whose product is rational.

16. *Hint:* Can you think of any "nice" integers x and y that are greater than 1 and have the property that $x^2 = y^3$?

18. **a.** <u>Proof by contraposition:</u>

Let n be any integer such that n is not even. *[We must show that n^3 is not even.]*

Because n is not even, it is odd, and so n^3 is also odd *[because it is a product of odd integers].*

Hence n^3 is not even *[as was to be shown].*

b. <u>Proof by contradiction:</u>

Suppose not. That is, suppose $\sqrt[3]{2}$ is rational. *[We must show that this supposition leads logically to a contradiction.]*

By definition of rational,

$$\sqrt[3]{2} = \frac{a}{b} \text{ for some integers } a \text{ and } b \text{ with } b \neq 0.$$

By canceling common factors if necessary, we may assume that a and b have no common factors. Cubing both sides of equation $\sqrt[3]{2} = \dfrac{a}{b}$ gives

$$2 = \frac{a^3}{b^3}, \quad \text{and so} \quad 2b^3 = a^3.$$

It follows that a^3 is even, and so, by part (a), a is even. Thus $a = 2k$ for some integer k.
By substitution

$$a^3 = (2k)^3 = 8k^3 = 2b^3,$$

and so

$$b^3 = 4k^3 = 2(2k^3).$$

Therefore b^3 is even, and hence, by part (a), b is even.

Thus both a and b are even which contradicts the assumption that a and b have no common factor. Therefore, the supposition is false, and so $\sqrt[3]{2}$ is irrational.

19. **a.** <u>Proof by contradiction</u>: Suppose not. That is, suppose there is an integer n such that

$$n = 3q_1 + r_1 = 3q_2 + r_2,$$

where q_1, q_2, r_1, and r_2 are integers, $0 \le r_1 < 3$, $0 \le r_2 < 3$, and $r_1 \ne r_2$.
[We must show that this supposition leads logically to a contradiction.]
By interchanging the labels for r_1 and r_2 if necessary, we may assume that $r_2 > r_1$. Then

$$3(q_1 - q_2) = r_2 - r_1 > 0,$$

and because both r_1 and r_2 are less than 3, either $r_2 - r_1 = 1$ or $r_2 - r_1 = 2$. So either $3(q_1 - q_2) = 1$ or $3(q_1 - q_2) = 2$.

The first case implies that $3 \mid 1$, and hence, by Theorem 4.4.1, that $3 \le 1$, and the second case implies that $3 \mid 2$, and hence, by Theorem 4.4.1, that $3 \le 2$.

These results contradict the fact that 3 is greater than both 1 and 2.

Thus in either case we have reached a contradiction, which shows that the supposition is false and the given statement is true.

b. <u>Proof by contradiction</u>: Suppose not. That is, suppose there is an integer n such that n^2 is divisible by 3 and n is not divisible by 3. *[We must deduce a contradiction.]*

By definition of divisibility, $n^2 = 3q$ for some integer q, and by the quotient- remainder theorem and part (a), $n = 3k + 1$ or $n = 3k + 2$ for some integer k.

Case 1 ($n = 3k + 1$ for some integer k): In this case

$$n^2 = (3k + 1)^2 = 9k^2 + 6k + 1 = 3(3k^2 + 2k) + 1.$$

Let $s = 3k^2 + 2k$. Then $n^2 = 3s + 1$, and s is an integer because it is a sum of products of integers. Thus $n^2 = 3q = 3s + 1$ for some integers q and s, which contradicts the result of part (a).

Case 2 ($n = 3k + 2$ for some integer k): In this case

$$n^2 = (3k + 2) = 9k^2 + 12k + 4 = 3(3k^2 + 4k + 1) + 1.$$

Let $t = 3k^2 + 4k + 1$. Then $n^2 = 3t + 1$, and t is an integer because it is a sum of products of integers. Thus there are integers q and t so that $n^2 = 3q = 3t + 1$, which contradicts the result of part (a).

Thus in either case 1 or case 2, a contradiction is reached, which shows that the supposition is false and the given statement is true.

c. <u>Proof by contradiction</u>: Suppose not. That is, suppose $\sqrt{3}$ is rational. By definition of rational, $\sqrt{3} = \dfrac{a}{b}$ for some integers a and b with $b \ne 0$. Without loss of generality, assume that a and b have no common factor. (If not, divide both a and b by their greatest common factor to obtain integers a' and b' with the property that a' and b' have no common factor and

$\sqrt{3} = \dfrac{a'}{b'}$. Then redefine $a = a'$ and $b = b'$.) Squaring both sides of $\sqrt{3} = \dfrac{a}{b}$ gives $3 = \dfrac{a^2}{b^2}$, and multiplying both sides by b^2 gives

$$3b^2 = a^2 \quad (*).$$

Thus a^2 is divisible by 3, and so, by part (b), a is also divisible by 3. By definition of divisibility, then, $a = 3k$ for some integer k, and so

$$a^2 = 9k^2 \quad (**).$$

Substituting equation $(**)$ into equation $(*)$ gives $3b^2 = 9k^2$, and dividing both sides by 3 yields

$$b^2 = 3k^2.$$

Hence b^2 is divisible by 3, and so, by part (b), b is also divisible by 3. Consequently, both a and b are divisible by 3, which contradicts the assumption that a and b have no common factor. Thus the supposition is false, and so $\sqrt{3}$ is irrational.

21. <u>Proof</u> :

Suppose that a and d are integers with $d > 0$ and that q_1, q_2, r_1, and r_2 are integers such that

$$a = dq_1 + r_1 \quad \text{and} \quad a = dq_2 + r_2, \quad \text{where} \quad 0 \le r_1 < d \quad \text{and} \quad 0 \le r_2 < d.$$

[We will show that $r_1 = r_2$ and $q_1 = q_2$.]

Then $\qquad\qquad\qquad\qquad\qquad\qquad dq_1 + r_1 = dq_2 + r_2,$

and so $\qquad\qquad\qquad\qquad\qquad r_2 - r_1 = dq_1 - dq_2 = d(q_1 - q_2).$

This implies that $\qquad\qquad\qquad\qquad d \mid (r_2 - r_1)$

because $q_1 - q_2$ is an integer (since it is a difference of integers).

But both r_1 and r_2 lie between 0 and d, and thus the difference $r_2 - r_1$ lies between $-d$ and d.

[This is justified by applying properties T23 and T26 of Appendix A. Multiplying $0 \le r_1 < d$ by -1 gives $0 \ge -r_1 > -d$, or, equivalently, $-d < -r_1 \le 0$, and adding $-d < -r_1 \le 0$ and $0 \le r_2 < d$ gives $-d < r_2 - r_1 < d$.]

Since $r_2 - r_1$ is a multiple of d and yet lies between $-d$ and d, the only possibility is that $r_2 - r_1 = 0$, or, equivalently, that $r_1 = r_2$.

Substituting back into the original expressions for a and equating the two gives

$$dq_1 + r_1 = dq_2 + r_1 \qquad \text{because } r_1 = r_2.$$

Subtracting r_1 from both sides gives $dq_1 = dq_2$, and since $d \neq 0$, we have that $q_1 = q_2$.

Hence, $r_1 = r_2$ and $q_1 = q_2$, *[as was to be shown]*.

22. *Hint:* First prove that for all integers a, if 5 divides a squared then 5 divides a. The rest of the proof is similar to the solution for exercise 19(c).

23. *Hint:* The statement is true. If $a^2 - 3 = 9b$, then $a^2 = 9b + 3 = 3(3b + 1)$, and so a^2 is divisible by 3. Hence, by exercise 19(b), a is divisible by 3. Thus $a^2 = (3c)^2$ for some integer c.

24. <u>Proof by contradiction</u>:

Suppose not. That is, suppose $\sqrt{2}$ is rational. *[We will show that this supposition leads to a contradiction.]*

By definition of rational, we may write $\sqrt{2} = a/b$ for some integers a and b with $b \neq 0$. Then $2 = a^2/b^2$, and so $a^2 = 2b^2$.

Consider the prime factorizations for a^2 and for $2b^2$. By the unique factorization of integers theorem, these factorizations are unique except for the order in which the factors are written.

Now because every prime factor of a occurs twice in the prime factorization of a^2, the prime factorization of a^2 contains an even number of 2's. (If 2 is a factor of a, then this even number is positive, and if 2 is not a factor of a, then this even number is 0.)

On the other hand, because every prime factor of b occurs twice in the prime factorization of b^2, the prime factorization of $2b^2$ contains an odd number of 2's.

Therefore, the equation $a^2 = 2b^2$ cannot be true. So the supposition is false, and hence $\sqrt{2}$ is irrational.

26. *Hint:* One solution uses only Theorem 4.8.1. Another uses the result of exercise 25 that $\sqrt{6}$ is irrational.

27. <u>Proof by contradiction :</u>

Suppose not. That is, suppose that $\log_5(2)$ is rational. *[We will show that this supposition leads logically to a contradiction.]*

By definition of rational,

$$\log_5(2) = \frac{a}{b} \quad \text{for some integers} \quad a \text{ and } b \text{ with } b \neq 0.$$

Since logarithms are always positive, we may assume that a and b are both positive.

By definition of logarithm,

$$5^{\frac{a}{b}} = 2, \quad \text{and so} \quad (5^{\frac{a}{b}})^b = 2^b \quad \text{or, equivalently,} \quad 5^a = 2^b$$

Let

$$N = 5^a = 2^b.$$

Now $b \geq 1$ because b is a positive integer, and so $N \geq 2^1 = 2$. Thus we may consider the prime factorization of N.

Because $N = 5^a$, the prime factors of N are all 5. On the other hand, because $N = 2^b$, the prime factors of N are all 2.

This contradicts the unique factorization of integers theorem, which states that the prime factors of any integer greater than 1 are unique except for the order in which they are written.

Hence the supposition is false, and so $\log_5(2)$ is irrational.

28. *Hint:* Divide $2 \cdot 3 \cdot 5 \cdot 7 + 1$ by each of 2, 3, 5, and 7, using the quotient-remainder theorem.

29. *Hint:* You can deduce that $p = 3$.

30. **a.** $N_1 = 2 + 1 = 3$, $N_2 = 2 \cdot 3 + 1 = 7$, $N_3 = 2 \cdot 3 \cdot 5 + 1 = 31$, $N_4 = 2 \cdot 3 \cdot 5 \cdot 7 + 1 = 211$, $N_5 = 2 \cdot 3 \cdot 5 \cdot 7 \cdot 11 + 1 = 2311$, and $N_6 = 2 \cdot 3 \cdot 5 \cdot 7 \cdot 11 \cdot 13 + 1 = 30031$.

b. Each of N_1, N_2, N_3, N_4, and N_5 is itself prime, but $N_6 = 30031 = 59 \cdot 509$, and so N_6 is not prime. However since both of its factors, 59 and 509, are prime and since both are greater than 13, this shows that the largest prime factor in the construction of N_6 could not be the largest prime number.

32. *Hint:* By Theorem 4.2.4 (divisibility by a prime) there is a prime number p such that $p|(n!-1)$. Show that the supposition that $p \leq n$ leads to a contradiction. It will then follow that $n < p < n!$.

33. Proof: Let $p_1, p_2, \ldots, p_n$ be distinct prime numbers with $p_1 = 2$ and $n > 1$. *[We must show that $p_1 p_2 \cdots p_n + 1 = 4k + 3$ for some integer k.]* Let

$$N = p_1 p_2 \cdots p_n + 1.$$

By the quotient-remainder theorem, there is an integer k such that

$$N = 4k, \; 4k + 1, \; 4k + 2, \; \text{or } 4k + 3.$$

But N is odd (because $p_1 = 2$); hence

$$N = 4k + 1 \quad \text{or} \quad 4k + 3$$

Suppose $N = 4k + 1$. *[We will show that this supposition leads to a contradiction.]*
By substitution,
$$4k + 1 = p_1 p_2 \cdots p_n + 1 \quad \text{and so} \quad 4k = p_1 p_2 \cdots p_n.$$
Thus
$$4 \mid p_1 p_2 \cdots p_n.$$

But $p_1 = 2$ and all of $p_2, p_3, \ldots, p_n$ are odd (being prime numbers that are greater than 2). Consequently, there is only one factor of 2 in the prime factorization of $p_1 p_2 \cdots p_n$, and so

$$4 \nmid p_1 p_2 \cdots p_n,$$

which results in a contradiction. Therefore the supposition that $N = 4k + 1$ for some integer k is false, and so *[by elimination]* $N = 4k + 3$ for some integer k *[as was to be shown]*.

34. **a.** *Hint:* Prove the contrapositive: If for some integer $n > 2$ that is not a power of 2, $x^n + y^n = z^n$ has a positive integer solution, then for some prime number $p > 2, x^p + y^p = z^p$ has a positive integer solution. Note that if $n = kp$, then $x^n = x^{kp} = (x^k)^p$.

35. Existence proof: When $n = 2$, then $n^2 - 1 = 3$, which is prime. Hence there exists a prime number of the form $n^2 - 1$, where n is an integer and $n \geq 2$.

Uniqueness proof (by contradiction): By the existence proof above, we know that when $n = 2$, then $n^2 - 1$ is prime. Suppose that m is another integer such that $m > 2$ and $m^2 - 1$ is prime. *[We must derive a contradiction.]*

Factor $m^2 - 1$ to obtain $m^2 - 1 = (m - 1)(m + 1)$. Since $m > 2$, both $m - 1 > 1$ and $m + 1 > 1$.

Hence $m^2 - 1$ is a product of two integers both greater than 1 and thus is not prime. This is a contradiction.

[The contradiction shows that the supposition is false, and so there is no other integer $m > 2$ such that $m^2 - 1$ is prime.]

Uniqueness proof (direct): Suppose m is any integer such that $m \geq 2$ and $m^2 - 1$ is prime. *[We must show that $m = 2$.]*

Now $m^2 - 1 = (m - 1)(m + 1)$ by factoring, and since $m^2 - 1$ is prime, either $m - 1 = 1$ or $m + 1 = 1$.

Since $m > 2$, then $m + 1 \geq 2 + 1 = 3$. Hence, by elimination, $m - 1 = 1$, and so $m = 2$.

36. Existence Proof: When $n = 2$, then

$$n^2 + 2n - 3 = 2^2 + 2 \cdot 2 - 3 = 5,$$

which is prime. Thus there is a prime number of the form $n^2 + 2n - 3$, where n is a positive integer.

Uniqueness Proof (by contradiction): By the existence proof above, we know that when $n = 2$, then $n^2 + 2n - 3$ is prime. Suppose there is another positive integer m, not equal to 2, such that $m^2 + 2m - 3$ is prime. *[We will show that this supposition leads logically to a contradiction.]* By factoring, we see that

$$m^2 + 2m - 3 = (m + 3)(m - 1).$$

Now $m \neq 1$ because otherwise $m^2 + 2m - 3 = 0$, which is not prime. Also $m \neq 2$ by supposition. Thus $m > 2$. Consequently, $m + 3 > 5$ and $m - 1 > 1$, and so $m^2 + 2m - 3$ can be written as a product of two positive integers neither of which is 1 (namely $m + 3$ and $m - 1$). This contradicts the supposition that $m^2 + 2m - 3$ is prime. Hence the supposition is false: there is no integer m other than 2 such that $m^2 + 2m - 3$ is prime.

Uniqueness Proof (direct): Suppose m is any positive integer such that $m^2 + 2m - 3$ is prime. *[We will show that $m = 2$.]* By factoring,

$$m^2 + 2m - 3 = (m + 3)(m - 1).$$

Since $m^2 + 2m - 3$ is prime, either $m + 3 = 1$ or $m - 1 = 1$. Now $m + 3 \neq 1$ because m is positive and if $m + 3 = 1$ then $m = -2$. Thus $m - 1 = 1$, which implies that $m = 2$ *[as was to be shown.]*

37. Proof (by contradiction):

Suppose not. That is, suppose there are two distinct real numbers a_1 and a_2 such that for every real number r,

$$(1)\ a_1 + r = r \quad \text{and} \quad (2)\ a_2 + r = r.$$

Then

$$a_1 + a_2 = a_2 \quad \text{by (1) with} \quad r = a_2$$

and

$$a_2 + a_1 = a_1 \quad \text{by (2) with} \quad r = a_1.$$

It follows that

$$a_2 = a_1 + a_2 = a_2 + a_1 = a_1,$$

which implies that $a_2 = a_1$. But this contradicts the supposition that a_1 and a_2 are distinct. *[Thus the supposition is false and there is at most one real number a such that $a + r = r$ for every real number r.]*

Proof (direct): Suppose a_1 and a_2 are real numbers such that for every real number r,

$$(1)\ a_1 + r = r \quad \text{and} \quad (2)\ a_2 + r = r.$$

Then

$$a_1 + a_2 = a_2 \quad \text{by (1) with} \quad r = a_2$$

and

$$a_2 + a_1 = a_1 \quad \text{by (2) with} \quad r = a_1.$$

It follows that

$$a_2 = a_1 + a_2 = a_2 + a_1 = a_1.$$

Hence $a_2 = a_1$. *[Thus there is at most one real number a such that $a + r = r$ for every real number r.]*

Section 4.9

1.

vertex	v_1	v_2	v_3	v_4	v_5	v_6
degree	3	2	4	2	1	0

Total degree $= 3 + 2 + 4 + 2 + 1 = 12$

# of edges $= 6 = \left(\dfrac{1}{2}\right) 12 =$ one-half of the total degree

3. The total degree of the graph is $0+2+2+3+9 = 16$, so, by the handshake theorem (Theorem 4.9.1), the number of edges is $16/2 = 8$.

5. One such graph is

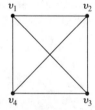

6. If there were a graph with four vertices of degree 1, 2, 3, and 3, then its total degree would be 9, which is odd. But by Corollary 4.9.2, the total degree of the graph must be even. *[This is a contradiction.]* Hence there is no such graph. (Alternatively, if there were such a graph, it would have an odd number of vertices of odd degree, and, by Proposition 4.9.3 this is impossible.)

9. Suppose there were a simple graph with four vertices of degrees 1, 2, 3, and 4. Then the vertex of degree 4 would have to be connected by edges to four distinct vertices other than itself because of the assumption that the graph is simple (and hence has no loops or parallel edges). This contradicts the assumption that the graph has four vertices in total. Hence there is no simple graph with four vertices of degrees 1, 2, 3, and 4.

12.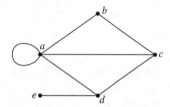

14. **a.** Define a graph G by letting each vertex represent a person at the party and drawing an edge between each pair of people who knew each other before the party. Let x be the number of people who knew three other people before the party.

    Then the total degree of the graph $= 2 \cdot 1 + 5 \cdot 2 + x \cdot 3 = 12 + 3x$

    because 2 people knew 1 other person before the party, 5 people knew 2 other people before the party, and x people knew 3 other people before the party. In addition, since a total of 15 pairs of people knew each other before the party, the graph has 15 edges. By the handshake theorem (Theorem 4.9.1), the total degree is twice the number of edges. Hence the total degree of the graph $= 2 \cdot 15 = 30$.

    It follows that $12 + 3x = 30$. Thus

$$3x = 30 - 12 = 18, \quad \text{and so} \quad x = \frac{18}{3} = 6.$$

In other words, 6 people at the party knew 3 other people before the party.

b. Now the total number of people at the party is the sum of the number who knew 1 other person before the party, plus the number who knew 2 other people before the party, plus the number who knew 3 other people before the party. Therefore, the number of people at the party $= 2 + 5 + 6 = 13$.

15. **a.** Define a graph G by letting each vertex represent a person in the social network and drawing an edge between each pair of network friends. Let x be the number of people who are network friends with three other people in the network. Then

$$\text{the total degree of the graph} = 3 \cdot 6 + 1 \cdot 5 + 5 \cdot 4 + x \cdot 3 = 43 + 3x$$

because 3 people are friends with 6 others in the network, 1 person is friends with 5 others in the network, 5 people are friends with 4 others in the network, and x people are friends with 3 others in the network. In addition, since the network contains a total of 41 pairs of network friends, the graph has 41 edges. By the handshake theorem (Theorem 4.9.1), the total degree is twice the number of edges. Hence

$$\text{the total degree of the graph} = 2 \cdot 41 = 82.$$

It follows that
$$43 + 3x = 82.$$

Thus
$$3x = 82 - 43 = 39, \quad \text{and so} \quad x = \frac{39}{3} = 13.$$

In other words, 13 people are friends with 6 others in the network.

b. Now the total number of people in the network is the sum of the number who are friends with 6 others, plus the number who are friends with 5 others, plus the number who are friends with 4 others, plus the number who are friends with 3 others. Therefore

$$\text{the number of people in the network} = 3 + 1 + 5 + 13 = 22.$$

16. **a.** Suppose that, in a group of 15 people, each person had exactly three friends. Then you could draw a graph representing each person by a vertex and connecting two vertices by an edge if the corresponding people were friends. But such a graph would have 15 vertices, each of degree 3, for a total degree of 45. This would contradict the fact that the total degree of any graph is even. Hence the supposition must be false, and in a group of 15 people it is not possible for each to have exactly three friends.

18. Yes. For example, the graph shown below satisfies this condition.

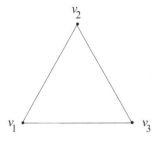

19. *Hint:* Let t be the total degree of the graph, let d_{min} be the minimum degree of any vertex in G, and let d_{max} be the maximum degree of any vertex in G.

21. **a.** Yes. Let G be a simple graph with n vertices and let v be a vertex of G. Since G has no parallel edges, v can be joined by at most a single edge to each of the $n - 1$ other vertices of G, and since G has no loops, v cannot be joined to itself. Therefore, the maximum degree of v is $n - 1$.

b. No. Suppose there is a simple graph with four vertices, all of which have different degrees. By part (a), no vertex can have degree greater than three, and of course, no vertex can have degree less than 0. Therefore, the only possible degrees of the vertices are 0, 1, 2, and 3. Since all four vertices have different degrees, there is one vertex with each degree. But then the vertex of degree 3 is connected to all the other vertices, which contradicts the fact that one of the vertices has degree 0. Hence the supposition is false, and there is no simple graph with four vertices each of which has a different degree.

c. No. Suppose there is a simple graph with n vertices (where $n \geq 2$) all of which have different degrees. By part (a), no vertex can have degree greater than $n-1$, and, of course, no vertex can have degree less than 0. Therefore, the only possible degrees of the vertices are 0, 1, 2 ,..., $n-1$. Since the vertices all have different degrees and there are n vertices, and since there are n integers from 0 to $n-1$ inclusive, there is one vertex with each degree. But then the vertex of degree $n-1$ is connected to all the other vertices, which contradicts the fact that one of the vertices has degree 0. This contradiction shows that the supposition is false, and there is no simple graph with n vertices each of which has a different degree.

22. *Hint:* Use the result of exercise 21, part (c).

23. **a.** $K_{4,2}$:

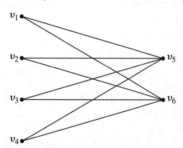

d. If $n \neq m$, the vertices of $K_{m,n}$ are divided into two groups: one of size m and the other of size n. Every vertex in the group of size m has degree n because each is connected to every vertex in the group of size n. So $K_{m,n}$ has m vertices of degree n. Similarly, every vertex in the group of size n has degree m because each is connected to every vertex in the group of size m. So $K_{m,n}$ has n vertices of degree m. Note that if $n = m$, then all $n + m = 2n$ vertices have the same degree, namely, n.

24. **a.** The given graph is bipartite, as shown by the following diagram.

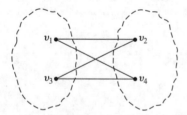

b. Suppose the given graph is bipartite. Then the vertex set can be partitioned into two mutually disjoint subsets such that vertices in each subset are connected by edges only to vertices in the other subset and not to vertices in the same subset. Now v_1 is in one subset of the partition, say, V_1. Since v_1 is connected by edges to v_2 and v_3 both v_2 and v_3 must be in the other subset, V_2. But v_2 and v_3 are connected by an edge to each other. This contradicts the fact that no vertices in V_2 are connected by edges to other vertices in V_2. Hence the supposition is false, and so the graph is not bipartite.

c. The given graph is bipartite, as shown by the following diagram.

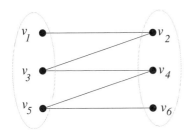

d. Suppose the given graph were bipartite with disjoint vertex sets V_1 and V_2, where no vertices within either V_1 or V_2 are connected by edges. Then v_1 would be in one of the sets, say V_1, and so v_2 and v_6 would be in V_2 (because each is connected by an edge to v_1). Furthermore, v_3, v_4, and v_5 would be in V_1 (because all are connected by edges to v_2). But v_4 is connected by an edge to v_5, and so both cannot be in V_1. This contradiction shows that the supposition is false, and so the graph is not bipartite.

e. The given graph is bipartite, as shown by the following diagram.

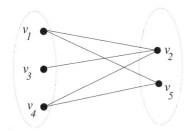

f. Suppose the given graph were bipartite with disjoint vertex sets V_1 and V_2, where no vertices within either V_1 or V_2 are connected by edges. Then v_1 would be in one of the sets, say V_1, and so v_2 and v_5 would be in V_2 (because each is connected by an edge to v_1). Furthermore, v_3 and v_4 would be in V_1 (because v_3 is connected by an edge to v_2 and v_4 is connected by an edge to v_5). But v_3 is connected by an edge to v_4, and so both cannot be in V_1. This contradiction shows that the supposition is false, and so the graph is not bipartite.

Section 4.10

1. $z = 0$

3. **a.** $z = 18$ **b.** $z = 6$

4. Trace table:

i	0	1	2	3
a	2	7	22	67

After execution: $a = 67$

6.

	Iteration Number			
	0	1	2	3
a	26			
d	7			
q	0	1	2	3
r	26	19	12	5

8. **a.**

	Iteration Number			
	0	1	2	3
A	69	19	9	
q	2			
d		1		
n			1	
p				4

9. $\gcd(27, 72) = 9$

10. $\gcd(5, 9) = 1$

12. *Solution 1*: $\gcd(48, 54) = \gcd(6 \cdot 8, 6 \cdot 9) = 6$

   Solution 2: $\gcd(48, 54) = \gcd(2^4 \cdot 3, 2 \cdot 3^3) = 2 \cdot 3 = 6$

13. Divide the larger number, 1,188, by the smaller, 385, to obtain a quotient of 3 and a remainder of 33. Next divide 385 by 33 to obtain a quotient of 11 and a remainder of 22. Then divide 33 by 22 to obtain a quotient of 1 and a remainder of 11. Finally, divide 22 by 11 to obtain a quotient of 2 and a remainder of 0. Thus, by Lemma 4.10.2, $\gcd(1188, 385) = \gcd(385, 33) = \gcd(33, 22) = \gcd(22, 11) = \gcd(11, 0)$, and by Lemma 4.10.1, $\gcd(11, 0) = 11$. So $\gcd(1188, 385) = 11$.

14. Divide the larger number, 1,177, by the smaller, 509, to obtain a quotient of 2 and a remainder of 159. Next divide 509 by 159 to obtain a quotient of 3 and a remainder of 32. Next divide 159 by 32 to obtain a quotient of 4 and a remainder of 31. Then divide 32 by 31 to obtain a quotient of 1 and a remainder of 1. Finally, divide 31 by 1 to obtain a quotient of 31 and a remainder of 0. Thus, by Lemma 4.10.2, $\gcd(1177, 509) = \gcd(509, 159) = \gcd(159, 32) = \gcd(32, 31) = \gcd(31, 1) = \gcd(1, 0)$, and by Lemma 4.10.1, $\gcd(1, 0) = 1$. So $\gcd(1177, 509) = 1$.

15.
$$
\begin{array}{r}
13 \\
832\overline{)10933} \\
10816 \\
\hline
117
\end{array}
$$
So $10933 = 832 \cdot 13 + 117$, and hence $\gcd(10933, 832) = \gcd(832, 117)$

$$
\begin{array}{r}
7 \\
117\overline{)832} \\
819 \\
\hline
13
\end{array}
$$
So $832 = 117 \cdot 7 + 13$, and hence $\gcd(832, 117) = \gcd(117, 13)$

$$
\begin{array}{r}
9 \\
13\overline{)117} \\
117 \\
\hline
0
\end{array}
$$
So $117 = 13 \cdot 9 + 0$, and hence $\gcd(117, 13) = \gcd(13, 0)$

Thus, by Lemma 4.10.2, $\gcd(10933, 832) = \gcd(832, 117) = \gcd(117, 13) = \gcd(13, 0)$, and by Lemma 4.10.1, $\gcd(13, 0) = 13$. So $\gcd(10933, 832) = 13$.

17.

A	1,001					
B	871					
r	871	130	91	39	13	0
b	871	130	91	39	13	0
a	1,001	871	130	91	39	13
gcd						13

18.

A	5859						
B	1232						
r	1232	931	301	28	21	7	0
b	1232	931	301	28	21	7	0
a	5859	1232	931	301	28	21	7
gcd							7

20.

A	4,617							
B	2,563							
a	4,617	2,563	2,054	509	18	5	3	1
b	2,563	2,054	509	18	5	3	2	0
r	2,563	2,054	509	18	5	3	2	0
gcd								1

The table shows that the greatest common divisor of 4,617 and 2,563 is 1, and so these integers are relatively prime.

21.

A	34,391							
B	6,728							
a	34,391	6,728	751	720	31	7	3	1
b	6,728	751	720	31	7	3	1	0
r	6,728	751	720	31	7	3	1	0
gcd								1

The table shows that the greatest common divisor of 34,319 and 6,728 is 1, and so these integers are relatively prime.

22. *Hint:* Divide the proof into two parts.

In part 1 suppose a and b are any positive integers such that $a \mid b$, and derive the conclusion that $\gcd(a, b) = a$. To do this, note that because it is also the case that $a \mid a$, a is a common divisor of a and b. Thus, by definition of greatest common divisor, a is less than or equal to the greatest common divisor of a and b. In symbols, $a \leq \gcd(a, b)$. Then show that $a \geq \gcd(a, b)$ by using Theorem 4.4.1.

In part 2 of the proof, suppose a and b are positive integers such that $\gcd(a, b) = a$ and deduce that $a \mid b$.

24. Proof:

Suppose a and b are any integers with $b \neq 0$, and suppose q and r are any integers such that

$$a = bq + r.$$

We must show that

$$\gcd(b, r) \leq \gcd(a, b).$$

***Step 1 (proof that any common divisor of b and r is also a common divisor of a and b)*:**

Let c be a common divisor of b and r. Then $c \mid b$ and $c \mid r$, and so by definition of divisibility, there are integers n and m so that

$$b = nc \quad \text{and} \quad r = mc$$

Substitute these values into the equation $a = bq + r$ to obtain

$$a = (nc)q + mc = c(nq + m).$$

But $nq + m$ is an integer, and so by definition of divisibility $c \mid a$. Because we already know that $c \mid b$, we can conclude that c is a common divisor of a and b.

Step 2 (proof that $\gcd(b, r) \leq \gcd(a, b)$):

By step 1, every common divisor of b and r is a common divisor of a and b. It follows that the greatest common divisor of b and r is a common divisor of a and b. But then $\gcd(b, r)$ (being one of the common divisors of a and b) is less than or equal to the greatest common divisor of a and b:

$$\gcd(b, r) \leq \gcd(a, b)$$

[as was to be shown].

25. **a.** *Hint 1:* If $a = dq - r$, then $-a = -dq + r = -dq - d + d - r = d(-q - 1) + (d - r)$.

Hint 2: If $0 \leq r < d$, then $0 \geq -r > -d$. Add d to all parts of this inequality and see what results.

26. **a.** <u>Proof</u>: Suppose a, d, q, and r are integers such that $a = dq + r$ and $0 \leq r < d$. *[We must show that $q = \left\lfloor \frac{a}{d} \right\rfloor$ and $r = a - d\left\lfloor \frac{a}{d} \right\rfloor$.]*

Solving $a = dq + r$ for r gives $r = a - dq$, and substituting into $0 \leq r < d$ gives $0 \leq a - dq < d$. Add dq to both sides to obtain

$$dq \leq a < d + dq = d(q + 1).$$

Then divide through by d to obtain

$$q \leq \frac{a}{d} < q + 1.$$

Therefore, by definition of floor, $\left\lfloor \frac{a}{d} \right\rfloor = q$.

Finally, substitution into $a = dq + r$ gives $a = d\left\lfloor \frac{a}{d} \right\rfloor + r$, and subtracting $d\left\lfloor \frac{a}{d} \right\rfloor$ from both sides yields $r = a - d\left\lfloor \frac{a}{d} \right\rfloor$ *[as was to be shown]*.

27. **a.** <u>Proof</u>: Suppose a and b are integers and $a \geq b > 0$.

Part 1 (proof that every common divisor of a and b is a common divisor of b and $a - b$):

Suppose

$$d \mid a \quad \text{and} \quad d \mid b.$$

Then, by exercise 16 of Section 4.4,

$$d \mid (a - b).$$

Hence d is a common divisor of a and $a - b$.

Part 2 (proof that every common divisor of b and $a - b$ is a common divisor of a and b):

Suppose

$$d \mid b \quad \text{and} \quad d \mid (a - b).$$

Then, by exercise 15 of Section 4.4,

$$a \mid [b + (a - b)].$$

But $b + (a - b) = a$, and so

$$d \mid a.$$

Hence d is a common divisor of a and b.

Part 3 (end of proof):

Because every common divisor of a and b is a common divisor of b and $a - b$, the greatest common divisor of a and b is a common divisor of b and $a - b$ and so is less than or equal to the greatest common divisor of a and $a - b$. Thus

$$\gcd(a, b) \leq \gcd(b, a - b).$$

By similar reasoning,

$$\gcd(b, a - b) \leq \gcd(a, b) \quad \text{and, therefore,} \quad \gcd(a, b) = \gcd(b, a - b).$$

b.

	Iteration Number									
	0	1	2	3	4	5	6	7	8	9
A	630									
B	336									
a	630	294		252	210	168	126	84	42	0
b	336		42							
gcd										42

c.

	Iteration Number												
	0	1	2	3	4	5	6	7	8	9	10	11	12
A	768												
B	348												
a	768	420	72					12					0
b	348			276	204	132	60		48	36	24	12	
gcd													12

28. **a.** $\operatorname{lcm}(12, 18) = 36$

29. <u>Proof</u>:

Part 1:

Let a and b be positive integers, and suppose $d = \gcd(a, b) = \operatorname{lcm}(a, b)$.

By definition of greatest common divisor and least common multiple, $d > 0$, $d \mid a$, $d \mid b$, $a \mid d$, and $b \mid d$.

Thus, in particular, $a = dm$ and $d = an$ for some integers m and n.

By substitution, $\quad\quad\quad\quad a = dm = (an)m = anm.$

Dividing both sides by a gives $\quad\quad 1 = nm.$

But the only divisors of 1 are 1 and -1 (Theorem 4.4.2), and so $m = n = \pm 1$.

Since both a and d are positive, $m = n = 1$, and hence

$$a = dm = d \cdot 1 = d.$$

Similar reasoning shows that $b = d$ also, and so $a = b$.

Part 2:

Given any positive integers a and b such that $a = b$, we have

$$\gcd(a, b) = \gcd(a, a) = a \quad \text{and} \quad \operatorname{lcm}(a, b) = \operatorname{lcm}(a, a) = a.$$

Hence $\gcd(a, b) = \operatorname{lcm}(a, b)$.

30. <u>Proof</u>: Let a and b be any positive integers.

    Part 1 (proof that if $\text{lcm}(a,b) = b$ then $a \mid b$): Suppose that

    $$\text{lcm}(a,b) = b.$$

    By definition of least common multiple,

    $$a \mid \text{lcm}(a,b),$$

    and so by substitution, $a \mid b$.

    Part 2 (proof that if $a \mid b$ then $\text{lcm}(a,b) = b$): Suppose that

    $$a \mid b.$$

    Then since it is also the case that

    $$b \mid b,$$

    b is a common multiple of a and b. Moreover, because b divides any common multiple of both a and b,

    $$\text{lcm}(a,b) = b.$$

32. *Hint:* Divide the proof into two parts. In part 1, suppose a and b are any positive integers, and deduce that

    $$\gcd(a,b) \cdot \text{lcm}(a,b) \le ab.$$

    Derive this result by showing that $\text{lcm}(a,b) \le \frac{ab}{\gcd(a,b)}$. To do this, show that $\frac{ab}{\gcd(a,b)}$ is a multiple of both a and b. For instance, to see that $\frac{ab}{\gcd(a,b)}$ is a multiple of b, note that because $\gcd(a,b)$ divides a, $a = \gcd(a,b) \cdot k$ for some integer k, and thus $ab = \gcd(a,b) \cdot kb$. Divide both sides by $\gcd(a,b)$ to obtain $\frac{ab}{\gcd(a,b)} = kb$. But since k is an integer, this equation implies that $\frac{ab}{\gcd(a,b)}$ is a multiple of b. The argument that $\frac{ab}{\gcd(a,b)}$ is a multiple of a is almost identical. In part 2 of the proof, use the definition of least common multiple to show that $\frac{ab}{\text{lcm}(a,b)} \mid a$ and $\frac{ab}{\text{lcm}(a,b)} \mid b$. Conclude that $\frac{ab}{\text{lcm}(a,b)} \le \gcd(a,b)$ and hence that $ab \le \gcd(a,b) \cdot \text{lcm}(a,b)$.

    <u>Proof</u>: Let a and b be any positive integers.

    Part 1 (proof that $\gcd(a,b) \cdot \text{lcm}(a,b) \le ab$):

    By definition of greatest common divisor, $\gcd(a,b) \mid a$. Hence by definition of divisibility, there is an integer k such that

    $$a = gcd(a,b) \cdot k.$$

    Multiplying both sides by b gives

    $$ab = \gcd(a,b) \cdot k \cdot b,$$

    and so

    $$\frac{ab}{\gcd(a,b)} = bk.$$

    It follows by definition of divisibility that

    $$b \quad \text{divides} \quad \left\lfloor \frac{ab}{\gcd(a,b)} \right\rfloor.$$

    An almost exactly identical sequence of steps shows that

    $$a \quad \text{divides} \quad \left\lfloor \frac{ab}{\gcd(a,b)} \right\rfloor.$$

Thus, by definition of least common multiple,

$$\text{lcm}(a,b) \leq \frac{ab}{\gcd(a,b)} \quad \text{or, equivalently,} \quad \gcd(a,b) \cdot \text{lcm}(a,b) \leq ab$$

because $\gcd(a,b) > 0$).

Part 2 (proof that $ab \leq \gcd(a,b) \cdot \text{lcm}(a,b))$):

By definition of least common multiple, $a \mid \text{lcm}(a,b)$. Hence by definition of divisibility, there is an integer k such that

$$\text{lcm}(a,b) = ak.$$

Multiplying both sides by b gives

$$b \cdot \text{lcm}(a,b) = b \cdot ak,$$

and dividing both sides of the result by $\text{lcm}(a,b)$ gives

$$b = \left[\frac{ab}{\text{lcm}(a,b)}\right] \cdot k.$$

It follows by definition of divisibility that

$$\left[\frac{ab}{\text{lcm}(a,b)}\right] \;\Big|\; b.$$

An almost exactly identical sequence of steps shows that

$$\left[\frac{ab}{\text{lcm}(a,b)}\right] \;\Big|\; a.$$

Thus, by definition of greatest common divisor,

$$\gcd(a,b) \geq \frac{ab}{\text{lcm}(a,b)} \quad \text{or, equivalently,} \quad \gcd(a,b) \cdot \text{lcm}(a,b) \geq ab.$$

because $\text{lcm}(a,b) > 0$.

Conclusion:

By part 1, $\gcd(a,b) \cdot \text{lcm}(a,b) \leq ab$, and by part 2, $ab \leq \gcd(a,b) \cdot \text{lcm}(a,b)$. Therefore, $\gcd(a,b) \cdot \text{lcm}(a,b) = ab$.

Review Guide: Chapter 4

Definitions

- Why is the phrase "if, and only if" used in a definition?
- How are the following terms defined?
  - even integer
  - odd integer
  - prime number
  - composite number
  - rational number
  - divisibility of one integer by another
  - *n div d* and *n mod d*
  - the floor of a real number
  - the ceiling of a real number
  - greatest common divisor of two integers

Proving an Existential Statement/Disproving a Universal Statement

- How do you determine the truth of an existential statement?
- What does it mean to "disprove" a statement?
- What is disproof by counterexample?
- How do you establish the falsity of a universal statement?

Proving a Universal Statement/Disproving an Existential Statement

- If a universal statement is defined over a small, finite domain, how do you use the method of exhaustion to prove that it is true?
- What is the method of generalizing from the generic particular?
- If you use the method of direct proof to prove a statement of the form "$\forall x$, if $P(x)$ then $Q(x)$," what do you suppose and what do you have to show?
- What are the guidelines for writing proofs of universal statements?
- What are some common mistakes people make when writing mathematical proofs?
- How do you disprove an existential statement?
- What is the method of proof by division into cases?
- What is the triangle inequality?
- If you use the method of proof by contradiction to prove a statement, what do you suppose and what do you have to show?
- If you use the method of proof by contraposition to prove a statement of the form "$\forall x$, if $P(x)$ then $Q(x)$," what do you suppose and what do you have to show?
- Are you able to use the various methods of proof and disproof to establish the truth or falsity of statements about odd and even integers , prime numbers , rational and irrational numbers, divisibility of integers, absolute value , and the floor and ceiling of a real number ?

Some Important Theorems and Algorithms

- What is the transitivity of divisibility theorem?
- What is the theorem about divisibility by a prime number?
- What is the unique factorization of integers theorem? (This theorem is also called the fundamental theorem of arithmetic.)
- What is the quotient-remainder theorem? Can you apply it to specific situations?

- What is the theorem about the irrationality of the square root of 2? Can you prove this theorem?
- What is the theorem about the infinitude of the prime numbers? Can you prove this theorem?
- What is the division algorithm ?
- What is the Euclidean algorithm?
- How do you use the Euclidean algorithm to compute the greatest common divisor of two positive integers?

The Handshake Theorem

- What is the total degree of a graph?
- What is the statement of the handshake theorem? Can you prove this theorem?
- What can you say about the parity of the total degree of a graph?
- What can you say about the number of odd degree vertices in a graph?

Notation for Algorithms

- How is an assignment statement executed?
- How is an **if-then** statement executed?
- How is an **if-then-else** statement executed?
- How are the statements **do** and **end do** used in an algorithm?
- How is a **while** loop executed?
- How is a **for-next** loop executed?
- How do you construct a trace table for a segment of an algorithm?

Fill-in-the-Blank Review Questions

Section 4.1

1. An integer is even if, and only if, _____.
2. An integer is odd if, and only if, _____.
3. An integer n is prime if, and only if, _____.
4. The most common way to disprove a universal statement is to find _____.
5. According to the method of generalizing from the generic particular, to show that every element of a domain satisfies a certain property, suppose x is a _____, and show that _____.
6. To use the method of direct proof to prove a statement of the form, "For every x in a domain D, if $P(x)$ then $Q(x)$," one supposes that _____ and one shows that _____.

Section 4.2

1. The meaning of every variable used in a proof should be explained within ____ .
2. Proofs should be written in sentences that are ____ and ____ .
3. Every assertion in a proof should be supported by a ____ .
4. The following are some useful "little words and phrases" that clarify the reasoning in a proof: ____, ____, ____, ____, and ____ .
5. A new thought or fact that does not follow as an immediate consequence of the preceding statement can be introduced by writing ____, ____, ____, ____, or ____ .
6. To introduce a new variable that is defined in terms of previous variables, use the word ____ .
7. Displaying equations and inequalities increases the ____ of a proof.

8. Some proof-writing mistakes are ____, ____, ____, ____, ____, ____,and ____.

Section 4.3

1. To show that a real number is rational, we must show that we can write it as _____.
2. An irrational number is a _____ that is _____.
3. Zero is a rational number because _____.

Section 4.4

1. To show that a nonzero integer d divides an integer n, we must show that _____.
2. To say that d divides n means the same as saying that ___ is divisible by ___.
3. If a and b are positive integers and $a \mid b$, then _____ is less than or equal to _____.
4. For all integers n and d with $d \neq 0$, $d \nmid n$ if, and only if, _____.
5. If a and b are integers, the notation $a \mid b$, denotes _____ and the notation a/b denotes _____.
6. The transitivity-of-divisibility theorem says that for all integers a, b, and c, if _____ then _____.
7. The divisibility-by-a-prime theorem says that every integer greater than 1 is _____.
8. The unique-factorization-of-integers theorem says that any integer greater than 1 is either _____ or can be written as _____ in a way that is unique except possibly for the _____ in which the numbers are written.

Section 4.5

1. The quotient-remainder theorem says that for all integers n and d with $d \geq 0$, there exist _____ q and r such that _____ and _____.
2. If n and d are integers with $d \geq 0$, n div d is _____ and n mod d is _____.
3. The parity of an integer indicates whether the integer is _____.
4. According to the quotient-remainder theorem, if an integer n is divided by an integer d, the possible remainders are _____. This implies that n can be written in one of the forms _____ for some integer q.
5. To prove a statement of the form "If A_1 or A_2 or A_3, then C," prove _____ and _____ and _____.
6. The triangle inequality says that for all real numbers x and y, _____.

Section 4.6

1. Given any real number x, the floor of x is the unique integer n such that _____.
2. Given any real number x, the ceiling of x is the unique integer n such that _____.

Section 4.7

1. To prove a statement by contradiction, you suppose that _____ and you show that _____.
2. A proof by contraposition of a statement of the form "$\forall x \in D$, if $P(x)$ then $Q(x)$" is a direct proof of _____.
3. To prove a statement of the form "$\forall x \in D$, if $P(x)$ then $Q(x)$" by contraposition, you suppose that _____ and you show that _____.

Section 4.8

1. The ancient Greeks discovered that in a right triangle where both legs have length 1, the ratio of the length of the hypotenuse to the length of one of the legs is not equal to a ratio of ____.

2. One way to prove that $\sqrt{2}$ is an irrational number is to assume that $\sqrt{2} = a/b$ for some integers a and b that have no common factor greater than 1, use the lemma that says that if the square of an integer is even then ____, and eventually show that a and b ____.

3. One way to prove that there are infinitely many prime numbers is to assume that there are only finitely many prime numbers 2, 3, 5, 7, 11, . . . , p, construct the number ____, and then show that this number has to be divisible by a prime number that is greater than ____.

Section 4.9

1. The total degree of a graph is defined as ____.

2. The handshake theorem says that the total degree of a graph is ____.

3. In any graph the number of vertices of odd degree is ____.

4. A simple graph is ____.

5. A complete graph on n vertices is a ____.

6. A complete bipartite graph on (m, n) vertices is a simple graph whose vertices can be divided into two distinct sets, say V with m vertices and W with n vertices, in such a way that (1) there is ____ from each vertex of V to each vertex of W, (2) there is ____ from any one vertex of V to any other vertex of V, and (3) there is ____ from any one vertex of W to any other vertex of W.

Section 4.10

1. When an algorithm statement of the form $x := e$ is executed, the expression e is ____.

2. Consider an algorithm statement of the following form:
   if (*condition*)
      then s_1
      else s_2
   When such a statement is executed, the truth or falsity of the *condition* is evaluated. If *condition* is true, ____. If *condition* is false, ____.

3. Consider an algorithm statement of the following form:
   while (*condition*)
      [statements that make up the body of the loop]
   end while
   When such a statement is executed, the truth or falsity of the *condition* is evaluated. If *condition* is true, ____. If *condition* is false, ____.

4. Consider an algorithm statement of the following form:
   for *variable := initial expression* **to** *final expression*
      [statements that make up the body of the loop]
   next *(same) variable*
   When such a statement is executed, *variable* is set equal to the value of the *initial expression*, and a check is made to determine whether the value of *variable* is less than or equal to the value of *final expression*. If so, ____. If not, ____.

5. Given a nonnegative integer a and a positive integer d, the division algorithm computes ____.

6. Given integers a and b, not both zero, $\gcd(a, b)$ is the integer d that satisfies the following two conditions: ____ and ____.

7. If r is a positive integer, then $\gcd(r, 0) = $ ____.

8. If a and b are integers with $b \neq 0$ and if q and r are nonnegative integers such that $a = bq + r$, then $\gcd(a, b) = $ _____.

9. Given positive integers A and B with $A > B$, the Euclidean algorithm computes _____.

Answers for Fill-in-the-Blank Review Questions

Section 4.1

1. it equals twice some integer

2. it equals twice some integer plus 1

3. n is greater than 1 and if n equals the product of any two positive integers, then one of the integers equals 1 and the other equals n.

4. a counterexample

5. particular but arbitrarily chosen element of the domain; x satisfies the given property

6. x is a particular but arbitrarily chosen element of the domain D that makes $P(x)$ true; x makes $Q(x)$ true

Section 4.2

1. the body of the proof (or: the proof itself)

2. complete; grammatically correct

3. reason (or explanation)

4. Because; Since; Then; Thus; So; Hence; Therefore; Consequently; It follows that; By substitution

5. Observe that; Note that; Recall that; But; Now

6. Let

7. readability

8. Arguing from examples; Using the same letter to mean two different things; Jumping to a conclusion; Circular reasoning; Confusion between what is known and what is still to be shown; Use of *any* when the correct word is *some*; Misuse of the word *if*

9. To show that a real number is rational, we must show that we can write it as

10. An irrational number is a _____ that is _____.

11. Zero is a rational number because _____.

Section 4.3

1. a ratio of integers with a nonzero denominator

2. real number; not rational

3. $0 = \dfrac{0}{1}$

Section 4.4

1. n equals d times some integer (*Or:* there is an integer r such that $n = dr$)

2. n; d

3. a; b

4. $\dfrac{n}{d}$ is not an integer

5. the sentence "a divides b"; the number obtained when a is divided by b

6. a divides b and b divides c; a divides c

7. divisible by some prime number
8. prime; a product of prime numbers; order

Section 4.5

1. integers; $n = dq + r$; $0 \leq r < d$
2. the quotient obtained when n is divided by d; the remainder obtained when n is divided by d
3. odd or even
4. $0, 1, 2, \ldots, (d-1)$; $dq, dq+1, dq+2, \ldots, dq+(d-1)$
5. If A_1, then C; If A, then C; If A_3, then C
6. $|x + y| \leq |x| + |y|$

Section 4.6

1. $n \leq x < n + 1$
2. $n - 1 < x \leq n$

Section 4.7

1. the statement is false; this supposition leads to a contradiction
2. $\forall x \in D$, if $\sim Q(x)$ then $\sim P(x)$
3. x is any *[particular but arbitrarily chosen]* element of D for which $Q(x)$ is false;
   $P(x)$ is false

Section 4.8

1. two integers
2. the integer is even; have a common factor greater than 1
3. $2 \cdot 3 \cdot 5 \cdot 7 \cdot 11 \cdots p + 1$; p

Section 4.9

1. the sum of the degrees of all the vertices of the graph
2. equal to twice the number of edges of the graph
3. an even number
4. a graph with no loops or parallel edges
5. simple graph with n vertices whose set of edges contains exactly one edge for each pair of vertices
6. one edge; no edge; no edge

Section 4.10

1. evaluated (using the current values of all the variables in the expression), and this value is placed in the memory location corresponding to x (replacing any previous contents of the location)
2. statement s_1 is executed;   statement s_2 is executed
3. all statements in the body of the loop are executed in order and then execution moves back to the beginning of the loop and the process repeats;
   execution passes to the next algorithm statement following the loop
4. the statements in the body of the loop are executed in order, *variable* is increased by 1, and execution returns to the top of the loop;
   execution passes to the next algorithm statement following the loop
5. integers q and r with the property that $n = dq + r$ and $0 \leq r < d$
6. d divides a and d divides b; if c is a common divisor of both a and b, then $c \leq d$
7. r
8. $\gcd(b, r)$
9. the greatest common divisor of A and B

Chapter 5: Sequences and Mathematical Induction

The first section of this chapter introduces the notation for sequences, summations, products, and factorial. The section is intended to help you learn to recognize patterns so as to be able, for instance, to transform expanded versions of sums into summation notation, and to handle subscripts, particularly to change variables for summations and to distinguish index variables from variables that are constant with respect to a summation.

The second, third, and fourth sections of the chapter treat mathematical induction. The ordinary form is discussed in Sections 5.2 and 5.3 and the strong form in Section 5.4. Because of the importance of mathematical induction in discrete mathematics, a wide variety of examples is given to help you become comfortable with using the technique in many different situations. Section 5.5 then shows how to use a variation of mathematical induction to prove the correctness of an algorithm. Sections 5.6-5.8 deal with recursively defined sequences, both how to analyze a situation using recursive thinking to obtain a sequence that describes the situation and how to find an explicit formula for the sequence once it has been defined recursively. Section 5.9 applies recursive thinking to the question of defining a set, and it describes the technique of structural induction, which is the variation of mathematical induction that can be used to verify properties of a recursively defined set. Section 5.9 also introduces the concept of a recursively defined function.

The logic of ordinary mathematical induction can be described by relating it to the logic discussed in Chapters 2 and 3. The main point is that the inductive step establishes the truth of a sequence of if-then statements. Together with the basis step, this sequence gives rise to a chain of inferences that lead to the desired conclusion. More formally:

Suppose
1. $P(1)$ is true; and
2. for all integers $k \geq 1$, if $P(k)$ is true then $P(k+1)$ is true.

The truth of statement (2) implies, according to the law of universal instantiation, that no matter what particular integer $k \geq 1$ is substituted in place of k, the statement "If $P(k)$ then $P(k+1)$" is true. The following argument, therefore, has true premises, and so by modus ponens it has a true conclusion:

	If $P(1)$ then $P(2)$.	by 2 and universal instantiation
	$P(1)$	by 1
$\therefore$	$P(2)$	by modus ponens

Similar reasoning gives the following chain of arguments, each of which has a true conclusion by modus ponens:

	If $P(2)$ then $P(3)$.
	$P(2)$
$\therefore$	$P(3)$
	If $P(3)$ then $P(4)$.
	$P(3)$
$\therefore$	$P(4)$
	If $P(4)$ then $P(5)$.
	$P(4)$
$\therefore$	$P(5)$
	And so forth.

Thus no matter how large a positive integer n is specified, the truth of $P(n)$ can be deduced as the final conclusion of a (possibly very long) chain of arguments continuing those shown above.

Section 5.1

1. $\dfrac{1}{11}, \dfrac{2}{12}, \dfrac{3}{13}, \dfrac{4}{14}$

3. $1, -\dfrac{1}{3}, \dfrac{1}{9}, -\dfrac{1}{27}$

5. $0, 0, 2, 2$

6. $f_1 = \left\lfloor \dfrac{1}{4} \right\rfloor \cdot 4 = 0 \cdot 4 = 0, \quad f_2 = \left\lfloor \dfrac{2}{4} \right\rfloor \cdot 4 = 0 \cdot 4 = 0, \quad f_3 = \left\lfloor \dfrac{3}{4} \right\rfloor \cdot 4 = 0 \cdot 4 = 4,$

  $f_4 = \left\lfloor \dfrac{4}{4} \right\rfloor \cdot 4 = 1 \cdot 4 = 4$

8. $g_1 = \lfloor \log_2 1 \rfloor = 0$

  $g_2 = \lfloor \log_2 2 \rfloor = 1, \quad g_3 = \lfloor \log_2 3 \rfloor = 1$

  $g_4 = \lfloor \log_2 4 \rfloor = 2, \quad g_5 = \lfloor \log_2 5 \rfloor = 2$

  $g_6 = \lfloor \log_2 6 \rfloor = 2, \quad g_7 = \lfloor \log_2 7 \rfloor = 2$

  $g_8 = \lfloor \log_2 8 \rfloor = 3, \quad g_9 = \lfloor \log_2 9 \rfloor = 3$

  $g_{10} = \lfloor \log_2 10 \rfloor = 3, \quad g_{11} = \lfloor \log_2 11 \rfloor = 3$

  $g_{12} = \lfloor \log_2 12 \rfloor = 3, \quad g_{13} = \lfloor \log_2 13 \rfloor = 3$

  $g_{14} = \lfloor \log_2 14 \rfloor = 3, \quad g_{15} = \lfloor \log_2 15 \rfloor = 3$

  When n is an integral power of 2, g_n is the exponent of that power. For instance, $8 = 2^3$ and $g_8 = 3$. More generally, if $n = 2^k$, where k is an integer, then $g_n = k$. All terms of the sequence from g_{2^k} up to, but not including, $g_{2^{k+1}}$ have the same value, namely k. For instance, all terms of the sequence from g_8 through g_{15} have the value 3.

9. $\begin{aligned}
h_1 &= 1 \cdot \lfloor \log_2 1 \rfloor &= 1 \cdot 0 \\
h_2 &= 2 \cdot \lfloor \log_2 2 \rfloor &= 2 \cdot 1 \\
h_3 &= 3 \cdot \lfloor \log_2 3 \rfloor &= 3 \cdot 1 \\
h_4 &= 4 \cdot \lfloor \log_2 4 \rfloor &= 4 \cdot 2 \\
h_5 &= 5 \cdot \lfloor \log_2 5 \rfloor &= 5 \cdot 2 \\
h_6 &= 6 \cdot \lfloor \log_2 6 \rfloor &= 6 \cdot 2 \\
h_7 &= 7 \cdot \lfloor \log_2 7 \rfloor &= 7 \cdot 2 \\
h_8 &= 8 \cdot \lfloor \log_2 8 \rfloor &= 8 \cdot 3 \\
h_9 &= 9 \cdot \lfloor \log_2 9 \rfloor &= 9 \cdot 3 \\
h_{10} &= 10 \cdot \lfloor \log_2 10 \rfloor &= 10 \cdot 3 \\
h_{11} &= 11 \cdot \lfloor \log_2 11 \rfloor &= 11 \cdot 3 \\
h_{12} &= 12 \cdot \lfloor \log_2 12 \rfloor &= 12 \cdot 3 \\
h_{13} &= 13 \cdot \lfloor \log_2 13 \rfloor &= 13 \cdot 3 \\
h_{14} &= 14 \cdot \lfloor \log_2 14 \rfloor &= 14 \cdot 3 \\
h_{15} &= 15 \cdot \lfloor \log_2 15 \rfloor &= 15 \cdot 3
\end{aligned}$

  When n is an integral power of 2, h_n is n times the exponent of that power. For instance, $8 = 2^3$ and $h_8 = 8 \cdot 3$. If m and n are integers and $2^m \le n < 2^{m+1}$, then $h_n = n \cdot m$.

Exercises 10-16 have more than one correct answer.

10. $a_n = (-1)^n$, where n is an integer and $n \ge 1$

11. $a_n = (n-1)(-1)^n$, where n is an integer and $n \ge 1$

12. $a_n = \dfrac{n}{(n+1)^2}$, where n is an integer and $n \ge 1$

14. $a_n = \dfrac{n^2}{3^n}$, where n is an integer and $n \geq 1$

15. $a_n = (-1)^{n-1}\left(\dfrac{n-1}{n}\right)$ for every integer $n \geq 1$

18. **a.** $2 + 3 + (-2) + 1 + 0 + (-1) + (-2) = 1$
    b. $a_0 = 2$
    c. $a_2 + a_4 + a_6 = -2 + 0 + (-2) = -4$
    d. $2 \cdot 3 \cdot (-2) \cdot 1 \cdot 0 \cdot (-1) \cdot (-2) = 0$
    e. $\displaystyle\prod_{k=2}^{2} a_k = a_2 = -2$

19. $2 + 3 + 4 + 5 + 6 = 20$

20. $2^2 \cdot 3^2 \cdot 4^2 = 576$

21. $\displaystyle\sum_{k=1}^{3}(k^2 + 1) = (1^2 + 1) + (2^2 + 1) + (3^2 + 1) = 2 + 5 + 10 = 17$

23. $1(1 + 1) = 2$

24. $\displaystyle\sum_{j=0}^{0}(j + 1) \cdot 2^j = (0 + 1) \cdot 2^0 = 1 \cdot 1 = 1$

27. $\left(\dfrac{1}{1} - \dfrac{1}{2}\right) + \left(\dfrac{1}{2} - \dfrac{1}{3}\right) + \left(\dfrac{1}{3} - \dfrac{1}{4}\right) + \left(\dfrac{1}{4} - \dfrac{1}{5}\right) + \left(\dfrac{1}{5} - \dfrac{1}{6}\right) + \dfrac{1}{6} - \dfrac{1}{7} = 1 - \dfrac{1}{7} = \dfrac{6}{7}$

29. $(-2)^1 + (-2)^2 + (-2)^3 + \cdots + (-2)^n = -2 + 2^2 - 2^3 + \cdots + (-1)^n 2^n$

Exercises 30-31 have more than one correct answer.

30. $\displaystyle\sum_{j=1}^{n} j(j + 1) = 1 \cdot 2 + 2 \cdot 3 + 3 \cdot 4 + \cdots + n \cdot (n + 1)$

31. $\displaystyle\sum_{k=0}^{n+1} \dfrac{1}{k!} = \dfrac{1}{0!} + \dfrac{1}{1!} + \dfrac{1}{2!} + \cdots + \dfrac{1}{(n+1)!}$

33. $\dfrac{1}{1^2} = 1$

35. $\left(\dfrac{1}{1+1}\right)\left(\dfrac{2}{2+1}\right)\left(\dfrac{3}{3+1}\right) = \left(\dfrac{1}{2}\right)\left(\dfrac{2}{3}\right)\left(\dfrac{3}{4}\right) = \dfrac{1}{4}$

36. $\left(\dfrac{1 \cdot 2}{3 \cdot 4}\right) = \dfrac{1}{6}$

37. $\displaystyle\sum_{i=1}^{k} i^3 + (k + 1)^3 = \sum_{i=1}^{k+1} i^3$

39. $\displaystyle\sum_{m=0}^{n}(m + 1)2^m + (n + 2)2^{n+1} = \sum_{m=0}^{n}(m + 1)2^m + ((n + 1) + 1)2^{n+1} = \sum_{m=0}^{n+1}(m + 1)2^m$

40. $\displaystyle\sum_{i=1}^{k+1} i(i!) = \sum_{i=1}^{k} i(i!) + (k + 1)(k + 1)!$

42. $\displaystyle\sum_{m=1}^{n+1} m(m+1) = \sum_{m=1}^{n} m(m+1) + (n+1)((n+1)+1)$

43. $\displaystyle\sum_{k=1}^{7} (-1)^{k+1} k^2$ or $\displaystyle\sum_{k=0}^{6} (-1)^k (k+1)^2$

Exercises 44-52 have more than one correct answer.

45. $\displaystyle\prod_{i=2}^{4} (i^2 - 1)$

46. $\displaystyle\sum_{j=2}^{6} \frac{(-1)^j j}{(j+1)(j+2)}$ or $\displaystyle\sum_{k=3}^{7} \frac{(-1)^{k+1}(k-1)}{k(k+1)}$

47. $\displaystyle\sum_{i=0}^{5} (-1)^i r^i$

48. $\displaystyle\prod_{j=1}^{4} (1 - t^j)$

49. $\displaystyle\sum_{k=1}^{n} k^3$

51. $\displaystyle\sum_{i=0}^{n-1} (n - i)$

53. When $k = 0$, then $i = 1$. When $k = 5$, then $i = 6$. Since $i = k + 1$, then $k = i - 1$. Thus,

$$k(k-1) = (i-1)((i-1)-1) = (i-1)(i-2),$$

and so

$$\sum_{k=0}^{5} k(k-1) = \sum_{i=1}^{6} (i-1)(i-2)$$

54. When $k = 1$, $i = 1 + 1 = 2$. When $k = n$, $i = n + 1$. Since $i = k + 1$, then $k = i - 1$. So

$$\frac{k}{k^2+4} = \frac{(i-1)}{(i-1)^2+4} = \frac{i-1}{i^2-2i+1+4} = \frac{i-1}{i^2-2i+5}.$$

Therefore,

$$\prod_{k=1}^{n} \left(\frac{k}{k^2+4}\right) = \prod_{i=2}^{n+1} \left(\frac{i-1}{i^2-2i+5}\right).$$

55. When $i = 1$, then $j = 0$. When $i = n + 1$, then $j = n$. Since $j = i - 1$, then $i = j + 1$. Thus,

$$\frac{(i-1)^2}{i \cdot n} = \frac{((j+1)-1)^2}{(j+1) \cdot n} = \frac{j^2}{jn+n}.$$

(Note that n has the same value in each term of the sum.)

$$\text{So } \sum_{i=1}^{n+1} \frac{(i-1)^2}{i \cdot n} = \sum_{j=0}^{n} \frac{j^2}{jn+n}.$$

56. When $i = 3$, then $j = 2$. When $i = n$, then $j = n - 1$. Since $j = i - 1$, then $i = j + 1$. Thus,

$$\sum_{i=3}^{n} \frac{i}{i + n - 1} = \sum_{j=2}^{n-1} \frac{j+1}{(j+1) + n - 1}$$

$$= \sum_{j=2}^{n-1} \frac{j+1}{j+n}.$$

57. When $i = 1$, $j = 1 - 1 = 0$. When $i = n - 1$, $j = n - 2$. Since $j = i - 1$, then $i = j + 1$. So

$$\frac{i}{(n-i)^2} = \frac{j+1}{(n-(j+1))^2} = \frac{j+1}{(n-j-1)^2}.$$

Therefore,

$$\sum_{i=1}^{n-1} \left(\frac{i}{(n-i)^2} \right) = \sum_{j=0}^{n-2} \left(\frac{j+1}{(n-j-1)^2} \right).$$

(Note that n has the same value in each term of the sum.)

59. $\sum_{k=1}^{n} [3(2k - 3) + (4 - 5k)] = \sum_{k=1}^{n} [(6k - 9) + (4 - 5k)] = \sum_{k=1}^{n} (k - 5)$

60. By Theorem 5.1.1,

$$2 \cdot \sum_{k=1}^{n} (3k^2 + 4) + 5 \cdot \sum_{k=1}^{n} (2k^2 - 1) \quad = \quad \sum_{k=1}^{n} 2(3k^2 + 4) + \sum_{k=1}^{n} 5(2k^2 - 1)$$

$$= \quad \sum_{k=1}^{n} (6k^2 + 8) + \sum_{k=1}^{n} (10k^2 - 5)$$

$$= \quad \sum_{k=1}^{n} (6k^2 + 8 + 10k^2 - 5)$$

$$= \quad \sum_{k=1}^{n} (16k^2 + 3).$$

62. $\dfrac{4 \cdot \cancel{3} \cdot \cancel{2} \cdot \cancel{1}}{\cancel{3} \cdot \cancel{2} \cdot \cancel{1}} = 4$

63. $\dfrac{6!}{8!} = \dfrac{6!}{8 \cdot 7 \cdot 6!} = \dfrac{1}{56}$

65. $\dfrac{n(n-1)(n-2) \cdots 3 \cdot 2 \cdot 1}{(n-1)(n-2) \cdots 3 \cdot 2 \cdot 1} = n$

66. $\dfrac{(n-1)(n-2) \cdots 3 \cdot 2 \cdot 1}{(n+1)n(n-1)(n-2) \cdots 3 \cdot 2 \cdot 1} = \dfrac{1}{n(n+1)}$

68. $\dfrac{[(n+1)n(n-1)(n-2) \cdots 3 \cdot 2 \cdot 1]^2}{[n(n-1)(n-2) \cdots 3 \cdot 2 \cdot 1]^2} = (n+1)^2$

69. $\dfrac{n(n-1)(n-2) \cdots (n-k+1)(n-k)(n-k-1) \cdots 2 \cdot 1}{(n-k)(n-k-1) \cdots 2 \cdot 1} = n(n-1)(n-2) \cdots (n-k+1)$

71. $\dbinom{5}{3} = \dfrac{5!}{(3!)(5-3)!} = \dfrac{5 \cdot 4 \cdot 3 \cdot 2 \cdot 1}{(3 \cdot 2 \cdot 1)(2 \cdot 1)} = 10$

72. $\dbinom{7}{4} = \dfrac{7!}{4!(7-4)!} = \dfrac{7 \cdot 6 \cdot 5 \cdot 4 \cdot 3 \cdot 2 \cdot 1}{(4 \cdot 3 \cdot 2 \cdot 1)(3 \cdot 2 \cdot 1)} = \dfrac{7 \cdot 6 \cdot 5}{(3 \cdot 2 \cdot 1)} = 35$

73. $\begin{pmatrix} 3 \\ 0 \end{pmatrix} = \dfrac{3!}{(0!)(3-0)!} = \dfrac{3!}{(1)(3!)} = 1$

75. $\begin{pmatrix} n \\ n-1 \end{pmatrix} = \dfrac{n!}{(n-1)!(n-(n-1))!} = \dfrac{n(n-1)!}{(n-1)!(n-n+1)!} = \dfrac{n}{1} = n$

77. **a.** <u>Proof</u>: Let n be an integer such that $n \geq 2$. By definition of factorial,

$$n! = \begin{cases} 2 \cdot 1 & \text{if } n = 2 \\ 3 \cdot 2 \cdot 1 & \text{if } n = 3 \\ n \cdot (n-1) \cdot \cdots \cdot 2 \cdot 1 & \text{if } n > 3. \end{cases}$$

In each case, $n!$ has a factor of 2, and so $n! = 2k$ for some integer k. Then

$$\begin{aligned} n! + 2 &= 2k + 2 & \text{by substitution} \\ &= 2(k+1) & \text{by factoring out the 2.} \end{aligned}$$

Since $k+1$ is an integer, $n! + 2$ is divisible by 2 *[as was to be shown]*.

b. *Hint:* Consider the sequence $m! + 2, m! + 3, m! + 4, \ldots, m! + m$.

78. <u>Proof</u>: Suppose n and r are nonnegative integers with $r + 1 \leq n$. The right-hand side of the equation to be shown is

$$\begin{aligned} \frac{n-r}{r+1} \cdot \begin{pmatrix} n \\ r \end{pmatrix} &= \frac{n-r}{r+1} \cdot \frac{n!}{r!(n-r)!} \\ &= \frac{n-r}{r+1} \cdot \frac{n!}{r!(n-r) \cdot (n-r-1)!} \\ &= \frac{n!}{(r+1)! \cdot (n-r-1)!} \\ &= \frac{n!}{(r+1)! \cdot (n-(r+1))!} \\ &= \begin{pmatrix} n \\ r+1 \end{pmatrix}, \end{aligned}$$

which is the left-hand side of the equation to be shown.

80. **a.** $m-1,\ sum + a[i+1]$ **b.** $m+1;\ sum + a[j-1]$

81.

$$
\begin{array}{r r l}
& 0 & \text{remainder} = r[6] = 1 \\
2\,\lfloor 1 & & \text{remainder} = r[5] = 0 \\
2\,\lfloor 2 & & \text{remainder} = r[4] = 1 \\
2\,\lfloor 5 & & \text{remainder} = r[3] = 1 \\
2\,\lfloor 11 & & \text{remainder} = r[2] = 0 \\
2\,\lfloor 22 & & \text{remainder} = r[1] = 1 \\
2\,\lfloor 45 & & \text{remainder} = r[0] = 0 \\
2\,\lfloor 90 & &
\end{array}
$$

Hence $90_{10} = 1011010_2$.

84.

a	23					
i	0	1	2	3	4	5
q	23	11	5	2	1	0
$r[0]$		1				
$r[1]$			1			
$r[2]$				1		
$r[3]$					0	
$r[4]$						1

87. Let a nonnegative integer a be given. Divide a by 16 using the quotient-remainder theorem to obtain a quotient $q[0]$ and a remainder $r[0]$. If the quotient is nonzero, divide by 16 again to obtain a quotient $q[1]$ and a remainder $r[1]$. Continue this process until a quotient of 0 is obtained. The remainders calculated in this way are the hexadecimal digits of a:

$$a_{10} = (r[k]r[k-1]\ldots r[2]r[1]r[0])_{16}.$$

88.

$$
\begin{array}{r}
0 \\
\hline
16\,|\;\;1 \\
\hline
16\,|\;\;17 \\
\hline
16\,|\;\;287
\end{array}
\qquad
\begin{array}{l}
\text{remainder } 1 \;= r[2] = 1_{16} \\
\text{remainder } 1 \;= r[1] = 1_{16} \\
\text{remainder } 15 = r[0] = F_{16}
\end{array}
$$

Hence $287_{10} = 11F_{16}$.

90.

$$
\begin{array}{r}
0 \\
\hline
16\,|\;\;8 \\
\hline
16\,|\;\;143 \\
\hline
16\,|\;\;2301
\end{array}
\qquad
\begin{array}{l}
\text{remainder } 8 \;= r[2] = 8_{16} \\
\text{remainder } 15 = r[1] = F_{16} \\
\text{remainder } 13 = r[1] = D_{16}
\end{array}
$$

Hence $2301_{10} = 8FD_{16}$.

Section 5.2

1. **a.** The statement in part (a) is true because

$$\text{if } \left(1 - \frac{1}{2}\right)\left(1 - \frac{1}{3}\right) = \frac{1}{3}, \quad \text{then} \quad \left(1 - \frac{1}{2}\right)\left(1 - \frac{1}{3}\right)\left(1 - \frac{1}{4}\right) = \frac{1}{3}\left(1 - \frac{1}{4}\right) = \frac{1}{3}\cdot\frac{3}{4} = \frac{1}{4}.$$

2. **a.** $P(1)$ is the equation $1 = 1^2$, which is true.

   b. $P(k)$ is the equation $1 + 3 + 5 + \cdots + (2k - 1) = k^2$.

   c. $P(k + 1)$ is the equation $1 + 3 + 5 + \cdots + (2(k + 1) - 1) = (k + 1)^2$

   Or, equivalently, $P(k + 1)$ is $1 + 3 + 5 + \cdots + (2k + 1) = (k + 1)^2$

   [because $2(k + 1) - 1 = 2k + 2 - 1 = 2k + 1$].

   d. In the inductive step, show that if k is any integer for which $k \geq 1$ and

$$1 + 3 + 5 + \cdots + (2k - 1) = k^2 \text{ is true,}$$

   then

$$1 + 3 + 5 + \cdots + (2k + 1) = (k + 1)^2 \text{ is also true.}$$

3. **a.** $P(1)$ is the equation $1^2 = \dfrac{1 \cdot (1 + 1) \cdot (2 \cdot 1 + 1)}{6}$. $P(1)$ is true because

   the left-hand side equals $1^2 = 1$

   and the right-hand side equals $\dfrac{1 \cdot (1 + 1) \cdot (2 + 1)}{6} = \dfrac{2 \cdot 3}{6} = 1$ also.

   b. $P(k)$ is the equation $1^2 + 2^2 + 3^2 + \cdots + k^2 = \dfrac{k(k + 1)(2k + 1)}{6}$.

   c. $P(k + 1)$ is the equation

$$1^2 + 2^2 + 3^2 + \cdots + (k + 1)^2 = \frac{(k + 1)[(k + 1) + 1][(2(k + 1) + 1)]}{6}.$$

Or, by simplifying the right-hand side, $P(k+1)$ is

$$1^2 + 2^2 + 3^2 + \cdots + (k+1)^2 = \frac{(k+1)(k+2)(2k+3)}{6}.$$

d. In the inductive step, show that if k is any integer such that $k \geq 1$ and

$$1^2 + 2^2 + 3^2 + \cdots + k^2 = \frac{k(k+1)(2k+1)}{6}$$

is true, then

$$1^2 + 2^2 + 3^2 + \cdots + (k+1)^2 = \frac{(k+1)(k+2)(2k+3)}{6}$$

is also true.

5. **a.** 1^2

   b. k^2

   c. $1 + 3 + 5 + \cdots + [(2(k+1) - 1]$

   d. $(k+1)^2$

   e. the odd integer just before $2k+1$ is $2k-1$

   f. inductive hypothesis

6. <u>Proof</u>: For the given statement, the property $P(n)$ is the equation

$$2 + 4 + 6 + \cdots + 2n = n^2 + n. \qquad \leftarrow \ P(n)$$

Show that $P(1)$ is true:

To prove $P(1)$, we must show that when 1 is substituted into the equation in place of n, the left-hand side equals the right-hand side. But when 1 is substituted for n, the left-hand side is the sum of all the even integers from 2 to $2 \cdot 1$, which is just 2, and the right-hand side is $1^2 + 1$, which also equals 2. Thus $P(1)$ is true.

Show that for every integer $k \geq 1$, if $P(k)$ is true then $P(k+1)$ is true:

Let k be any integer with $k \geq 1$, and suppose $P(k)$ is true. That is, suppose

$$2 + 4 + 6 + \cdots + 2k = k^2 + k. \qquad \leftarrow \ \begin{array}{l} P(k) \\ \text{inductive hypothesis} \end{array}$$

We must show that $P(k+1)$ is true. That is, we must show that

$$2 + 4 + 6 + \ldots + 2(k+1) = (k+1)^2 + (k+1).$$

Because $(k+1)^2 + (k+1) = k^2 + 2k + 1 + k + 1 = k^2 + 3k + 2$, this is equivalent to showing that

$$2 + 4 + 6 + \cdots + 2(k+1) = k^2 + 3k + 2. \quad \leftarrow \ P(k+1)$$

Now the left-hand side of $P(k+1)$ is

$$
\begin{array}{lll}
2 + 4 + 6 + \cdots + 2(k+1) & = & 2 + 4 + 6 + \cdots + 2k + 2(k+1) \quad \text{by making the next-to-last} \\
& & \hspace{5.5cm} \text{term explicit} \\
& = & (k^2 + k) + 2(k+1) \hspace{1.8cm} \text{by substitution from the} \\
& & \hspace{5.5cm} \text{inductive hypothesis} \\
& = & k^2 + 3k + 2 \hspace{3cm} \text{by algebra,}
\end{array}
$$

and this is the right-hand side of $P(k+1)$. Hence $P(k+1)$ is true.

[Since both the basis step and the inductive step have been proved, $P(n)$ is true for every integer $n \geq 1$.]

8. Proof: For the given statement, the property $P(n)$ is the equation

$$1 + 2 + 2^2 + \cdots + 2^n = 2^{n+1} - 1. \qquad \leftarrow P(n)$$

Show that $P(0)$ is true:

The left-hand side of $P(0)$ is 1, and the right-hand side is $2^{0+1} - 1 = 2 - 1 = 1$ also. Thus $P(0)$ is true.

Show that for every integer $k \geq 0$, if $P(k)$ is true then $P(k+1)$ is true:

Let k be any integer with $k \geq 0$, and suppose $P(k)$ is true. That is, suppose

$$1 + 2 + 2^2 + \cdots + 2^k = 2^{k+1} - 1. \qquad \leftarrow \begin{array}{l} P(k) \\ \text{inductive hypothesis} \end{array}$$

We must show that $P(k+1)$ is true. That is, we must show that

$$1 + 2 + 2^2 + \cdots + 2^{k+1} = 2^{(k+1)+1} - 1.$$

or, equivalently,

$$1 + 2 + 2^2 + \cdots + 2^{k+1} = 2^{k+2} - 1. \qquad \leftarrow P(k+1)$$

Now the left-hand side of $P(k+1)$ is

$$
\begin{aligned}
1 + 2 + 2^2 + \cdots + 2^{k+1} &= 1 + 2 + 2^2 + \cdots + 2^k + 2^{k+1} && \text{by making the next-to-last} \\
&&& \text{term explicit} \\
&= (2^{k+1} - 1) + 2^{k+1} && \text{by substitution from the} \\
&&& \text{inductive hypothesis} \\
&= 2 \cdot 2^{k+1} - 1 && \text{by combining like terms} \\
&= 2^{k+2} - 1 && \text{by the laws of exponents,}
\end{aligned}
$$

and this is the right-hand side of $P(k+1)$. Hence the property is true for $n = k+1$.

[Since both the basis step and the inductive step have been proved, $P(n)$ is true for every integer $n \geq 0$.]

9. Proof (by mathematical induction): Let the property $P(n)$ be the equation

$$4^3 + 4^4 + 4^5 + \cdots + 4^n = \frac{4(4^n - 16)}{3}. \qquad \leftarrow P(n)$$

We will show that $P(n)$ is true for every integer $n \geq 3$.

Show that $P(3)$ is true: $P(3)$ is true because its left-hand side is $4^3 = 64$ and its right-hand side is $\frac{4(4^3 - 16)}{3} = \frac{4(64 - 16)}{3} = \frac{4 \cdot 48}{3} = 64$ also.

Show that for every integer $k \geq 3$, if $P(k)$ is true then $P(k+1)$ is true: Let k be any integer with $k \geq 3$, and suppose that

$$4^3 + 4^4 + 4^5 + \cdots + 4^k = \frac{4(4^k - 16)}{3}. \qquad \leftarrow \begin{array}{l} P(k) \\ \text{inductive hypothesis} \end{array}$$

We must show that

$$4^3 + 4^4 + 4^5 + \cdots + 4^{k+1} = \frac{4(4^{k+1} - 16)}{3}. \qquad \leftarrow P(k+1)$$

Now the left-hand side of $P(k+1)$ is

$$
\begin{aligned}
4^3 + 4^4 + 4^5 + \cdots + 4^{k+1} \;=\;& 4^3 + 4^4 + 4^5 + \cdots + 4^k + 4^{k+1} \\
& \quad\text{by making the next-to-last term explicit} \\
=\;& \frac{4(4^k - 16)}{3} + 4^{k+1} \\
& \quad\text{by inductive hypothesis} \\
=\;& \frac{4^{k+1} - 64}{3} + \frac{3 \cdot 4^{k+1}}{3} \\
& \quad\text{by creating a common denominator} \\
=\;& \frac{4 \cdot 4^{k+1} - 64}{3} \\
& \quad\text{by adding the fractions} \\
=\;& \frac{4(4^{k+1} - 16)}{3} \\
& \quad\text{by factoring out the 4,}
\end{aligned}
$$

and this is the right-hand side of $P(k+1)$ *[as was to be shown]*.

[Since both the basis and the inductive steps have been proved, we conclude that $P(n)$ is true for every integer $n \geq 3$.]

10. <u>Proof</u>: For the given statement, the property is the equation

$$
1^2 + 2^2 + 3^2 + \cdots n^2 = \frac{n(n+1)(2n+1)}{6}. \qquad \leftarrow\ P(n)
$$

Show that $P(1)$ is true:

The left-hand side of $P(1)$ is $1^2 = 1$, and the right-hand side is $\frac{1(1+1)(2\cdot1+1)}{6} = \frac{2\cdot3}{6} = 1$ also. Thus $P(1)$ is true.

Show that for every integer $k \geq 1$, if $P(k)$ is true then $P(k+1)$:

Let k be any integer with $k \geq 1$, and suppose $P(k)$ is true. That is, suppose

$$
1^2 + 2^2 + 3^2 + \cdots + k^2 = \frac{k(k+1)(2k+1)}{6}. \qquad \leftarrow\ \begin{array}{l} P(k) \\ \text{inductive hypothesis} \end{array}
$$

We must show that $P(k+1)$ is true. That is, we must show that

$$
1^2 + 2^2 + 3^2 + \cdots + (k+1)^2 = \frac{(k+1)((k+1)+1)(2(k+1)+1)}{6},
$$

or, equivalently,

$$
1^2 + 2^2 + 3^2 + \cdots + (k+1)^2 = \frac{(k+1)(k+2)(k+3)}{6}. \qquad \leftarrow\ P(k+1)
$$

Now the left-hand side of $P(k + 1)$ is

$$1^2 + 2^2 + 3^2 + \cdots + (k+1)^2 \quad = \quad 1^2 + 2^2 + 3^2 + \cdots + k^2 + (k+1)^2$$

by making the next-to-last term explicit

$$= \quad \frac{k(k+1)(2k+1)}{6} + (k+1)^2$$

by inductive hypothesis

$$= \quad \frac{k(k+1)(2k+1)}{6} + \frac{6(k+1)^2}{6}$$

by creating a common denominator

$$= \quad \frac{k(k+1)(2k+1) + 6(k+1)^2}{6}$$

by adding the fractions

$$= \quad \frac{(k+1)[k(2k+1) + 6(k+1)]}{6}$$

by factoring out $(k+1)$

$$\frac{(k+1)(2k^2 + 7k + 6)}{6}$$

by multiplying out and combining like terms

$$\frac{(k+1)(k+2)(2k+3)}{6}$$

because $(k+2)(2k+3) = 2k^2 + 7k + 6,$

and this is the right-hand side of $P(k+1)$. Hence the property is true for $n = k + 1$.

[Since both the basis step and the inductive step have been proved, $P(n)$ is true for every integer $n \geq 1$].

12. <u>Proof (by mathematical induction)</u>: Let the property $P(n)$ be the equation

$$\frac{1}{1 \cdot 2} + \frac{1}{2 \cdot 3} + \cdots + \frac{1}{n(n+1)} = \frac{n}{n+1}. \qquad \leftarrow P(n)$$

We will show that $P(n)$ is true for every integer $n \geq 1$.

Show that $P(1)$ is true: $P(1)$ is true because the left-hand side equals $\dfrac{1}{1 \cdot 2} = \dfrac{1}{2}$ and the right-hand side equals $\dfrac{1}{1+1} = \dfrac{1}{2}$ also.

Show that for every integer $k \geq 1$, if $P(k)$ is true then $P(k+1)$ is true: Let k be any integer with $k \geq 1$ and suppose that

$$\frac{1}{1 \cdot 2} + \frac{1}{2 \cdot 3} + \cdots + \frac{1}{k(k+1)} = \frac{k}{k+1}. \qquad \leftarrow \begin{array}{l} P(k) \\ \text{inductive hypothesis} \end{array}$$

We must show that

$$\frac{1}{1 \cdot 2} + \frac{1}{2 \cdot 3} + \cdots + \frac{1}{(k+1)((k+1)+1)} = \frac{k+1}{(k+1)+1},$$

or, equivalently,

$$\frac{1}{1 \cdot 2} + \frac{1}{2 \cdot 3} + \cdots + \frac{1}{(k+1)(k+2)} = \frac{k+1}{k+2}. \qquad \leftarrow P(k+1)$$

The left-hand side of $P(k+1)$ is

$$\frac{1}{1 \cdot 2} + \frac{1}{2 \cdot 3} + \cdots + \frac{1}{(k+1)(k+2)}$$

$$= \frac{1}{1 \cdot 2} + \frac{1}{2 \cdot 3} + \cdots + \frac{1}{k(k+1)} + \frac{1}{(k+1)(k+2)}$$

by making the next-to-last term explicit

$$= \frac{k}{k+1} + \frac{1}{(k+1)(k+2)}$$

by inductive hypothesis

$$= \frac{k(k+2)}{(k+1)(k+2)} + \frac{1}{(k+1)(k+2)}$$

by creating a common denominator

$$= \frac{k^2 + 2k + 1}{(k+1)(k+2)}$$

by adding the fractions

$$= \frac{(k+1)^2}{(k+1)(k+2)}$$

because $k^2 + 2k + 1 = (k+1)^2$

$$= \frac{k+1}{k+2}$$

by canceling $(k+1)$ from numerator and denominator,

and this is the right-hand side of $P(k+1)$ *[as was to be shown]*.

13. <u>Proof</u>: For the given statement, the property $P(n)$ is the equation

$$\sum_{i=1}^{n=1} i(i+1) = \frac{n(n-1)(n+1)}{3}. \qquad \leftarrow P(n)$$

Show that $P(2)$ is true:

The left-hand side of $P(2)$ is $\sum_{i=1}^{1} i(i+1) = 1 \cdot (1+1) = 2$, and the right-hand side is

$$\frac{2(2-1)(2+1)}{3} = \frac{6}{3} = 2 \quad \text{also.}$$

Thus $P(2)$ is true.

Show that for every integer $k \geq 2$, if $P(k)$ is true then $P(k+1)$ is true:

Let k be any integer with $k \geq 2$, and suppose $P(k)$ is true. That is, suppose

$$\sum_{i=1}^{k=1} i(i+1) = \frac{k(k-1)(k+1)}{3} \qquad \leftarrow \begin{array}{l} P(k) \\ \text{inductive hypothesis} \end{array}$$

We must show that $P(k+1)$ is true. That is, we must show that

$$\sum_{i=1}^{(k+1)-1} i(i+1) = \frac{(k+1)((k+1)-1)((k+1)+1)}{3},$$

or, equivalently,

$$\sum_{i=1}^{k} i(i+1) = \frac{(k+1)k(k+2)}{3}. \qquad \leftarrow P(k+1)$$

Now the left-hand side of $P(k+1)$ is

$$\sum_{i=1}^{i} i(i+1) = \sum_{i=1}^{k-1} i(i+1) + k(k+1) \qquad \text{by writing the last term separately}$$

$$= \frac{k(k-1)(k+1)}{3} + k(k+1) \qquad \text{by substitution from the inductive hypothesis}$$

$$= \frac{k(k-1)(k+1)}{3} + \frac{3k(k+1)}{3} \qquad \text{because } \frac{3}{3} = 1$$

$$= \frac{k(k-1)(k+1) + 3k(k+1)}{3} \qquad \text{by adding the fractions}$$

$$= \frac{k(k+1)[(k-1)+3]}{3} \qquad \text{by factoring out } k(k+1)$$

$$= \frac{k(k+1)(k+2)}{3} \qquad \text{by algebra,}$$

and this is the right-hand side of $P(k+1)$. Hence $P(k+1)$ is true.

[Since both the basis step and the inductive step have been proved, $P(n)$ is true for every integer $n \ge 2$.].ide of $P(k+1)$ [as was to be shown].

15. <u>Proof (by mathematical induction):</u> Let the property $P(n)$ be the equation

$$\sum_{i=1}^{n} i(i!) = (n+1)! - 1. \qquad \leftarrow \; P(n)$$

We will show that $P(n)$ is true for every integer $n \ge 1$.

Show that $P(1)$ is true: We must show that $\sum_{i=1}^{1} i(i!) = (1+1)! - 1$. Now the left-hand side of this equation is $\sum_{i=1}^{1} i(i!) = 1 \cdot (1!) = 1$ and the right-hand side is $(1+1)! - 1 = 2! - 1 = 2 - 1 = 1$ also. So $P(1)$ is true.

Show that for every integer $k \ge 1$, if $P(k)$ is true then $P(k+1)$ is true: Let k be any integer with $k \ge 1$, and suppose that

$$\sum_{i=1}^{k} i(i!) = (k+1)! - 1. \qquad \leftarrow \begin{array}{l} P(k) \\ \text{inductive hypothesis} \end{array}$$

We must show that

$$\sum_{i=1}^{k+1} i(i!) = ((k+1)+1)! - 1,$$

or, equivalently,

$$\sum_{i=1}^{k+1} i(i!) = (k+2)! - 1. \qquad \leftarrow \; P(k+1)$$

The left-hand side of $P(k+1)$ is

$$\sum_{i=1}^{k+1} i(i!) = \sum_{i=1}^{k} i(i!) + (k+1)((k+1)!) \qquad \text{by writing the } (k+1)\text{st term separately}$$

$$= [(k+1)! - 1] + (k+1)((k+1)!) \qquad \text{by inductive hypothesis}$$

$$= ((k+1)!)(1 + (k+1)) - 1 \qquad \text{by combining the terms with the common factor } (k+1)!$$

$$= (k+1)!(k+2) - 1$$

$$= (k+2)! - 1 \qquad \text{by algebra,}$$

and this is the right-hand side of $P(k+1)$ *[as was to be shown].*

18. Proof (by mathematical induction): Let the property $P(n)$ be the equation

$$\prod_{i=2}^{n}\left(1-\frac{1}{i}\right)=\frac{1}{n}. \qquad \leftarrow P(n)$$

We will show that $P(n)$ is true for every integer $n \geq 2$.

Show that $P(2)$ is true: $P(2)$ is true because the left-hand side equals

$$\prod_{i=2}^{2}\left(1-\frac{1}{i}\right)=1-\frac{1}{2}=\frac{1}{2},$$

which equals the right-hand side of $P(n)$ when $n = 2$.

Show that for every integer $k \geq 2$, if $P(k)$ is true then $P(k+1)$ is true: Let k be any integer with $k \geq 2$ and suppose that

$$\prod_{i=2}^{k}\left(1-\frac{1}{i}\right)=\frac{1}{k}. \qquad \leftarrow \begin{array}{l}P(k)\\ \text{inductive hypothesis}\end{array}$$

We must show that

$$\prod_{i=2}^{k+1}\left(1-\frac{1}{i}\right)=\frac{1}{k+1}. \qquad \leftarrow P(k+1)$$

The left-hand side of $P(k+1)$ is

$$
\begin{aligned}
\prod_{i=2}^{k+1}\left(1-\frac{1}{i}\right) &= \prod_{i=2}^{k}\left(1-\frac{1}{i}\right)\left(1-\frac{1}{k+1}\right) && \text{by separating off the final term}\\
&= \frac{1}{k}\left(1-\frac{1}{i}\right) && \text{by inductive hypothesis}\\
&= \frac{1}{k}\left(\frac{k+1}{k+1}-\frac{1}{k+1}\right) && \text{by creating a common denominator}\\
&= \frac{1}{k}\cdot\frac{k}{k+1} && \text{by subtracting fractions}\\
&= \frac{1}{k+1} && \text{by multiplying out,}
\end{aligned}
$$

and this is the right-hand side of $P(k+1)$ *[as was to be shown]*.

20. $4 + 8 + 12 + 16 + \cdots + 200 = 4(1 + 2 + 3 + \cdots + 50) = 4\left(\dfrac{50 \cdot 51}{2}\right) = 5{,}100.$

21. $5 + 10 + 15 + 20 + \cdots + 300 = 5(1 + 2 + 3 + \cdots + 60) = 5\left(\dfrac{60 \cdot 61}{2}\right) = 9{,}150.$

22. **a.** $3 + 4 + 5 + 6 + \cdots + 1000 = (1 + 2 + 3 + 4 + \cdots + 1000) - (1 + 2)$

$= \left(\dfrac{1000 \cdot 1001}{2}\right) - 3 = 500{,}497.$

b. $3 + 4 + 5 + 6 + \cdots + m = (1 + 2 + 3 + 4 + \cdots + m) - (1 + 2)$

$= \dfrac{m(m+1)}{2} - 3 = \dfrac{m^2 + m}{2} - \dfrac{6}{2} = \dfrac{m^2 + m - 6}{2}$

24. $1 + 2 + 3 + \cdots + (k-1) = \dfrac{(k-1)((k-1)+1)}{2} = \dfrac{k(k-1)}{2}$

25. **a.** $\dfrac{2^{26} - 1}{2 - 1} = 2^{26} - 1 = 67{,}108{,}863$

b.

$$2 + 2^2 + 2^3 + \cdots + 2^{26} \quad = \quad 2(1 + 2 + 2^2 + \cdots + 2^{25})$$

$$= \quad 2 \cdot (67{,}108{,}863) \qquad \text{by part (a)}$$

$$= \quad 134{,}217{,}726$$

c. *Solution 1:*

$$2 + 2^2 + 2^3 + \cdots + 2^n \quad = \quad 2(1 + 2 + 2^2 + 2^3 + \cdots + 2^{n-1})$$

$$= \quad 2\left(\frac{2^{(n-1)+1} - 1}{2 - 1}\right)$$

$$= \quad \left(\frac{2^n - 1}{2 - 1}\right) - 2$$

$$= \quad 2^{n+1} - 2$$

Solution 2:

$$2 + 2^2 + 2^3 + \cdots + 2^n \quad = \quad 2(1 + 2 + 2^2 + 2^3 + \cdots + 2^n) - 1$$

$$= \quad \frac{2^{n+1} - 1}{2 - 1} - 1$$

$$= \quad 2^{n+1} - 2$$

27. *Solution 1:*

$$5^3 + 5^4 + 5^5 + \cdots + 5^k \quad = \quad 5^3(1 + 5 + 5^2 + \cdots + 5^{k-3})$$

$$= \quad 5^3\left(\frac{5^{(k-3)+1} - 1}{5 - 1}\right) = \frac{5^3(5^{k-2} - 1)}{4}$$

Solution 2:

$$5^3 + 5^4 + 5^5 + \cdots + 5^k \quad = \quad 1 + 5 + 5^2 + \cdots + 5^k - (1 + 5 + 5^2)$$

$$= \quad \frac{5^{k+1} - 1}{5 - 1} - 31 = \frac{5^{k+1} - 1}{4} - 31$$

Note that the expression obtained in solution 2 can be transformed into the one obtained in solution 1:

$$\frac{5^{k+1} - 1}{4} - 31 = \frac{5^{k+1} - 1}{4} - \frac{31 \cdot 4}{4} = \frac{5^{k+1} - 125}{4} = \frac{5^{k+1} - 5^3}{4} = \frac{5^3(5^{k-2} - 1)}{4}.$$

28.

$$\frac{\left(\dfrac{1}{2}\right)^{n+1} - 1}{\dfrac{1}{2} - 1} \quad = \quad \frac{\dfrac{1}{2^{n+1}} - 1}{-\dfrac{1}{2}}$$

$$= \quad \left(\frac{1}{2^{n+1}} - 1\right)(-2)$$

$$= \quad -\frac{2}{2^{n+1}} + 2$$

$$= \quad 2 - \frac{1}{2^n}$$

30. *General formula:* For every integer $n \geq 1$, $\dfrac{1}{1 \cdot 3} + \dfrac{1}{3 \cdot 5} + \cdots + \dfrac{1}{(2n-1)(2n+1)} = \dfrac{n}{2n+1}.$

Proof (by mathematical induction): Let the property $P(n)$ be the equation

$$\frac{1}{1 \cdot 3} + \frac{1}{3 \cdot 5} + \cdots + \frac{1}{(2n-1)(2n+1)} = \frac{n}{2n+1}. \qquad \leftarrow P(n)$$

Show that $P(1)$ is true:

The left-hand side of $P(1)$ equals $\frac{1}{1 \cdot 3}$, and the right-hand side equals $\frac{1}{2 \cdot 1 + 1}$. Since both sides equal $\frac{1}{3}$, $P(1)$ is true.

Show that for each integer $k \geq 1$, if $P(k)$ is true then $P(k+1)$ is true:

Suppose that k is any integer with $k \geq 1$, and

$$\frac{1}{1 \cdot 3} + \frac{1}{3 \cdot 5} + \cdots + \frac{1}{(2k-1)(2k+1)} = \frac{k}{2k+1}. \qquad \begin{array}{l} \leftarrow \quad P(k) \\ \quad \text{inductive hypothesis} \end{array}$$

We must show that

$$\frac{1}{1 \cdot 3} + \frac{1}{3 \cdot 5} + \cdots + \frac{1}{(2(k+1)-1)(2(k+1)+1)} = \frac{k+1}{2(k+1)+1},$$

or, equivalently,

$$\frac{1}{1 \cdot 3} + \frac{1}{3 \cdot 5} + \cdots + \frac{1}{(2k+1)(2k+3)} = \frac{k+1}{2k+3}. \qquad \leftarrow P(k+1)$$

Now the left-hand side of $P(k+1)$ is

$$\frac{1}{1 \cdot 3} + \frac{1}{3 \cdot 5} + \cdots + \frac{1}{(2k+1)(2k+3)}$$

$$= \frac{1}{1 \cdot 3} + \frac{1}{3 \cdot 5} + \cdots + \frac{1}{(2k+1)(2k+1)} + \frac{1}{(2k+1)(2k+3)}$$

$$= \frac{k}{2k+1} + \frac{1}{(2k+1)(2k+3)} \qquad \text{by inductive hypothesis}$$

$$= \frac{k(2k+3)}{(2k+1)(2k+3)} + \frac{1}{(2k+1)(2k+3)}$$

$$= \frac{2k^2 + 3k + 1}{(2k+1)(2k+3)}$$

$$= \frac{(2k+1)(k+1)}{(2k+1)(2k+3)}$$

$$= \frac{k+1}{(2k+3)} \qquad \text{by algebra.}$$

and this is the right-hand side of $P(k+1)$ *[as was to be shown]*.

32. *Hint 1:* The general formula is

$$\begin{aligned} & 1 - 4 + 9 - 16 + \cdots + (-1)^{n-1} n^2 \\ & = (-1)^{n-1}(1 + 2 + 3 + \cdots + n) \qquad \text{in expanded form} \\ & Or: \sum_{i=1}^{n} (-1)^{i-1} i^2 = (-1)^{n-1} \left(\sum_{i=1}^{n} i \right) \qquad \text{in summation notation.} \end{aligned}$$

Hint 2: In the proof, use the fact that

$$1 + 2 + 3 + \cdots + n = \sum_{j=1}^{n} i = \frac{n(n+1)}{2}.$$

33. $(a + md) + (a + (m + 1)d) + (a + (m + 2)d) + \cdots + (a + (m + n)d)$

$$= (a + md) + (a + md + d) + (a + md + 2d) + \cdots + (a + md + nd)$$

$$= \underbrace{((a + md) + (a + md) + \cdots + (a + md))}_{n + 1 \text{ terms}} + d(1 + 2 + 3 + \cdots + n)$$

$$= (n + 1)(a + md) + d\left(\frac{n(n + 1)}{2}\right) \qquad \text{by Theorem 5.2.1}$$

$$= (a + md + \frac{n}{2}d)(n + 1)$$

$$= [a + (m + \frac{n}{2})d](n + 1)$$

Any one of the last three equations or their algebraic equivalents could be considered a correct answer.

36. In the inductive step, both the inductive hypothesis and what is to be shown are wrong. The inductive hypothesis should be

Suppose that k is any integer with $k \geq 1$ such that

$$1^2 + 2^2 + \cdots + k^2 = \frac{k(k + 1)(2k + 1)}{6}.$$

And what is to be shown should be

$$1^2 + 2^2 + \cdots + (k + 1)^2 = \frac{(k + 1)((k + 1) + 1)(2(k + 1) + 1)}{6}.$$

37. *Hint:* See the Caution note in Section 5.1, page 262.

38. *Hint:* See the subsection Proving an Equality on page 284 in Section 5.2.

39. Proof: Suppose m and n are any positive integers such that m is odd. By definition of odd, $m = 2q + 1$ for some integer k, and so, by Theorems 5.1.1 and 5.2.2,

$$\sum_{k=0}^{m-1} (n + k) = \sum_{k=0}^{(2q+1)-1} (n + k) = \sum_{k=0}^{2q} (n + k) = \sum_{k=0}^{2q} n + \sum_{k=0}^{2q} k = (2q + 1)n + \sum_{k=1}^{2q} k$$

$$= (2q + 1)n + \frac{2q(2q + 1)}{2} = (2q + 1)n + q(2q + 1) = (2q + 1)(n + q) = m(n + q).$$

Now $n + q$ is an integer because it is a sum of integers. Hence, by definition of divisibility, $\sum_{k=0}^{m-1} (n + k)$ is divisible by m.

Note: If m is even, the property is no longer true. For example, if $n = 1$ and $m = 2$, then $\sum_{k=0}^{m-1} (n + k) = \sum_{k=0}^{2-1} (1 + k) = 1 + 2 = 3$, and 3 is not divisible by 2.

40. *Hint:* Form the sum $n^2 + (n + 1)^2 + (n + 2)^2 + \cdots + (n + (p - 1))^2$, and show that it equals

$$pn^2 + 2n(1 + 2 + 3 + \cdots + (p - 1)) + (1 + 4 + 9 + 16 + \cdots + (p - 1)^2).$$

Section 5.3

1. Proof (by mathematical induction): Let the property $P(n)$ be the sentence

n cents can be obtained by using 3-cent and 8-cent coins.

We will show that $P(n)$ is true for every integer $n \geq 14$.

Show that $P(14)$ is true: $P(14)$ is true because fourteen cents can be obtained by using two 3-cent coins and one 8-cent coin.

Show that for every integer $k \geq 14$, if $P(k)$ is true then $P(k+1)$ is also true:

Suppose k is any integer with $k \geq 14$ such that

k cents can be obtained using 3-cent and 8-cent coins. $\leftarrow$ $\begin{array}{c} P(k) \\ \text{inductive hypothesis} \end{array}$

We must show that

$k+1$ cents can be obtained using 3-cent and 8-cent coins. $\leftarrow P(k+1)$

If the k cents includes an 8-cent coin, replace it by three 3-cent coins to obtain a total of $k+1$ cents. Otherwise the k cents consists of 3-cent coins exclusively, and so there must be least five 3-cent coins (since the total amount is at least 14 cents). In this case, replace five of the 3-cent coins by two 8-cent coin to obtain a total of $k+1$ cents. Thus, in either case, $k+1$ cents can be obtained using 3-cent and 8-cent coins. *[This is what we needed to show.]*

[Since we have proved the basis step and the inductive step, we conclude that the given statement is true for every integer $n \geq 14$.]

3. **a.** $5 = 1 \cdot 5$, $8 = 1 \cdot 8$, $10 = 2 \cdot 5$, $13 = 5 + 8$, $15 = 3 \cdot 5$, $16 = 2 \cdot 8$, $20 = 4 \cdot 5$, $21 = 2 \cdot 8 + 1 \cdot 5$, $24 = 3 \cdot 8$, $25 = 5 \cdot 5$

b. <u>Proof (by mathematical induction):</u> Let the property $P(n)$ be the sentence

n stamps can be obtained by buying a collection
of 5-stamp packages and 8-stamp packages. $\leftarrow P(n)$

We will show that $P(n)$ is true for every integer $n \geq 28$.

Show that $P(28)$ is true: $P(28)$ is true because twenty-eight stamps can be obtained by buying four 5-stamp packages and one 8-stamp package.

Show that for any integer $k \geq 28$, if $P(k)$ is true then $P(k+1)$ is true:

Let k be any integer with $k \geq 28$, and suppose that

k stamps can be obtained by buying a collection
of 5-stamp packages and 8-stamp packages. $\leftarrow$ $\begin{array}{c} P(k) \\ \text{inductive hypothesis} \end{array}$

We must show that

$k+1$ stamps can be obtained by buying a collection
of 5-stamp packages and 8-stamp packages. $\leftarrow P(k+1)$

There are two possible cases: either the collection of k stamps includes three 5-stamp packages or it does not.

Case 1 (the collection of k stamps includes three 5-stamp packages): In this case replace three 5-stamp packages by two 8-stamp packages to make a collection of $k+1$ stamps.

Case 2 (the collection of k stamps does not include three 5-stamp packages): In this case the collection includes at most two 5-stamp packages. Since there are at least 28 stamps and at most 10 come from 5-stamp packages, at least 18 stamps come from 8-stamp packages. And because the smallest multiple of 8 that is at least 18 is 24, at least three 8-stamp packages must be in the collection. Replace three 8-stamp packages by five 5-stamp packages to make a collection of $k+1$ stamps.

Thus in either of the two possible cases a collection of $k+1$ stamps can be obtained by buying a collection of 5-stamp packages and 8-stamp packages *[as was to be shown]*.

[Since both the basis and the inductive steps have been proved, we conclude that $P(n)$ is true for every integer $n \geq 28$.]

4. **a.** $P(0)$ is the sentence "$5^0 - 1$ is divisible by 4." $P(0)$ is true because $5^0 - 1 = 0$, which is divisible by 4.

b. $P(k)$ is the sentence $5^k - 1$ is divisible by 4."

c. $P(k+1)$ is the sentence $5^k - 1$ is divisible by 4."

d. *Must show:* If k is any integer such that $k \geq 0$ and $5^k - 1$ is divisible by 4, then $5^{k+1} - 1$ is divisible by 4.

6. For each positive integer n, let $P(n)$ be the sentence

   Any checkerboard with dimensions $2 \times 3n$ can be completely covered with L-shaped trominoes.

   a. $P(1)$ is the sentence "Any checkerboard with dimensions 2×3 can be completely covered with L-shaped trominoes." The following diagram shows that $P(1)$ is true:

   b. $P(k)$ is the sentence "Any checkerboard with dimensions $2 \times 3k$ can be completely covered with L-shaped trominoes."

   c. $P(k+1)$ is the sentence "Any checkerboard with dimensions $2 \times 3(k+1)$ can be completely covered with L-shaped trominoes."

   d. The inductive step requires showing that for every integer $k \geq 1$, if any checkerboard with dimensions $2 \times 3k$ can be completely covered with L-shaped trominoes, then any checkerboard with dimensions $2 \times 3(k+1)$ can be completely covered with L-shaped trominoes.

8. <u>Proof (by mathematical induction)</u>: For the given statement, the property is the sentence

$$5^n - 1 \text{ is divisible by 4.} \qquad \leftarrow \ P(n)$$

Show that $P(0)$ is true:

$P(0)$ is the sentence "$5^0 - 1$ is divisible by 4." Now $5^0 - 1 = 1 - 1 = 0$, and 0 is divisible by 4 because $0 = 4 \cdot 0$. Thus $P(0)$ is true.

Show that for every integer $k \geq 0$, if $P(k)$ is true then $P(k+1)$ is true:

Let k be any integer with $k \geq 0$, and suppose $P(k)$ is true. That is, suppose

$$5^k - 1 \text{ is divisible by 4.} \qquad \leftarrow \quad \begin{array}{l} P(k) \\ \text{inductive hypothesis} \end{array}$$

We must show that $P(k+1)$ is true. That is, we must show that

$$5^{k+1} - 1 \text{ is divisible by 4.} \qquad \leftarrow \ P(k+1)$$

Now

$$5^{k+1} - 1 = 5^k \cdot 5 - 1 = 5^k \cdot (4+1) - 1 = 5^k \cdot 4 + (5^k - 1). \ (*)$$

By the inductive hypothesis, $5^k - 1$ is divisible by 4, and so $5^k - 1 = 4r$ for some integer r. Substitute $4r$ in place of $5^k - 1$ in equation (*), to obtain

$$5^{k+1} - 1 = 5^k \cdot 4 + 4r = 4(5^k + r).$$

But $5^k + r$ is an integer because k and r are integers. Hence, by definition of divisibility, $5^{k+1} - 1$ is divisible by 4 *[as was to be shown]*.

An alternative proof of the inductive step goes as follows:

Let k be any integer with $k \geq 0$, and suppose that $5^k - 1$ is divisible by 4. Then $5^k - 1 = 4r$ for some integer r, and hence $5^k = 4r + 1$.

It follows by substitution that

$$5^{k+1} = 5^k \cdot 5 = (4r + 1) \cdot 5 = 20r + 5.$$

Subtracting 1 from both sides gives that

$$5^{k+1} - 1 = 20r + 4 = 4(5r + 1).$$

Now since $5r + 1$ is an integer, by definition of divisibility, $5^{k+1} - 1$ is divisible by 4.

9. <u>Proof (by mathematical induction)</u>: Let the property $P(n)$ be the sentence

$$7^n - 1 \text{ is divisible by 6.} \qquad \leftarrow \quad P(n)$$

We will prove that $P(n)$ is true for every integer $n \geq 0$.

Show that $P(0)$ is true: $P(0)$ is true because because $7^0 - 1 = 1 - 1 = 0$ and 0 is divisible by 6 (since $0 = 0 \cdot 6$).

Show that for every integer $k \geq 0$, if $P(k)$ is true then $P(k+1)$ is true:

Let k be any integer with $k \geq 0$, and suppose

$$7^k - 1 \text{ is divisible by 6.} \qquad \leftarrow \quad \begin{array}{l} P(k) \\ \text{inductive hypothesis} \end{array}$$

We must show that

$$7^{k+1} - 1 \text{ is divisible by 6.} \qquad \leftarrow \quad P(k+1)$$

By definition of divisibility, the inductive hypothesis is equivalent to the statement

$$7^k - 1 = 6r$$

for some integer r. Then

$$\begin{aligned} 7^{k+1} - 1 &= 7 \cdot 7^k - 1 \\ &= (6+1)7^k - 1 \\ &= 6 \cdot 7^k + (7^k - 1) \quad \text{by algebra} \\ &= 6 \cdot 7^k + 6r \quad \text{by inductive hypothesis} \\ &= 6(7^k + r) \quad \text{by algebra.} \end{aligned}$$

Now $7^k + r$ is an integer because products and sums of integers are integers. Thus, by definition of divisibility, $7^{k+1} - 1$ is divisible by 6 *[as was to be shown]*.

11. <u>Proof (by mathematical induction)</u>: For the given statement, the property $P(n)$ is the sentence

$$3^{2n} - 1 \text{ is divisible by 8.} \qquad \leftarrow \quad P(n)$$

Show that $P(0)$ is true:

$P(0)$ is the sentence "$3^{2 \cdot 0} - 1$ is divisible by 8." Observe that $3^{2 \cdot 0} - 1 = 1 - 1 = 0$, and 0 is divisible by 8 because $0 = 8 \cdot 0$. Thus $P(0)$ is true.

Show that for every integer $k \geq 0$, if $P(k)$ is true then $P(k + 1)$ is true:

Let k be any integer with $k \geq 0$, and suppose $P(k)$ is true. That is, suppose

$$3^{2k} - 1 \text{ is divisible by 8.} \qquad \leftarrow \quad \begin{array}{c} P(k) \\ \text{inductive hypothesis} \end{array}$$

We must show that $P(k + 1)$ is true. That is, we must show that

$$3^{2(k+1)} - 1 \text{ is divisible by 8.}$$

or, equivalently, that

$$3^{2k+2} - 1 \text{ is divisible by 8.} \qquad \leftarrow \quad P(k+1)$$

Now

$$
\begin{aligned}
3^{2k+2} - 1 &= 3^{2k} \cdot 3^2 - 1 \\
&= 3^{2k} \cdot 9 - 1 \\
&= 3^{2k} \cdot (8 + 1) - 1 \\
&= 3^{2k} \cdot 8 + (3^{2k} - 1). \ (*) \quad \text{by algebra}
\end{aligned}
$$

By the inductive hypothesis $3^{2k} - 1$ is divisible by 8, and so $3^{2k} - 1 = 8r$ for some integer r. Thus substitution into equation $(*)$ gives

$$3^{2k+2} - 1 = 3^{2k} \cdot 8 + 8r = 8(3^{2k} + r).$$

Now $3^{2k} + r$ is an integer because k and r are integers, and hence, by definition of divisibility, $3^{2k+2} - 1$ is divisible by 8 *[as was to be shown]*.

12. <u>Proof (by mathematical induction)</u>: Let the property $P(n)$ be the sentence

$$7^n - 2^n \text{ is divisible by 5.} \qquad \leftarrow \quad P(n)$$

We will prove that $P(n)$ is true for every integer $n \geq 0$.

Show that $P(0)$ is true: $P(0)$ is true because $7^0 - 2^0 = 0 - 0 = 0$ and 0 is divisible by 5 (since $0 = 5 \cdot 0$).

Show that for every integer $k \geq 0$, if $P(k)$ is true then $P(k + 1)$ is true:

Let k be any integer with $k \geq 0$, and suppose

$$7^k - 2^k \text{ is divisible by 5.} \qquad \leftarrow \quad \begin{array}{c} P(k) \\ \text{inductive hypothesis} \end{array}$$

We must show that

$$7^{k+1} - 2^{k+1} \text{ is divisible by 5.} \qquad \leftarrow \quad P(k+1)$$

By definition of divisibility, the inductive hypothesis is equivalent to the statement $7^k - 2^k = 5r$ for some integer r. Then

$$
\begin{aligned}
7^{k+1} - 2^{k+1} &= 7 \cdot 7^k - 2 \cdot 2^k \\
&= (5 + 2) \cdot 7^k - 2 \cdot 2^k \\
&= 5 \cdot 7^k + 2 \cdot 7^k - 2 \cdot 2^k \\
&= 5 \cdot 7^k + 2(7^k - 2^k) \qquad \text{by algebra} \\
&= 5 \cdot 7^k + 2 \cdot 5r \qquad \text{by inductive hypothesis} \\
&= 5(7^k + 2r) \qquad \text{by algebra.}
\end{aligned}
$$

Now $7^k + 2r$ is an integer because products and sums of integers are integers. Therefore, by definition of divisibility, $7^{k+1} - 2^{k+1}$ is divisible by 5 *[as was to be shown]*.

13. *Hint:*
$$x^{k+1} - y^{k+1} = x^{k+1} - x \cdot y^k + x \cdot y^k - y^{k+1}$$
$$= x \cdot (x^k - y^k) + y^k \cdot (x - y)$$

14. *Hint 1:*
$$(k+1)^3 - (k+1) = k^3 + 3k^2 + 3k + 1 - k - 1$$
$$= (k^3 - k) + 3k^2 + 3k$$
$$= (k^3 - k) + 3k(k+1)$$

Hint 2: $k(k+1)$ is a product of two consecutive integers. By Theorem 4.5.2, one of these must be even.

15. <u>Proof (by mathematical induction)</u>: Let the property $P(n)$ be the sentence

$$n(n^2 + 5) \text{ is divisible by 6.} \qquad \leftarrow \ P(n)$$

We will prove that $P(n)$ is true for every integer $n \geq 0$.

Show that $P(0)$ is true: $P(0)$ is true because $0(0^2 + 5) = 0$ and 0 is divisible by 6.

Show that for every integer $k \geq 0$, if $P(k)$ is true then $P(k+1)$ is true:

Let k be any integer with $k \geq 0$, and suppose

$$k(k^2 + 5) \text{ is divisible by 6.} \qquad \leftarrow \ \begin{matrix} P(k) \\ \text{inductive hypothesis} \end{matrix}$$

We must show that

$$(k+1)((k+1)^2 + 5) \text{ is divisible by 6.} \qquad \leftarrow P(k+1)$$

By definition of divisibility $k(k^2 + 5) = 6r$ for some integer r. Then

$$
\begin{aligned}
(k+1)((k+1)^2 + 5) &= (k+1)(k^2 + 2k + 1 + 5) \\
&= (k+1)(k^2 + 2k + 6) \\
&= k^3 + 2k^2 + 6k + k^2 + 2k + 6 \\
&= k^3 + 3k^2 + 8k + 6 \\
&= (k^3 + 5k) + (3k^2 + 3k + 6) \\
&= k(k^2 + 5) + (3k^2 + 3k + 6) \qquad \text{by algebra} \\
&= 6r + 3(k^2 + k) + 6, \qquad \text{by inductive hypothesis.}
\end{aligned}
$$

Now $k(k+1)$ is a product of two consecutive integers. By Theorem 4.4.2 one of these is even, and so *[by properties 1 and 4 of Example 4.3.3]* the product $k(k+1)$ is even. Hence $k(k+1) = 2s$ for some integer s. Thus

$$6r + 3(k^2 + k) + 6 = 6r + 3(2s) + 6 = 6(r + s + 1).$$

By substitution, then,
$$(k+1)((k+1)^2 + 5) = 6(r + s + 1),$$

which is divisible by 6 because $r + s + 1$ is an integer. Therefore, $(k+1)((k+1)^2 + 5)$ is divisible by 6 *[as was to be shown]*.

16. <u>Proof (by mathematical induction)</u>: For the given statement, let the property $P(n)$ be the inequality
$$2^n < (n+1)!. \qquad \leftarrow \ P(n)$$

Show that $P(2)$ is true:

$P(2)$ says that $2^2 < (2+1)!$. The left-hand side is $2^2 = 4$ and the right-hand side is $3! = 6$. So, because $4 < 6$, $P(2)$ is true.

Show that for every integer $k \geq 2$, if $P(k)$ is true then $P(k+1)$ is true:
Let k be any integer with $k \geq 2$, and suppose $P(k)$ is true. That is, suppose

$$2^k < (k+1)!. \qquad \leftarrow \begin{array}{l} P(k) \\ \text{inductive hypothesis} \end{array}$$

We must show that $P(k+1)$ is true. That is, we must show that

$$2^{k+1} < ((k+1)+1)!,$$

or, equivalently, that

$$2^{k+1} < (k+2)!. \qquad \leftarrow P(k+1)$$

$2^{k+1} < ((k+1)+1)$, or, equivalently, that $2^{k+1} < (k+2)!$. By the laws of exponents and the inductive hypothesis,

$$2^{k+1} = 2 \cdot 2^k < 2(k+1)!. \, (*)$$

Since $k \geq 2$, then $2 < k+2$, and so

$$2(k+1)! < (k+2)(k+1)! = (k+2)!. \, (**)$$

Combining inequalities $(*)$ and $(**)$ gives

$$2^{k+1} < (k+2)!$$

[as was to be shown].

18. <u>Proof (by mathematical induction)</u>: Let the property $P(n)$ be the inequality

$$5^n + 9 < 6^n. \qquad \leftarrow P(n)$$

We will prove that $P(n)$ is true for every integer $n \geq 2$.

Show that $P(2)$ is true: $P(2)$ is true because the left-hand side is $5^2 + 9 = 25 + 9 = 34$ and the right-hand side is $6^2 = 36$, and $34 < 36$.

Show that for every integer $k \geq 2$, if $P(k)$ is true then $P(k+1)$ is true:
Let k be any integer with $k \geq 2$, and suppose

$$5^k + 9 < 6^k. \qquad \leftarrow \begin{array}{l} P(k) \\ \text{inductive hypothesis} \end{array}$$

We must show that

$$5^{k+1} + 9 < 6^{k+1}. \qquad \leftarrow P(k+1)$$

Multiplying both sides of the inequality in the inductive hypothesis by 5 gives

$$5(5^k + 9) < 5 \cdot 6^k. \, (*)$$

Note that

$$5^{k+1} + 9 < 5^{k+1} + 45 = 5(5^k + 9) \quad \text{and} \quad 5 \cdot 6^k < 6^{k+1}. \, (**)$$

Thus, by by the transitive property of order, $(*)$, and $(**)$,

$$5^{k+1} + 9 < 5(5^k + 9) \quad \text{and} \quad 5(5^k + 9) < 5 \cdot 6^k \quad \text{and} \quad 5 \cdot 6^k < 6^{k+1}.$$

So, by the transitive property of order,

$$5^{k+1} + 9 < 6^{k+1}$$

[as was to be shown].

19. <u>Proof (by mathematical induction)</u>: For the given statement, let the property $P(n)$ be the inequality

$$n^2 < 2^n. \qquad \leftarrow \ P(n)$$

Show that $P(5)$ is true:

$P(5)$ says that $5^2 < 2^5$. But $5^2 = 25$ and $2^5 = 32$, and $25 < 32$. Hence $P(5)$ is true.

Show that for any integer $k \geq 5$, if $P(k)$ is true then $P(k+1)$ is true:

Let k be any integer with $k \geq 5$, and suppose $P(k)$ is true. That is, suppose

$$k^2 < 2^k. \qquad \leftarrow \quad \begin{array}{l} P(k) \\ \text{inductive hypothesis} \end{array}$$

We must show that $P(k+1)$ is true. That is, we must show that

$$(k+1)^2 < 2^{k+1}. \qquad \leftarrow P(k+1)$$

Now

$$(k+1)^2 = k^2 + 2k + 1 < 2^k + 2k + 1 \quad \text{by inductive hypothesis.}$$

Also, by Proposition 5.3.2,

$$2k + 1 < 2^k.$$

[Prop. 5.3.2 applies since $k \geq 5 \geq 3$.] Putting these inequalities together gives

$$(k+1)^2 < 2^k + 2k + 1 < 2^k + 2^k = 2^{k+1}$$

[as was to be shown].

21. <u>Proof (by mathematical induction)</u>: Let the property $P(n)$ be the inequality

$$\sqrt{n} < \frac{1}{\sqrt{1}} + \frac{1}{\sqrt{2}} + \frac{1}{\sqrt{3}} + \cdots + \frac{1}{\sqrt{n}}. \qquad \leftarrow \ P(n)$$

We will prove that $P(n)$ is true for every integer $n \geq 2$.

Show that $P(2)$ is true: To show that $P(2)$ is true we must show that

$$\sqrt{2} < \frac{1}{\sqrt{1}} + \frac{1}{\sqrt{2}}.$$

Now this inequality is true if, and only if,

$$2 < \sqrt{2} + 1$$

(by multiplying both sides by $\sqrt{2}$). And this is true if, and only if,

$$1 < \sqrt{2}$$

(by subtracting 1 on both sides). Now $1 < \sqrt{2}$, and so $P(2)$ is true.

Show that for every integer $k \geq 2$, if $P(k)$ is true then $P(k+1)$ is true:

Let k be any integer with $k \geq 2$, and suppose

$$\sqrt{k} < \frac{1}{\sqrt{1}} + \frac{1}{\sqrt{2}} + \frac{1}{\sqrt{3}} + \cdots + \frac{1}{\sqrt{k}}. \qquad \leftarrow \quad \begin{array}{l} P(k) \\ \text{inductive hypothesis} \end{array}$$

We must show that

$$\sqrt{k+1} < \frac{1}{\sqrt{1}} + \frac{1}{\sqrt{2}} + \frac{1}{\sqrt{3}} + \cdots + \frac{1}{\sqrt{k+1}}. \qquad \leftarrow P(k+1)$$

Now for each integer $k \geq 2$,

$$\sqrt{k} < \sqrt{k+1} \ (*),$$

and multiplying both sides of (*) by $\sqrt{k}$ gives

$$k < \sqrt{k} \cdot \sqrt{k+1}.$$

Adding 1 to both sides gives

$$k+1 < \sqrt{k} \cdot \sqrt{k+1} + 1,$$

and dividing both sides by $\sqrt{k+1}$ gives

$$\sqrt{k+1} < \sqrt{k} + \frac{1}{\sqrt{k+1}}.$$

By substitution from the inductive hypothesis, then,

$$\sqrt{k+1} < \frac{1}{\sqrt{1}} + \frac{1}{\sqrt{2}} + \frac{1}{\sqrt{3}} + \cdots + \frac{1}{\sqrt{k}} + \frac{1}{\sqrt{k+1}}$$

[as was to be shown].

(*) *Note*: Strictly speaking, the reason for this claim is that $k < k+1$ and for all positive real numbers a and b, if $a < b$, then $\sqrt{a} < \sqrt{b}$.

24. <u>Proof (by mathematical induction)</u>: For the given statement, let the property $P(n)$ be the equation

$$a_n = 3 \cdot 7^{n-1}. \qquad \leftarrow \ P(n)$$

Show that $P(1)$ is true:

The left-hand side of $P(1)$ is a_1, which equals 3 by definition of the sequence. The right-hand side is $3 \cdot 7^{1-1} = 3$ also. Thus $P(1)$ is true.

Show that for every integer $k \geq 1$, if $P(k)$ is true then $P(k+1)$ is true:

Let k be any integer with $k \geq 1$, and suppose $P(k)$ is true. That is, suppose

$$a_k = 3 \cdot 7^{k-1}. \qquad \leftarrow \quad \begin{array}{l} P(k) \\ \text{inductive hypothesis} \end{array}$$

We must show that $P(k+1)$ is true. That is, we must show that

$$a_{k+1} = 3 \cdot 7^{(k+1)-1},$$

or, equivalently, that

$$a_{k+1} = 3 \cdot 7^k. \qquad \leftarrow P(k+1)$$

But the left-hand side of $P(k+1)$ is

$$\begin{array}{rll} a_{k+1} &= 7a_k & \text{by definition of the sequence } a_1, a_2, a_3, \ldots \\ &= 7(3 \cdot 7^{k-1}) & \text{by inductive hypothesis} \\ &= 3 \cdot 7^k & \text{by the laws of exponents,} \end{array}$$

and this is the right-hand side of $P(k+1)$ *[as was to be shown].*

25. <u>Proof (by mathematical induction)</u>: According to the definition of $b_0, b_1, b_2, \ldots$, we have that $b_0 = 5$ and $b_k = 4 + b_{k-1}$ for every integer $k \geq 1$. Let the property $P(n)$ be the inequality

$$b_n > 4n. \qquad \leftarrow \ P(n)$$

We will prove that $P(n)$ is true for each integer $n \geq 0$.

Show that $P(0)$ is true: To show that $P(0)$ is true we must show that $b_0 > 4 \cdot 0$. But $4 \cdot 0 = 0, b_0 = 5$ *[by definition of $b_0, b_1, b_2, \ldots$],* and $5 > 0$. So $P(0)$ is true.

Show that for every integer $k \geq 0$, if $P(k)$ is true then $P(k + 1)$ is true: Let k be any integer with $k \geq 0$, and suppose that

$$b_k > 4k. \qquad \leftarrow \begin{array}{c} P(k) \\ \text{inductive hypothesis} \end{array}$$

We must show that

$$b_{k+1} > 4(k + 1). \qquad \leftarrow P(k + 1)$$

Now

$$
\begin{array}{rcll}
b_{k+1} & = & 4 + b_k & \text{by definition of } b_1, b_2, b_3, \ldots \\
& > & 4 + 4k & \text{because } b_k > 4k \text{ by inductive hypothesis} \\
& > & 4(1 + k) & \text{by factoring out a 4} \\
& > & 4(k + 1) & \text{by the commutative law of addition}
\end{array}
$$

[as was to be shown].

27. <u>Proof (by mathematical induction):</u> According to the definition of $d_1, d_2, d_3, \ldots$, we have that $d_1 = 2$ and $d_k = \dfrac{d_{k-1}}{k}$ for every integer $k \geq 2$. Let the property $P(n)$ be the equation

$$d_n = \frac{2}{n!}. \qquad \leftarrow P(n)$$

We will prove that $P(n)$ is true for every integer $n \geq 1$.

Show that $P(1)$ is true: To show that $P(1)$ is true we must show that $d_1 = \frac{2}{1!}$. Now $\frac{2}{1!} = 2$ and $d_1 = 2$ (by definition of $d_1, d_2, d_3, \ldots$). So the property holds for $n = 1$.

Show that for every integer $k \geq 1$, if $P(k)$ is true then $P(k + 1)$ is true:

Let k be any integer with $k \geq 1$, and suppose that

$$d_k = \frac{2}{k!}. \qquad \leftarrow \begin{array}{c} P(k) \\ \text{inductive hypothesis} \end{array}$$

We must show that

$$d_{k+1} = \frac{2}{(k + 1)!}. \qquad \leftarrow P(k + 1)$$

Now the left-hand side of this equation is

$$
\begin{array}{rcll}
d_{k+1} & = & \dfrac{d_k}{k + 1} & \text{by definition of } d_1, d_2, d_3, \ldots \\[3mm]
& = & \dfrac{\frac{2}{k!}}{k + 1} & \text{by inductive hypothesis} \\[3mm]
& = & \dfrac{2}{(k + 1)k!} & \\[3mm]
& = & \dfrac{2}{(k + 1)!} & \text{by the algebra of fractions,}
\end{array}
$$

which is the right-hand side of the equation. *[This is what was to be shown].*

29. Proof (by mathematical induction):

A set L consists of strings obtained by juxtaposing one or more of *abb*, *bab*, and *bba*. Let the property $P(n)$ be the sentence

> If a string s in L has length $3n$, then
> s contains an even number of b's.

Show that $P(1)$ is true: $P(1)$ is the statement that a string s in L of length 3 contains an even number of b's. The only strings in L that have length 3 are *abb*, *bab*, and *bba*, and each of these strings has an even number of b's. So $P(1)$ is true.

Show that for every integer $k \geq 1$, if $P(k)$ is true then $P(k+1)$ is true: Let k be any integer with $k \geq 1$ and suppose that

> If a string s in L has length $3k$, then $\leftarrow$ $P(k)$
> s contains an even number of b's   inductive hypothesis

We must show that

> If a string s in L has length $3(k+1)$, then
> s contains an even number of b's. $\leftarrow$ $P(k+1)$

So, suppose s is a string in L that has length $3(k+1)$. Now $3(k+1) = 3k+3$ and the strings in L are obtained by juxtaposing strings already in L with one of *abb*, *bab*, or *bba*. Thus, either the initial or the final three characters in s are *abb*, *bab*, or *bba*. Moreover, the other $3k$ characters in s are also in L by definition of L, and so, by inductive hypothesis, the other $3k$ characters in s contain an even number, say m, of b's. Because each of *abb*, *bab,* and *bba* contains 2 b's, the total number of b's in s is $m+2$, which is a sum of even integers and hence is even *[as was to be shown]*.

30. Proof (by mathematical induction):

A set S consists of strings obtained by juxtaposing one or more copies of 1110 and 0111. Let the property $P(n)$ be the sentence

> If a string s in S has length $4n$, then the
> number of 1's in s is a multiple of 3.

Show that $P(1)$ is true: $P(1)$ is the statement that the number of 1's in a string of length 4 is a multiple of 3. The only strings in S that have length 4 are 1110 and 0111, and the number of 1's in each of these strings is a multiple of 3. So $P(1)$ is true.

Show that for every integer $k \geq 1$, if $P(k)$ is true then $P(k+1)$ is true:

Let k be any integer with $k \geq 1$ and suppose that

> If a string s in S has length $4k$, then the $\leftarrow$ $P(k)$
> number of 1's in s is a multiple of 3.   inductive hypothesis

We must show that

> If a string s in S has length $4(k+1)$, then
> the number of 1's in s is a multiple of 3. $\leftarrow$ $P(k+1)$

Suppose s is a string in S that has length $4(k+1)$. Now all the strings in S are obtained by juxtaposing strings of the form 1110 or 0111. Thus, since s has length $4(k+1) = 4k+4$, the initial or the final four characters in s are either 1110 or 0111, and the other part of s is a string of $4k$ characters, which also consists only of juxtapositions of 1110 and 0111. So the other part of s is also a string in S. It follows by inductive hypothesis that the number of 1's in the other part of s is a multiple of 3 and, therefore, has the form $3a$ for some integer a. Because each of 1110 and 0111 contains three 1's, the total number of 1's in s is $3a+3$, which equals $3(a+1)$ and is a multiple of 3. *[This is what was to be shown.]*

32. *Hint:* Consider the problem of trying to cover a 3×3 checkerboard with trominoes. Place a checkmark in certain squares as shown in the following figure.

Observe that no two squares containing checkmarks can be covered by the same tromino. Since there are four checkmarks, four trominoes would be needed to cover these squares. But, since each tromino covers three squares, four trominoes would cover twelve squares, not the nine squares in this checkerboard. It follows that such a covering is impossible.

33.

34. **a.** *Hint:* For the inductive step, note that a $2 \times 3(k+1)$ checkerboard can be split into a $2 \times 3k$ checkerboard and a 2×3 checkerboard.

35. **b.** *Hint:* Consider a 3×5 checkerboard, and refer to the hint for exercise 32. Figure out a way to place six checkmarks in squares so that no two of the squares that contain checkmarks can be covered by the same tromino.

36. <u>Proof by mathematical induction:</u> Let the property $P(n)$ be the sentence

> In any round-robin tournament involving n teams, it is possible to label the teams $T_1, T_2, T_3, \ldots, T_n$ so that for each $i = 1, 2, 3, \ldots, n-1$, T_i beats T_{i+1}. $\leftarrow P(n)$

We will prove that $P(n)$ is true for every integer $n \geq 2$.

Show that $P(2)$ is true: Consider any round-robin tournament involving two teams. By definition of round-robin tournament, these teams play each other exactly once. Let T_1 be the winner and T_2 the loser of this game. Then T_1 beats T_2, and so the labeling is as required for $P(2)$ to be true.

Show that for every integer $k \geq 2$, if $P(k)$ is true then $P(k+1)$ is true:

Let k be any integer with $k \geq 2$ and suppose that

> In any round-robin tournament involving k teams, it is possible to label the teams $T_1, T_2, T_3, \ldots, T_k$ so that for each $i = 1, 2, 3, \ldots, k-1$, T_i beats T_{i+1}. $\leftarrow$ $P(k)$ inductive hypothesis

We must show that

> In any round-robin tournament involving $k+1$ teams, it is possible to label the teams $T_1, T_2, T_3, \ldots, T_{k+1}$ so that for each $i = 1, 2, 3, \ldots, k$, T_i beats T_{i+1}. $\leftarrow P(k+1)$

Consider any round-robin tournament with $k+1$ teams. Pick one and call it T'. Temporarily remove T' and consider the remaining k teams. Since each of these teams plays each other team exactly once, the games played by these k teams form a round-robin tournament. It follows by inductive hypothesis that these k teams may be labeled $T_1, T_2, T_3, \ldots, T_k$ where T_i beats T_{i+1} for all $i = 1, 2, 3, \ldots, k-1$.

Case 1 (T' beats T_1): In this case, relabel each T_i to be T_{i+1}, and let $T_1 = T'$. Then T_1 beats the newly labeled T_2 (because T' beats the old T_1), and T_i beats T_{i+1} for all $i = 2, 3, \ldots, k$ (by inductive hypothesis).

Case 2 (T' loses to $T_1, , T_2, T_3, \ldots, T_m$ and beats T_{m+1} where $1 \leq m \leq k-1$): In this case, relabel teams $T_{m+1}, T_{m+2}, \ldots, T_k$ to be $T_{m+2}, T_{m+3}, \ldots, T_{k+1}$ and let $T_{m+1} = T'$.

Then for each i with $1 \leq i \leq m - 1$, T_i beats T_{i+1} (by inductive hypothesis), T_m beats T_{m+1} (because T_m beats T'), T_{m+1} beats T_{m+2} (because T' beats the old T_{m+1}), and for each i with $m + 2 \leq i \leq k$, T_i beats T_{i+1} (by inductive hypothesis).

Case 3 (T' loses to T_i for all $i = 1, 2, \ldots, k$): In this case, let $T_{k+1} = T'$. Then for all $i = 1, 2, \ldots, k-1$, T_i beats T_{i+1} (by inductive hypothesis) and T_k beats T_{k+1} (because T_k beats T').

Thus in all three cases the teams may be relabeled in the way specified [as was to be shown].

37. *Hint:* Use proof by contradiction. If the statement is false, then there exists some ordering of the integers from 1 to 30, say, $x_1, x_2, \ldots, x_{30}$, such that $x_1 + x_2 + x_3 < 45$, $x_2 + x_3 + x_4 < 45, \ldots$, and $x_{30} + x_1 + x_2 < 45$. Evaluate the sum of all these inequalities using the fact that $\sum_{i=1}^{30} x_i = \sum_{i=1}^{30} i$ and Theorem 5.2.1.

38. *Hint:* Given $k + 1$ a's and $k + 1$ b's arrayed around the outside of the circle, there has to be at least one location where an a is followed by a b as one travels in the clockwise direction. In the inductive step, temporarily remove such an a and the b that follows it, and apply the inductive hypothesis.

39. Proof (by mathematical induction): Let the property $P(n)$ be the sentence

> The interior angles of any n-sided convex $\leftarrow$ $P(n)$
> polygon add up to $180(n - 2)$ degrees.

We will prove that $P(n)$ is true for every integer $n \geq 3$.

Show that $P(3)$ is true: $P(3)$ is true because any convex 3-sided polygon is a triangle, the sum of the interior angles of any triangle is 180 degrees, and $180(3 - 2) = 180$. So the angles of any 3-sided convex polygon add up to $180(3 - 2)$ degrees.

Show that for every integer $k \geq 3$, if $P(k)$ is true then $P(k + 1)$ is true:

Let k be any integer with $k \geq 3$ and suppose that

> The interior angles of any k-sided convex $P(k)$
> polygon add up to $180(k - 2)$ degrees. $\leftarrow$ inductive hypothesis

We must show that

> The interior angles of any $(k + 1)$-sided convex polygon
> add up to $180((k + 1) - 2) = 180(k - 1)$ degrees. $\leftarrow$ $P(k + 1)$

Let p be any $(k + 1)$-sided convex polygon. Label adjacent vertices of p as $v_1, v_2, v_3, \ldots, v_k, v_{k+1}$, and draw a straight line from v_1 to v_3. Because the angles at v_1, v_2, and v_3 are all less than 180 degrees, this line lies entirely inside the polygon. Thus polygon p is split in two pieces: (1) the polygon p' obtained from p by using all of its vertices except v_2, and (2) triangle t with vertices v_1, v_2, v_3. This situation is illustrated in the diagram below.

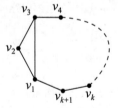

Note that p' has k vertices. In addition, it is convex because the sizes of the angles at v_1 and v_3 in p' are less than their sizes in p [since angle v_1 in p is the sum of an angle in p' plus an angle of triangle t, and similarly for angle v_3 in p]. Because p' is a convex polygon with

k vertices, by inductive hypothesis, the sum of its interior angles is $180(k-2)$ degrees. Now polygon p is obtained by joining p' and t, and since the sum of the interior angles in t is 180 degrees,

$$
\begin{aligned}
\text{the sum of the interior angles in } p \quad &= \quad \text{the sum of the interior angles in } p' \\
&\qquad + \text{ the sum of the interior angles in } t \\
&= \quad 180(k-2) \text{ degrees } + 180 \text{ degrees} \\
&= \quad 180(k-2+1) \text{ degrees} \\
&= \quad 180(k-1) \text{ degrees,}
\end{aligned}
$$

as was to be shown.

40. **b.** *Hint:* In the inductive step, imagine dividing a $2(k+1) \times 2(k+1)$ checkerboard into two sections: a center checkerboard of dimensions $2k \times 2k$ and an outer perimeter of single, adjacent squares. Then examine three cases: case 1 is where both removed squares are in the central $2k \times 2k$ checkerboard, case 2 is where one removed square is in the central $2k \times 2k$ checkerboard and the other is on the perimeter, and case 3 is where both removed squares are on the perimeter.

41. *Hint:* Let $P(n)$ be the sentence: If (1) $2n+1$ people are all positioned so that the distance between any two people is different from the distance between any two other people, and if (2) each person sends a message to their nearest neighbor, then there is at least one person who does not receive a message from anyone. Use mathematical induction to prove that $P(n)$ is true for each integer $n \geq 1$.

42. *One solution of several:* Let n be any positive even integer. Then $n = 2k$ for some positive integer k. Label a set of n people $1, 2, \ldots, n$; and position them so that for each $i = 1, 2, \ldots, k$, persons i and $k+i$ are i meters apart, and both person i and person $k+i$ are at least $k+1$ meters from every other person in the group. *[One way to achieve this is to surround each person by a circle of radius greater than $k+1$, next surround each pair of circles around persons i and $k+i$ by larger circles, and finally position the resulting circles so that none of them overlap.]* It follows that the nearest neighbor for person i is person $i+k$, and the nearest neighbor for person $i+k$ is person i, and so, for each $i = 1, 2, \ldots, k$, person $i+k$ receives a message from person i, and person i receives a message from person $i+k$. Also neither receives a message from anyone else in the group because, for each $i = 1, 2, \ldots, k$, since $i < k+1$, the distance between i and $i+k$ *[i meters]* is less than the distance between both i and $i+k$ and every other person in the group *[more than $k+1$ meters]*. Hence each person in the group receives a message from another person in the group.

43. **a.**   *Hint:*

Two Balls		
WW	→	B
WB	→	W
BB	→	B

Summary			
Start		End	
W	B	W	B
2	0	0	1
1	1	1	0
0	2	0	1

b. *Hint:* In all three cases when the urn initially contains an odd number of white balls, there is one white ball in the urn at the end of the game, and when the urn initially contains an even number of white balls, there is one black ball (i.e., zero white balls) in the urn at the end of the game.

44. *Hint:* Given a graph G satisfying the given condition, form a new graph G' by deleting one vertex v of G and all the edges that are incident on v. Then apply the inductive hypothesis to G'.

45. The inductive step fails for going from $n = 1$ to $n = 2$ because when $k = 1$, $A = \{a_1, a_2\}$ and sets B and C each have just one element. If, for instance, $B = \{a_1\}$, then $C = \{a_1\}$ because $B = C$. Hence, a_2 is not in either B or C, and so we cannot deduce any information about whether $a_2 = a_1$ or $a_2 \neq a_1$. Thus the inductive step fails for going from $n = 1$ to $n = 2$. This breaks the sequence of inductive steps, and so none of the statements for $n > 2$ is proved either.

In terms of the domino analogy, here is an explanation for why the given "proof" fails. The first domino *is* tipped backward (in other words, the basis step is proved), and, if any domino from the second onward tips backward it tips over the one behind it (the inductive step works for $n \geq 2$). However, when the first domino is tipped backward, it does *not* tip the second domino backward (the inductive step fails for going from $n = 1$ to $n = 2$). So only the first domino falls down; the rest remain standing.

46. *Hint:* Is the basis step true?

Section 5.4

1. Proof (by strong mathematical induction): Let the property $P(n)$ be the sentence

$$a_n \text{ is odd.} \qquad \leftarrow P(n)$$

Show that $P(1)$ and $P(2)$ are true: By definition of $a_1, a_2, a_3, \ldots$, we have that $a_1 = 1$ and $a_2 = 3$ and both 1 and 3 are odd. Thus $P(1)$ and $P(2)$ are true.

Show that for every integer $k \geq 2$, if $P(i)$ is true for each integer i with $1 \leq i \leq k$, then $P(k + 1)$ is true:

Let k be any integer with $k \geq 2$, and suppose

$$a_i \text{ is odd for each integer } i \text{ with } 1 \leq i \leq k. \qquad \leftarrow \text{ inductive hypothesis}$$

We must show that

$$a_{k+1} \text{ is odd.}$$

We know that

$$a_{k+1} = a_{k-1} + 2a_k$$

by definition of $a_1, a_2, a_3, \ldots$. Moreover, $k - 1$ is less than $k + 1$ and is greater than or equal to 1 (because $k \geq 2$). Thus, by inductive hypothesis, a_{k-1} is odd.

Also, every term of the sequence is an integer (being a sum of products of integers), and so $2a_k$ is even by definition of even.

It follows that a_{k+1} is the sum of an odd integer and an even integer and hence is odd by Theorem 4.1.2 (exercise 30, Section 4.1). *[This is what was to be shown.]*

3. Proof (by strong mathematical induction): Let the property $P(n)$ be the sentence

$$c_n \text{ is even.}$$

We will prove that $P(n)$ is true for every integer $n \geq 0$.

Show that $P(0)$, $P(1)$, and $P(2)$ are true: By definition of $c_0, c_1, c_2, \ldots$, we have that $c_0 = 2$, $c_1 = 2$, and $c_2 = 6$ and 2, 2, and 6 are all even. So $P(0)$, $P(1)$, and $P(2)$ are all true.

Show that for every integer $k \geq 2$, if $P(i)$ is true for each integer i from 0 through k, then $P(k + 1)$ is true: Let k be any integer with $k \geq 2$, and suppose

$$c_i \text{ is even for every integer } i \text{ with } 0 \leq i \leq k \qquad \leftarrow \text{ inductive hypothesis}$$

We must show that
$$c_{k+1} \text{ is even.}$$

Now by definition of $c_0, c_1, c_2, \ldots, c_{k+1} = 3c_{k-2}$. Since $k \geq 2$, we have that $0 \leq k - 2 \leq k$, and so, by inductive hypothesis, c_{k-2} is even. Now the product of an even integer with any integer is even *[properties 1 and 4 of Example 4.2.3]*, and hence $3c_{k-2}$, which equals c_{k+1}, is also even *[as was to be shown]*.

[Since both the basis and the inductive steps have been proved, we conclude that $P(n)$ is true for every integer $n \geq 0$.]

4. <u>Proof (by strong mathematical induction)</u>: Let the property $P(n)$ be the inequality
$$d_n \leq 1.$$

Show that $P(1)$ and $P(2)$ are true: Observe that $d_1 = \dfrac{9}{10}$ and $d_2 = \dfrac{10}{11}$ and both $\dfrac{9}{10} \leq 1$ and $\dfrac{10}{11} \leq 1$. Thus $P(1)$ and $P(2)$ are true.

Show that for every integer $k \geq 2$, if $P(i)$ is true for each integer i with $1 \leq i \leq k$, then $P(k+1)$ is true:

Let k be any integer with $k \geq 2$, and suppose
$$d_i \leq 1 \text{ for each integer } i \text{ with } 1 \leq i \leq k. \qquad \leftarrow \text{inductive hypothesis}$$

We must show that
$$d_{k+1} \leq 1.$$

Now, by definition of $d_1, d_2, d_3, \ldots,$ we know that $d_{k+1} = d_k \cdot d_{k-1}$. Moreover $d_k \leq 1$ and $d_{k-1} \leq 1$ by inductive hypothesis because both $k - 1$ and k are less than or equal to k.

Consequently,
$$d_{k+1} = d_k \cdot d_{k-1} \leq 1$$

because if two positive numbers are each less than or equal to 1, then their product is less than or equal to 1. *[To see why this is so, note that if $0, a < 1 \leq 1$ and $0 < b \leq 1$, then multiplying $a \leq 1$ by b gives $ab \leq b$, and since $b \leq 1$, then, by transitivity of order, $ab \leq 1$.]*

Thus the inductive step has been proved. *[Since we have proved both the basis step and the inductive step, we conclude that $d_n \leq 1$ for every integer $n \geq 1$.]*

5. <u>Proof (by strong mathematical induction)</u>: Let the property $P(n)$ be the equation
$$e_n = 5 \cdot 3^n + 7 \cdot 2^n.$$

Show that $P(0)$ and $P(1)$ are true: We must show that $e_0 = 5 \cdot 3^0 + 7 \cdot 2^0$ and $e_1 = 5 \cdot 3^1 + 7 \cdot 2^1$.

The left-hand side of the first equation is 12 (by definition of $e_0, e_1, e_2, \ldots$), and its right-hand side is $5 \cdot 1 + 7 \cdot 1 = 12$ also.

The left-hand side of the second equation is 29 (by definition of $e_0, e_1, e_2, \ldots$), and its right-hand side is $5 \cdot 3 + 7 \cdot 2 = 29$ also.

Thus $P(0)$ and $P(1)$ are true.

Show that for every integer $k \geq 1$, if $P(i)$ is true for each integer i with $0 \leq i \leq k$, then $P(k+1)$ is true:

Let k be any integer with $k \geq 1$, and suppose
$$e_i = 5 \cdot 3^i + 7 \cdot 2^i \text{ for each integer } i \text{ with } 1 \leq i \leq k. \qquad \leftarrow \text{inductive hypothesis}$$

We must show that
$$e_{k+1} = 5 \cdot 3^{k+1} + 7 \cdot 2^{k+1}.$$

Now

$$
\begin{aligned}
e_{k+1} &= 5e_k - 6e_{k-1} &&\text{by definition of } e_0, e_1, e_2, \ldots \\
&= 5(5 \cdot 3^k + 7 \cdot 2^k) - 6(5 \cdot 3^{k-1} + 7 \cdot 2^{k-1}) &&\text{by inductive hypothesis} \\
&= 25 \cdot 3^k + 35 \cdot 2^k - 30 \cdot 3^{k-1} - 42 \cdot 2^{k-1} \\
&= 25 \cdot 3^k + 35 \cdot 2^k - 10 \cdot 3 \cdot 3^{k-1} - 21 \cdot 2 \cdot 2^{k-1} \\
&= 25 \cdot 3^k + 35 \cdot 2^k - 10 \cdot 3^k - 21 \cdot 2^k \\
&= (25 - 10) \cdot 3^k + (35 - 21) \cdot 2^k \\
&= 15 \cdot 3^k + 14 \cdot 2^k \\
&= 5 \cdot 3 \cdot 3^k + 7 \cdot 2 \cdot 2^k \\
&= 5 \cdot 3^{k+1} + 7 \cdot 2^{k+1} &&\text{by algebra,}
\end{aligned}
$$

[as was to be shown].

6. <u>Proof (by strong mathematical induction)</u>: Let the property $P(n)$ be the equation
$$f_n = 3 \cdot 2^n + 2 \cdot 5^n.$$

We will prove that $P(n)$ is true for every integer $n \geq 0$.

Show that $P(0)$ and $P(1)$ are true: By definition of $f_0, f_1, f_2, \ldots$, we have that $f_0 = 5$ and $f_1 = 16$. Since $3 \cdot 2^0 + 2 \cdot 5^0 = 3 + 2 = 5$ and $3 \cdot 2^1 + 2 \cdot 5^1 = 6 + 10 = 16$, both $P(0)$ and $P(1)$ are true.

Show that for every integer $k \geq 1$, if $P(i)$ is true for each integer i from 0 through k, then $P(k + 1)$ is true: Let k be any integer with $k \geq 1$, and suppose

$$f_i = 3 \cdot 2^i + 2 \cdot 5^i \text{ for every integer } i \text{ with } 0 \leq i \leq k. \qquad \leftarrow \text{inductive hypothesis}$$

We must show that
$$f_{k+1} = 3 \cdot 2^{k+1} + 2 \cdot 5^{k+1}.$$

Now

$$
\begin{aligned}
f_{k+1} &= 7f_k - 10f_{k-1} &&\text{by definition of } f_0, f_1, f_2, \ldots \\
&= 7(3 \cdot 2^k + 2 \cdot 5^k) - 10(3 \cdot 2^{k-1} + 2 \cdot 5^{k-1}) &&\text{by inductive hypothesis} \\
&= 7(6 \cdot 2^{k-1} + 10 \cdot 5^{k-1}) - 10(3 \cdot 2^{k-1} + 2 \cdot 5^{k-1}) &&\text{since } 2^k = 2 \cdot 2^{k-1} \text{ and } 5^k = 5 \cdot 5^{k-1} \\
&= (42 \cdot 2^{k-1} + 70 \cdot 5^{k-1}) - (30 \cdot 2^{k-1} + 20 \cdot 5^{k-1}) \\
&= (42 - 30) \cdot 2^{k-1} + (70 - 20) \cdot 5^{k-1} \\
&= 12 \cdot 2^{k-1} + 50 \cdot 5^{k-1} \\
&= 3 \cdot 2^2 \cdot 2^{k-1} + 2 \cdot 5^2 \cdot 5^{k-1} \\
&= 3 \cdot 2^{k+1} + 2 \cdot 5^{k+1} &&\text{by algebra,}
\end{aligned}
$$

[as was to be shown].

[Since both the basis and the inductive steps have been proved, we conclude that $P(n)$ is true for every integer $n \geq 0$.]

9. <u>Proof (by strong mathematical induction)</u>: Let the property $P(n)$ be the inequality
$$a_n \leq \left(\frac{7}{4}\right)^n.$$

We will prove that $P(n)$ is true for every integer $n \geq 1$.

Show that $P(1)$ and $P(2)$ are true: By definition of $a_1, a_2, a_3, \ldots$, we have that $a_1 = 1$ and $a_2 = 3$. Now

$$\frac{7}{4} > 1 \quad \text{and} \quad \left(\frac{7}{4}\right)^2 = \frac{49}{16} = 3 + \frac{1}{16} > 3$$

So $a_1 \leq \frac{7}{4}$ and $a_2 \leq \left(\frac{7}{4}\right)^2$, and thus $P(1)$ and $P(2)$ are both true.

Show that for every integer $k \geq 2$, if $P(i)$ is true for each integer i from 1 through k, then $P(k+1)$ is true: Let k be any integer with $k \geq 2$, and suppose

$$a_i \leq \left(\frac{7}{4}\right)^i \quad \text{for every integer } i \text{ with } 0 \leq i \leq k. \qquad \leftarrow \text{ inductive hypothesis}$$

We must show that $\qquad\qquad\qquad a_{k+1} \leq \left(\frac{7}{4}\right)^{k+1}.$

Since $k \geq 2$,

$$
\begin{aligned}
a_{k+1} &= a_k + a_{k-1} &&\text{by definition of } a_1, a_2, a_3, \ldots \\
&\leq \left(\frac{7}{4}\right)^k + \left(\frac{7}{4}\right)^{k-1} &&\text{by inductive hypothesis} \\
&\leq \left(\frac{7}{4}\right)^{k-1}\left(\frac{7}{4}+1\right) &&\text{by factoring out } \left(\frac{7}{4}\right)^{k-1} \\
&\leq \left(\frac{7}{4}\right)^{k-1}\left(\frac{11}{4}\right) &&\text{by adding } \frac{7}{4} \text{ and } 1 \\
&\leq \left(\frac{7}{4}\right)^{k-1}\left(\frac{44}{16}\right) &&\text{by multiplying numerator and denominator of } \frac{11}{4} \text{ by 4} \\
&\leq \left(\frac{7}{4}\right)^{k-1}\left(\frac{49}{16}\right) &&\text{because } \frac{44}{16} < \frac{49}{16} \\
&\leq \left(\frac{7}{4}\right)^{k-1}\left(\frac{7}{4}\right)^2 &&\text{because } \left(\frac{49}{16}\right) = \left(\frac{7}{4}\right)^2 \\
&\leq \left(\frac{7}{4}\right)^{k+1} &&\text{by a law of exponents.}
\end{aligned}
$$

Thus, by transitivity of order, $a_{k+1} \leq \left(\frac{7}{4}\right)^{k+1}$ *[as was to be shown].*

10. *Hint:* In the basis step, show that $P(14)$, $P(15)$, and $P(16)$ are all true. For the inductive step, note that $k+1 = [(k+1)-3]+3$, and if $k \geq 16$, then $(k+1)-3 \geq 14$.

11. <u>Proof (by strong mathematical induction):</u> Let the property $P(n)$ be the sentence

A jigsaw puzzle consisting of n pieces takes $n-1$ steps to put together.

Show that $P(1)$ is true: A jigsaw puzzle consisting of just one piece does not take any steps to put together. Hence it is correct to say that it takes zero steps to put together.

Show that for every integer $k \geq 1$, if $P(i)$ is true for each integer i with $1 \leq i \leq k$, then $P(k+1)$ is true:

Let k be any integer with $k \geq 1$ and suppose that

For every integer i with $1 \leq i \leq k$,
a jigsaw puzzle consisting of n pieces takes $n-1$ steps to put together. $\qquad \leftarrow$ inductive hypothesis

We must show that

A jigsaw puzzle consisting of $k+1$ pieces takes k steps to put together.

Consider assembling a jigsaw puzzle consisting of $k+1$ pieces. The last step involves fitting together two blocks. Suppose one of the blocks consists of r pieces and the other consists of s pieces. Then $r+s = k+1$, and $1 \leq r \leq k$ and $1 \leq s \leq k$. Thus, by the inductive hypothesis, the numbers of steps required to assemble the blocks are $r - 1$ and $s - 1$, respectively. Then the total number of steps required to assemble the puzzle is

$$(r - 1) + (s - 1) + 1 = (r + s) - 1 = (k + 1) - 1 = k$$

[as was to be shown].

12. *Note*: This problem can be solved with ordinary mathematical induction.

    Proof (by mathematical induction): Let the property $P(n)$ be the sentence

    > Given any sequence of n cans of gasoline, deposited around a circular track in such a way that the total amount of gasoline is enough for a car to make one complete circuit of the track, it is possible to find an inital location for the car so that it will be able to traverse the entire track by using the various amounts of gasoline in the cans that it encounters along the way.

    We will prove that $P(n)$ is true for every integer $n \geq 1$.

    Show that $P(1)$ is true: When there is just one can, the car should be placed next to it. By hypothesis, the can contains enough gasoline to enable the car to make one complete circuit of the track. Hence $P(1)$ is true.

    Show that for every integer $k \geq 1$, if $P(k)$ is true then $P(k+1)$ is true: Let k be any integer with $k \geq 1$, and suppose that

    > For every integer i with $1 \leq i \leq k$, given any sequence of n cans of gasoline, deposited around a circular track in such a way that the total amount of gasoline is enough for a car to make one complete circuit of the track, it is possible to find an inital location for the car so that it will be able to traverse the entire track by using the various amounts of gasoline in the cans that it encounters along the way. $\leftarrow$ inductive hypothesis

    We must show that

    > Given any sequence of $k + 1$ cans of gasoline, deposited around a circular track in such a way that the total amount of gasoline is enough for a car to make one complete circuit of the track, it is possible to find an inital location for the car so that it will be able to traverse the entire track by using the various amounts of gasoline in the cans that it encounters along the way.

    Now, because the total amount of gasoline in all $k + 1$ cans is enough for a car to make one complete circuit of the track, there must be at least one can, call it C, that contains enough gasoline to enable the car to reach the next can, say D, in the direction of travel along the track. Imagine pouring all the gasoline from D into C. The result would be k cans deposited around the track in such a way that the total amount of gasoline would be enough for a car to make one complete circuit of the track. By inductive hypothesis, it is possible to find an initial location for the car so that it could traverse the entire track by using the various amounts of gasoline in the cans that it encounters along the way. Use that location as the starting point for the car. When the car reaches can C, the amount of gasoline in C is enough to enable it to reach can D, and once the car reaches D, the additional amount of gasoline in D enables it to complete the circuit. *[This is what was to be shown.]*

13. *Sketch of proof:* Given any integer $k > 1$, either k is prime or k is a product of two smaller positive integers, each greater than 1. In the former case, the property is true. In the latter case, the inductive hypothesis ensures that both factors of k are products of primes and hence that k is also a product of primes.

14. Proof (by strong mathematical induction): Let the property $P(n)$ be the sentence

Any product of n odd integers is odd.

Show that $P(1)$ and $P(2)$ are true: $P(1)$ is true because any "product" of a single odd integer is odd. $P(2)$ is true because any product of two odd integers is odd (exercise 20 of Section 4.2).

Show that for every integer $k \geq 2$, if $P(i)$ is true for each integer i with $1 \leq i \leq k$, then $P(k+1)$ is true:

Let k be any integer with $k \geq 1$, and suppose that

For every integer i from 1 through k, $\leftarrow$ inductive hypothesis
any product of i odd integers is odd.

Consider any product M of $k+1$ odd integers. Some multiplication is the final one that is used to obtain M. Thus, there are integers A and B such that $M = AB$, and each of A and B is a product of between 1 and k odd integers. (For instance, if $M = ((a_1 a_2)a_3)a_4$, then $A = (a_1 a_2)a_3$ and $B = a_4$.)

By inductive hypothesis, each of A and B is odd, and, as in the basis step, we know that any product of two odd integers is odd. Hence $M = AB$ is odd.

15. Proof (by strong mathematical induction): Let the property $P(n)$ be the sentence

Any sum of n even integers is even.

We will prove that $P(n)$ is true for every integer $n \geq 1$.

Show that $P(1)$ and $P(2)$ are true: $P(1)$ is true because any "sum" of a single even integer is even. $P(2)$ is true because any sum of two even integers is even (Theorem 4.1.1).

Show that for every integer $k \geq 2$, if $P(i)$ is true for each integer i from 1 through k, then $P(k+1)$ is true: Let k be any integer with $k \geq 1$, and suppose that

For every integer i from 1 through k, $\leftarrow$ inductive hypothesis
any sum of i even integers is even.

We must show that

any sum of $k+1$ even integers is even.

Consider any sum S of $k+1$ even integers. Some addition is the final one that is used to obtain S. Thus there are integers A and B such that $S = A + B$, and each of A and B is a sum of between 1 and k even integers.

By inductive hypothesis, both A and B are even, and hence $S = A + B$ is even because, as in the basis step, it is a sum of two even integers.

16. _Hint:_ Let the property $P(n)$ be the sentence "If n is even, then any sum of n odd integers is even, and if n is odd, then any sum of n odd integers is odd." For the inductive step, consider any sum S of $k+1$ odd integers. Some addition is the final one that is used to obtain S. Thus, there are integers A and B such that $S = A + B$, and A is a sum of r odd integers and B is a sum of $(k+1) - r$ odd integers. Consider the two cases where $k+1$ is even and $k+1$ is odd, and for each case consider the two subcases where r is even and where r is odd.

17. $4^1 = 4$, $4^2 = 16$, $4^3 = 64$, $4^4 = 256$, $4^5 = 1024$, $4^6 = 4096$, $4^7 = 16384$, and $4^8 = 65536$.

Conjecture: The units digit of 4^n equals 4 if n is odd and equals 6 if n is even.

Proof by strong mathematical induction: Let the property $P(n)$ be the sentence

The units digit of 4^n is 4 if n is even and is 6 if n is odd.

Show that P(1) and P(2) are true: When $n = 1, 4^n = 4^1 = 4$, and so the units digit is 4. When $n = 2$, then $4^n = 4^2 = 16$, and so the units digit is 6. Thus, $P(1)$ and $P(2)$ are true.

Show that for every integer $k \geq 2$, if $P(i)$ is true for each integer i with $1 \leq i \leq k$, then $P(k+1)$ is true: Let k by any integer with $k \geq 2$, and suppose that

For every integer i from 1 through k,
$$\text{the units digit of } 4^i = \begin{cases} 4 \text{ if } i \text{ is even} \\ 6 \text{ if } i \text{ is odd} \end{cases} \quad \leftarrow \quad \text{inductive hypothesis}$$

We must show that the units digit of 4^{k+1} equals 4 if $k+1$ is odd and equals 6 if $k+1$ is even.

Case 1 ($k+1$ is odd): In this case, k is even, and so, by inductive hypothesis, the units digits of 4^k is 6. Thus $4^k = 10q + 6$ for some nonnegative integer q. It follows that

$$4^{k+1} = 4^k \cdot 4 = (10q+6) \cdot 4 = 40q + 24 = 10(4q+2) + 4.$$

Thus, the units digit of 4^{k+1} is 4.

Case 2 ($k+1$ is even): In this case, k is odd, and so, by inductive hypothesis, the units digit of 4^k is 4. Thus $4^k = 10q + 4$ for some nonnegative integer q. It follows that

$$4^{k+1} = 4^k \cdot 4 = (10q+4) \cdot 4 = 40q + 16 = 10(4q+1) + 6.$$

Thus, the units digit of 4^{k+1} is 6.

Conclusion: Because cases 1 and 2 are the only possibilities and 4^{k+1} has one of the required forms in each case, we have shown that $P(k+1)$ is true.

18. $9^1 = 9, 9^2 = 81, 9^3 = 729, 9^4 = 6561,$ and $9^5 - 59049.$

Conjecture: For every integer $n \geq 0$, the units digit of 9^n is 1 if n is even and is 9 if n is odd.

Proof (by strong mathematical induction): Let the property $P(n)$ be the sentence

The units digit of 9^n is 1 if n is even and is 9 if n is odd.

We will prove that $P(n)$ is true for every integer $n \geq 0$.

Show that $P(0)$ and $P(1)$ are true: $P(0)$ is true because 0 is even and the units digit of $9^0 = 1$. $P(1)$ is true because 1 is odd and the units digit of $9^1 = 9$.

Show that for every integer $k \geq 1$, if $P(i)$ is true for each integer i from 0 through k, then $P(k+1)$ is true: Let k be any integer with $k \geq 1$, and suppose:

For every integer i from 0 through k,
$$\text{the units digit of } 9^i = \begin{cases} 1 \text{ if } i \text{ is even} \\ 9 \text{ if } i \text{ is odd} \end{cases} \quad \leftarrow \quad \text{inductive hypothesis}$$

We must show that

$$\text{the units digit of } 9^{k+1} = \begin{cases} 1 \text{ if } i \text{ is even} \\ 9 \text{ if } i \text{ is odd} \end{cases} \quad \leftarrow \quad P(k+1)$$

Case 1 ($\leq$ is even): In this case k is odd, and so, by inductive hypothesis, the units digit of 9^k is 9. This implies that there is an integer a so that $9^k = 10a + 9$, and hence

$$\begin{aligned} 9^{k+1} &= 9^1 \cdot 9^k && \text{by algebra (a law of exponents)} \\ &= 9(10a+9) && \text{by substitution} \\ &= 90a + 81 \\ &= 90a + 80 + 1 \\ &= 10(9a+8) + 1 && \text{by algebra.} \end{aligned}$$

Because $9a + 8$ is an integer, it follows that the units digit of 9^{k+1} is 1.

Case 2 (k + 1 is odd): In this case k is even, and so, by inductive hypothesis, the units digit of 9^{k-1} is 1. This implies that there is an integer a so that $9^k = 10a + 1$, and hence

$$
\begin{aligned}
9^{k+1} &= 9^1 \cdot 9^k && \text{by algebra (a law of exponents)} \\
&= 9(10a + 1) && \text{by substitution} \\
&= 90a + 9 \\
&= 10(9a) + 9 && \text{by algebra.}
\end{aligned}
$$

Because $9a$ is an integer, it follows that the units digit of 9^{k+1} is 9.

Hence in both cases the units digit of 9^{k+1} is as specified in $P(k + 1)$ *[as was to be shown]*.

19. Proof (by strong mathematical induction): Let $a_1, a_2, a_3, \ldots$ be a sequence that satisfies the recurrence relation $a_k = 2 \cdot a_{\lfloor k/2 \rfloor}$ for every integer $k \geq 2$, with initial condition $a_1 = 1$, and let the property $P(n)$ be the inequality

$$a_n \leq n. \qquad \leftarrow P(n)$$

We will show that $P(n)$ is true for each integer $n \geq 1$.

Show that P(1) is true: $a_1 = 1$ and $1 \leq 1$. So $P(1)$ is true.

Show that for every integer $k \geq 1$, if $P(i)$ is true for each integer i from 1 through k, then $P(k + 1)$ is true: Let k be any integer with $k \geq 1$, and suppose that

$$a_i \leq i \quad \text{for each integer } i \text{ with } 1 \leq i \leq k \quad \leftarrow \quad \text{inductive hypothesis}$$

We must show that

$$a_{k+1} \leq k + 1. \qquad \leftarrow \quad P(k + 1)$$

Now

$$
\begin{aligned}
a_{k+1} &= 2 \cdot a_{\lfloor (k+1)/2 \rfloor} && \text{by definition of } a_1, a_2, a_3, \ldots \\
&\leq 2 \cdot \lfloor (k+1)/2 \rfloor \\
&\leq \begin{cases} 2 \cdot ((k+1)/2) & \text{if } k \text{ is odd} \\ 2 \cdot (k/2) & \text{if } k \text{ is even} \end{cases} && \text{by inductive hypothesis} \\
&\leq \begin{cases} k + 1 & \text{if } k \text{ is odd} \\ k & \text{if } k \text{ is even} \end{cases} \\
&\leq k + 1 && \text{because both } k \leq k + 1 \text{ and } k + 1 \leq k + 1.
\end{aligned}
$$

Thus $a_{k+1} \leq k + 1$ *[as was to be shown]*.

21. Proof (by strong mathematical induction): Let $c_0, c_1, c_2, \ldots$ be a sequence that satisfies the recurrence relation $c_k = c_{\lfloor k/2 \rfloor} + c_{\lceil k/2 \rceil}$ for every integer $k \geq 2$, with initial conditions $c_0 = 1$ and $c_1 = 1$, and let the property $P(n)$ be the equation

$$c_n = n. \qquad \leftarrow P(n)$$

We will show that $P(n)$ is true for each integer $n \geq 1$.

Show that P(1) is true: $c_1 = 1$ and $1 = 1$. So $P(1)$ is true.

Show that for every integer $k \geq 1$, if $P(i)$ is true for each integer i from 1 through k, then $P(k + 1)$ is true: Let k be any integer with $k \geq 1$, and suppose that

$$c_i = i \quad \text{for each integer } i \text{ with } 1 \leq i \leq k \quad \leftarrow \quad \text{inductive hypothesis}$$

We must show that

$$c_{k+1} = k+1.$$

Now

$$c_{k+1} = c_{\lfloor (k+1)/2 \rfloor} + c_{\lceil (k+1)/2 \rceil} \qquad \text{by definition of } c_0, c_1, c_2, \ldots$$

$$= \left\lfloor \frac{k+1}{2} \right\rfloor + \left\lceil \frac{k+1}{2} \right\rceil \qquad \text{by inductive hypothesis because when}$$
$$\text{$k \geq 1$ then $(k+1) \geq 2$ and so $(k+1)/2 \geq 1$}$$

$$= \begin{cases} \dfrac{k}{2} + \dfrac{k+2}{2} & \text{if } k \text{ is odd} \\[2mm] \dfrac{k+1}{2} + \dfrac{k+1}{2} & \text{if } k \text{ is even} \end{cases} \qquad \text{by Theorem 4.6.2 and exercise 21, Section 4.6}$$

$$= \begin{cases} \dfrac{2k+2}{2} & \text{if } k \text{ is odd} \\[2mm] \dfrac{2k+2}{2} & \text{if } k \text{ is even} \end{cases} \qquad \text{by combining like terms}$$

$$= k+1 \qquad \text{by basic algebra.}$$

Thus $c_{k+1} = k+1$ [as was to be shown].

22. Proof (by strong mathematical induction): Let $P(n)$ be the sentence

> In this version of NIM, if both piles initially contain n objects, the player who goes second can always win.

We will prove that $P(n)$ is true for every integer $n \geq 0$.

Show that $P(0)$ is true:

If neither pile contains any objects, the player who goes first automatically loses because of not being able to make a move. So the second player wins the game by default. Thus $P(0)$ is true.

Show that for every integer $k \geq 1$, if $P(i)$ is true for each integer i from 0 through k, then $P(k+1)$ is true:

Let k be any integer with $k \geq 0$, and suppose:

> In this version of NIM, for every integer i with $0 \leq i \leq k$, if both piles initially contain i objects, the player who goes second can always win.   ← inductive hypothesis

We must show that

> In this version of NIM, if both piles initially contain $k+1$ objects, the player who goes second can always win.   ← $P(k+1)$

So suppose both piles contain $k+1$ objects, where $0 \leq i \leq k$, and suppose the first player removes r objects from pile #1, where $1 \leq r \leq 3$. If the second player removes r objects from pile #2, then both piles will have the same number of objects, namely $(k+1) - r$. Now $(k+1) - r \leq k$ because $r \geq 1$, and thus, by inductive hypothesis, the second player will win.

23. **a.**

			10							
	3			7			$3 \cdot 7$		21	
1		2		4		3	$1 \cdot 2 + 4 \cdot 3$		14	
	1	1	2		2	1	2	$1 \cdot 1 + 2 \cdot 2 + 1 \cdot 2$	7	
		1	1	1	1		1	1	$1 \cdot 1 + 1 \cdot 1 + 1 \cdot 1$	3

TOTAL   45

24. **a.** The results are shown both diagrammatically and in a table.

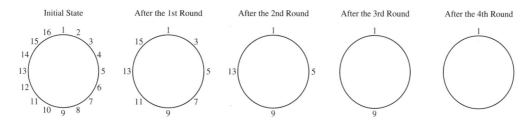

Round	Eliminated	Remaining
1	2, 4, 6, 8, 10, 12, 14, 16	1, 3, 5, 7, 9, 11, 13
2	3, 7, 11, 15	1, 5, 9, 13
3	5, 13	1, 9
4	9	1

b. <u>Proof (by ordinary mathematical induction)</u>: Let the property $P(n)$ be the sentence

> Given a set of 2^n people arranged in a circle and numbered consecutively from 1 to 2^n clockwise around the circle, if every second person starting with person #2 is eliminated until only one person remains, then the last person remaining is person #1.

We will prove that $P(n)$ is true for every integer $n \geq 1$.

Show that $P(1)$ is true: When $n = 1$, $2^n = 2^1 = 2$, and so the circle contains only two people. Person 2 is eliminated and person 1 remains. So $P(1)$ is true.

Show that for every integer $k \geq 1$, if $P(k)$ is true then $P(k+1)$ is true: Let k be any integer with $k \geq 2$, and suppose that

> Given a set of 2^k people arranged in a circle and numbered consecutively from 1 to 2^k clockwise around the circle, if every second person starting with person #2 is eliminated until only one person remains, then the last person remaining is person #1.

$P(k)$
← inductive hypothesis

We must show that

> Given a set of 2^{k+1} people arranged in a circle and numbered consecutively from 1 to 2^{k+1} clockwise around the circle, if every second person starting with person #2 is eliminated until only one person remains, then the last person remaining is person #1.

← $P(k+1)$

Suppose 2^{k+1} people are arranged in a circle and numbered consecutively from 1 to 2^{k+1} clockwise around the circle. If every second person starting with person #2 is eliminated, then in the first time around the circle all the even numbered people are eliminated, i.e., the people numbered 2, 4, 6, ... , 2^{k+1}. There are 2^k even-numbered people and 2^k odd-numbered people (since $2^k + 2^k = 2 \cdot 2^k = 2^{k+1}$). This is now the same situation as in the inductive hypothesis because 2^k people remain after the first time around the circle, and the second round starts by eliminating the next person in the circle after person #1. Thus, by inductive hypothesis, when only one person remains, it is person #1. *[This is what was to be shown.]*

c. <u>Proof</u>: Suppose m and n are any nonnegative integers with $2^n \leq 2^n + m < 2^{n+1}$, and suppose r people, numbered consecutively 1 through r, are arranged in a circle, where

$$r = 2^n + m.$$

Suppose that one starts from person #1 and goes repeatedly around the circle successively eliminating every second person. During the first round of eliminations, the first people to be eliminated are the m people in positions $2, 4, 6, \ldots, 2m$. At that point, the number of people remaining is

$$r - m = (2^n + m) - m = 2^n.$$

This situation is illustrated below for $n = 3$ and $m = 2$.

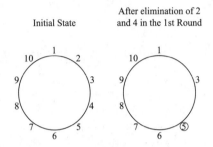

The elimination process continues by eliminating the next person in the circle after person #$(2m + 1)$. Thus, in this situation, person #$(2m + 1)$ plays the same role as person #1 in part (b), and, therefore, after all the rounds are complete and when only one person remains, it is person #$(2m + 1)$.

26. <u>Proof:</u> Let n be any integer greater than 1. Consider the set S of all positive integers other than 1 that divide n. Since $n \mid n$ and $n > 1$, there is at least one element in S. Hence, by the well-ordering principle for the integers, S has a smallest element; call it p.

    We claim that p is prime. For suppose p is not prime. Then there are integers a and b with $1 < a < p$, $1 < b < p$, and $p = ab$. And so, by definition of divides, $a \mid p$. Also $p \mid n$ because p is in S and every element in S divides n. Therefore, $a \mid p$ and $p \mid n$, and so, by transitivity of divisibility, $a \mid n$. Consequently, $a \in S$.

    But this contradicts the fact that $a < p$, and p is the smallest element of S. *[This contradiction shows that the supposition that p is not prime is false.]*

    Hence p is prime, and we have shown the existence of a prime number that divides n.

27. <u>Proof by contradiction:</u> Suppose not. That is, suppose that there exists an integer that is greater than 1, that is not prime, and that is not a product of primes. *[We will show that this supposition leads to a contradiction.]*

    Let S be the set of all integers that are greater than 1, are not prime, and are not a product of primes. That is,

    $$S = \{n \in \mathbf{Z} \mid n > 1, n \text{ is not prime, and } n \text{ is not a product of primes }\}.$$

    Then, by supposition, S has one or more elements. By the well-ordering principle for the integers, S has a least element; call it m. Then m is greater than 1, is not prime, and is not a product of primes.

    Now because m is greater than 1 and is not prime, $m = rs$ for some integers r and s with $1 < r < m$ and $1 < s < m$. Also, because both r and s are less than m, which is the least element of S, neither r nor s is in S. Thus both r and s are either prime or products of primes.

    But this implies that m is a product of primes because m is a product of r and s. Thus m is not in S. So m is in S and m is not in S, which is a contradiction. *[Hence the supposition is false, and so every integer greater than 1 is either prime or a product of primes.]*

28. **a.** <u>Proof:</u> Suppose r is any rational number. *[We need to show that there is an integer n such that $r < n$.]*

    Case 1 ($r \leq 0$): In this case, take $n = 1$. Then $r < n$.

    Case 2 ($r > 0$): In this case, $r = \frac{a}{b}$ for some positive integers a and b *[by definition of rational and because r is positive].* Note that $r = \frac{a}{b} < n$ if, and only if, $a < nb$. Let $n = 2a$. Multiply both sides of the inequality $1 < 2$ by a to obtain $a < 2a$, and multiply both sides of the inequality $1 < b$ by $2a$ to obtain $2a < 2ab = nb$. Thus $a < 2a < nb$, and so, by transitivity of order, $a < nb$. Dividing both sides by b gives that $\frac{a}{b} < n$, or, equivalently, that $r < n$.

    Hence, in, both cases, $r < n$ *[as was to be shown].*

29. *Hint:* If r is any rational number, let S be the set of all integers n such that $r < n$. Use the results of exercises 28(a), 28(c), and the well-ordering principle for the integers to show that S has a least element, say v, and then show that $v - 1 \leq r < v$..

30. <u>Proof:</u> Let S be the set of all integers r such that $n = 2^i \cdot r$ for some integer i. Then $n \in S$ because $n = 2^0 \cdot n$, and so $S \neq \varnothing$. Also, since $n \geq 1$, each r in S is positive, and so, by the well-ordering principle, S has a least element m. This means that $n = 2^k \cdot m$ (*) for some nonnegative integer k, and $m \leq r$ for every r in S. We claim that m is odd. The reason is that if m is even, then $m = 2p$ for some integer p. Substituting into equation (*) gives

    $$n = 2^k \cdot m = 2^k \cdot 2p = (2^k \cdot 2)p = 2^{k+1} \cdot p.$$

    It follows that $p \in S$ and $p < m$, which contradicts the fact that m is the *least* element of S. Hence m is odd, and so $n = m \cdot 2^k$ for some odd integer m and nonnegative integer k.

33. Suppose $P(n)$ is a property that is defined for integers n and suppose the following statement can be proved using strong mathematical induction:

    $$P(n) \text{ is true for every integer } n \geq a.$$

    Then for some integer $b \geq a$ the following two statements are true:

    1. $P(a), P(a+1), P(a+2), \ldots, P(b)$ are all true.

    2. For any integer $k \geq b$, if $P(i)$ is true for every integer i from a through k, then $P(k+1)$ is true.

    We will show that we can reach the conclusion that $P(n)$ is true for every integer $n \geq a$ using ordinary mathematical induction.

    <u>Proof by ordinary mathematical induction:</u> Let $Q(n)$ be the property

    $$P(j) \text{ is true for every integer } j \text{ from } a \text{ through } n.$$

    Show that $Q(b)$ is true: For $n = b$, the property is "$P(j)$ is true for every integer j with $a \leq j \leq b$." And this is true by (1) above.

    Show that for every integer $k \geq b$, if $Q(k)$ is true then $Q(k+1)$ is true: Let k be any integer with $k \geq b$, and suppose that $Q(k)$ is true. In other words, suppose that

    $$P(j) \text{ is true for every integer } j \text{ from } a \text{ through } k. \quad \leftarrow \text{ inductive hypothesis}$$

    We must show that $Q(k+1)$ is true. In other words, we must show that

    $$P(j) \text{ is true for every integer } j \text{ from } a \text{ through } k+1.$$

    Since, by inductive hypothesis, $P(j)$ is true for every integer j from a through k, it follows from (2) above that $P(k+1)$ is also true. Hence $P(j)$ is true for every integer j from a through $k+1$, *[as was to be shown].*

It follows by the principle of ordinary mathematical induction that $P(j)$ is true for every integer j from a through n for every integer $n \geq b$. From this and from (1) above, we conclude that $P(n)$ is true for every integer $n \geq a$.

34. *Hint:* In the inductive step, divide into cases depending on whether k can be written as $k = 3x$ or $k = 3x + 1$ or $k = 3x + 2$ for some integer x.

35. *Hint:* In the inductive step, let an integer $k \geq 0$ be given and suppose that there exist integers q' and r' such that $k = dq' + r'$ and $0 \leq r' < d$. You must show that there exist integers q and r such that
$$k + 1 = dq + r \quad \text{and} \quad 0 \leq r < d.$$

To do this, consider the two cases $r' < d - 1$ and $r' = d - 1$.

36. <u>Proof:</u> Let $P(n)$ be a property that is defined for integers n, and let a be a fixed integer. Suppose the statement "$P(n)$ is true for every integer $n \geq a$" can be proved using ordinary mathematical induction. Then the following two statements are true:

(1) $P(a)$ is true;

(2) For every integer $k \geq a$, if $P(k)$ is true then $P(k + 1)$ is true.

We will use proof by contradiction and the well-ordering principle for the integers to deduce the truth of the statement

$$P(n) \text{ is true for every integer } n \geq a.$$

Let S be the set of all integers greater than or equal to a for which $P(n)$ is false. Suppose S has one or more elements. *[We will show that this supposition leads logically to a contradiction.]* By the well-ordering principle for the integers, S has a least element, b, and by definition of S,

$$P(b) \text{ is false.}$$

Now

$$b - 1 \geq a$$

because S consists entirely of integers greater than or equal to a and

$$b \neq a$$

because $P(a)$ is true and $P(b)$ is false. Also

$$P(b - 1) \text{ is true}$$

because $b - 1 < b$ and b is the *least* element greater than or equal to a for which $P(n)$ is false. Thus $b - 1 \geq a$ and $P(b - 1)$ is true, and so by (2) above,

$$P((b - 1) + 1) = P(b) \text{ is true.}$$

Hence $P(b)$ is both true and false, which is a contradiction. This contradiction shows that the supposition is false, and so S has no elements, which means that $P(n)$ is true for every integer $n \geq a$.

37. *Hint:* Suppose S is a set containing one or more integers, all of which are greater than or equal to some integer a, and suppose that S does not have a least element. Let the property $P(n)$ be the sentence "$i \notin S$ for any integer i with $a \leq i \leq n$." Use mathematical induction to prove that $P(n)$ is true for every integer $n \geq a$, and explain how this result contradicts the supposition that S does not have a least element.

Section 5.5

1. <u>Proof</u>: Suppose the predicate $m + n = 100$ is true before entry to the loop. Then

$$m_{old} + n_{old} = 100.$$

After execution of the loop,

$$m_{\text{new}} = m_{\text{old}} + 1 \text{ and } n_{\text{new}} = n_{\text{old}} - 1,$$

So

$$m_{\text{new}} + n_{\text{new}} = (m_{\text{old}} + 1) + (n_{\text{old}} - 1)$$
$$= m_{\text{old}} + n_{\text{old}} = 100.$$

3. <u>Proof</u>: Suppose the predicate $m^3 > n^2$ is true before entry to the loop. Then

$$m_{\text{old}}^3 > n_{\text{old}}^2.$$

After execution of the loop,

$$m_{\text{new}} = 3 \cdot m_{\text{old}} \text{ and } n_{\text{new}} = 5 \cdot n_{\text{old}},$$

so

$$m_{\text{new}}^3 = (3 \cdot m_{\text{old}})^3 = 27 \cdot \text{m}_{\text{old}}^3 > 27 \cdot n_{\text{old}}^2.$$

Now since $n_{\text{new}} = 5 \cdot n_{old}$, then $n_{old} = \frac{1}{5} n_{\text{new}}$. Hence

$$m_{\text{new}}^3 > 27 \cdot n_{\text{old}}^2 = 27 \cdot \left(\frac{1}{5} n_{\text{new}}\right)^2 = 27 \cdot \frac{1}{25} n_{\text{new}}^2$$
$$= \frac{27}{25} \cdot n_{\text{new}}^2 > n_{\text{new}}^2.$$

6. <u>Proof</u>: *[The wording of this proof is almost the same as that of Example 5.5.2.]*

I. Basis Property: *[$I(0)$ is true before the first iteration of the loop.]*

$I(0)$ is "$exp = x^0$ and $i = 0$." According to the precondition, before the first iteration of the loop, $exp = 1$ and $i = 0$. Since $x^0 = 1$, $I(0)$ is evidently true.

II. Inductive Property: *[If $G \wedge I(k)$ is true before a loop iteration (where $k \geq 0$), then $(k \mid 1)$ is true after the loop iteration.]*

Suppose k is any nonnegative integer such that $G \wedge I(k)$ is true before an iteration of the loop. Then as execution reaches the top of the loop, $i \neq m, exp = x^k$, and $i = k$. Since $i \neq m$, the guard is passed and statement 1 is executed. Now before execution of statement 1,

$$exp_{\text{old}} = x^k,$$

so execution of statement 1 has the following effect:

$$exp_{\text{new}}^{\cdot} = exp_{\text{old}} \cdot x = x^k \cdot x = x^{k+1}.$$

Similarly, before statement 2 is executed,

$$i_{\text{old}} = k,$$

so after execution of statement 2,

$$i_{\text{new}} = i_{\text{old}} + 1 = k + 1$$

Hence after the loop iteration, the two statements $exp = x_{k+1}$ and $i = k + 1$ are true, and so $I(k + 1)$ is true.

III. Eventual Falsity of Guard: *[After a finite number of iterations of the loop, G becomes false.]*

The guard G is the condition $i \neq m$, and m is a nonnegative integer. By I and II, it is known that

(a) for every integer $n \geq 0$, if the loop is iterated n times, then $exp = x^n$ and $i = n$.

So after m iterations of the loop, $i = m$. Thus G becomes false after m iterations of the loop.

IV. Correctness of the Post-Condition: *[If N is the least number of iterations after which G is false and $I(N)$ is true, then the value of the algorithm variables will be as specified in the post-condition of the loop.]*

According to the post-condition, the value of exp after execution of the loop should be x^m. But when G is false, $i = m$. And when $I(N)$ is true, $i = N$ and $exp = x^N$. Since *both* conditions (G false and $I(N)$ true) are satisfied, $m = i = N$ and $exp = x^m$, as required.

8. Proof:

I. Basis Property: $I(0)$ is "$i = 1$ and $sum = A[1]$." According to the pre-condition, this statement is true.

II. Inductive Property: Suppose k is a nonnegative integer such that $G \wedge I(k)$ is true before an iteration of the loop. Then as execution reaches the top of the loop,

$$i \neq m, \ i = k + 1, \ \text{and} \ sum = A[1] + A[2] + \cdots + A[k+1].$$

Since $i \neq m$, the guard is passed and statement 1 is executed. Now before execution of statement 1, $i_{\text{old}} = k + 1$. So after execution of statement 1,

$$i_{\text{new}} = i_{old} + 1 = (k+1) + 1 = k + 2.$$

Also before statement 2 is executed,

$$sum_{\text{old}} = A[1] + A[2] + \cdots + A[k+1].$$

Execution of statement 2 adds $A[k+2]$ to this sum, and so after statement 2 is executed,

$$sum_{\text{new}} = A[1] + A[2] + \cdots + A[k+1] + A[k+2].$$

Thus after the loop iteration, $I(k+1)$ is true.

III. Eventual Falsity of Guard: The guard G is the condition $i \neq m$, where m is a positive integer. By I and II, it is known that for every integer $n \geq 1$, after n iterations of the loop, $I(n)$ is true. Hence, after $m - 1$ iterations of the loop, $I(m)$ is true, which implies that $i = m$ and G is false.

IV. Correctness of the Post-Condition: Suppose that N is the least number of iterations after which G is false and $I(N)$ is true. Then (since G is false)

$$i = m \ \text{and (since} \ I(N) \ \text{is true)} \ i = N + 1 \ \text{and} \ sum = A[1] + A[2] + \cdots + A[N+1].$$

Putting these together gives $m = N + 1$, and so

$$sum = A[1] + A[2] + \cdots + A[m],$$

which is the post-condition.

9. Proof

I. Basis Property: $I(0)$ is the statement

both a and A are even integers or both are odd integers and, in either case, $a \geq -1$.

According to the pre-condition this statement is true.

II. Inductive Property: The guard condition G is $a_{\text{old}} > 0$. Suppose k is any nonnegative integer such that $G \wedge I(k)$ is true before an iteration of the loop. Then when execution comes to the top of the loop, $a_{\text{old}} > 0$ and

both a_{old} and A are even integers or both are odd integers and, in either case, $a_{\text{old}} \geq -1$.

Execution of statement 1 sets a_{new} equal to $a_{\text{old}} - 2$. Hence a_{new} has the same parity as a_{old} which is the same as A. Also since $a_{\text{old}} > 0$, then

$$a_{\text{new}} = a_{\text{old}} - 2 > 0 - 2 = -2.$$

Now since $a_{\text{new}} > -2$ and since a_{new} is an integer, $a_{\text{new}} \geq -1$. Hence after the loop iteration, $I(k+1)$ is true.

III. Eventual Falsity of Guard: The guard G is the condition $a > 0$. After each iteration of the loop,

$$a_{\text{new}} = a_{\text{old}} - 2 < a_{\text{old}},$$

and so successive iterations of the loop give a strictly decreasing sequence of integer values of a which eventually becomes less than or equal to zero, at which point G becomes false.

IV. Correctness of the Post-Condition: Suppose that N is the least number of iterations after which G is false and $I(N)$ is true. Then (since G is false) $a \leq 0$ and (since $I(N)$ is true) both a and A are even integers or both are odd integers, and $a \geq -1$. Putting the inequalities together gives

$$-1 \leq a \leq 0,$$

and so, since a is an integer, $a = -1$ or $a = 0$. Since a and A have the same parity, then, $a = 0$ if A is even and $a = -1$ if A is odd. This is the post-condition.

10. *Hint:* Assume $G \wedge I(k)$ is true for a nonnegative integer k. Then $a_{\text{old}} \neq 0$ and $b_{\text{old}} \neq 0$ and

(**1**) a_{old} and b_{old} are nonnegative integers with $\gcd(a_{\text{old}}, b_{\text{old}}) = \gcd(A, B)$.

(**2**) At most one of a_{old} and b_{old} equals 0.

(**3**) $0 \leq a_{\text{old}} + b_{\text{old}} \leq A + B - k$.

It must be shown that $I(k+1)$ is true after the loop iteration. That means it is necessary to show that

(**1**) a_{new} and b_{new} are nonnegative integers with $\gcd(a_{\text{new}}, b_{\text{new}}) = \gcd(A, B)$.

(**2**) At most one of a_{new} and b_{new} equals 0.

(**3**) $0 \leq a_{\text{new}} + b_{\text{new}} \leq A + B - (k+1)$.

To show (3), observe that

$$a_{\text{new}} + b_{\text{new}} = \begin{cases} a_{\text{old}} - b_{\text{old}} + b_{\text{old}} & \text{if } a_{\text{old}} \geq b_{\text{old}} \\ b_{\text{old}} - a_{\text{old}} + a_{\text{old}} & \text{if } a_{\text{old}} < b_{\text{old}} \end{cases}.$$

[The reason for this is that when $a_{old} \geq b_{old}$, then $a_{new} = a_{old} - b_{old}$ and $b_{new} = b_{old}$, and when $a_{old} < b_{old}$, then $b_{new} = b_{old} - a_{old}$ and $a_{new} = a_{old}$.]

$$a_{\text{new}} + b_{\text{new}} = \begin{cases} a_{\text{old}} & \text{if } a_{\text{old}} \geq b_{\text{old}} \\ b_{\text{old}} & \text{if } a_{\text{old}} < b_{\text{old}} \end{cases}.$$

Now since $a_{\text{old}} \neq 0$ and $b_{\text{old}} \neq 0$, and since a_{old} and b_{old} are nonnegative integers, then $a_{\text{old}} \geq 1$ and $b_{\text{old}} \geq 1$. Hence, $a_{\text{old}} - 1 \geq 0$ and $b_{\text{old}} - 1 \geq 0$, and so $a_{\text{old}} \leq a_{\text{old}} + b_{\text{old}} - 1$ and $b_{\text{old}} \leq b_{\text{old}} + a_{\text{old}} - 1$. It follows that $a_{\text{new}} + a_{\text{new}} \leq a_{\text{old}} + b_{\text{old}} - 1 \leq (A + B - k) - 1$ by noting that (3) is true when going into the kth iteration. Thus, $a_{\text{new}} + b_{\text{new}} < A + B - (k+1)$ by algebraic simplification.

12. **a.** Suppose the following condition is satisfied before entry to the loop: "there exist integers u, v, s, and t such that $a = uA + vB$ and $b = sA + tB$." Then

$$a_{\text{old}} = u_{\text{old}}A + v_{\text{old}}B \quad \text{and} \quad b_{\text{old}} = s_{\text{old}}A + t_{\text{old}}B,$$

for some integers u_{old}, v_{old}, s_{old}, and t_{old}. Observe that $b_{\text{new}} = r_{\text{new}} = a_{\text{old}} \bmod b_{\text{old}}$. So by the quotient-remainder theorem, there exists a unique integer q_{new} with $a_{\text{old}} = b_{\text{old}} \cdot q_{\text{new}} + r_{\text{new}} = b_{\text{old}} \cdot q_{\text{new}} + b_{\text{new}}$. Solving for b_{new} gives

$$
\begin{aligned}
b_{\text{new}} &= a_{\text{old}} - b_{\text{old}} \cdot q_{\text{new}} \\
&= (u_{\text{old}}A + v_{\text{old}}B) - (s_{\text{old}}A + t_{\text{old}}B)q_{\text{new}} \\
&= (u_{\text{old}} - s_{\text{old}}q_{\text{new}})A + (v_{\text{old}} - t_{\text{old}}q_{\text{new}})B.
\end{aligned}
$$

Therefore, let

$$s_{\text{new}} = u_{\text{old}} - s_{\text{old}}q_{\text{new}} \quad \text{and} \quad t_{\text{new}} = v_{\text{old}} - t_{\text{old}}q_{\text{new}}.$$

Also since

$$a_{\text{new}} = b_{\text{old}} = s_{\text{old}}A + t_{\text{old}}B,$$

let

$$u_{\text{new}} = s_{\text{old}} \quad \text{and} \quad v_{\text{new}} = t_{\text{old}}.$$

Hence

$$a_{\text{new}} = u_{\text{new}} \cdot A + v_{\text{new}} \cdot B \quad \text{and} \quad b_{\text{new}} = s_{\text{new}} \cdot A + t_{\text{new}} \cdot B,$$

and so the condition is true after each iteration of the loop and hence after exit from the loop.

b. Initially $a = A$ and $b = B$. Let $u = 1$, $v = 0$, $s = 0$, and $t = 1$. Then before the first iteration of the loop,

$$a = uA + vB \quad \text{and} \quad b = sA + tB,$$

as was to be shown.

c. By part (b) there exist integers u, v, s, and t such that before the first iteration of the loop,

$$a = uA + vB \quad \text{and} \quad b = sA + tB.$$

So by part (a), after each subsequent iteration of the loop, there exist integers u, v, s, and t such that

$$a = uA + vB \quad \text{and} \quad b = sA + tB.$$

Now after the final iteration of the **while** loop in the Euclidean algorithm, the variable gcd is given the current value of a. (See page 224.) Now by the correctness proof for the Euclidean algorithm, $gcd = \gcd(A, B)$. Hence there exist integers u and v such that

$$\gcd(A, B) = uA + vB.$$

d. The method discussed in part (a) gives the following formulas for u, v, s, and t:

$$u_{\text{new}} = s_{\text{old}}, \quad v_{\text{new}} = t_{\text{old}}, \quad s_{\text{new}} = u_{\text{old}} - s_{\text{old}}q_{\text{new}}, \quad \text{and} \quad t_{\text{new}} = v_{\text{old}} - t_{\text{old}}q_{\text{new}},$$

where in each iteration q_{new} is the quotient obtained by dividing a_{old} by b_{old}. The trace table below shows the values of a, b, r, q, gcd, and u, v, s, and t for the iterations of the **while** loop from the Euclidean algorithm *Algorithm 4.10.2]*. By part (b) the initial values of u, v, s, and t are $u = 1$, $v = 0$, $s = 0$, and $t = 1$.

r		18	12	6	0
q		2	8	1	2
a	330	156	18	12	6
b	156	18	12	6	0
gcd					6
u	1	0	1	-8	9
v	0	1	-2	17	-19
s	0	1	-8	9	-26
t	1	-2	17	-19	55

Since the final values of gcd, u, and v are 6, 9 and -19 and since $A = 330$ and $B = 156$, we have $\gcd(330, 156) = 6 = 330u + 156v = 330 \cdot 9 + 156 \cdot (-19)$, which is true.

Section 5.6

1. $a_1 = 1$, $a_2 = 2a_1 + 2 = 2 \cdot 1 + 2 = 4$, $a_3 = 2a_2 + 3 = 2 \cdot 4 + 3 = 11$, $a_4 = 2a_3 + 4 = 2 \cdot 11 + 4 = 26$

3. $c_0 = 1$, $c_1 = 1 \cdot (c_0)^2 = 1 \cdot (1)^2 = 1$, $c_2 = 2(c_1)^2 = 2 \cdot (1)^2 = 2$, $c_3 = 3(c_2)^2 = 3 \cdot (2)^2 = 12$

5. $s_0 = 1$, $s_1 = 1$, $s_2 = s_1 + 2s_0 = 1 + 2 \cdot 1 = 3$, $s_3 = s_2 + 2s_1 = 3 + 2 \cdot 1 = 5$

6. $t_0 = -1$, $t_1 = 2$, $t_2 = t_1 + 2 \cdot t_0 = 2 + 2 \cdot (-1) = 0$, $t_3 = t_2 + 2 \cdot t_1 = 0 + 2 \cdot 2 = 4$

7. $u_1 = 1$, $u_2 = 1$, $u_3 = 3u_2 - u_1 = 3 \cdot 1 - 1 = 2$, $u_4 = 4u_3 - u_2 = 4 \cdot 2 - 1 = 7$

9. By definition of $a_0, a_1, a_2, \ldots$, $a_n = 3n + 1$ for every integer $n \geq 1$. Substitute k and $k - 1$ in place of n to get
$$a_k = 3k + 1 \text{ (*)} \quad \text{and} \quad a_{k-1} = 3(k - 1) + 1 \text{ (**)}$$
for each integer $k \geq 1$. Then

$$\begin{aligned} a_{k-1} + 3 &= 3(k-1) + 1 + 3 &&\text{by substitution from (**)} \\ &= 3k - 3 + 1 + 3 \\ &= 3k + 1 &&\text{by basic algebra} \\ &= a_k &&\text{by substitution from (*).} \end{aligned}$$

Thus $a_k = a_{k-1} + 3$ for every integer $k \geq 1$.

11. By definition of $c_0, c_1, c_2, \ldots$, $c_n = 2^n - 1$, for each integer $n \geq 0$. Substitute k and $k - 1$ in place of n to get
$$c_k = 2^k - 1 \text{ (*)} \quad \text{and} \quad c_{k-1} = 2^{k-1} - 1 \text{ (**)}$$
for every integer $k \geq 1$. Then

$$\begin{aligned} 2c_{k-1} + 1 &= 2(2^{k-1} - 1) + 1 &&\text{by substitution from (**)} \\ &= 2^k - 2 + 1 \\ &= 2^k - 1 &&\text{by basic algebra} \\ &= c_k &&\text{by substitution from (*).} \end{aligned}$$

Thus $c_k = 2c_{k-1} + 1$ for every integer $k \geq 1$.

12. For every integer $n \geq 0$, $s_n = \dfrac{(-1)^n}{n!}$. Thus for each integer k with $k \geq 1$,

$$s_k = \frac{(-1)^k}{k!} \text{ (*)} \quad \text{and} \quad s_{k-1} = \frac{(-1)^{k-1}}{(k-1)!} \text{ (**).}$$

It follows that for each integer k with $k \geq 1$,

$$\frac{-s_{k-1}}{k} = \frac{-\frac{(-1)^{k-1}}{(k-1)!}}{k} \qquad \text{by substitution from (**)}$$

$$= \frac{-(-1)^{k-1}}{k(k-1)!}$$

$$= \frac{(-1)^k}{k!}$$

$$= s_k \qquad \text{by substitution from (*)}$$

Thus $s_k = \dfrac{-s_{k-1}}{k}$ for every integer $k \geq 1$.

13. By definition of $t_0, t_1, t_2, \ldots, t_n = 2 + n$, for each integer $n \geq 0$. Substitute k, $k - 1$, and $k - 2$ in place of n to get

$$t_k = 2 + k, \ (*) \quad t_{k-1} = 2 + (k - 1), \ (**) \quad \text{and} \quad t_{k-2} = 2 + (k - 2) \ (***)$$

for each integer $k \geq 2$. Then

$$\begin{aligned} 2t_{k-1} - t_{k-2} &= 2(2 + (k - 1)) - (2 + (k - 2)) \quad \text{by substitution from (**) and (***)} \\ &= 2(k + 1) - k \\ &= 2 + k \qquad\qquad\qquad\qquad \text{by basic algebra} \\ &= t_k \qquad\qquad\qquad\qquad\quad \text{by substitution from (*).} \end{aligned}$$

Thus $t_k = 2 + k$ for every integer $k \geq 2$.

15. <u>Proof</u> : Let n be any integer with $n \geq 1$. Then

$$\begin{aligned} \frac{1}{4n + 2}\binom{2n + 2}{n + 1} &= \left(\frac{1}{2(2n + 1)}\right)\left(\frac{(2n + 2)!}{(n + 1)!((2n + 2) - (n + 1))!}\right) \\[2mm] &= \left(\frac{1}{2(2n + 1)}\right)\left(\frac{(2n + 2)!}{(n + 1)!(n + 1)!}\right) \\[2mm] &= \left(\frac{1}{2(2n + 1)}\right)\left(\frac{(2n + 2)(2n + 1)(2n)!}{(n + 1) \cdot n! \cdot (n + 1) \cdot n!}\right) \\[2mm] &= \frac{1}{2}\left(\frac{2(n + 1)}{(n + 1) \cdot (n + 1)}\right)\left(\frac{(2n)!}{n! \cdot n!}\right) \\[2mm] &= \frac{1}{n + 1}\binom{2n}{n} \\[2mm] &= C_n. \end{aligned}$$

Thus $C_n = \dfrac{1}{4n + 2}\dbinom{2n + 2}{n + 1}$.

17. **a.** $a_1 = 2$

$a_2 = 2$ (moves to move the top disk from pole A to pole C)

$\quad + 1$ (move to move the bottom disk from pole A to pole B)

$\quad + 2$ (moves to move top disk from pole C to pole A)

$\quad + 1$ (move to move the bottom disk from pole B to pole C)

$\quad + 2$ (moves to move top disk from pole A to pole C)

$= 8$

$a_3 = 8 + 1 + 8 + 1 + 8 = 26$

b. $a_4 = 26 + 1 + 26 + 1 + 26 = 80$

c. For every integer $k \geq 2$,

$a_k = a_{k-1}$ (moves to move the top $k-1$ disks from pole A to pole C)

$\qquad + 1$ (move to move the bottom disk from pole A to pole B)

$\qquad + a_{k-1}$ (moves to move the top disk from pole C to pole A)

$\qquad + 1$ (move to move the bottom disks from pole B to pole C)

$\qquad + a_{k-1}$ (moves to move the top disks from pole A to pole C)

$\quad = 3a_{k-1} + 2.$

18. **a.** $b_1 = 1, \quad b_2 = 1 + 1 + 1 + 1 = 4, \quad b_3 = 4 + 4 + 1 + 4 = 13$

b. $b_4 = 40$

c. Note that it takes just as many moves to move a stack of disks from the middle pole to an outer pole as from an outer pole to the middle pole: the moves are the same except that their order and direction are reversed. For every integer $k \geq 2$,

$b_k \quad = \quad a_{k-1} \qquad$ (moves to transfer the top $k-1$ disks from pole A to pole C)

$\qquad\quad +1 \qquad$ (move to transfer the bottom disk from pole A to pole B)

$\qquad\qquad +b_{k-1} \qquad$ (moves to transfer the top $k-1$ disks from pole C to pole B).

$\quad = \quad a_{k-1} + 1 + b_{k-1}.$

d. One way to transfer a tower of k disks from pole A to pole B is first to transfer the top $k-1$ disks from pole A to pole B *[this requires b_{k-1} moves]*, then transfer the top $k-1$ disks from pole B to pole C *[this also requires b_{k-1} moves]*, then transfer the bottom disk from pole A to pole B *[this requires one move]*, and finally transfer the top $k-1$ disks from pole C to pole B *[this again requires b_{k-1} moves]*. This sequence of steps need not necessarily, however, result in a minimum number of moves. Therefore, at this point, all we can say for sure is that for each integer $k \geq 2$,

$$b_k \leq b_{k-1} + b_{k-1} + 1 + b_{k-1} = 3b_{k-1} + 1.$$

e. Proof (by mathematical induction): Let the property $P(k)$ be the equation

$$b_k = 3b_{k-1} + 1.$$

We will show that $P(k)$ is true for every integer $n \geq 2$.

Show that $P(2)$ is true: The property is true for $k = 2$ because for $k = 2$ the left-hand side is 4 (by part (a)) and the right-hand side is $3 \cdot 1 + 1 = 4$ also.

Show that for every integer $i \geq 2$, if $P(i)$ is true then $P(i+1)$ is true: Let i be any integer with $i \geq 2$, and suppose that

$$b_i = 3b_{i-1} + 1. \qquad \leftarrow \text{inductive hypothesis}$$

We must show that

$$b_{i+1} = 3b_i + 1.$$

Now the left-hand side of $P(i+1)$ is

$$
\begin{aligned}
b_{i+1} \quad &= \quad a_i + 1 + b_i & \text{by part (c)} \\
&= \quad a_i + 1 + 3b_{i-1} + 1 & \text{by inductive hypothesis} \\
&= \quad (3a_{i-1} + 2) + 1 + 3b_{i-1} + 1 & \text{by exercise 17 (c)} \\
&= \quad 3a_{i-1} + 3 + 3b_{i-1} + 1 & \\
&= \quad 3(a_{i-1} + 1 + b_{i-1}) + 1 & \text{by algebra} \\
&= \quad 3b_i + 1 & \text{by part (c) of this exercise,}
\end{aligned}
$$

which is the right-hand side of $P(i+1)$ *[as was to be shown]*.

19. **a.** $s_1 = 1, s_2 = 1 + 1 + 1 = 3,$

$s_3 = s_1 + (1 + 1 + 1) + s_1 = 5$

b. $s_4 = s_2 + (1 + 1 + 1) + s_2 = 9$

c. Name the poles A, B, C, and D going from left to right. Because disks can be moved from one pole to any other pole, the number of moves needed to transfer a tower of disks from any one pole to any other pole is the same for any two poles. One way to transfer a tower of k disks from pole A to pole D is to first transfer the top $k - 2$ disks from pole A to pole B, next transfer the second largest disk from pole A to pole C, next transfer the largest disk from pole A to pole D, next transfer the second largest disk from pole C to pole D, and finally transfer the top $k - 2$ disks from pole B to pole D. This might not result in a minimal number of moves, but it does justify that for each integer $k \geq 3$,

$$s_k \quad \leq \quad s_{k-2} \qquad \text{(moves to transfer the top } k - 2 \text{ disks from pole } A \text{ to pole } B)$$

$$+1 \qquad \text{(move to transfer the second largest disk from pole } A \text{ to pole } C)$$

$$+1 \qquad \text{(move to transfer the largest disk from pole } A \text{ to pole } D)$$

$$+1 \qquad \text{(move to transfer the second largest disk from pole } C \text{ to pole } D)$$

$$+s_{k-2} \qquad \text{(moves to transfer the top } k - 2 \text{ disks from pole } B \text{ to pole } D)$$

$$\leq \quad 2s_{k-2} + 3.$$

20. **a.** Let c_n be the minimum number of moves needed to transfer a pile of n disks from one pole to the next adjacent pole in the clockwise direction. Name the poles A, B, and C going in a clockwise direction around the circle. Because disks can be moved from one pole to any other pole, the number of moves needed to transfer a tower of disks from any one pole to any other pole is the same for any two poles. One way to transfer a tower of k disks from pole A to pole B is to first transfer the top $k - 1$ disks from pole A to pole B, next transfer the top $k - 1$ disks from pole B to pole C, next transfer the largest disk from pole A to pole B, then transfer the top $k - 1$ disks from pole C to pole A, and finally transfer the top $k - 1$ disks from pole A to pole B. This might not result in a minimal number of moves, but it does justify that for each integer $k \geq 2$,

$$c_k \quad \leq \quad c_{k-1} \qquad \text{(moves to transfer the top } k - 1 \text{ disks from pole } A \text{ to pole } B)$$

$$+c_{k-1} \qquad \text{(moves to transfer the top } k - 1 \text{ disks from pole } B \text{ to pole } C)$$

$$+1 \qquad \text{(move to transfer the largest disk from pole } A \text{ to pole } B)$$

$$+c_{k-1} \qquad \text{(moves to transfer the top } k - 1 \text{ disks from pole } C \text{ to pole } A)$$

$$+c_{k-1} \qquad \text{(moves to transfer the top } k - 1 \text{ disks from pole } A \text{ to pole } B)$$

$$\leq \quad 4c_{k-1} + 1.$$

b. Call the poles A, B, and C. Compute c_2 by using the following sequence of steps to transfer two disks from A to B:

1 (move to transfer the top disk from A to B)

　+ 1 (move to transfer the top disk from B to C)

　+ 1 (move to transfer the bottom disk from A to B)

　+ 1 (move to transfer the top disk from C to A)

　+ 1 (move to transfer the top disk from A to B)

　= 5.

This sequence of steps is the least possible, and so $c_2 = 5$.

To compute c_3, use the following sequence of steps to start the transfer of three disks from A to B:

1 (move to transfer the top disk from A to B)

    $+ 1$ (move to transfer the top disk from B to C)

    $+ 1$ (move to transfer the middle disk from A to B)

    $+ 1$ (move to transfer the top disk from C to A)

    $+ 1$ (move to transfer the middle disk from B to C)

    $+ 1$ (move to transfer the top disk from A to B)

    $+ 1$ (move to transfer the top disk from B to C).

After these 7 steps have been completed, the bottom disk can be transferred from A to B. At that point the top two disks are on C, and a modified version of the initial seven steps can be used to transfer them from C to B. Thus the total number of steps is $c_3 = 7 + 1 + 7 = 15$, and $15 < 21 = 4c_2 + 1$.

21. **a.** $t_1 = 2,\quad t_2 = 2 + 2 + 2 = 6$

    b. $t_3 = 14$

    c. For every integer $k \geq 2$,

$$
\begin{aligned}
t_k \;=\;\; & t_{k-1} \quad \text{(moves to transfer the top } 2k - 2 \text{ disks from pole } A \text{ to pole } B) \\
& +2 \quad \text{(moves to transfer the bottom two disks from pole } A \text{ to pole } C) \\
& +t_{k-1} \quad \text{(moves to transfer the top } 2k - 2 \text{ disks from pole } B \text{ to pole } C) \\
=\;\; & 2t_{k-1} + 2.
\end{aligned}
$$

Note that transferring the stack of $2k$ disks from pole A to pole C requires at least two transfers of the top $2(k-1)$ disks: one to transfer them off the bottom two disks to free the bottom disks so that they can be moved to pole C and another to transfer the top $2(k-1)$ disks back on top of the bottom two disks. Thus at least $2t_{k-1}$ moves are needed to effect these two transfers. Two more moves are needed to transfer the bottom two disks from pole A to pole C, and this transfer cannot be effected in fewer than two moves. It follows that the sequence of moves indicated in the description of the equation above is, in fact, minimal.

22. **b.** $r_0 = 1 \quad r_1 = 1 \quad r_2 = 1 + 4 \cdot 1 = 5 \quad r_3 = 5 + 4 \cdot 1 = 9 \quad r_4 = 9 + 4 \cdot 5 = 29$
$r_5 = 29 + 4 \cdot 9 = 65 \quad r_6 = 65 + 4 \cdot 29 = 181$

23. **c.** There are 904 rabbit pairs, or 1,808 rabbits, after 12 months.

24. $F_{13} = F_{12} + F_{11} = 233 + 144 = 377, \quad F_{14} = F_{13} + F_{12} = 377 + 233 = 610$

25. **a.** Each term of the Fibonacci sequence beyond the second equals the sum of the previous two. For any integer $k \geq 1$, the two terms previous to F_{k+1} are F_k and F_{k-1}. Hence, for every integer $k \geq 1$, $F_{k+1} = F_k + F_{k-1}$.

26. By repeated use of definition of the Fibonacci sequence, for each integer $k \geq 4$,

$$
\begin{aligned}
F_k \;&=\; F_{k-1} + F_{k-2} = (F_{k-2} + F_{k-3}) + (F_{k-3} + F_{k-4}) \\
&=\; ((F_{k-3} + F_{k-4}) + F_{k-3}) + (F_{k-3} + F_{k-4}) \\
&=\; 3F_{k-3} + 2 + F_{k-4}.
\end{aligned}
$$

27. For each integer $k \geq 1$,

$$
\begin{aligned}
F_k^2 - F_{k-1}^2 \;&=\; (F_k - F_{k-1})(F_k + F_{k-1}) \quad &&\text{by basic algebra (difference of two squares)} \\
&=\; (F_k - F_{k-1})F_{k+1} \quad &&\text{by definition of the Fibonacci sequence.} \\
&=\; F_k F_{k+1} - F_{k-1}F_{k+1} \quad &&\text{by basic algebra.}
\end{aligned}
$$

30. <u>Proof (by mathematical induction)</u>: Let the property $P(n)$ be the equation

$$F_{n+2}F_n - F_{n+1}^2 = (-1)^n. \quad \leftarrow P(n)$$

We will show that $P(n)$ is true for every integer $n \geq 0$.

Show that $P(0)$ is true: The left-hand side of $P(0)$ is $F_{0+2}F_0 - F_1{}^2 = 2 \cdot 1 - 1^2 = 1$, and the right-hand side is $(-1)^0 = 1$ also. So $P(0)$ is true.

Show that for every integer $k \geq 0$, if $P(k)$ is true then $P(k+1)$ is true: Let k be any integer with $k \geq 0$, and suppose that

$$F_{k+2}F_k - F_{k+1}^2 = (-1)^k. \quad \leftarrow P(k) \text{ inductive hypothesis}$$

We must show that

$$F_{k+3}F_{k+1} - F_{k+2}^2 = (-1)^{k+1}. \quad \leftarrow P(k+1)$$

Now by inductive hypothesis,

$$F_{k+1}^2 = F_{k+2}F_k - (-1)^k = F_{k+2}F_k + (-1)^{k+1}. \quad (*)$$

Hence,

$$
\begin{aligned}
F_{k+3}&F_{k+1} - F_{k+2}^2 \\
&= (F_{k+1} + F_{k+2})F_{k+1} - F_{k+2}^2 && \text{by definition of the Fibonacci sequence} \\
&= F_{k+1}^2 + F_{k+2}F_{k+1} - F_{k+2}^2 \\
&= F_{k+2}F_k + (-1)^{k+1} + F_{k+2}F_{k+1} - F_{k+2}^2 && \text{by substitution from equation } (*) \\
&= F_{k+2}(F_k + F_{k+1} - F_{k+2}) + (-1)^{k+1} && \text{by factoring out } F_{k+2} \\
&= F_{k+2}(F_{k+2} - F_{k+2}) + (-1)^{k+1} && \text{by definition of the Fibonacci sequence} \\
&= F_{k+2} \cdot 0 + (-1)^{k+1} \\
&= (-1)^{k+1}.
\end{aligned}
$$

32. _Hint:_ Use mathematical induction. In the inductive step, use Lemma 4.10.2 and the fact that $F_{k+2} = F_{k+1} + F_k$ to deduce that

$$\gcd(F_{k+2}, F_{k+1}) = \gcd(F_{k+1}, F_k).$$

33. Let $F_0, F_1, F_2, \ldots$ be the sequence defined as follows: for each integer $n \geq 0$,

$$F_n = \frac{1}{\sqrt{5}}\left[\left(\frac{1+\sqrt{5}}{2}\right)^{n+1} - \left(\frac{1-\sqrt{5}}{2}\right)^{n+1}\right].$$

We will show that $F_k = F_{k-1} + F_{k-2}$ for every integer $k \geq 2$. To do so, suppose k is any integer with $k \geq 2$. Then by definition of $F_0, F_1, F_2, \ldots$ and algebraic manipulation,

$$
\begin{aligned}
F_k &= \frac{1}{\sqrt{5}}\left[\left(\frac{1+\sqrt{5}}{2}\right)^{k+1} - \left(\frac{1-\sqrt{5}}{2}\right)^{k+1}\right] \\
&= \frac{1}{\sqrt{5}}\left[\left(\frac{1+\sqrt{5}}{2}\right)^{k-1}\left(\frac{1+\sqrt{5}}{2}\right)^2 - \left(\frac{1-\sqrt{5}}{2}\right)^{k-1}\left(\frac{1-\sqrt{5}}{2}\right)^2\right] \\
&= \frac{1}{\sqrt{5}}\left[\left(\frac{1+\sqrt{5}}{2}\right)^{k-1}\left(\frac{1+2\sqrt{5}+5}{4}\right) - \left(\frac{1-\sqrt{5}}{2}\right)^{k-1}\left(\frac{1-2\sqrt{5}+5}{4}\right)\right] \\
&= \frac{1}{\sqrt{5}}\left[\left(\frac{1+\sqrt{5}}{2}\right)^{k-1}\left(\frac{6+2\sqrt{5}}{4}\right) - \left(\frac{1-\sqrt{5}}{2}\right)^{k-1}\left(\frac{6-2\sqrt{5}}{4}\right)\right] \\
&= \frac{1}{\sqrt{5}}\left[\left(\frac{1+\sqrt{5}}{2}\right)^{k-1}\left(\frac{3+\sqrt{5}}{2}\right) - \left(\frac{1-\sqrt{5}}{2}\right)^{k-1}\left(\frac{3-\sqrt{5}}{2}\right)\right].
\end{aligned}
$$

On the other hand, also by definition of $F_0, F_1, F_2, \ldots$ and algebraic manipulation,

$F_{k-1} + F_{k-2}$

$$= \frac{1}{\sqrt{5}} \left[\left(\frac{1+\sqrt{5}}{2} \right)^k - \left(\frac{1-\sqrt{5}}{2} \right)^k \right] + \frac{1}{\sqrt{5}} \left[\left(\frac{1+\sqrt{5}}{2} \right)^{k-1} - \left(\frac{1-\sqrt{5}}{2} \right)^{k-1} \right]$$

$$= \frac{1}{\sqrt{5}} \left[\left(\frac{1+\sqrt{5}}{2} \right)^{k-1} \left(\frac{1+\sqrt{5}}{2} \right) - \left(\frac{1-\sqrt{5}}{2} \right)^{k-1} \left(\frac{1-\sqrt{5}}{2} \right) \right] + \frac{1}{\sqrt{5}} \left[\left(\frac{1+\sqrt{5}}{2} \right)^{k-1} - \left(\frac{1-\sqrt{5}}{2} \right)^{k-1} \right]$$

$$= \frac{1}{\sqrt{5}} \left[\left(\frac{1+\sqrt{5}}{2} \right)^{k-1} \left[\left(\frac{1+\sqrt{5}}{2} \right) + 1 \right] - \left(\frac{1-\sqrt{5}}{2} \right)^{k-1} \left[\left(\frac{1-\sqrt{5}}{2} \right) + 1 \right] \right]$$

$$= \frac{1}{\sqrt{5}} \left[\left(\frac{1+\sqrt{5}}{2} \right)^{k-1} \left(\frac{1+\sqrt{5}}{2} + \frac{2}{2} \right) - \left(\frac{1-\sqrt{5}}{2} \right)^{k-1} \left(\frac{1-\sqrt{5}}{2} + \frac{2}{2} \right) \right]$$

$$= \frac{1}{\sqrt{5}} \left[\left(\frac{1+\sqrt{5}}{2} \right)^{k-1} \left(\frac{3+\sqrt{5}}{2} \right) - \left(\frac{1-\sqrt{5}}{2} \right)^{k-1} \left(\frac{3-\sqrt{5}}{2} \right) \right].$$

Thus, both F_k and $F_{k-1} + F_{k-2}$ are equal to the same quantity, and so they are equal to each other.

34. *Hint:* Let $= \lim\limits_{n \to \infty} \frac{F_{n+1}}{F_n}$ and show that $L = \frac{1}{L} + 1$. Deduce that $L = \frac{1+\sqrt{5}}{2}$.

35. *Hint:* Use the result of exercise 30 to prove that the infinite sequence $\frac{F_0}{F_1}, \frac{F_2}{F_3}, \frac{F_4}{F_5}, \ldots$ is strictly decreasing and that the infinite sequence $\frac{F_1}{F_2}, \frac{F_3}{F_4}, \frac{F_5}{F_6}, \ldots$ is strictly increasing. The first sequence is bounded below by 0, and the second sequence is bounded above by 1. Deduce that the limits of both sequences exist, and show that they are equal.

36. Let $L = \lim_{n \to \infty} x_n$. By definition of $x_0, x_1, x_2, \ldots$ and by the continuity of the square root function,

$$L = \lim_{n \to \infty} x_n = \lim_{n \to \infty} \sqrt{2 + x_{n-1}} = \sqrt{2 + \lim_{n \to \infty} x_{n-1}} = \sqrt{2 + L}.$$

Hence $L^2 = 2 + L$, and so $L^2 - L - 2 = 0$. Factoring gives $(L - 2)(L + 1) = 0$, and so $L = 2$ or $L = -1$. Now $L \geq 0$ because each $x_i \geq 0$. Thus $L = 2$.

37. **a.** Because the 4% annual interest is compounded quarterly, the quarterly interest rate is $(4\%)/4 = 1\%$. Thus $R_k = R_{k-1} + 0.01 R_{k-1} - 1.01 R_{k-1}$.

b. Because one year equals four quarters, the amount on deposit at the end of one year is $R_4 = \$5,203.02$ (rounded to the nearest cent).

c. The annual percentage yield (APY) for the account is

$$\frac{\$5203.02 - \$5000.00}{\$5000.00} = 4.0604\%.$$

39. When one is climbing a staircase consisting of n stairs, the last step taken is either a single stair or two stairs together. The number of ways to climb the staircase and have the final step be a single stair is c_{n-1}; the number of ways to climb the staircase and have the final step be two stairs is c_{n-2}. Therefore, $c_n = c_{n-1} + c_{n-2}$. Note also that $c_1 = 1$ and $c_2 = 2$ *[because either the two stairs can be climbed one by one or they can be climbed as a unit]*.

40. Imagine a tower of height k cm. If the bottom block has height 1 cm, then the remaining blocks make up a tower of height $(k - 1)$ cm. There are t_{k-1} such towers. If the bottom block has height 2 cm, then the remaining blocks make up a tower of height $(k - 2)$ cm. There are t_{k-2} such towers. If the bottom block has height 4 cm, then the remaining blocks make up a tower of height $(k - 4)$ cm. There are t_{k-4} such towers. Therefore, $t_k = t_{k-1} + t_{k-2} + t_{k-4}$ for each integer $k \geq 5$.

41. <u>Proof (by mathematical induction)</u>: Let the property, $P(n)$, be the equation

$$\sum_{i=1}^{n} ca_i = \sum_{i=1}^{n} a_i, \text{ where } a_1, a_2, a_3, \ldots, a_n$$
$$\text{and } c \text{ are any real numbers.} \qquad \leftarrow P(n)$$

Show that $P(1)$ is true:

Let a_1 and c be any real numbers. By the recursive definition of sum, $\sum_{i=1}^{1}(ca_i) = ca_1$ and $\sum_{i=1}^{1} a_i = a_1$. Therefore, $\sum_{i=1}^{1}(ca_i) = c\sum_{i=1}^{1} a_i$, and so $P(1)$ is true.

Show that for every integer $k \geq 1$, if $P(k)$ is true, then $P(k+1)$ is true:

Let k be any integer with $k \geq 1$. Suppose that

$$\sum_{i=1}^{k} ca_i = \sum_{i=1}^{k} a_i, \text{ where } a_1, a_2, a_3, \ldots, a_k$$
$$\text{and } c \text{ are any real numbers.} \qquad \leftarrow P(k) \text{ inductive hypothesis}$$

We must show that

$$\sum_{i=1}^{k+1} ca_i = \sum_{i=1}^{k+1} a_i, \text{ where } a_1, a_2, a_3, \ldots, a_{k+1}$$
$$\text{and } c \text{ are any real numbers.} \qquad \leftarrow P(k+1)$$

Let $a_1, a_2, a_3, \ldots, a_{k+1}$ and c be any real numbers. Then

$$
\begin{aligned}
\sum_{i=1}^{k+1} ca_i &= \sum_{i=1}^{k} ca_i + ca_{k+1} && \text{by the recursive definition for } \Sigma \\
&= c\sum_{i=1}^{k} a_i + ca_{k+1} && \text{by inductive hypothesis} \\
&= c\left(\sum_{i=1}^{k} a_i + a_{k+1}\right) && \text{by the distributive law for the real numbers} \\
&= c\sum_{i=1}^{k+1} a_i && \text{by the recursive definition for } \Sigma.
\end{aligned}
$$

42. <u>Proof (by mathematical induction)</u>: Let the property $P(n)$ be the sentence

$$\text{If } a_1, a_2, \ldots, a_n \text{ and } b_1, b_2, \ldots, b_n \text{ are any real}$$
$$\text{numbers, then } \prod_{i=1}^{n}(a_ib_i) = \left(\prod_{i=1}^{n} a_i\right)\left(\prod_{i=1}^{n} b_i\right). \qquad \leftarrow P(n)$$

We will show that $P(n)$ is true for every integer $n \geq 1$.

Show that $P(1)$ is true: Let a_1 and b_1 be any real numbers. By the recursive definition of product,

$$\prod_{i=1}^{1}(a_ib_i) = a_1b_1, \prod_{i=1}^{1} a_i = a_1, and \prod_{i=1}^{1} b_i = b_1.$$

Show that for every integer $k \geq 1$, if $P(k)$ is true then $P(k+1)$ is true: Let k be any integer with $k \geq 1$, and suppose that

$$\text{If } a_1, a_2, \ldots, a_k \text{ and } b_1, b_2, \ldots, b_k \text{ are any real}$$
$$\text{numbers, then } \prod_{i=1}^{k}(a_ib_i) = \left(\prod_{i=1}^{k} a_i\right)\left(\prod_{i=1}^{k} b_i\right). \qquad \leftarrow \begin{array}{l} P(k) \text{ inductive} \\ \text{hypothesis} \end{array}$$

We must show that

$$\text{If } a_1, a_2, \ldots, a_{k+1} \text{ and } b_1, b_2, \ldots, b_{k+1} \text{ are any real}$$
$$\text{numbers, then } \prod_{i=1}^{k+1}(a_ib_i) = \left(\prod_{i=1}^{k+1} a_i\right)\left(\prod_{i=1}^{k+1} b_i\right). \qquad \leftarrow P(k+1)$$

So suppose $a_1, a_2, \ldots, a_{k+1}$ and $b_1, b_2, \ldots, b_{k+1}$ are any real numbers. Then

$\prod_{i=1}^{k+1}(a_i b_i)$

$$
\begin{aligned}
&= \left(\prod_{i=1}^{k}(a_i b_i)\right)(a_{k+1} b_{k+1}) && \text{by the recursive definition of product} \\
&= \left(\left(\prod_{i=1}^{k} a_i\right)\left(\prod_{i=1}^{k} b_i\right)\right)(a_{k+1} b_{k+1}) && \text{by substitution from the inductive hypothesis} \\
&= \left(\left(\prod_{i=1}^{k} a_i\right) a_{k+1}\right)\left(\left(\prod_{i=1}^{k} b_i\right) b_{k+1}\right) && \text{by the associative and commutative laws of algebra} \\
&= \left(\prod_{i=1}^{k+1} a_i\right)\left(\prod_{i=1}^{k+1} b_i\right) && \text{by the recursive definition of product.}
\end{aligned}
$$

[This is what was to be shown.]

45. We give two proofs for the given statement, one less formal and the other more formal.

Proof 1 (by mathematical induction): For the basis step observe that any "sum" of one even integer is the integer itself, which is even. For the inductive step we suppose that for an arbitrarily chosen even integer $r \geq 1$, the sum of any r even integers is even. Then we must show that any sum of $r + 1$ even integers is even. But any sum of $r + 1$ even integers is equal to a sum of r even integers, which is even (by inductive hypothesis), plus another even integer. The result is a sum of two even integers, which is even (by Theorem 4.1.1) *[as was to be shown]*.

Proof 2 (by mathematical induction): Let $P(n)$ be the sentence:

$$\text{If } a_1, a_2, a_3, \ldots, a_n \text{ are any even integers, then } \sum_{i=1}^{n} a_i \text{ is even.} \qquad \leftarrow P(n)$$

We will show that $P(n)$ is true for every integer $n \geq 1$.

Show that $P(1)$ is true:

Suppose a_1 is any even integer. Then $\sum_{i=1}^{1} a_i = a_1$, which is even. So P(1) is true.

Show that for every integer $k \geq 1$, if $P(k)$ is true, then $P(k+1)$ is true:

Let k be any integer with $k \geq 1$, and suppose that

$$\text{If } a_1, a_2, \ldots, a_k \text{ are any even integers, then } \sum_{i=1}^{k} a_i \text{ is even.} \qquad \leftarrow P(k) \text{ inductive hypothesis}$$

We must show that

$$\text{If } a_1, a_2, \ldots, a_{k+1} \text{ are any even integers, then } \sum_{i=1}^{k+1} a_i \text{ is even.} \qquad \leftarrow P(k+1)$$

So suppose $a_1, a_2, a_3, \ldots, a_{k+1}$ are any even integers, then

$$\sum_{i=1}^{k+1} a_i = \sum_{i=1}^{k} a_i + a_{k+1}$$

by writing the final term of the sum separately. Now, by inductive hypothesis, $\sum_{i=1}^{k} a_i$ is even, and, by assumption, a_{k+1} is even. Therefore, $\sum_{i=1}^{k+1} a_i$ is the sum of two even integers, which is even (by Theorem 4.1.1).*[This is what was to be shown.]*.

47. *Hint:* Use proof by contradiction or proof by contraposition.

Section 5.7

1. a. $1 + 2 + 3 + \cdots + (k-1) = \dfrac{(k-1)((k-1)+1)}{2} = \dfrac{(k-1)k}{2}$

b. $5 + 2 + 4 + 6 + 8 + \cdots + 2n = 5 + 2(1 + 2 + 3 + \cdots + n)$

$= 5 + 2\dfrac{n(n+1)}{2} = 5 + n(n+1) = n^2 + n + 5$

2. a. $1 + 2 + 2^2 + \cdots + 2^{i-1} = \dfrac{2^{(i-1)+1} - 1}{2 - 1} = 2^i - 1$

c. $2^n + 2^{n-2} \cdot 3 + 2^{n-3} \cdot 3 + \cdots + 2^2 \cdot 3 + 2 \cdot 3 + 3$

$\quad = \quad 2^n + 3(2^{n-2} + 2^{n-3} + \cdots + 2^2 + 2 + 1)$

$\quad = \quad 2^n + 3(1 + 2 + 2^2 + \cdots + 2^{n-3} + 2^{n-2})$

$\quad = \quad 2^n + 3\left(\dfrac{2^{(n-2)+1} - 1}{2 - 1}\right)$

$\quad = \quad 2^n + 3(2^{n-1} - 1)$

$\quad = \quad 5 \cdot 2^{n-1} - 3$

d. Note that $\dfrac{1}{(-1)^n} = (-1)^n$ and $\dfrac{1}{(-1)^n} \cdot (-1)^k = \dfrac{1}{(-1)^{n-k}} = (-1)^{n-k}$. Thus

$2^n - 2^{n-1} + 2^{n-2} - 2^{n-3} + \cdots + (-1)^{n-2} \cdot 2^2 + (-1)^{n-1} \cdot 2 + (-1)^n$

$\quad = \quad \dfrac{1}{(-1)^n}((-1)^n 2^n + (-1)^{n-1} 2^{n-1} + (-1)^{n-2} 2^{n-2} + \cdots + (-1)^2 \cdot 2^2 + (-1)^1 \cdot 2^1 + 1)$

$\quad = \quad \dfrac{1}{(-1)^n}(1 - 2 + 2^2 + \cdots + (-1)^{n-1} 2^{n-1} + (-1)^n 2^n)$

$\quad = \quad (-1)^n(1 + (-2) + (-2)^2 + \cdots + (-2)^{n-1} + (-2)^n)$

$\quad = \quad (-1)^n \left(\dfrac{(-2)^{n+1} - 1}{(-2) - 1}\right)$

$\quad = \quad (-1)^n \left(\dfrac{1 - (-2)^{n+1}}{3}\right)$

$\quad = \quad \dfrac{(-1)^n + 2^{n+1}}{3}.$

3.

$a_0 = 1$

$a_1 = 1 \cdot a_0 = 1 \cdot 1 = 1$

$a_2 = 2a_1 = 2 \cdot 1$

$a_3 = 3a_2 = 3 \cdot 2 \cdot 1$

$a_4 = 4a_3 = 4 \cdot 3 \cdot 2 \cdot 1$

$\vdots$

Guess: $a_n = n(n-1) \cdots 3 \cdot 2 \cdot 1 = n!$ for every integer $n \geq 1$.

5.

$$c_1 = 1$$
$$c_2 = 3c_1 + 1 = 3 \cdot 1 + 1 = 3 + 1$$
$$c_3 = 3c_2 + 1 = 3 \cdot (3 + 1) + 1 = 3^2 + 3 + 1$$
$$c_4 = 3c_3 + 1 = 3 \cdot (3^2 + 3 + 1) + 1$$
$$= 3^3 + 3^2 + 3 + 1$$
$$\vdots$$

Guess: $c_n = 3^{n-1} + 3^{n-2} + \cdots + 3^3 + 3^2 + 3 + 1$

$$= \frac{3^n - 1}{3 - 1} \qquad \textit{[by Theorem 5.2.2 with } r = 3\textit{]}$$

$$= \frac{3^n - 1}{2} \text{ for every integer } n \geq 1.$$

6.

$$d_1 = 2$$
$$d_2 = 2d_1 + 3 = 2 \cdot 2 + 3 = 2^2 + 3$$
$$d_3 = 2d_2 + 3 = 2(2^2 + 3) + 3 = 2^3 + 2 \cdot 3 + 3$$
$$d_4 = 2d_3 + 3 = 2(2^3 + 2 \cdot 3 + 3) + 3 = 2^4 + 2^2 \cdot 3 + 2 \cdot 3 + 3$$
$$d_5 = 2d_4 + 3 = 2(2^4 + 2^2 \cdot 3 + 2 \cdot 3 + 3) + 3 = 2^5 + 2^3 \cdot 3 + 2^2 \cdot 3 + 2 \cdot 3 + 3$$
$$\vdots$$

Guess: $d_n = 2^n + 2^{n-2} \cdot 3 + 2^{n-3} \cdot 3 + \cdots + 2^2 \cdot 3 + 2 \cdot 3 + 3$
$$= 2^n + 3(2^{n-2} + 2^{n-3} + \cdots + 2^2 + 2 + 1)$$
$$= 2^n + 3\left(\frac{2^{(n-2)+1} - 1}{2 - 1}\right) \quad \textit{[by Theorem 5.2.2 with } r = 2\textit{]}$$
$$= 2^n + 3(2^{n-1} - 1)$$
$$= 2^{n-1}(2 + 3) - 3 = 5 \cdot 2^{n-1} - 3 \text{ for every integer } n \geq 1$$

9.

$$g_1 = 1$$
$$g_2 = \frac{g_1}{g_1 + 2} = \frac{1}{1 + 2}$$

$$g_3 = \frac{g_2}{g_2 + 2} = \frac{\frac{1}{1+2}}{\frac{1}{1+2} + 2} = \frac{1}{1 + 2(1+2)} = \frac{1}{1 + 2 + 2^2}$$

$$g_4 = \frac{g_3}{g_3 + 2} = \frac{\frac{1}{1+2+2^2}}{\frac{1}{1+2+2^2} + 2} = \frac{1}{1 + 2(1 + 2 + 2^2)} = \frac{1}{1 + 2 + 2^2 + 2^3}$$

$$g_5 = \frac{g_4}{g_4 + 2} = \frac{\frac{1}{1+2+2^2+2^3}}{\frac{1}{1+2+2^2+2^3} + 2} = \frac{1}{1 + 2(1 + 2 + 2^2 + 2^3)} = \frac{1}{1 + 2 + 2^2 + 2^3 + 2^4}$$

$$\vdots$$

Guess: $g_n = \dfrac{1}{1 + 2 + 2^2 + 2^3 + \cdots + 2^{n-1}}$

$$= \frac{1}{2^n - 1} \quad \textit{[by Theorem 5.2.2 with } r = 2\textit{]} \quad \text{for every integer } n \geq 1$$

10.
$$h_0 = 1$$
$$h_2 = 2^2 - h_1 = 2^2 - (2^1 - 1) = 2^2 - 2^1 + 1$$
$$h_3 = 2^3 - h_2 = 2^3 - (2^2 - 2^1 + 1) = 2^3 - 2^2 + 2^1 - 1$$
$$h_4 = 2^4 - h_3 = 2^4 - (2^3 - 2^2 + 2^2 - 1) = 2^4 - 2^3 + 2^2 - 2^1 + 1$$
$$\vdots$$

Guess:
$$\begin{aligned}
h_n &= 2^n - 2^{n-1} + \cdots + (-1)^n \cdot 1 \\
&= (-1)^n[1 - 2 + 2^2 - \cdots + (-1)^n \cdot 2^n] \\
&= (-1)^n[1 + (-2) + (-2)^2 - \cdots + (-2)^n] \qquad \text{\textit{by basic algebra}} \\
&= (-1)^n\left[\frac{(-2)^{n+1} - 1}{(-2) - 1}\right] \qquad\qquad \text{\textit{by Theorem 5.2.2 with } } r = -2 \\
&= \frac{(-1)^{n+1} \cdot [(-2)^{n+1} - 1]}{(-1) \cdot (-3)} \\
&= \frac{2^{n+1} - (-1)^{n+1}}{3} \qquad \text{for every integer } n \geq 0.
\end{aligned}$$

12.
$$s_0 = 3$$
$$s_1 = s_0 + 2 \cdot 1 = 3 + 2 \cdot 1$$
$$s_2 = s_1 + 2 \cdot 2 = [3 + 2 \cdot 1] + 2 \cdot 2 = 3 + 2 \cdot (1 + 2)$$
$$s_3 = s_2 + 2 \cdot 3 = [3 + 2 \cdot (1 + 2)] + 2 \cdot 3 = 3 + 2 \cdot (1 + 2 + 3)$$
$$s_4 = s_3 + 2 \cdot 4 = [3 + 2 \cdot (1 + 2 + 3)] + 2 \cdot 4 = 3 + 2 \cdot (1 + 2 + 3 + 4)$$
$$\vdots$$

Guess:
$$\begin{aligned}
s_n &= 3 + 2 \cdot (1 + 2 + 3 + \cdots + (n - 1) + n) \\
&= 3 + 2 \cdot \frac{n(n+1)}{2} \qquad\qquad \text{\textit{by Theorem 5.2.1}} \\
&= 3 + n(n + 1) \quad \text{for every integer } n \geq 0.
\end{aligned}$$

14.
$$x_1 = 1$$
$$x_2 = 3x_1 + 2 = 3 + 2$$
$$x_3 = 3x_2 + 3 = 3(3 + 2) + 3 = 3^2 + 3 \cdot 2 + 3$$
$$x_4 = 3x_3 + 4 = 3(3^2 + 3 \cdot 2 + 3) + 4 = 3^3 + 3^2 \cdot 2 + 3 \cdot 3 + 4$$
$$x_5 = 3x_4 + 5 = 3(3^3 + 3^2 \cdot 2 + 3 \cdot 3 + 4) + 5 = 3^4 + 3^3 \cdot 2 + 3^2 \cdot 3 + 3 \cdot 4 + 5$$
$$x_6 = 3x_5 + 6 = 3(3^4 + 3^3 \cdot 2 + 3^2 \cdot 3 + 4 \cdot 3 + 5) + 6$$
$$\quad = 3^5 + 3^4 \cdot 2 + 3^3 \cdot 3 + 3^2 \cdot 4 + 3 \cdot 5 + 6$$
$$\vdots$$

Guess:
$$x_n = 3^{n-1} + 3^{n-2} \cdot 2 + 3^{n-3} \cdot 3 + \cdots + 3(n - 1) + n$$

$$= 3^{n-1} + \underbrace{3^{n-2} + 3^{n-2}}_{2 \text{ times}} + \underbrace{3^{n-3} + 3^{n-3} + 3^{n-3}}_{3 \text{ times}} + \underbrace{3 + 3 + \cdots + 3}_{(n-1) \text{ times}} + \underbrace{1 + 1 + \cdots + 1}_{n \text{ times}}$$

$$= (3^{n-1} + 3^{n-2} + \cdots + 3^2 + 3 + 1) + (3^{n-2} + 3^{n-3} + \cdots$$
$$\qquad\qquad + 3^2 + 3 + 1) + \cdots + (3^2 + 3 + 1) + (3 + 1) + 1$$

$$= \frac{3^n - 1}{2} + \frac{3^{n-1} - 1}{2} + \cdots + \frac{3^3 - 1}{2} + \frac{3^2 - 1}{2} + \frac{3 - 1}{2}$$

$$= \frac{1}{2}[(3^n + 3^{n-1} + \cdots + 3^2 + 3) - n]$$

$$= \frac{1}{2}[3(3^{n-1} + 3^{n-2} + \cdots + 3 + 1) - n]$$

$$= \frac{1}{2}\left(3\left(\frac{3^n - 1}{3 - 1}\right) - n\right)$$

$$= \frac{1}{4}(3^{n+1} - 3 - 2n)$$

15.
$$y_1 = 1$$
$$y_2 = y_1 + 2^2 = 1 + 2^2$$
$$y_3 = y_2 + 3^2 = (1 + 2^2) + 3^2 = 1 + 2^2 + 3^2$$
$$y_4 = y_3 + 4^2 = (1 + 2^2 + 3^2) + 4^2 = 1^2 + 2^2 + 3^2 + 4^2$$
$$\vdots$$

Guess:

$$y_n = 1^2 + 2^2 + 3^2 + \cdots + n^2 = \frac{n(n+1)(2n+1)}{6} \text{ by exercise 10 of Section 5.2}$$

18. Proof (by mathematical induction): Let d be any fixed constant, and let $a_0, a_1, a_2, \ldots$ be the sequence defined recursively by $a_k = a_{k-1} + d$ for each integer $k \geq 1$.

The property $P(n)$ is the equation

$$a_n = a_0 + nd. \qquad \leftarrow \; P(n)$$

We show by mathematical induction that $P(n)$ is true for every integer $n \geq 0$.

Show that $P(0)$ is true:

When $n = 0$, the left-hand side of the equation is a_0, and the right-hand side is $a_0 + 0 \cdot d = a_0$, which equals the left-hand side. Thus $P(0)$ is true.

Show that for every integer $k \geq 0$, if $P(k)$ is true, then $P(k+1)$ is true:

Suppose k is any integer such that $k \geq 0$ and

$$a_k = a_0 + kd. \qquad \leftarrow P(k) \text{ inductive hypothesis}$$

We must show that

$$a_{k+1} = a_0 + (k+1)d. \qquad \leftarrow \; P(k+1)$$

Now the left-hand side of $P(k+1)$ is

$$
\begin{aligned}
a_{k+1} \; &= \quad a_k + d && \text{by definition of } a_0, a_1, a_2, \ldots \\
&= \quad [a_0 + kd] + d && \text{by substitution from the inductive hypothesis} \\
&= \quad a_0 + (k+1)d && \text{by basic algebra}
\end{aligned}
$$

which is the right-hand side of $P(k+1)$ [as was to be shown].

19. Let $U_n =$ the number of units produced on day n. Then

$$U_k = U_{k-1} + 2 \qquad \text{for each integer } k \geq 1,$$
$$U_0 = 170.$$

Hence $U_0, U_1, U_2, \ldots$ is an arithmetic sequence with fixed constant 2. It follows that when $n = 30$,

$$U_n = U_0 + n \cdot 2 = 170 + 2n = 170 + 2 \cdot 30$$
$$= 230 \text{ units.}$$

21. Proof (by mathematical induction): Let r be a fixed constant and $a_0, a_1, a_2, \ldots$. a sequence that satisfies the recurrence relation $a_k = ra_{k-1}$ for each integer $k \geq 1$ with initial condition $a_0 = a$. Let the property $P(n)$ be the equation

$$a_n = ar^n. \qquad \leftarrow \; P(n)$$

We will show that $P(n)$ is true for every integer $n \geq 0$.

Show that $P(0)$ is true: The right-hand side of $P(0)$ is $ar^0 = a \cdot 1 = a$, which is the left-hand side of $P(0)$. So $P(0)$ is true.

Show that for every integer $k \geq 0$, if $P(k)$ is true then $P(k+1)$ is true: Let k be any integer with $k \geq 0$, and suppose that

$$a_k = ar^k. \qquad \leftarrow P(k) \text{ inductive hypothesis}$$

We must show that

$$a_{k+1} = ar^{k+1}. \qquad \leftarrow P(k+1)$$

The left-hand side of $P(k+1)$ is

$$
\begin{aligned}
a_{k+1} &= ra_k & \text{by definition of } a_0, a_1, a_2, \ldots \\
&= r(ar^k) & \text{by inductive hypothesis} \\
&= ar^{k+1} & \text{by the laws of exponents,}
\end{aligned}
$$

which is the right-hand side of $P(k+1)$, *[as to be shown]*.

24. $\sum_{k=0}^{20} 5^k = \dfrac{5^{21}-1}{4} \cong 1.192 \times 10^{14} \cong 119{,}200{,}000{,}000{,}000 \cong 119$ trillion people (This is about 20,000 times the current population of the earth!)

26. **b.** *Hint*: Before simplification, $A_n = 1{,}000(1.0025)^n + 200[(1.0025)^{n-1} + (1.0025)^{n-1} + \cdots + (1.0025)^2 + 1.0025 + 1]$.

    d. $A_{240} \cong \$67{,}481.15$, $A_{480} \cong \$188{,}527.05$

    e. *Hint*: Use logarithms to solve the equation $A_n = 10{,}000$, where A_n is the expression found after simplifying the result in part (b).

27. **a.** Let the original balance in the account be A dollars, and let A_n be the amount owed in month n assuming the balance is not reduced by making payments during the year. The annual interest rate is 18%, and so the monthly interest rate is $(18/12)\% = 1.5\% = 0.015$. The sequence $A_0, A_1, A_2, \ldots$ satisfies the recurrence relation

$$A_k = A_{k-1} + 0.015A_{k-1} = 1.015A_{k-1}.$$

Thus
$$
\begin{aligned}
A_1 &= 1.015A_0 = 1.015A \\
A_2 &= 1.015A_1 = 1.015(1.015A) = (1.015)^2 A \\
&\;\;\vdots \\
A_{12} &= 1.015A_{11} = 1.015(1.015)^{11}A = (1.015)^{12}A.
\end{aligned}
$$

So the amount owed at the end of the year is $(1.015)^{12}A$. It follows that the APY is

$$\frac{(1.015)^{12}A - A}{A} = \frac{A((1.015)^{12} - 1)}{A} = (1.015)^{12} - 1 \cong 19.6\%.$$

Note: Because $A_k = 1.015A_{k-1}$ for each integer $k \geq 1$, we could have immediately concluded that the sequence is geometric and, therefore, satisfies the equation $A_n = A_0(1.015)^n = A(1.015)^n$.

b. Because the person pays \$150 per month to pay off the loan, the balance at the end of month k is $B_k = 1.015B_{k-1} - 150$. We use iteration to find an explicit formula for $B_0, B_1, B_2, \ldots$.

$B_0 = 3000$

$B_1 = (1.015)B_0 - 150 = 1.015(3000) - 150$

$B_2 = (1.015)B_1 - 150 = (1.015)[1.015(3000) - 150] - 150$

$\quad = 3000(1.015)^2 - 150(1.015) - 150$

$B_3 = (1.015)B_2 - 150 = (1.015)[3000(1.015)^2 - 150(1.015) - 150] - 150$

$\quad = 3000(1.015)^3 - 150(1.015)^2 - 150(1.015) - 150$

$B_4 = (1.015)B_3 - 150$

$\quad = (1.015)[3000(1.015)^3 - 150(1.015)^2 - 150(1.015) - 150] - 150$

$\quad = 3000(1.015)^4 - 150(1.015)^3 - 150(1.015)^2 - 150(1.015) - 150$

$\quad\vdots$

Guess: $B_n = 3000(1.015)^n + [150(1.015)^{n-1} - 150(1.015)^{n-2} + \ldots$

$$-150(1.015)^2 - 150(1.015) - 150]$$

$\quad = 3000(1.015)^n - 150[(1.015)^{n-1} + (1.015)^{n-2} + \cdots + (1.015)^2 + 1.015 + 1]$

$\quad = 3000(1.015)^n - 150 \left(\dfrac{(1.015)^n - 1}{1.015 - 1} \right)$

$\quad = (1.015)^n(3000) - \dfrac{150}{0.015}((1.015)^n - 1)$

$\quad = (1.015)^n(3000) - 10000((1.015)^n - 1)$

$\quad = (1.015)^n(3000 - 10000) + 10000$

$\quad = (-7000)(1.015)^n + 10000$

So it appears that $B_n = (-7000)(1.015)^n + 10000$. We use mathematical induction to confirm this guess.

Proof (by mathematical induction): Let $B_0, B_1, B_2, \ldots.$ be a sequence that satisfies the recurrence relation $B_k = (1.015)B_{k-1} - 150$ for each integer $k \geq 1$, with initial condition $B_0 = 3000$, and let the property $P(n)$ be the equation

$$B_n = (-7000)(1.015)^n + 10000. \qquad \leftarrow P(n)$$

We will show that $P(n)$ is true for every integer $n \geq 0$.

Show that $P(0)$ is true: The right-hand side of $P(0)$ is $(-7000)(1.015)^0 + 10000 = 3000$, which equals B_0, the left-hand side of $P(0)$. So $P(0)$ is true.

Show that for every integer $k \geq 0$, if $P(k)$ is true then $P(k+1)$ is true: Let k be any integer with $k \geq 0$, and suppose that

$$B_k = (-7000)(1.015)^k + 10000. \qquad \leftarrow P(k) \text{ inductive hypothesis}$$

We must show that

$$B_{k+1} = (-7000)(1.015)^{k+1} + 10000. \qquad \leftarrow P(k+1)$$

The left-hand side of $P(k+1)$ is

$\begin{aligned} B_{k+1} &= (1.015)B_k - 150 & \text{by definition of } B_0, B_1, B_2, \ldots \\ &= (1.015)[(-7000)(1.015)^k + 10000] - 150 & \text{by substitution from} \\ & & \text{the inductive hypothesis} \\ &= (-7000)(1.015)^{k+1} + 10150 - 150 \\ &= (-7000)(1.015)^{k+1} + 10000 & \text{by the laws of algebra,} \end{aligned}$

which is the right-hand side of $P(k+1)$ *[as to be shown]*.

c. By part (b), $B_n = (-7000)(1.015)^n + 10000$, and so we need to find the value of n for which

$$(-7000)(1.015)^n + 10000 = 0.$$

Now this equation holds

$$\Leftrightarrow \qquad 7000(1.015)^n = 10000$$

$$\Leftrightarrow \qquad (1.015)^n = \frac{10000}{7000} = \frac{10}{7}$$

$$\Leftrightarrow \quad \log_{10}(1.015)^n = \log_{10}\left(\frac{10}{7}\right) \qquad \text{by a property of logarithms}$$

$$\Leftrightarrow \quad n\log_{10}(1.015) = \log_{10}\left(\frac{10}{7}\right) \qquad \text{by a property of logarithms}$$

$$\Leftrightarrow \qquad n = \frac{\log_{10}(10/7)}{\log_{10}(1.015)} \cong 24.$$

So $n \cong 24$ months $= 2$ years. It will require approximately 2 years to pay off the balance, assuming that payments of \$150 are made each month and the balance is not increased by any additional purchases.

d. Assuming that the person makes no additional purchases and pays \$150 each month, the person will have made 24 payments of \$150 each, for a total of \$3600 to pay off the initial balance of \$3000.

28. Proof (by mathematical induction): Let $a_0, a_1, a_2, \ldots$ be the sequence defined recursively by $a_0 = 1$ and $a_k = ka_{k-1}$ for each integer $k \geq 1$, and let the property $P(n)$ be the equation

$$a_n = n!. \qquad \leftarrow P(n)$$

We show by mathematical induction that $P(n)$ is true for every integer $n \geq 0$.

Show that $P(0)$ is true: When $n = 0$, the right-hand side of the equation is $0! = 1$, and by definition of a_0, a_1, a_2, the left-hand side of the equation, a_0, is also 1. Thus the property is true for $n = 0$.

Show that for every integer $k \geq 0$, if $P(k)$ is true, then $P(k+1)$ is true: Suppose k is any integer with $k \geq 0$ and

$$a_k = k!. \qquad \leftarrow P(k) \text{ inductive hypothesis}$$

We must show that

$$a_{k+1} = (k+1)!. \qquad \leftarrow \ P(k+1)$$

The left-hand side of $P(k+1)$ is

$$\begin{aligned} a_{k+1} \ \ &= \ \ (k+1)\cdot a_k \quad \text{by definition of } a_0, a_1, a_2, \ldots \\ &= \ \ (k+1)\cdot k! \quad \text{by inductive hypothesis} \\ &= \ \ (k+1)! \qquad \text{by definition of factorial,} \end{aligned}$$

which is the right-hand side of $P(k+1)$. *[Hence if P(k) is true, then P(k+1) is true, as was to be shown.]*

30. Proof (by mathematical induction): Let $c_1, c_2, c_3, \ldots$ be the sequence defined recursively by $c_1 = 1$ and $c_k = 3c_{k-1} + 1$ for each integer $k \geq 2$.

Let the property $P(n)$ be the equation

$$c_n = \frac{3^n - 1}{2}. \qquad \leftarrow P(n)$$

We show by mathematical induction that $P(n)$ is true for every integer $n \geq 1$.

Show that $P(1)$ is true: When $n = 1$, the right-hand side of the equation is $\dfrac{3^1 - 1}{2} = \dfrac{3 - 1}{2} = 1$, and by definition of $c_1, c_2, c_3, \ldots$, the left-hand side of the equation, c_1, is also 1. Thus the property is true for $n = 1$.

Show that for every integer $k \geq 1$, if $P(k)$ is true, then $P(k + 1)$ is true: Suppose that k is any integer with $k \geq 1$ and

$$c_k = \frac{3^k - 1}{2}. \qquad \leftarrow P(k) \text{ inductive hypothesis}$$

We must show that

$$c_{k+1} = \frac{3^{k+1} - 1}{2}. \qquad \leftarrow P(k+1)$$

The left-hand side of $P(k + 1)$ is

$$
\begin{aligned}
c_{k+1} &= 3c_k + 1 && \text{by definition of } c_1, c_2, c_3, \ldots \\
&= 3\left(\frac{3^k - 1}{2}\right) + 1 && \text{by substitution from the inductive hypothesis} \\
&= \frac{3^{k+1} - 3}{2} + \frac{2}{2} \\
&= \frac{3^{k+1} - 1}{2} && \text{by basic algebra,}
\end{aligned}
$$

which is the right-hand side of $P(k + 1)$ *[as to be shown]*.

33. Proof (by mathematical induction): Let $f_1, f_2, f_3, \ldots$ be a sequence that satisfies the recurrence relation $f_k = f_{k-1} + 2^k$ for each integer $k \geq 2$, with initial condition $f_1 = 1$, and let the property $P(n)$ be the equation

$$f_n = 2^{n+1} - 3. \qquad \leftarrow P(n)$$

We will show that $P(n)$ is true for every integer $n \geq 1$.

Show that $P(1)$ is true: The right-hand side of $P(1)$ is $2^{1+1} - 3 = 2^2 - 3 = 1$, which equals f_1, the left-hand side of $P(1)$. So $P(1)$ is true.

Show that for every integer $k \geq 1$, if $P(k)$ is true then $P(k + 1)$ is true: Let k be any integer with $k \geq 1$, and suppose that

$$f_k = 2^{k+1} - 3. \qquad \leftarrow P(k) \text{ inductive hypothesis}$$

We must show that

$$f_{k+1} = 2^{(k+1)+1} - 3$$

or, equivalently,

$$f_{k+1} = 2^{k+2} - 3. \qquad \leftarrow P(k+1)$$

The left-hand side of $P(k + 1)$ is

$$
\begin{aligned}
f_{k+1} &= f_k + 2^{k+1} && \text{by definition of } f_1, f_2, f_3, \ldots \\
&= 2^{k+1} - 3 + 2^{k+1} && \text{by inductive hypothesis} \\
&= 2 \cdot 2^{k+1} - 3 \\
&= 2^{k+2} - 3 && \text{by the laws of algebra,}
\end{aligned}
$$

which is the right-hand side of $P(k + 1)$ *[as to be shown]*.

35. *Hint:*
$$2^{k+1} - \frac{2^{k+1} - (-1)^{k+1}}{3} = \frac{3 \cdot 2^{k+1}}{3} - \frac{2^{k+1} - (-1)^{k+1}}{3} = \frac{2 \cdot 2^{k+1} + (-1)^{k+1}}{3} = \frac{2^{k+2} - (-1)^{k+2}}{3}$$

36. Proof (by mathematical induction): Let $p_1, p_2, p_3, \ldots$ be a sequence that satisfies the recurrence relation $p_k = p_{k-1} + 2 \cdot 3^k$ for each integer $k \geq 2$, with initial condition $p_1 = 2$, and let the property $P(n)$ be the equation

$$p_n = 3^{n+1} - 7. \qquad \leftarrow P(n)$$

We will show that $P(n)$ is true for every integer $n \geq 1$.

Show that $P(1)$ *is true*: The right-hand side of $P(1)$ is $3^{1+1} - 7 = 3^2 - 7 = 9 - 7 = 2$, which equals p_1, the left-hand side of $P(1)$. So $P(1)$ is true.

Show that for every integer $k \geq 1$, if $P(k)$ is true then $P(k + 1)$ is true: Let k be any integer with $k \geq 1$, and suppose that

$$p_k = 3^{k+1} - 7. \qquad \leftarrow P(k) \text{ inductive hypothesis}$$

We must show that
$$p_{k+1} = 3^{(k+1)+1} - 7,$$

or, equivalently, that

$$p_{k+1} = 3^{k+2} - 7. \qquad \leftarrow P(k + 1)$$

The left-hand side of $P(k + 1)$ is

$$
\begin{aligned}
p_{k+1} &= p_k + 2 \cdot 3^{k+1} & \text{by definition of } p_1, p_2, p_3, \ldots \\
&= (3^{k+1} - 7) + 2 \cdot 3^{k+1} & \text{by inductive hypothesis} \\
&= 3^{k+1}(1 + 2) - 7 & \\
&= 3 \cdot 3^{k+1} - 7 & \\
&= 3^{k+2} - 7 & \text{by the laws of algebra,}
\end{aligned}
$$

which is the right-hand side of $P(k + 1)$ *[as to be shown].*

37. *Hint:* $[3 + k(k + 1)] + 2(k + 1) = 3 + k^2 + k + 2k + 2 = 3 + [k^2 + 3k + 2]$
$$= 3 + (k + 1)(k + 2) = 3 + (k + 1)[(k + 1) + 1] = 3 + (k + 1)(k + 2)$$

39. Proof (by mathematical induction): Let $x_1, x_2, x_3, \ldots$ be the sequence defined recursively by $x_1 = 1$ and $x_k = 3x_{k-1} + k$ for each integer $k \geq 2$. Let the property, $P(n)$, be the equation

$$x_n = \frac{3^{n+1} - 2n - 3}{4}. \qquad \leftarrow P(n)$$

We will show by mathematical induction that $P(n)$ is true for every integer $n \geq 1$.

Show that $P(1)$ *is true:* When $n = 1$, the right-hand side of the equation is $\dfrac{3^{1+1} - 2 \cdot 1 - 3}{4} = \dfrac{3^2 - 2 - 3}{4} = 1$, and by definition of $x_1, x_2, x_3, \ldots$, the left-hand side of the equation, x_1, is also 1. Thus $P(1)$ is true.

Show that for every integer $k \geq 1$, if $P(k)$ is true then $P(k + 1)$ is true:

Suppose that k is any integer with $k \geq 0$ and

$$x_k = \frac{3^{k+1} - 2k - 3}{4}. \qquad \leftarrow P(k) \text{ inductive hypothesis}$$

We must show that

$$x_{k+1} = \frac{3^{(k+1)+1} - 2(k+1) - 3}{4},$$

or, equivalently, that

$$x_{k+1} = \frac{3^{k+2} - 2k - 5}{4}. \qquad \leftarrow \; P(k+1)$$

The left-hand side of $P(k+1)$ is

$$
\begin{aligned}
x_{k+1} \;&=\; 3x_k + k && \text{by definition of } y_1, y_2, y_3, \ldots \\[4pt]
&=\; 3\left(\frac{3^{k+1} - 2k - 3}{4}\right) + k + 1 && \text{by inductive hypothesis} \\[4pt]
&=\; \frac{3 \cdot 3^{k+1} - 3 \cdot 2k - 3 \cdot 3}{4} + \frac{4(k+1)}{4} \\[4pt]
&=\; \frac{3^{k+2} - 6k - 9 + 4k + 4}{4} \\[4pt]
&=\; \frac{3^{k+2} - 2k - 5}{4} && \text{by the laws of algebra,}
\end{aligned}
$$

which is the right-hand side of $P(k+1)$ *[as to be shown]*.

42. Proof (by mathematical induction): Let $t_1, t_2, t_3, \ldots$. be a sequence that satisfies the recurrence relation $t_k = 2t_{k-1} + 2$ for each integer $k \geq 2$, with initial condition $t_1 = 2$, and let the property $P(n)$ be the equation $t_n = 2^{n+1} - 2$.

$$t_n = 2^{n+1} - 2. \qquad \leftarrow \; P(n)$$

We will show that $P(n)$ is true for every integer $n \geq 1$.

Show that $P(1)$ is true: The right-hand side of $P(1)$ is $2^2 - 2 = 2$, which equals t_1, the left-hand side of $P(1)$. So $P(1)$ is true.

Show that for every integer $k \geq 1$, if $P(k)$ is true then $P(k+1)$ is true: Let k be any integer with $k \geq 1$, and suppose that

$$t_k = 2^{k+1} - 2. \qquad \leftarrow \; P(k) \text{ inductive hypothesis}$$

We must show that

$$t_{k+1} = 2^{(k+1)+1} - 2$$

or, equivalently, that

$$t_{k+1} = 2^{k+2} - 2. \qquad \leftarrow \; P(k+1)$$

The left-hand side of $P(k+1)$ is

$$
\begin{aligned}
t_{k+1} \;&=\; 2t_k + 2 && \text{by definition of } t_1, t_2, t_3, \ldots \\
&=\; 2(2^{k+1} - 2) + 2 && \text{by inductive hypothesis} \\
&=\; 2^{k+2} - 4 + 2 \\
&=\; 2^{k+2} - 2 && \text{by the laws of algebra,}
\end{aligned}
$$

which is the right-hand side of $P(k+1)$ *[as to be shown]*.

43. a.

$$a_0 = 2$$

$$a_1 = \frac{a_0}{2a_0 - 1} = \frac{2}{2 \cdot 2 - 1} = \frac{2}{3}$$

$$a_2 = \frac{a_1}{2a_1 - 1} = \frac{\frac{2}{3}}{2 \cdot \frac{2}{3} - \frac{3}{3}} = \frac{\frac{2}{3}}{\frac{1}{3}} = 2$$

$$a_3 = \frac{a_2}{2a_2 - 1} = \frac{2}{2 \cdot 2 - 1} = \frac{2}{3}$$

$$a_4 = \frac{a_3}{2a_3 - 1} = \frac{\frac{2}{3}}{2 \cdot \frac{2}{3} - \frac{3}{3}} = \frac{\frac{2}{3}}{\frac{1}{3}} = 2$$

$$\vdots$$

Guess: $a_n = \begin{cases} 2 & \text{if } n \text{ is even} \\ \dfrac{2}{3} & \text{if } n \text{ is odd.} \end{cases}$

b. Proof (by strong mathematical induction): Let $a_0, a_1, a_2, \ldots$ be the sequence defined recursively by

$$a_0 = 2 \quad \text{and} \quad a_k = \frac{a_{k-1}}{2a_{k-1} - 1} \quad \text{for each integer } k \geq 1.$$

Let the property, $P(n)$, be the equation

$$a_n = \begin{cases} 2 & \text{if } n \text{ is even} \\ \dfrac{2}{3} & \text{if } n \text{ is odd.} \end{cases} \quad \leftarrow P(n)$$

We show by strong mathematical induction that $P(n)$ is true for every integer $n \geq 1$.

Show that $P(0)$ and $P(1)$ are true:

The results of part (a) show that $P(0)$ and $P(1)$ are true.

Show that if $k \geq 1$ and $P(i)$ is true for every integer i from 0 through k, then $P(k + 1)$ is true: Let k be any integer with $k \geq 0$, and suppose that

$$a_i = \begin{cases} 2 & \text{if } i \text{ is even} \\ \dfrac{2}{3} & \text{if } i \text{ is odd.} \end{cases} \quad \text{for every integer } i \text{ with } 0 \leq i \leq k. \quad \leftarrow \begin{array}{l} \text{inductive} \\ \text{hypothesis} \end{array}$$

We must show that

$$a_{k+1} = \begin{cases} 2 & \text{if } k \text{ is even} \\ \dfrac{2}{3} & \text{if } k \text{ is odd.} \end{cases} \quad \leftarrow P(k + 1)$$

Now

$$a_{k+1} = \frac{a_k}{2a_k - 1}$$ by definition of $a_0, a_1, a_2, \ldots$

$$= \begin{cases} \dfrac{2}{2 \cdot 2 - 1} & \text{if } k \text{ is even} \\[3mm] \dfrac{\frac{2}{3}}{2 \cdot \frac{2}{3} - 1} & \text{if } k \text{ is odd} \end{cases}$$ by inductive hypothesis

$$= \begin{cases} \dfrac{2}{3} & \text{if } k \text{ is even} \\[3mm] -\dfrac{\frac{2}{3}}{\frac{1}{3}} & \text{if } k \text{ is odd} \end{cases}$$

$$= \begin{cases} \dfrac{2}{3} & \text{if } k + 1 \text{ is odd} \\[3mm] 2 & \text{if } k + 1 \text{ is even} \end{cases}$$ because $k + 1$ is odd when k is even

and $k + 1$ is even when k is odd

[as was to be shown].

45. a.

$$v_1 = 1$$

$$v_2 = v_{\lfloor 2/2 \rfloor} + v_{\lfloor 3/2 \rfloor} + 2 = v_1 + v_1 + 2 = 1 + 1 + 2$$

$$v_3 = v_{\lfloor 3/2 \rfloor} + v_{\lfloor 4/2 \rfloor} + 2 = v_1 + v_2 + 2 = 1 + (1 + 1 + 2) + 2 = 3 + 2 \cdot 2$$

$$v_4 = v_{\lfloor 4/2 \rfloor} + v_{\lfloor 5/2 \rfloor} + 2 = v_2 + v_2 + 2 = (1 + 1 + 2) + (1 + 1 + 2) + 2 = 4 + 3 \cdot 2$$

$$v_5 = v_{\lfloor 5/2 \rfloor} + v_{\lfloor 6/2 \rfloor} + 2 = v_2 + v_3 + 2 = (3 + 2 \cdot 2) + (1 + 1 + 2) + 2 = 5 + 4 \cdot 2$$

$$v_6 = v_{\lfloor 6/2 \rfloor} + v_{\lfloor 7/2 \rfloor} + 2 = v_3 + v_3 + 2 = (3 + 2 \cdot 2) + (3 + 2 \cdot 2) + 2 = 6 + 5 \cdot 2$$

$$\vdots$$

Guess: $v_n = n + 2(n - 1) = 3n - 2$ for every integer $n \geq 1$

Proof (by strong mathematical induction): Let $v_1, v_2, v_3, \ldots$ be the sequence defined recursively by $v_1 = 1$ and $v_k = v_{\lfloor k/2 \rfloor} + v_{\lfloor (k+1)/2 \rfloor} + 2$ for each integer $k \geq 1$. Let the property, $P(n)$, be the equation

$$v_n = 3n - 2. \qquad \leftarrow P(n)$$

We show by strong mathematical induction that $P(n)$ is true for every integer $n \geq 1$.

Show that $P(1)$ is true: When $n = 1$, the right-hand side of the equation is $3 \cdot 1 - 2 = 1$, which equals v_1 by definition of $v_1, v_2, v_3, \ldots$ Thus $P(1)$ is true.

Show that if $k \geq 1$ and $P(i)$ is true for every integer i from 0 through k, then $P(k + 1)$ is true: Let k be any integer with $k \geq 1$, and suppose that

$$v_i = 3i - 2 \text{ for each integer } i \text{ with } 1 \leq i \leq k. \qquad \leftarrow \text{ inductive hypothesis}$$

We must show that

$$v_{k+1} = 3(k + 1) - 2 = 3k + 1 \qquad \leftarrow P(k + 1)$$

The left-hand side of $P(k+1)$ is

$$v_{k+1} = v_{\lfloor (k+1)/2 \rfloor} - v_{\lfloor (k+2)/2 \rfloor} = 2 \qquad \text{by definition of } v_1, v_2, v_3, \ldots$$

$$= \left(3\left\lfloor \frac{k+1}{2} \right\rfloor - 2\right) + \left(3\left\lfloor \frac{k+2}{2} \right\rfloor - 2\right) + 2$$

$$= 3\left(\left\lfloor \frac{k+1}{2} \right\rfloor + \left\lfloor \frac{k+2}{2} \right\rfloor\right) - 2$$

$$= \begin{cases} 3\left(\dfrac{k}{2} + \dfrac{k+2}{2}\right) - 2 & \text{if } k \text{ is even} \\[2ex] 3\left(\dfrac{k+1}{2} + \dfrac{k+1}{2}\right) - 2 & \text{if } k \text{ is odd} \end{cases}$$

$$= 3\left(\frac{2k+2}{2}\right) - 2$$

$$= 3(k+1) - 2$$

$$= 3k + 1 \qquad \text{by the laws of algebra}$$

and this is the right-hand side of $P(k+1)$ *[as was to be shown]*.

46. *Hint:* Show that for every integer $n \geq 0, s_{2n} = 2^n$ and $s_{2n+1} = 2^{n+1}$. Then combine these formulas using the ceiling function to obtain $s_n = 2^{\lceil n/2 \rceil}$.

48. **a.**

$$w_1 - 1$$
$$w_2 = 2$$
$$w_3 = w_1 + 3 = 1 + 3$$
$$w_4 = w_2 + 4 = 2 + 4$$
$$w_5 = w_3 + 5 = 1 + 3 + 5$$
$$w_6 = w_4 + 6 = 2 + 4 + 6$$
$$w_7 = w_5 + 7 = 1 + 3 + 5 + 7$$
$$\vdots$$

$$\text{Guess:} \quad w_n = \begin{cases} 1 + 3 + 5 + \cdots + n & \text{if } n \text{ is odd} \\[1ex] 2 + 4 + 6 + \cdots + n & \text{if } n \text{ is even} \end{cases}$$

$$= \begin{cases} \left(\dfrac{n+1}{2}\right)^2 & \text{if } n \text{ is odd} \\[2ex] 2\left(1 + 2 + 3 + \cdots + \dfrac{n}{2}\right) & \text{if } n \text{ is even} \end{cases} \qquad \text{by exercise 5 of Section 5.2}$$

$$= \begin{cases} \left(\dfrac{n+1}{2}\right)^2 & \text{if } n \text{ is odd} \\[2ex] 2\left(\dfrac{\dfrac{n}{2}\left(\dfrac{n}{2}+1\right)}{2}\right) & \text{if } n \text{ is even} \end{cases} \qquad \text{by Theorem 5.2.1}$$

$$= \begin{cases} \dfrac{(n+1)^2}{4} & \text{if } n \text{ is odd} \\[2ex] \dfrac{n(n+2)}{4} & \text{if } n \text{ is even} \end{cases} \qquad \text{by the laws of algebra.}$$

b. Proof (by strong mathematical induction): Let $w_1, w_2, w_3, \ldots$ be a sequence that satisfies the recurrence relation $w_k = w_{k-2} + k$ for each integer $k \geq 3$, with initial conditions $w_1 = 1$ and $w_2 = 2$, and let the property $P(n)$ be the equation

$$w_n = \begin{cases} \dfrac{(n+1)^2}{4} & \text{if } n \text{ is odd} \\ \dfrac{n(n+2)}{4} & \text{if } n \text{ is even} \end{cases} \qquad \text{for all integers } n \geq 1. \qquad \leftarrow P(n)$$

We will show that $P(n)$ is true for every integer $n \geq 1$.

Show that $P(1)$ and $P(2)$ are true: For $n = 1$ and $n = 2$ the right-hand sides of $P(n)$ are

$$\frac{(1+1)^2}{4} = 1 \quad \text{and} \quad \frac{2(2+2)}{4} = 2,$$

which equal w_1 and w_2 respectively. So $P(1)$ and $P(2)$ are true.

Show that if $k \geq 2$ and $P(i)$ is true for every integer i from 1 through k, then $P(k+1)$ is true: Let k be any integer with $k \geq 2$, and suppose that

$$w_i = \begin{cases} \dfrac{(i+1)^2}{4} & \text{if } i \text{ is odd} \\ \dfrac{i(i+2)}{4} & \text{if } i \text{ is even} \end{cases} \qquad \text{for all integers } i \text{ with } 1 \leq i < k. \quad \leftarrow \begin{array}{l}\text{inductive} \\ \text{hypothesis}\end{array}$$

We must show that

$$w_{k+1} = \begin{cases} \dfrac{(k+2)^2}{4} & \text{if } k+1 \text{ is odd} \\ \dfrac{(k+1)(k+3)}{4} & \text{if } k+1 \text{ is even} \end{cases} \qquad \leftarrow P(k+1)$$

Now

$$w_{k+1} = w_{k-1} + (k+1) \qquad \text{by definition of } w_1, w_2, w_3, \ldots$$

$$= \begin{cases} \dfrac{((k-1)+1)^2}{4} + (k+1) & \text{if } k-1 \text{ is odd} \\ \dfrac{(k-1)[(k-1)+2]}{4} + (k+1) & \text{if } k-1 \text{ is even} \end{cases} \qquad \text{by inductive hypothesis}$$

$$= \begin{cases} \dfrac{k^2}{4} + \dfrac{4(k+1)}{4} & \text{if } k+1 \text{ is odd} \\ \dfrac{(k-1)(k+1)}{4} + \dfrac{4(k+1)}{4} & \text{if } k+1 \text{ is even} \end{cases} \qquad \begin{array}{l}\text{because } k-1 \text{ and } k+1 \text{ have} \\ \text{the same parity}\end{array}$$

$$= \begin{cases} \dfrac{k^2 + 4k + 4}{4} & \text{if } k+1 \text{ is odd} \\ \dfrac{k^2 + 4k + 3}{4} & \text{if } k+1 \text{ is even} \end{cases}$$

$$= \begin{cases} \dfrac{(k+2)^2}{4} & \text{if } k+1 \text{ is odd} \\ \dfrac{(k+1)(k+3)}{4} & \text{if } k+1 \text{ is even} \end{cases} \qquad \text{by the laws of algebra.}$$

[This is what was to be shown.]

49. **a.** *Hint:* Express the answer using the Fibonacci sequence.

50. Performing the inductive step for a proof by mathematical induction of the formula, involves substituting $(k-2)^2$ in place of a_{k-1} in the expression $a_k = 2a_{k-1}+k-1$ in hopes of showing that a_k equals $(k-1)^2$. However, solving $2(k-2)^2 + k - 1 = (k-1)^2$ for k, gives that $k^2 - 5k + 6 = 0$, which implies that $k = 2$ or $k = 3$. It turns out that the sequence $a_1, a_2, a_3, \ldots$ does satisfy the given formula for $k = 2$ and $k = 3$, but when $k = 4, a_4 = 2 \cdot 4 + (4-1) = 11$, and $11 \neq (4-1)^2$. Hence the sequence does not satisfy the formula for $n = 4$.

51. Performing the inductive step for a proof by mathematical induction of the formula involves substituting $(k-2)^2$ in place of a_{k-1} in the expression $a_k = 4a_{k-1} - k + 3$ in hopes of showing that a_k equals $(k-1)^2$. However, if you set $4(k-2)^2 - k + 3 = (k-1)^2$ and solve for k, you find that $3k^2 - 9k + 1 = 0$, and this quadratic equation is true for only two values of k. You can show that $a_2 = (2-1)^2$ and $a_3 = (3-1)^2$, but $a_4 = 4 \cdot 4 - 4 + 3 = 15$ whereas $(4-1)^2 = 3^2 = 9$.

52. **a.** *Hint:* The maximum number of regions is obtained when each additional line crosses all the previous lines, but not at any point that is already the intersection of two lines. When a new line is added, it divides each region through which it passes into two pieces. The number of regions a newly added line passes through is one more than the number of lines it crosses.

54. **a.**

$$Y_1 = E + c + mY_0$$

$$Y_2 = E + c + mY_1 = E + c + m(E + c + mY_0) = (E + c) + m(E + c) + m^2 Y_0$$

$$Y_3 = E + c + mY_2 = E + c + m((E+c) + m(E+c) + m^2 Y_0) = (E+c) + m(E+c) + m^2(E+c) + m^3 Y_0$$

$$Y_4 = E + c + mY_3 = E + c + m((E+c) + m(E+c) + m^2(E+c) + m^3 Y_0)$$

$$= (E+c) + m(E+c) + m^2(E+c) + m^3(E+c) + m^4 Y_0$$

$$\vdots$$

Guess: $Y_n = (E+c) + m(E+c) + m^2(E+c) + \cdots + m^{n-1}(E+c) + m^n Y_0$

$$= (E+c)[1 + m + m^2 + \cdots + m^{n-1}] + m^n Y_0$$

$$= (E+c)\left(\frac{m^n - 1}{m - 1}\right) + m^n Y_0, \text{ for all integers } n \geq 1.$$

b. Suppose $0 < m < 1$. Then

$$\lim_{n \to \infty} Y_n = \lim_{n \to \infty}\left((E+c)\left(\frac{m^n - 1}{m - 1}\right) + m^n Y_0\right)$$

$$= (E+c)\left(\frac{\lim_{n \to \infty} m^n - 1}{m - 1}\right) + \lim_{n \to \infty} m^n Y_0$$

$$= (E+c)\left(\frac{0 - 1}{m - 1}\right) + 0 \cdot Y_0 \qquad\qquad \text{because when } 0 < m < 1,$$

$$\text{then } \lim_{n \to \infty} m^n = 0$$

$$= \frac{E+c}{1-m}.$$

Section 5.8

1. (a), (d), and (f)

3. **a.**

$$\left.\begin{cases} a_0 = C \cdot 2^0 + D = C + D = 1 \\ a_1 = C \cdot 2^1 + D = 2C + D = 3 \end{cases}\right\} \Leftrightarrow \left\{\begin{array}{l} D = 1 - C \\ 2C + (1 - C) = 3 \end{array}\right\} \Leftrightarrow \left\{\begin{array}{l} C = 2 \\ D = -1 \end{array}\right\}$$

Therefore, $a_2 = C \cdot 2^2 + D = 2 \cdot 2^2 + (-1) = 7$.

b.

$$\left.\begin{cases} a_0 = C \cdot 2^0 + D = C + D = 0 \\ a_1 = C \cdot 2^1 + D = 2C + D = 2 \end{cases}\right\} \Leftrightarrow \left\{\begin{array}{l} D = -C \\ 2C + (-C) = 2 \end{array}\right\} \Leftrightarrow \left\{\begin{array}{l} D = -C \\ C = 2 \end{array}\right\} \Leftrightarrow \left\{\begin{array}{l} C = 2 \\ D = -2 \end{array}\right\}$$

Therefore, $a_2 = C \cdot 2^2 + D = 2 \cdot 2^2 + (-2) = 6$

4. **a.**

$$\left.\begin{cases} b_0 = C \cdot 3^0 + D \cdot (-2)^0 = C + D = 0 \\ b_1 = C \cdot 3^1 + D \cdot (-2)^1 = 3C - 2D = 5 \end{cases}\right\} \Leftrightarrow \left\{\begin{array}{l} D = -C \\ 3C - 2(-C) = 5 \end{array}\right\} \Leftrightarrow \left\{\begin{array}{l} C = 1 \\ D = -1 \end{array}\right\}$$

Therefore, $b_2 = C \cdot 3^2 + D(-2)^2 = 2 \cdot 3^2 + 1 \cdot (-2)^2 = 18 + 4 = 22$

5. <u>Proof:</u> Given that $a_n = C \cdot 2^n + D$, then for any choice of C and D and an integer $k > 2$,

$$a_k = C \cdot 2^k + D$$
$$a_{k-1} = C \cdot 2^{k-1} + D$$
$$a_{k-2} = C \cdot 2^{k-2} + D.$$

Hence

$$\begin{aligned} 3a_{k-1} - 2a_{k-2} &= 3(C \cdot 2^{k-1} + D) - 2(C \cdot 2^{k-2} + D) \\ &= 3C \cdot 2^{k-1} + 3D - 2C \cdot 2^{k-2} - 2D \\ &= 3C \cdot 2^{k-1} - C \cdot 2^{k-1} + D \\ &= 2C \cdot 2^{k-1} + D \\ &= C \cdot 2^k + D = a_k. \end{aligned}$$

6. <u>Proof:</u> Given that $b_n = C \cdot 3^n + D(-2)^n$, then for any choice of C and D and integer $k \geq 2$,

$$b_k = C \cdot 3^k + D \cdot (-2)^k, \quad b_{k-1} = C \cdot 3^{k-1} + D \cdot (-2)^{k-1}, \quad \text{and} \quad b_{k-2} = C \cdot 3^{k-2} + D \cdot (-2)^{k-2}.$$

Hence,

$$\begin{aligned} b_{k-1} + 6b_{k-2} &= \left(C \cdot 3^{k-1} + D(-2)^{k-1}\right) + 6\left(C \cdot 3^{k-2} + D \cdot (-2)^{k-2}\right) \\ &= C \cdot \left(3^{k-1} + 6 \cdot 3^{k-2}\right) + D \cdot \left((-2)^{k-1} + 6 \cdot (-2)^{k-2}\right) \\ &= C \cdot 3^{k-2}(3 + 6) + D \cdot (-2)^{k-2}(-2 + 6) \\ &= C \cdot 3^{k-2} \cdot 3^2 + D \cdot (-2)^{k-2}2^2 \\ &= C \cdot 3^k + D \cdot (-2)^k \\ &= b_k. \end{aligned}$$

8. **a.** If for each integer $k \geq 2$, $t^k = 2t^{k-1} + 3t^{k-2}$ and $t \neq 0$, then $t^2 = 2t + 3$ [by dividing by t^2], and so $t^2 - 2t - 3 = 0$. In addition, $t^2 - 2t - 3 = (t - 3)(t + 1)$, and thus $t = 3$ or $t = -1$.

b. It follows from part (a) and the distinct roots theorem that for some constants C and D, the terms of the sequence $a_0, a_1, a_2, \ldots$ satisfy the equation

$$a_n = C \cdot 3^n + D \cdot (-1)^n \text{ for every integer } n \geq 0.$$

Since $a_0 = 1$ and $a_1 = 2$, then

$$\left\{ \begin{array}{l} a_0 = C \cdot 3^0 + D \cdot (-1)^0 = C + D = 1 \\ a_1 = C \cdot 3^1 + D \cdot (-1)^1 = 3C - D = 2 \end{array} \right\} \Leftrightarrow \left\{ \begin{array}{l} D = 1 - C \\ 3C - (1 - C) = 2 \end{array} \right\} \Leftrightarrow \left\{ \begin{array}{l} D = 1 - C \\ 4C - 1 = 2 \end{array} \right\}$$

$$\Leftrightarrow \left\{ \begin{array}{l} C = 3/4 \\ D = 1/4. \end{array} \right\}$$

Thus $a_n = \frac{3}{4}(3^n) + \frac{1}{4}(-1)^n$ for every integer $n \geq 0$.

9. **a.** If for each integer $k \geq 2$, $t^k = 7t^{k-1} - 10t^{k-2}$ and $t \neq 0$, then $t^2 = 7t - 10$ and so $t^2 - 7t + 10 = 0$. In addition, $t^2 - 7t + 10 = (t - 2)(t - 5)$. Thus $t = 2$ or $t = 5$.

b. It follows from part (a) and the distinct roots theorem that for some constants C and D, the terms of the sequence $b_0, b_1, b_2, \ldots$. satisfy the equation

$$b_n = C \cdot 2^n + D \cdot 5^n \quad \text{for every integer } n \geq 0.$$

Since $b_0 = 2$ and $b_1 = 2$, then

$$\left\{ \begin{array}{l} b_0 = C \cdot 2^0 + D \cdot 5^0 = C + D = 2 \\ b_1 = C \cdot 2^1 + D \cdot 5^1 = 2C + 5D = 2 \end{array} \right\} \Leftrightarrow \left\{ \begin{array}{l} D = 2 - C \\ 2C + 5(2 - C) = 2 \end{array} \right\} \Leftrightarrow \left\{ \begin{array}{l} D = 2 - C \\ C = 8/3 \end{array} \right\}$$

$$\Leftrightarrow \left\{ \begin{array}{l} D = 2 - (8/3) = -(2/3) \\ C = 8/3 \end{array} \right\}$$

Thus $b_n = \frac{8}{3} \cdot 2^n - \frac{2}{3} \cdot 5^n$ for every integer $n \geq 0$.

11. The characteristic equation is $t^2 - 4 = 0$. Since $t^2 - 4 = (t - 2)(t + 2)$, the roots are $t = 2$ and $t = -2$. By the distinct roots theorem, there exist constants C and D such that

$$e_n = C \cdot 2^n + D \cdot (-2)^n \quad \text{for every integer } n \geq 0.$$

Since $d_0 = 1$ and $d_1 = -1$, then

$$\left\{ \begin{array}{l} d_0 = C \cdot 2^0 + D \cdot (-2)^0 = C + D = 1 \\ d_1 = C \cdot 2^1 + D \cdot (-2)^1 = 2C - 2D = -1 \end{array} \right\} \Leftrightarrow \left\{ \begin{array}{l} D = 1 - C \\ 2C - 2(1 - C) = -1 \end{array} \right\}$$

$$\Leftrightarrow \left\{ \begin{array}{l} D = 1 - C \\ 2C - 2(1 - C) = -1 \end{array} \right\} \Leftrightarrow \left\{ \begin{array}{l} C = 1/4 \\ D = 3/4 \end{array} \right\}$$

Thus $d_n = \frac{1}{4}(2^n) + \frac{3}{4}(-2)^n$ for every integer $n \geq 0$.

12. The characteristic equation is $t^2 - 9 = 0$. Since $t^2 - 9 = (t - 3)(t + 3)$, the roots are $t = 3$ and $t = -3$. By the distinct roots theorem, there exist constants C and D such that

$$e_n = C \cdot 3^n + D \cdot (-3)^n \quad \text{for every integer } n \geq 0.$$

Since $e_0 = 0$ and $e_1 = 2$, then

$$\left\{ \begin{array}{l} e_0 = C \cdot 3^0 + D \cdot (-3)^0 = C + D = 0 \\ e_1 = C \cdot 3^1 + D \cdot (-3)^1 = 3C - 3D = 2 \end{array} \right\} \Leftrightarrow \left\{ \begin{array}{l} D = -C \\ 3C - 3(-C) = 2 \end{array} \right\} \Leftrightarrow \left\{ \begin{array}{l} D = -1/3 \\ C = 1/3 \end{array} \right\}$$

Thus $e_n = \frac{1}{3} \cdot 3^n - \frac{1}{3}(-3)^n = 3^{n-1} + (-3)^{n-1} = 3^{n-1}(1 + (-1)^{n-1}) = \left\{ \begin{array}{ll} 2 \cdot 3^{n-1} & \text{if } n \text{ is odd} \\ 0 & \text{if } n \text{ is even} \end{array} \right.$

for every integer $n \geq 0$.

13. The characteristic equation is $t^2 - 2t + 1 = 0$. Since $t^2 - 2t + 1 = (t - 1)^2$, there is only one root, $t = 1$. *[Alternatively, by the quadratic formula,*

$$t = \frac{2 \pm \sqrt{4 - 4 \cdot 1}}{2} = \frac{2}{2} = 1.]$$

By the single root theorem, there exist constants C and D such that

$$r_n = C \cdot (1^n) + D \cdot n(1^n) = C + nD \quad \text{for every integer } n \geq 0.$$

Since $r_0 = 1$ and $r_1 = 5$, then

$$\left\{ \begin{array}{l} r_0 = C + 0 \cdot D = C = 1 \\ sr_1 = C + 1 \cdot D = C + D = 4 \end{array} \right\} \Leftrightarrow \left\{ \begin{array}{l} C = 1 \\ 1 + D = 4 \end{array} \right\} \Leftrightarrow \left\{ \begin{array}{l} C = 1 \\ D = 3 \end{array} \right\}$$

Thus $r_n = 1 + 3n$ for every integer $n \geq 0$.

15. The characteristic equation is $t^2 - 6t + 9 = 0$. Since $t^2 - 6t + 9 = (t - 3)^2$, there is only one root, $t = 3$. By the single root theorem, there exist constants C and D such that

$$t_n = C \cdot 3^n + D \cdot n \cdot 3^n \quad \text{for every integer } n \geq 0.$$

Since $t_0 = 1$ and $t_1 = 3$, then

$$\left\{ \begin{array}{l} t_0 = C \cdot 3^0 + D \cdot 0 \cdot 3^0 = C = 1 \\ t_1 = C \cdot 3^1 + D \cdot 1 \cdot 3^1 = 3C + 3D = 3 \end{array} \right\} \Leftrightarrow \left\{ \begin{array}{l} C = 1 \\ C + D = 1 \end{array} \right\} \Leftrightarrow \left\{ \begin{array}{l} C = 1 \\ D = 0 \end{array} \right\}$$

Thus $t_n = 1 \cdot 3^n + 0 \cdot n \cdot 3^n = 3^n$ for every integer $n \geq 0$.

16. *Hint:* For every integer $n \geq 0$,

$$s_n = \frac{\sqrt{3} + 2}{2\sqrt{3}} (1 + \sqrt{3})^n + \frac{\sqrt{3} - 2}{2\sqrt{3}} (1 - \sqrt{3})^n.$$

18. <u>Proof:</u> Suppose that $s_0, s_1, s_2, \ldots$ and $t_0, t_1, t_2, \ldots$ are sequences such that for each integer $k \geq 2$,

$$s_k = 5s_{k-1} - 4s_{k-2} \quad \text{and} \quad t_k = 5t_{k-1} - 4t_{k-2}.$$

Then for each integer $k \geq 2$,

$$\begin{aligned} 5(2s_{k-1} + 3t_{k-1}) - 4(2s_{k-2} + 3t_{k-2}) &= (5 \cdot 2s_{k-1} - 4 \cdot 2s_{k-2}) + (5 \cdot 3t_{k-1} - 4 \cdot 3t_{k-2}) \\ &= 2(5s_{k-1} - 4s_{k-2}) + 3(5t_{k-1} - 4t_{k-2}) \\ &= 2s_k + 3t_k. \end{aligned}$$

[This is what was to be shown.]

19. <u>Proof:</u>Suppose r, s, a_0, and a_1 are numbers with $r \neq s$. Consider the system of equations

$$C + D = a_0$$
$$Cr + Ds = a_1.$$

By solving for D and substituting, we find that

$$D = a_0 - C \quad \text{and so} \quad Cr + (a_0 - C)s = a_1.$$

Hence

$$C(r - s) = a_1 - a_0 s.$$

Since $r \neq s$, both sides may be divided by $r - s$. Thus the given system of equations has the unique solution

$$C = \frac{a_1 - a_0 s}{r - s}$$

and

$$D = a_0 - C = a_0 - \frac{a_1 - a_0 s}{r - s} = \frac{a_0 r - a_0 s - a_1 + a_0 s}{r - s} = \frac{a_0 r - a_1}{r - s}.$$

Alternative solution: Since the determinant of the system is $1 \cdot s - r \cdot 1 = s - r$ and since $r \neq s$, the given system has nonzero determinant and therefore has a unique solution.

21. Let $a_0, a_1, a_2, \ldots$ be any sequence that satisfies the recurrence relation $a_k = Aa_{k-1} + Ba_{k-2}$ for some real numbers A and B with $B \neq 0$ and for each integer $k \geq 2$. Furthermore, suppose that the equation $t^2 - At - B = 0$ has a single real root r. First note that $r \neq 0$ because otherwise we would have $0^2 - A \cdot 0 - B = 0$, which would imply that $B = 0$ and contradict the hypothesis. Second, note that the following system of equations with unknowns C and D has a unique solution.

$$a_0 = Cr^0 + 0 \cdot Dr^0 = 1 \cdot C + 0 \cdot D$$
$$a_1 = Cr^1 + 1 \cdot Dr^1 = C \cdot r + D \cdot r$$

One way to reach this conclusion is to observe that the determinant of the system is $1 \cdot r - r \cdot 0 = r \neq 0$. Another way to reach the conclusion is to write the system as

$$a_0 = C$$
$$a_1 = Cr + Dr$$

and let $C = a_0$ and $D = (a_1 - a_0 r)/r$. It is clear by substitution that these values of C and D satisfy the system. Conversely, if any numbers C and D satisfy the system, then $C = a_0$, and substituting C into the second equation and solving for D yields $D = (a_1 - Cr)/r$.

Proof of the exercise statement by strong mathematical induction: Let $a_0, a_1, a_2, \ldots$ be any sequence that satisfies the recurrence relation $a_k = Aa_{k-1} + Ba_{k-2}$ for some real numbers A and B with $B \neq 0$ and for each integer $k \geq 2$. Furthermore, suppose that the equation $t^2 - At - B = 0$ has a single real root r. Let the property $P(n)$ be the equation

$$a_n = Cr^n + nDr^n \qquad \leftarrow \ P(n)$$

where C and D are the unique real numbers such that $a_0 = Cr^0 + 0 \cdot Dr^0$ and $a_1 = Cr^1 + 1 \cdot Dr^1$. We will show that $P(n)$ is true for every integer $n \geq 0$.

Show that $P(0)$ and $P(1)$ are true: The fact $P(0)$ and $P(1)$ are true is automatic because C and D are exactly those numbers for which $a_0 = Cr^0 + 0 \cdot Dr^0$ and $a_1 = C \cdot r^1 + 1 \cdot Dr^1$.

Show that if $k \geq 1$ and $P(i)$ is true for every integer i from 0 through k, then $P(k+1)$ is true: Let k be any integer with $k \geq 1$ and suppose that

$$a_i = Cr^i + iDr^i \ \text{ for every integer } i \text{ with } 1 \leq i \leq k. \quad \leftarrow \text{inductive hypothesis}$$

We must show that

$$a_{k+1} = Cr^{k+1} + (k+1)Dr^{k+1}.$$

Now by the inductive hypothesis,

$$a_k = Cr^k + kDr^k \quad \text{and} \quad a_{k-1} = Cr^{k-1} + (k-1)Dr^{k-1}.$$

So

$$
\begin{aligned}
a_{k+1} &= Aa_k + Ba_{k-1} & \text{by definition of } a_0, a_1, a_2, \ldots \\
&= A(Cr^k + kDr^k) + B(Cr^{k-1} + (k-1)Dr^{k-1}) \\
& & \text{by inductive hypothesis} \\
&= C(Ar^k + Br^{k-1}) + D(Akr^k + B(k-1)r^{k-1}) \\
& & \text{by algebra} \\
&= Cr^{k+1} + Dkr^{k+1} & \text{by Lemma 5.8.4.}
\end{aligned}
$$

[This is what was to be shown.]

22. The characteristic equation is $t^2 - 2t + 2 = 0$. By the quadratic formula, its roots are

$$t = \frac{2 \pm \sqrt{4 - 8}}{2} = \frac{2 \pm 2i}{2} = \begin{cases} 1 + i \\ 1 - i. \end{cases}$$

By the distinct roots theorem, for some constants C and D

$$a_n = C(1+i)^n + D(1-i)^n$$

for every integer $n \geq 0$.

Since $a_0 = 1$ and $a_1 = 2$, then

$$a_0 = C(1+i)^0 + D(1-i)^0 = C + D = 1$$

$$a_1 = C(1+i)^1 + D(1-i)^1$$
$$= C(1+i) + D(1-i) = 2$$

$$\Leftrightarrow \left\{ \begin{array}{l} D = 1 - C \\ C(1+i) + (1-C)(1-i) = 2 \end{array} \right\}$$

$$\Leftrightarrow \left\{ \begin{array}{l} D = 1 - C \\ C(1+i-1+i) + 1 - i = 2 \end{array} \right\}$$

$$\Leftrightarrow \left\{ \begin{array}{l} D = 1 - C \\ C(2i) = 1 + i \end{array} \right\}$$

$$\Leftrightarrow \left\{ \begin{array}{l} D = 1 - C \\ C = \dfrac{1+i}{2i} = \dfrac{1+i}{2i} \cdot \dfrac{i}{i} = \dfrac{i-1}{-2} = \dfrac{1-i}{2} \end{array} \right\}$$

$$\Leftrightarrow \left\{ \begin{array}{l} D = 1 - \dfrac{1-i}{2} = \dfrac{2-1+i}{2} = \dfrac{1+i}{2} \\ C = \dfrac{1-i}{2} \end{array} \right\}.$$

Thus for every integer, $n \geq 0$,

$$a_n = \left(\frac{1-i}{2} \right)(1+i)^n + \left(\frac{1+i}{2} \right)(1-i)^n.$$

24. **a.** If $\dfrac{\phi}{1} = \dfrac{1}{\phi - 1}$, then $\phi(\phi - 1) = 1$, or, equivalently, $\phi^2 - \phi - 1 = 0$ and so ϕ satisfies the equation $t^2 - t - 1 = 0$.

b. By the quadratic formula, the solutions to $t^2 - t - 1 = 0$ are

$$t = \frac{1 \pm \sqrt{1+4}}{2} = \left\{ \begin{array}{l} \dfrac{1+\sqrt{5}}{2} \\ \dfrac{1-\sqrt{5}}{2}. \end{array} \right.$$

Let

$$\phi_1 = \frac{1+\sqrt{5}}{2} \quad \text{and} \quad \phi_2 = \frac{1-\sqrt{5}}{2}.$$

c. $F_n = \dfrac{1}{\sqrt{5}} \cdot \phi_1^{n+1} - \dfrac{1}{\sqrt{5}} \cdot \phi_2^{n+1} = \dfrac{1}{\sqrt{5}}(\phi_1^{n+1} - \phi_2^{n+1})$

This equation is an alternative way to write equation (5.8.8).

Section 5.9

1. **a.** (1) p, q, r, and s are Boolean expressions by I.

(2) $\sim s$ is a Boolean expression by (1) and II(c).

(3) $r \vee \sim s$ is a Boolean expression by (1), (2), and II(b).

(4) $(r \vee \sim s)$ is a Boolean expression by (3) and II(d).

(5) $q \wedge (r \vee \sim s)$ is a Boolean expression by (1), (4), and II(a)

(6) $(q \wedge (r \vee \sim s))$ is a Boolean expression by (5) and II(d)

(7) $\sim p$ is a Boolean expression by (1) and II(c).

(8) $\sim p \vee (q \wedge (r \vee \sim s))$ is a Boolean expression by (6), (7), and II(b).

2. **a.** (1) ( ) is in C by I.

   (2) (( )) is in C by (1) and II(a).

   (3) ( )(( )) is in C by (1), (2), and II(b).

3. **a.** (1) By Theorem 5.9.1, a and b are strings in S because a and b are in A.

   (2) By (1) and part II(c) of the definition of string, $a(bc) = (ab)c$ because a and b are strings in S and c is in A.

   b. (1) By Theorem 5.9.1, a and b are strings in S.

   (2) By (1) and part II(c) of the definition of string, ab is a string in S because a is a string in S and b is in A.

   (3) By (2), and part II(c) of the definition of string, $(ab)c$ is a string in S because ab is a string in S and c is in A.

   (4) By (3) and part (a), $a(bc)$ is a string in S because $a(bc) = (ab)c$ and $(ab)c$ is a string in S.

4. **a.** (1) $M I$ is in the $M I U$ system by I.

   (2) $M I I$ is in the $M I U$ system by (1) and II(b).

   (3) $M I I I I$ is in the $M I U$ system by (3) and II(b).

   (4) $M I I I I I I I I$ is in the $M I U$ system by (3) and II(b).

   (5) $M I U I I I I$ is in the $M I U$ system by (4) and II(c).

   (6) $M I U U I$ is in the $M I U$ system by (5) and II(c).

   (7) $M I U I$ is in the $M I U$ system by (6) and II(d).

5. **a.** (1) 2, 0.3, 4.2, and 7 are arithmetic expressions by I.

   (2) $(0.3 - 4.2)$ is an arithmetic expression by (1) and II(d).

   (3) $(2 \cdot (0.3 - 4.2))$ is an arithmetic expression by (1), (2), and II(e).

   (4) (-7) is an arithmetic expression by (1) and II(b).

6. Proof (by structural induction): By the definition of S in exercise 6, the only integer in the base for S is 5, and the recursion rule states that for every integer n in S, $n + 4$ is in S. Given any integer n in S, let property $P(n)$ be the sentence, "$n \bmod 2 = 1$."

   Show that $P(n)$ is true for each integer n in the base for S:

   The only integer in the base for S is 5, and $P(5)$ is true because $5 \bmod 2 = 1$ since $5 = 2 \cdot 2 + 1$.

   Show that for each integer n in S, if $P(n)$ is true and if m is obtained from n by applying a rule from the recursion for S, then $P(m)$ is true:

   Suppose n is any integer in S such that $P(n)$ is true, or, in other words, $n \bmod 2 = 1$. *[This is the inductive hypothesis.]* The recursion for S consists of only one rule, and when the rule is applied to n, the result is $n + 4$. To complete the inductive step, we must show that $P(n+4)$ is true, or, equivalently, that $(n+4) \bmod 2 = 1$. Now since $n \bmod 2 = 1$, then

   $$n = 2k + 1 \text{ for some integer } k.$$

   Hence

   $$
   \begin{aligned}
   (n+4) \bmod 2 &= [(2k+1) + 4] \bmod 2 && \text{by substitution} \\
   &= [2(k+2) + 1] \bmod 2 && \text{by basic algebra} \\
   &= 1 && \text{because } k+2 \text{ is an integer.}
   \end{aligned}
   $$

   and so $P(n+4)$ is true *[as was to be shown]*.

   Conclusion: Because there are no integers in S other than those obtained from the base and the recursion for S, we conclude that every integer n in S satisfies the equation $n \bmod 2 = 1$.

7. Proof (by structural induction): By the definition of S in exercise 7, the only element in the base is 1, and the recursion rules II(a) and II(b) state that for every string s in S, $0s$ and $1s$ are in S. Given any string s in S, let property $P(s)$ be the sentence

$$s \text{ ends in a } 1. \quad \leftarrow \ P(s)$$

Show that $P(a)$ is true for each string a in the base for S: The only string in the base for S is 1, which ends in a 1, and so $P(1)$ is true.

Show that for each string x in S, if $P(x)$ is true and if y is obtained from x by applying a rule from the recursion for S, then $P(y)$ is true: The recursion for S consists of two rules: II(a) and II(b). Suppose s is any string in S such that $P(s)$ is true. In other words, suppose

$$x \text{ is any string in } S \text{ such that } x \text{ ends in a } 1. \quad \leftarrow \text{ inductive hypothesis}$$

To complete the inductive step, we must show that applying either of the two recursion rules to s also results in a string that ends in 1.

Now when rule II(a) is applied to x, the result is $0x$ and when rule II(b) is applied to x, the result is $1x$. Because x ends in a 1, so do $0x$ and $1x$, which means that $P(0x)$ and $P(1x)$ are true. This completes the inductive step.

Conclusion: Because there are no strings in S other than those obtained from the base and recursion for S, we conclude that every string in S ends in a 1.

9. Proof (by structural induction): By the definition of S in exercise 9, the only element in the base is λ, and the recursion rules II(a)–II(d) state that for every string s in S, bs, sb, saa, and aas are in S. Given any string, s in S, let property $P(s)$ be the sentence

$$s \text{ contains an even number of } a\text{'s} \quad \leftarrow \ P(s)$$

Show that $P(a)$ is true for each string a in the base for S: The only string in the base for S is λ, which contains 0 a's. Since 0 is an even number, $P(\lambda)$ is true.

Show that for each string x in S, if $P(x)$ is true and if y is obtained from x by applying a rule from the recursion for S, then $P(y)$ is true: Suppose x is any string in S such that $P(x)$ is true, or, in other words, suppose

$$x \text{ is any string in } S \text{ that has an even number of } a\text{'s.} \quad \leftarrow \ \textit{inductive hypothesis}$$

When either rule II(a) or II(b) is applied to x, the result is either bx or xb, each of which contains the same number of a's as x and hence an even number of a's. Thus both $P(bx)$ and $P(xb)$ are true.

When either rule II(c) or II(d) is applied to x, the result is either aax or xaa, each of which contains two more a's than does x. Because two more than an even number is an even number, both aax and xaa contain an even number of a's. Hence both $P(aax)$ and $P(xaa)$ are true. This completes the inductive step because II(a)–II(d) are the only rules in the recursion.

Conclusion: Because there are no strings in S other than those obtained from the base and the recursion for S, we conclude that every string in S contains an even number of a's.

10. *Hint:* For each string s in S, let property $P(s)$ be the sentence: "s represents an odd integer." In the decimal notation, a string represents an odd integer if, and only if, it ends in 1, 3, 5, 7, or 9.

11. *Hint:* By divisibility results from Chapter 4 (exercises 15 and 16 of Section 4.4), if both k and m are divisible by 5, then so are $k + m$ and $k - m$.

12. Proof (by structural induction):

By the definition of S in exercise 12, the only elements in the base are the integers 0 and 5, and the recursion rules II(a) and II(b) state that for all integers k and p in S, $k + p$ and $k - p$ are in S. Given any integer n in S, let property $P(n)$ be the sentence

$$n \text{ is divisible by } 5. \qquad \leftarrow P(n)$$

Show that P(a) is true for each integer a in the base for S: The integers in the base are 0 and 5. Both of these integers are divisible by 5, and thus both $P(0)$ and $P(5)$ are true.

Show that for each integer a in S, if P(a) is true and if b is obtained from a by applying a rule from the recursion for S, then P(b) is true: The recursion for S consists of two rules, denoted II(a) and II(b). Suppose k and p are any integers in S such that $P(k)$ and $P(p)$ are true. In other words, suppose

$$k \text{ and } p \text{ are any integers in } S \text{ such that both } k \text{ and } p \text{ are divisible by } 5. \qquad \leftarrow \begin{array}{l} \textit{inductive} \\ \textit{hypothesis} \end{array}$$

By rules II(a) and II(b), both $k + p$ and $k - p$ are in S, and, because k and p are divisible by 5, so are their sum and difference, $k + p$ and $k - p$ *[by exercises 15 and 16 from Section 4.4]*. Hence, both $P(k + p)$ and $P(k - p)$ are true.

This completes the inductive step because II(a) and II(b) are the only rules in the recursion for S

Conclusion: Because there are no integers in S other than those obtained from the base and the recursion for S, we conclude that all the integers in S are divisible by 5.

13. *Hint:* Can the number of I's in a string in the $M I U$ system be a multiple of 3? How do rules II(a)–II(d) affect the number of I's in a string?

15. **a.** The parenthesis structure ( )(( ) is not in C. To see why this is so, we will prove that every parenthesis structure in C has an equal number of left and right parentheses. It will follow that, since ( )(( ) has 3 left parentheses and 2 right parentheses, ( )(( ) cannot be in C.

Proof (by structural induction):

Define a function $f \colon C \to \mathbf{Z}$ as follows: For each parenthesis structure x in C, let

$$f(x) = \left[\begin{array}{l} \text{the number of left} \\ \text{parentheses in } x \end{array} \right] - \left[\begin{array}{l} \text{the number of right} \\ \text{parentheses in } x \end{array} \right].$$

Given any parenthesis structure x in C, let property $P(x)$ be the sentence,

$$f(x) = 0. \qquad \leftarrow P(x)$$

Show that P(a) is true for each parenthesis structure a in the base for C: The only parenthesis structure in the base for C is ( ), and $f[( )] = 0$ because ( ) has one left parenthesis and one right parenthesis. Hence $P[( )]$ is true.

Show that for each parenthesis structure x in C, if P(x) is true and if y is obtained from x by applying a rule from the recursion for C, then P(y) is true: The recursion for C consists of two rules, denoted II(a) and II(b). Suppose u and v are any parenthesis structures in C such that $P(u)$ and $P(v)$ are true. In other words, suppose that

$$u \text{ and } v \text{ are any parenthesis structures} \\ \text{in } C \text{ such that } f(u) = 0 \text{ and } f(v) = 0. \qquad \leftarrow \begin{array}{l} \textit{inductive} \\ \textit{hypothesis} \end{array}$$

We must show that when any of the recursion rules are applied to u and v, the results also satisfy the property. Let k and m be the numbers of left and right parentheses, respectively, in u, and let n and p be the numbers of left and right parentheses, respectively, in v.

When rule II(a) is applied to u, the result is (u). Since u is in C, $k - m = 0$, and, because (u) has one more left parenthesis and one more right parenthesis than u,

$$f[(u)] = (k + 1) - (m + 1) = (k - m) + (1 - 1) = 0 + 0 = 0.$$

Hence, $P(u)$ is true.

When rule II(b) is applied to u and v, the result is uv, and, because u and v are in C, $k - m = 0$ and $n - p = 0$. Now uv has $k + n$ left parentheses and $m + p$ right parentheses. So, by definition of f,

$$f(uv) = (k + n) - (m + p) = (k - m) + (n - p) = 0 + 0 = 0.$$

Hence, $f(uv) = 0$, and so $P(uv)$ is true.

Conclusion: Because there are no parenthesis structures in C other than those obtained from the base and the recursion for C, we conclude that given any parenthesis structure x in C, $f(x) = 0$. Therefore, every parenthesis structure in C has the same number of left and right parentheses.

b. Even though the number of its left parentheses equals the number of its right parentheses, this structure is not in C either. Roughly speaking, the reason is that given any parenthesis structure derived from the base for C by repeated application of the recursion rules, as you move from left to right along the structure, the total in the number of right parentheses you encounter will never be larger than the number of left parentheses you have already passed. But if you move along $(( )( )))(( )$ from left to right, you encounter an extra right parenthesis in the seventh position.

A more formal explanation is the following: Let A be the set of all finite sequences of integers and define a function $g\colon C \to A$ as follows:

For each parenthesis structure x in C, let

$$g[x] = (a_1, a_2, \ldots, a_n) \quad \text{where } a_i \text{ is the number of left parentheses in } x$$
$$\text{minus the number of right parentheses in } x$$
$$\text{as one counts from left to right through position } i.$$

For instance, if $x = (( )( ))$, then $g[x] = (1, 2, 1, 2, 1, 0)$.

We claim that for every parenthesis structure x in G, each component of $g[x]$ is nonnegative and the final component is 0. It will follow from the claim that if $(( )( )))(( )$ were in C, then all components of $g[(( )( )))(( )]$ would be nonnegative. But $g[(( )( )))(( )] = (1, 2, 1, 2, 1, 0, -1, 0, 1, 0)$, and one of these components is negative. Consequently, $(( )( )))(( )$ is not in C.

Proof of the claim (by structural induction):

Let the function g be defined as above. Given any parenthesis structure x in C, let property $P(x)$ be the sentence

Each component of $g[x]$ is nonnegative, and the final component is 0. $\leftarrow P(x)$

Show that $P(a)$ is true for each parenthesis structure a in the base for C: The only object in the base is $(\ )$, and $g[(\ )] = (1, 0)$. Both components of $g[(\ )]$ are nonnegative and the final component is 0. Hence $P[(\ )]$ is true.

Show that for each parenthesis structure x in C, if $P(x)$ is true and if y is obtained from x by applying a rule from the recursion for C, then $P(y)$ is true: The recursion for C consists of two rules, denoted II(a) and II(b).

Suppose u and v are any parenthesis structures in C such that $P(u)$ and $P(v)$ are true. In other words, suppose

all components of $g[u]$ and $g[v]$ are nonnegative, $\leftarrow$ *inductive*
with the final components of both $g[u]$ and $g[v]$ being 0.   *hypothesis*

Consider the effect of applying rules II(a) and II(b).

When rule II(a) is applied to u, the result is (u). Now the first component in $g[(u)]$ is 1 *[because (u) starts with a left parenthesis]*, and every subsequent component of $g[(u)]$ except the last is one more than a corresponding (nonnegative) component of $g[u]$. Hence the next-to-last component of $g[(u)]$ is 1 *(because the final component of $g[u]$ is 0)*. Thus the final right-parenthesis of (u) reduces the final component of $g[(u)]$ to 0. So each component of $g[(u)]$ is nonnegative and the final component is 0. Hence $P(u)$ is true.

When rule II(b) is applied to u and v, the result is uv. We must show that each component of $g[uv]$ is nonnegative, with the final component being 0. For concreteness, suppose $g[u]$ has m components and $g[v]$ has n components. The first through the mth components of $g[uv]$ are the same as the first through the mth components of u, which are all nonnegative, with the mth component being 0. Thus the $(m+1)$st through the $(m+n)$th components of $g[uv]$ are the same as the first through the nth components of v, which are all nonnegative, with the final component being 0. Because the final component of $g[uv]$ is the same as the final component of $g[v]$, all components of $g[uv]$ are nonnegative and the final component is 0. Hence $P(uv)$ is true.

Conclusion: Because there are no parenthesis structures in C other than those obtained from the base and the recursion for C, we conclude that given any parenthesis structure x in C, all components of $g[x]$ are nonnegative and the final component is 0.

16. Let S be the set of all strings of 0's and 1's with the same number of 0's and 1's. The following is a recursive definition for S.

    I. Base: The null string $\lambda \in S$.

    II. Recursion: If $s \in S$, then

    a. $01s \in S$ **b.** $s01 \subset S$ **c.** $10s \in S$ **d.** $s10 \in S$ **e.** $0s1 \in S$ **f.** $1s0 \in S$

    III. Restriction: There are no elements of S other that those obtained from the base and recursion for S.

18. Let T be the set of all strings of a's and b's that contain an odd number of a's. The following is a recursive definition of T.

    I. Base: $a \in T$.

    II. Recursion: If $t \in T$, then

    a. $bt \in T$ **b.** $tb \in T$ **c.** $aat \in T$ **d.** $ata \in T$ **e.** $taa \in T$

    III. Restriction: There are no elements of T other than those obtained from the base and recursion for T.

20. **a.** Suppose a is any character in A. *[We must show that $L(a) = 1$.]* Then

$$\begin{aligned} L(a) &= L(\lambda \cdot a) && \text{by part II(b) of the definition of string} \\ &= L(\lambda) + 1 && \text{by part (b) of the definition of the length function} \\ &= 0 + 1 && \text{by part (a) of the definition of the length function} \\ &= 1 && \text{by definition of 0.} \end{aligned}$$

21. <u>Proof (by structural induction):</u> Suppose S is the set of all strings over a finite set A, and suppose u and v are any strings in S. We will prove that for every string w in S, $u(vw) = (uv)w$. Let the property $P(w)$ be the sentence

$$\text{If } w \text{ is any string in } S, \text{ then } (uv)w = u(vw). \qquad \leftarrow \quad P(w)$$

Show that $P(a)$ is true for each string a in the base for S:

The only string in the base for the set of all strings is λ, and by property II(c) in the definition of S, $(uv)\lambda = u(v\lambda)$. Thus $P(\lambda)$ is true.

Show that for each string x in S, if P(x) is true and if y is obtained from x by applying a rule from the recursion for S, then P(y) is true:

The recursive definition for S consists of three rules denoted II(a), II(b), and (II(c). Suppose x is any string in S such that $P(x)$ is true. In other words, suppose

$$x \text{ is any string in } S \text{ such that } (uv)x = u(vx). \quad \leftarrow \text{ inductive hypothesis}$$

When rule II(a) is applied to x, the result is xc, where c is a character in A. We must show that $P(xc)$ is true, or, equivalently, that $u(v(xc)) = (uv)(xc)$. Now

$$
\begin{aligned}
u(v(xc)) &= u((vx)c) && \text{by part II(c) of the definition of string} \\
&= (u(vx))c && \text{by definition of string (since } u \text{ and } vx \text{ are in } S \text{ and } c \in A) \\
&= ((uv)x)c && \text{by inductive hypothesis} \\
&= (uv)(xc) && \text{by part II(c) of the definition of string.}
\end{aligned}
$$

Hence $P(xc)$ is true.

Conclusion: Because there are no strings in S other than those obtained from the base and the recursion for S, we conclude that if u and v are any strings in S, then for every string w in S, $u(vw) = (uv)w$.

22. *Hint:* If S is the set of all strings over a finite set A, then for any string u in S, let the property $P(v)$ be the sentence "If v is any string of length n, then $\text{Rev}(uv) = \text{Rev}(v)\text{Rev}(u)$." For the basis step you will show that $P(\lambda)$ is true by showing that $\text{Rev}(u\lambda) = \text{Rev}(\lambda)\text{Rev}(u)$. For the inductive step you will assume that x is any string for which $P(x)$ is true, and you will show that if y is the result of applying rule II(a) to x, then $P(y)$ is true.

23. **a.**
$$
\begin{aligned}
M(86) &= M(M(97)) && \text{since } 86 \leq 100 \\
&= M(M(M(108))) && \text{since } 97 \leq 100 \\
&= M(M(98)) && \text{since } 108 > 100 \\
&= M(M(M(109))) && \text{since } 98 < 100 \\
&= M(M(99)) && \text{since } 109 > 100 \\
&= M(91) && \text{by Example 5.9.7}
\end{aligned}
$$

24. Proof 1 (by a variation of strong mathematical induction):

Let the property $P(n)$ be the equation

$$M(n) = 91. \quad \leftarrow \ P(n)$$

Show that P(n) is true for every integer n from 91 through 101: This statement is proved in the solution to exercise 23.

Show that for every integer k, if $1 \leq k < 91$ and the property is true for every integer i from $k + 1$ through 101, then it is true for k: Let k be any integer such that $1 \leq k < 91$ and suppose

$$M(i) = 91 \text{ for every integer } i \text{ with } k < i \leq 101. \quad \leftarrow \text{ inductive hypothesis}$$

We must show that

$$M(k) = 91.$$

By definition of M, $M(k) = M(M(k+11))$. In addition, because $1 \leq k < 91$, we that that $12 \leq k + 11 < 102$, and so $k < k + 11 \leq 101$. Thus, by inductive hypothesis, $M(k+11) = 91$. It follows that $M(k) = M(91)$, which equals 91 by the basis step above. Hence $M(k) = 91$ *[as was to be shown]*.

Proof 2 (by contradiction):

Suppose not. That is, suppose there is at least one positive integer $k \le 101$ with $M(k) \ne 91$. Let q be the largest such integer. Then

$$1 \le q \le 101 \quad \text{and} \quad M(q) \ne 91.$$

Now by the solution to exercise 23, for each integer n with $91 \le n \le 101$, $M(n) = 91$. Thus $q \le 90$ and so $q + 11 \le 90 + 11 = 101$. Note, therefore, that $q + 11$ is a larger positive integer than q and is also less than or equal to 101.

Hence, because q is the largest positive integer less than or equal to 101 with $M(q) \ne 91$, we must have that $M(q + 11) = 91$. Now,, by definition of M, $M(q) = M(M(q + 11))$. So $M(q) = M(M(q + 11)) = M(91) = 91$ (by the solution to exercise 19). Therefore,

$$M(q) \ne 91 \quad \text{and} \quad M(q) = 91,$$

which is a contradiction. We conclude that the supposition is false and $M(n) = 91$ for all positive integers $k \le 101$.

25. **a.** $\begin{aligned} A(1,1) &= A(0, A(1,0)) && \text{by (5.9.3) with } m = 1 \text{ and } n = 1 \\ &= A(1,0) + 1 && \text{by (5.9.1) with } n = A(1,0) \\ &= A(0,1) + 1 && \text{by (5.9.2) with } m = 1 \\ &= (1 + 1) + 1 && \text{by (5.9.1) with } n = 1 \\ &= 3 \end{aligned}$

Alternative solution:

$\begin{aligned} A(1,1) &= A(0, A(1,0)) && \text{by (5.9.3) with } m = 1 \text{ and } n = 1 \\ &= A(0, A(0,1)) && \text{by (5.9.2) with } m = 1 \\ &= A(0,2) && \text{by (5.9.1) with } n = 1 \\ &= 3 && \text{by (5.9.1) with } n = 2 \end{aligned}$

26. **a.** <u>Proof by mathematical induction</u>: Let the property, $P(n)$, be the equation

$$A(1, n) = n + 2. \qquad \leftarrow P(n)$$

Show that $P(0)$ is true: When $n = 0$,

$$\begin{aligned} A(1, n) &= A(1,0) && \text{by substitution} \\ &= A(0,1) && \text{by (5.9.2)} \\ &= 1 + 1 && \text{by (5.9.1)} \\ &= 2. \end{aligned}$$

On the other hand, $n + 2 = 0 + 2$ also. Thus $A(1, n) = n + 2$ for $n = 0$.

Show that for every integer $k \ge 0$, if $P(k)$ is true, then $P(k + 1)$ is true: Let k be any integer with $k \ge 1$ and suppose $P(k)$ is true. In other words, suppose

$$A(1, k) = k + 2. \qquad \leftarrow P(k) \text{ inductive hypothesis}$$

We must show that $P(k + 1)$ is true. In other words, we must show that $A(1, k + 1) = (k + 1) + 2 = k + 3$. Now

$$\begin{aligned} A(1, k + 1) &= A(0, A(1, k)) && \text{by (5.9.3)} \\ &= A(1, k) + 1 && \text{by (5.9.1)} \\ &= (k + 2) + 1 && \text{by inductive hypothesis} \\ &= k + 3 \end{aligned}$$

[as was to be shown].

[Since both the basis and the inductive steps have been proved, we conclude that the equation holds for every nonnegative integer n.]

27. (1) $T(2) = T(1) = 1$

    (2) $T(3) = T(10) = T(5) = T(16) = T(8) = T(4) = T(2) = 1$

    (3) $T(4) = 1$ by (2)

    (4) $T(5) = 1$ by (2)

    (5) $T(6) = T(3) = 1$ by (2)

    (6) $T(7) = T(22) = T(11) = T(34) = T(17) = T(52) = T(26) = T(13) = T(40) = T(20)$
        $= T(10) = 1$ by (2)

28. Suppose F is a function. Then $F(1) = 1, F(2) = F(1) = 1, F(3) = 1 + F(5 \cdot 3 - 9) = 1 + F(6) = 1 + F(3)$. Subtracting $F(3)$ from the extreme left and extreme right of this sequence of equations gives $1 = 0$, which is false. Hence F is not a function. In other words, F is not well-defined.

Review Guide: Chapter 5

Sequences and Summations

- What is a method that can sometimes find an explicit formula for a sequence whose first few terms are given (provided a nice explicit formula exists!)?
- What is the expanded form for a sum that is given in summation notation?
- What is the summation notation for a sum that is given in expanded form?
- How do you evaluate $a_1 + a_2 + a_3 + \cdots a_n$ when n is small?
- What does it mean to "separate off the final term of a summation"?
- What is the product notation? *(p. 233)*
- What are some properties of summations and products?
- How do you transform a summation by making a change of variable?
- What is factorial notation?
- What is the n choose r notation?
- What is an algorithm for converting from base 10 to base 2?
- What does it mean for a sum to be written in closed form?

Mathematical Induction

- What do you show in the basis step and what do you show in the inductive step when you use (ordinary) mathematical induction to prove that a property involving an integer n is true for all integers greater than or equal to some initial integer?
- What is the inductive hypothesis in a proof by (ordinary) mathematical induction?
- Are you able to use (ordinary) mathematical induction to construct proofs involving various kinds of statements such as formulas, divisibility properties, inequalities, and other situations?

- Are you able to apply the formula for the sum of the first n positive integers?
- Are you able to apply the formula for the sum of the successive powers of a number, starting with the zeroth power?

Strong Mathematical Induction and The Well-Ordering Principle for the Integers

- What do you show in the basis step and what do you show in the inductive step when you use strong mathematical induction to prove that a property involving an integer n is true for all integers greater than or equal to some initial integer?
- What is the inductive hypothesis in a proof by strong mathematical induction?
- Are you able to use strong mathematical induction to construct proofs of various statements?

- What is the well-ordering principle for the integers?
- Are you able to use the well-ordering principle for the integers to prove statements, such as the existence part of the quotient-remainder theorem?
- How are ordinary mathematical induction, strong mathematical induction, and the well-ordering principle for the integers related?

Algorithm Correctness

- What are the pre-condition and the post-condition for an algorithm?

- What does it mean for a loop to be correct with respect to its pre- and post-conditions?
- What is a loop invariant?
- How do you use the loop invariant theorem to prove that a loop is correct with respect to its pre- and post-conditions?

Recursion

- What is an explicit formula for a sequence?
- What does it mean to define a sequence recursively?
- What is a recurrence relation with initial conditions?
- How do you compute terms of a recursively defined sequence?
- Can different sequences satisfy the same recurrence relation?
- What is the "recursive paradigm"?
- How do you develop a recurrence relation for the tower of Hanoi sequence?
- How do you develop a recurrence relations for the Fibonacci sequence?
- How do you develop recurrence relations for sequences that involve compound interest?
- How do you mathematical induction to prove properties of summations?

Solving Recurrence Relations

- What is the method of iteration for solving a recurrence relation?
- What is an arithmetic sequence?
- What is a geometric sequence?
- How do you use the formula for the sum of the first n integers and the formula for the sum of the first n powers of a real number r to simplify the answers you obtain when you solve recurrence relations?
- How is mathematical induction used to check that the solution to a recurrence relation is correct?
- What is a second-order linear homogeneous recurrence relation with constant coefficients?
- What is the characteristic equation for a second-order linear homogeneous recurrence relation with constant coefficients?
- What is the distinct-roots theorem? If the characteristic equation of a relation has two distinct roots, how do you solve the relation?
- What is the single-root theorem? If the characteristic equation of a relation has a single root, how do you solve the relation?

General Recursive Definitions

- When a set is defined recursively, what are the three parts of the definition?
- Given a recursive definition for a set, how can you show that a given element is in the set?
- Given a recursively defined set S and a property that objects in S may or may not satisfy, how do you use structural induction to prove that every element in S satisfies the property? Specifically, what do you show in the basis step and what do you show in the inductive step?
- What is a recursive function?
- What is McCarthy's 91 function?
- What is the Ackermann function

Fill-in-the-Blank Review Questions

Section 5.1

1. The notation $\displaystyle\sum_{k=m}^{n} a_k$ is read _____.

2. The expanded form of $\displaystyle\sum_{k=m}^{n} a_k$ is _____.

3. The value of $a_1 + a_2 + a_3 + \cdots + a_n$ when $n = 2$ is _____.

4. The notation $\displaystyle\prod_{k=m}^{n} a_k$ is read _____.

5. If n is a positive integer, then $n! = $ _____.

6. $\displaystyle\sum_{k=m}^{n} a_k + c \sum_{k=m}^{n} b_k = $ _____.

7. $\displaystyle\left(\prod_{k=m}^{n} a_k\right)\left(\prod_{k=m}^{n} b_k\right) = $ _____.

Section 5.2

1. Mathematical induction is a method for proving that a property defined by integers n is true for all values of n that are _____.

2. Let $P(n)$ be a property defined for integers n and consider constructing a proof by mathematical induction for the statement "$P(n)$ is true for all values of $n \geq a$."
   a. In the basis step one must show that _____.
   b. In the inductive step one supposes that _____ for some particular but arbitrarily chosen value of an integer $k \geq a$. This supposition is called the _____. One then has to show that _____.

Section 5.3

1. Mathematical induction differs from the kind of induction used in the natural sciences because it is actually a form of ____ reasoning.

2. Mathematical induction can be used to _____ conjectures that have been made using inductive reasoning.

Section 5.4

1. In a proof by strong mathematical induction the basis step may require checking a property $P(n)$ for more _____ value of n.

2. Suppose that in the basis step for a proof by strong mathematical induction the property $P(n)$ was checked for every integer n from a through b. Then in the inductive step one assumes that for any integer $k \geq b$, the property $P(n)$ is true for all values of i from _____ through _____ and one shows that _____ is true.

3. According to the well-ordering principle for the integers, if a set S of integers contains at least _____ and if there is some integer that is less than or equal to every _____, then _____.

Section 5.5

1. A pre-condition for an algorithm is _____ and a post-condition for an algorithm is _____.

2. A loop is defined as correct with respect to its pre- and post-conditions if, and only if, whenever the algorithm variables satisfy the pre-condition for the loop and the loop terminates after a finite number of steps, then _____.

3. For each iteration of a loop, if a loop invariant is true before iteration of the loop, then _____.

4. Given a **while** loop with guard G and a predicate $I(n)$ if the following four properties are true, then the loop is correct with respect to its pre- and post-conditions:
   (a) The pre-condition for the loop implies that _____ is true before the first iteration of the loop;
   (b) For every integer $k \geq 0$, if the guard G and the predicate $I(k)$ are both true before an iteration of the loop, then _____;
   (c) After a finite number of iterations of the loop, _____;
   (d) If N is the least number of iterations after which G is false and $I(N)$ is true, then the values of the algorithm variables will be as specified _____.

Section 5.6

1. A recursive definition for a sequence consists of a _____ and _____ .

2. A recurrence relation is an equation that defines each later term of a sequence by reference to _____ in the sequence.

3. Initial conditions for a recursive definition of a sequence consist of one or more of the _____ of the sequence.

4. To solve a problem recursively means to divide the problem into smaller subproblems of the same type as the initial problem, to suppose _____ , and to figure out how to use the supposition to _____ .

5. A crucial step for solving a problem recursively is to define a _____ in terms of which the recurrence relation and initial conditions can be specified.

Section 5.7

1. To use iteration to find an explicit formula for a recursively defined sequence, start with the _____ and use successive substitution into the _____ to look for a numerical pattern.

2. At every step of the iteration process, it is important to eliminate _____ .

3. If a single number, say a, is added to itself k times in one of the steps of the iteration, replace the sum by the expression _____ .

4. If a single number, say a, is multiplied by itself k times in one of the steps of the iteration, replace the product by the expression _____ .

5. A general arithmetic sequence with initial value a_0 and constant adder d satisfies the recurrence relation _____ and has the explicit formula _____ .

6. A general geometric sequence with initial value a_0 and constant multiplier r satisfies the recurrence relation _____ and has the explicit formula _____ .

7. When an explicit formula for a recursively defined sequence has been obtained by iteration, its correctness can be checked by _____ .

Section 5.8

1. A second-order linear homogeneous recurrence relation with constant coefficients is a recurrence relation of the form _____ for all integers $k \geq$ _____, where _____.

2. Given a recurrence relation of the form $a_k = Aa_{k-1} + Ba_{k-2}$ for all integers $k \geq 2$, the characteristic equation of the relation is _____.

3. If a sequence $a_1, a_2, a_3, \ldots$ is defined by a second-order linear homogeneous recurrence relation with constant coefficients and the characteristic equation for the relation has two distinct roots r and s (which could be complex numbers), then the sequence satisfies an explicit formula of the form _____.

4. If a sequence $a_1, a_2, a_3, \ldots$ is defined by a second-order linear homogeneous recurrence relation with constant coefficients and the characteristic equation for the relation has only a single root r, then the sequence satisfies an explicit formula of the form _____.

Section 5.9

1. The base for a recursive definition of a set is _____.

2. The recursion for a recursive definition of a set is _____.

3. The restriction for a recursive definition of a set is _____.

4. One way to show that a given element is in a recursively defined set is to start with an element or elements in the _____ and apply the rules from the _____ until you obtain the given element.

5. To use structural induction to prove that every element in a recursively defined set S satisfies a certain property, you show that _____ and that, for each rule in the recursion, if_____ then_____.

6. A function is said to be defined recursively if, and only if, _____.

Answers for Fill-in-the-Blank Review Questions

Section 5.1

1. the summation from k equals m to n of a-sub-k

2. $a_m + a_{m+1} + a_{m+2} + \cdots + a_n$

3. $a_1 + a_2$

4. the product from k equals m to n of a-sub-k

5. $n \cdot (n-1) \cdots 3 \cdot 2 \cdot 1$ (Or: $n \cdot (n-1)!$)

6. $\displaystyle\sum_{k=m}^{n} (a_k + cb_k)$

7. $\displaystyle\prod_{k=m}^{n} a_k b_k$

Section 5.2

1. greater than or equal to some initial value

2. **a.** $P(a)$ is true

   b. $P(k)$ is true; inductive hypothesis; $P(k+1)$ is true

Section 5.3

1. deductive

2. prove

Section 5.4

1. than one
2. a; k; $P(k+1)$
3. one integer; integer in S; S contains a least element

Section 5.5

1. a predicate that describes the initial state of the input variables for the algorithm; a predicate that describes the final state of the output variables for the algorithm
2. the algorithm variables satisfy the post-condition for the loop
3. it is true after iteration of the loop
4. (a) (0) is true
   (b) $I(k+1)$ is true after the iteration of the loop
   (c) the guard G becomes false
   (d) in the post-condition of the loop

Section 5.6

1. recurrence relation; initial conditions
2. earlier terms
3. values of the first few terms
4. that the smaller subproblems have already been solved; solve the initial problem
5. sequence

Section 5.7

1. initial conditions; recurrence relation
2. parentheses
3. $k \cdot a$
4. a^k
5. $a_k = a_{k-1} + d$; $a_n = a_0 + dn$
6. $a_k = ra_{k-1}$; $a_n = a_0 r^n$
7. mathematical induction

Section 5.8

1. $a_k = Aa_{k-1} + Ba_{k-2}$; 2; A and B are fixed real numbers with $B \neq 0$
2. $t^2 - At - B = 0$
3. $a_n = Cr^n + Ds^n$, where C and D are real or complex numbers
4. $a_n = Cr^n + Dnr^n$, where C and D are real numbers

Section 5.9

1. a statement that certain objects belong to the set
2. a collection of rules indicating how to form new set objects from those already known to be in the set
3. a statement that no objects belong to the set other than those coming from either the base or the recursion
4. base; recursion

5. each object in the base satisfies the property; the rule is applied to objects that satisfy the property; the objects defined by the rule also satisfy the property

6. its rule of definition refers to itself

Formats for Proving Formulas by Mathematical Induction

When using mathematical induction to prove a formula, it is sometimes tempting to use an invalid mode of proof because it is possible — validly — to deduce a true equation from a false one. Several formats can be used, besides the one shown most frequently in the textbook, to avoid this fallacy. A crucial point is this:

> If you are trying to prove that an equation is true but you haven't yet done so, you can avoid error by writing, "We must show that" or by putting a question mark above the equal sign This alerts both you and a reader that the truth of the equation is still in question.

Format 1 (the format used most often in the textbook for the inductive step): Start with the left-hand side (LHS) of the equation to be proved and successively transform it using definitions, known facts from basic algebra, and (for the inductive step) the inductive hypothesis, until you obtain the right-hand side (RHS) of the equation.

Format 2 (the format used most often in the textbook for the basis step): Transform the LHS and the RHS of the equation to be proved *independently*, one after the other, until both sides are shown to equal the same expression. Because two quantities equal to the same quantity are equal to each other, you can conclude that the two sides of the equation are equal to each other.

Format 3: This format is just like Format 2 except that the computations are done in parallel. But in order to avoid the fallacy of assuming what is to be proved, do NOT put an equal sign between the two sides of the equation until the very last step. Separate the two sides of the equation with a vertical line.

Format 4: This format is just like Format 3 except that the two sides of the equation are separated by an equal sign with a question mark on top: $\overset{?}{=}$

Format 5: Start by writing something like "We must show that" and the equation you want to prove true. In successive steps, indicate that this equation is true if, and only if, ($\Leftrightarrow$) various other equations are true. But be sure that both the directions of your "if and only if" claims are correct. In other words, be sure that the $\Leftarrow$ direction is just as true as the $\Rightarrow$ direction. If you finally get down to an equation that is known to be true, then because each subsequent equation is true *if, and only if*, the previous equation is true, you will have shown that the original equation is true.

Example: Prove that for each integer $n \geq 1$,

$$\boxed{1 + 3 + 5 + \cdots + (2n - 1) = n^2} \leftarrow \text{This is } P(n).$$

Proof that $P(1)$ is true:

Solution (Format 1):
When $n = 1$, the LHS of $P(1)$ equals 1, and the RHS equals 1^2 which also equals 1. So $P(1)$ is true.

Proof that for every $k \geq 1$, if $P(k)$ is true then $P(k + 1)$ is true:

Solution (Format 2):
Suppose that k is any integer with $k \geq 1$ such that $1 + 3 + 5 + \cdots + (2k - 1) = k^2$. *[This is the inductive hypothesis, $P(k)$.]* We must show that $P(k + 1)$ is true, where $P(k + 1)$ is the equation $1 + 3 + 5 + \cdots + (2k + 1) = (k + 1)^2$.

Now the LHS of $P(k+1)$ is

$$1 + 3 + 5 + \cdots + (2k+1) = 1 + 3 + 5 + \cdots + (2k-1) + (2k+1)$$

by making the next-to-last term explicit

$$= k^2 + (2k+1) \quad \text{by inductive hypothesis.}$$

And the RHS of $P(k+1)$ is

$$(k+1)^2 = k^2 + 2k + 1 \quad \text{by basic algebra.}$$

So the left-hand and right-hand sides of $P(k+1)$ equal the same quantity, and thus and thus $P(k+1)$ is true *[as was to be shown]*.

Solution (Format 3):

Suppose that k is any integer with $k \geq 1$ such that $1 + 3 + 5 + \cdots + (2k-1) = k^2$. *[This is the inductive hypothesis, $P(k)$.]* We must show that $P(k+1)$ is true, where $P(k+1)$ is the equation $1 + 3 + 5 + \cdots + (2k+1) = (k+1)^2$.

Consider the left-hand and right-hand sides of $P(k+1)$:

$1 + 3 + 5 + \cdots + (2k+1)$	$(k+1)^2$
$= 1 + 3 + 5 + \cdots + (2k-1) + (2k+1)$	
by making the next-to-last term explicit	
$= k^2 + (2k+1)$	
by inductive hypothesis	
$= k^2 + 2k + 1$	$= k^2 + 2k + 1$
by basic algebra	by basic algebra

So the left-hand and right-hand sides of $P(k+1)$ equal the same quantity, and thus and thus $P(k+1)$ is true *[as was to be shown]*.

Solution (Format 4):

Suppose that k is any integer with $k \geq 1$ such that $1 + 3 + 5 + \cdots + (2k-1) = k^2$. *[This is the inductive hypothesis, $P(k)$.]* We must show that $P(k+1)$ is true, where $P(k+1)$ is the equation $1 + 3 + 5 + \cdots + (2k+1) = (k+1)^2$.

Consider the left-hand and right-hand sides of $P(k+1)$:

$1 + 3 + 5 + \cdots + (2k+1)$	$\overset{?}{=} \quad (k+1)^2$
$1 + 3 + 5 + \cdots + (2k-1) + (2k+1)$	$\overset{?}{=} \quad k^2 + 2k + 1$
by making the next-to-last term explicit	by basic algebra
$k^2 + (2k+1)$	$\overset{?}{=} \quad k^2 + 2k + 1$
by inductive hypothesis	
$k^2 + 2k + 1$	$= \quad k^2 + 2k + 1$
by basic algebra	

So the left-hand and right-hand sides of $P(k+1)$ equal the same quantity, and thus $P(k+1)$ is true *[as was to be shown]*.

Solution (Format 5):

Suppose that k is any integer with $k \geq 1$ such that $1 + 3 + 5 + \cdots + (2k-1) = k^2$. *[This is the inductive hypothesis, $P(k)$.]* We must show that $P(k+1)$ is true, where $P(k+1)$ is the equation $1 + 3 + 5 + \cdots + (2k+1) = (k+1)^2$.

But $P(k+1)$ is true if, and only if, $(\Leftrightarrow)$

	$1 + 3 + 5 + \cdots + (2k-1) + (2k+1)$	$=$	$(k+1)^2$	by making the next-to-last term explicit
$\Leftrightarrow$	$k^2 + (2k+1)$	$=$	$(k+1)^2$	by inductive hypothesis
$\Leftrightarrow$	$k^2 + 2k + 1$	$=$	$(k+1)^2$	

which is true by basic algebra. Thus $P(k+1)$ is true *[as was to be shown]*.

Chapter 6: Set Theory

The first section of this chapter introduces additional terminology for sets and the concept of an element argument to prove that one set is a subset of another. The aim of this section is to provide a experience with a variety of types of sets and a basis for deriving the set properties discussed in the remainder of the chapter. The second and third sections show how to prove and disprove various proposed set properties of union, intersection, set difference and (general) complement using element arguments, algebraic arguments, and counterexamples. Section 6.4 introduces the concept of Boolean algebra, which generalizes both the algebra of sets with the operations of union and intersection and the properties of a set of statements with the operations of *or* and *and*. The section goes on to discuss Russell's paradox and shows that reasoning similar to Russell's can be used to prove an important property of computer algorithms.

One caution involves a tendency to interpret if-then statements as *and* statements. For instance, when asked to write what it means for A to be a subset of B, a common incorrect student response includes the words, "$x \in A$ and $x \in B$," instead of, "if $x \in A$ then $x \in B$." This mistake has an understandable basis: If it is true for a particular x that "if $x \in A$ then $x \in B$" and if it is also true that $x \in A$, then the statement "$x \in A$ and $x \in B$" is true. However, when we write "if $x \in A$ then $x \in B$", we can't assume that $x \in A$. The statement "if $x \in A$ then $x \in B$" doesn't give any information about x unless it is already known that x is in A. It might be in B or it might not be in B. When $x \in A$ is false, the statement "if $x \in A$ then $x \in B$" is true but the statement "$x \in A$ and $x \in B$" is false.

Section 6.1

1. **a.** $A = \{2, \{2\}, (\sqrt{2})^2\} = \{2, \{2\}, 2\} = \{2, \{2\}\}$ and $B = \{2, \{2\}, \{\{2\}\}\}$. So $A \subseteq B$ because every element in A is in B, but $B \not\subseteq A$ because $\{\{2\}\} \in B$ and $\{\{2\}\} \notin A$. Thus A is a proper subset of B.

   b. Note that $\sqrt{5^2 - 4^2} = \sqrt{25 - 16} = \sqrt{9} = 3$. Also 24 *mod* 7 = 3 because $24 - 3 = 21$ and 21 is divisible by 3. Thus

   $$A = \{3, \sqrt{5^2 - 4^2}, 24 mod\ 7\} = \{3, 3, 3\} = \{3\}.$$

   In addition, 8 *mod* 5 = 3 because $8 - 5 = 3$ and 3 is divisible by 3. Thus

   $$B = \{8 mod\ 5\} = \{3\},$$

   and so $A = B$. It follows that $A \subseteq B$ and $B \subseteq A$ and neither is a proper subset of the other.

   c. $A = \{\{1, 2\}, \{2, 3\}\}$ and $B = \{1, 2, 3\}$. So $A \not\subseteq B$ because $\{1, 2\} \in A$ and $\{1, 2\} \notin B$. Also $B \not\subseteq A$ because $1 \in B$ and $1 \notin A$.

   d. A has three elements: a, b, and c, and B has three different elements: $\{a\}$, $\{b\}$, and $\{c\}$. So A and B have no elements in common, and thus $A \not\subseteq B$ and $B \not\subseteq A$.

   e. $A = \{\sqrt{16}, \{4\}\} = \{4, \{4\}\}$ and $B = \{4\}$. Then $B \subseteq A$ because the only element in B is 4 and 4 is in A, but $A \not\subseteq B$ because $\{4\} \in A$ and $\{4\} \notin B$. Thus B is a proper subset of A.

   f. The only integer values taken by both the sine and the cosine function are 0, 1, and -1, and they occur at $x = 0, \dfrac{\pi}{2}, -\dfrac{\pi}{2}, \pi, -\pi, \dfrac{3\pi}{2}, -\dfrac{3\pi}{2}$, etc. Because all of these numbers can be expressed as $\dfrac{k\pi}{2}$ for some integer k,

   $$A = B = \{x \mid x = \frac{k\pi}{2} \text{ for some integer } k\}.$$

   It follows that $A \subseteq B$ and $B \subseteq A$ and neither is a proper subset of the other.

2. **Proof That $B \subseteq A$:**

Suppose x is a particular but arbitrarily chosen element of B.

   [We must show that $x \in A$. By definition of A, this means we must show that $x = 2 \cdot$ (some integer).]

By definition of B, there is an integer b such that $x = 2b - 2$.

   [Given that $x = 2b - 2$, can x also be expressed as $2 \cdot$ (some integer)? That is, is there an integer—say, a—such that $2b - 2 = 2a$? Solve for a to obtain $a = b - 1$. Check to see if this works.]

Let $a = b - 1$.

   [First check that a is an integer.]

We know that a is an integer because it is a difference of integers.

   [Then check that $x = 2a$.]

By substitution, $2a = 2(b - 1) = 2b - 2 = x$.

Thus, by definition of A, x is an element of A, *[as was to be shown]*.

3. **a.** $R \nsubseteq T$ because there are elements in R that are not in T. For example, the number 2 is in R but 2 is not in T since 2 is not divisible by 6.

   b. $T \subseteq R$ because every element in T is in R since every integer divisible by 6 is divisible by 2. To see why this is so, suppose n is any integer that is divisible by 6. Then $n = 6m$ for some integer m. Since $6m = 2(3m)$ and since $3m$ is an integer (being a product of integers), it follows that $n = 2 \cdot$ (*some integer*), and, hence, that n is divisible by 2.

   c. $T \subseteq S$ because every integer that is divisible by 6 is also divisible by 3. To see why this is so, suppose n is any integer that is divisible by 6. Then $n = 6m$ for some integer m. Since $6m = 3(2m)$ and since $2m$ is an integer (being a product of integers), it follows that $n = 3 \cdot$ (*some integer*), and, hence, that n is divisible by 3.

5. **a.** $C \subseteq D$ because every element in C is in D. To see why this is so, suppose n is any element of C. Then $n = 6r - 5$ for some integer r. Let $s = 2r - 2$. Then s is an integer (because products and differences of integers are integers), and

$$3s + 1 = 3(2r - 2) + 1 = 6r - 6 + 1 = 6r - 5,$$

   which equals n. Thus n satisfies the condition for being in D. Hence, every element in C is in D.

   b. $D \nsubseteq C$ because there are elements of D that are not in C. For example, 4 is in D because $4 = 3 \cdot 1 + 1$. But 4 is not in C because if it were, then $4 = 6r - 5$ for some integer r, which would imply that $9 = 6r$, or, equivalently, that $r = 3/2$, and this contradicts the fact that r is an integer.

6. **a.** $A \nsubseteq B$ because $2 \in A$ (because $2 = 5 \cdot 0 + 2$) but $2 \notin B$ (because if $2 = 10b - 3$ for some integer b, then $10b = 5$, so $b = 1/2$, which is not an integer).

   b. $B \subseteq A$

   <u>Proof:</u>

   Suppose y is a particular but arbitrarily chosen element of B.

   [We must show that y is in A. By definition of A, this means that we must show that $y = 5 \cdot$ (some integer) $+ 2$.]

   By definition of B, $y = 10b - 3$ for some integer **b**.

[Scratch work: Is there an integer, say a, such that $y = 5a + 2$? If so, then y will be in A. Now $5a + 2 = 10b - 3$, which implies that $5a = 10b - 5$, or, equivalently, that $a = 2b - 1$. So give this value to a and see if it works.]

Let $a = 2b - 1$. Then a is an integer, $5a + 2 \in B$, and $5a + 2 = 5(2b - 1) + 2 = 10b - 5 + 2 = 10b - 3 = y$. Thus y is in A *[as was to be shown]*.

c. $B = C$

<u>Proof:</u>

Part 1, Proof That $B \subseteq C$:

Suppose y is a particular but arbitrarily chosen element of B.

[We must show that y is in C. By definition of C, this means that we must show that y = 10·(some integer) + 7.]

By definition of B, $y = 10b - 3$ for some integer b.

[Scratch work: Is there an integer, say c, such that $y = 10c + 7$? If so, then $10c + 7 = 10b - 3$, which implies that $10c = 10b - 10$, or, equivalently, that $c = b - 1$. So give this value to c and see if it works.]

Let $c = b - 1$. Then c is an integer, and so $10c + 7 \in B$. Also $10c + 7 = 10(b - 1) + 7 = 10b - 10 + 7 = 10b - 3 = y$. Thus y satisfies the definition to be in C *[as was to be shown]*.

Part 2, Proof That $C \subseteq B$:

Suppose z is a particular but arbitrarily chosen element of C.

[We must show that z is in B. By definition of B, this means that we must show that z = 10·(some integer) − 3.]

By definition of C, $z = 10c + 7$ for some integer c.

[Scratch work: Is there an integer, say b, such that $z = 10b - 3$? If so, then $10b - 3 = 10c + 7$, which implies that $10b = 10c + 10$, or, equivalently, that $b = c + 1$. So give this value to b and see if it works.]

Let $b = c + 1$. Then b is an integer and so $10b - 3$ is in B. Also, by substitution, $10b - 3 = 10(c + 1) - 3 = 10c + 10 - 3 = 10c + 7 = z$, and hence z satisfies the definition to be in B *[as was to be shown]*.

8. **a.** *In words:* The set of all x in U such that x is in A and x is in B.

   In symbolic notation: $A \cap B$.

9. **a.** $x \notin A$ and $x \notin B$ **b.** $x \notin A$ or $x \notin B$ **c.** $x \notin A$ or $x \in B$

10. **a.** $A \cup B = \{1, 3, 5, 6, 7, 9\}$ **b.** $A \cap B = \{3, 9\}$ **c.** $A \cup C = \{1, 2, 3, 4, 5, 6, 7, 8, 9\}$
    d. $A \cap C = \emptyset$ **e.** $A - B = \{1, 5, 7\}$

11. **a.** $A \cup B = \{x \in \mathbf{R} \mid 0 < x < 4\}$ **b.** $A \cap B = \{x \in \mathbf{R} \mid 1 \le x \le 2\}$
    c. $A^c = \{x \in \mathbf{R} \mid x \le 0 \text{ or } x > 2\}$ **d.** $A \cup C = \{x \in \mathbf{R} \mid 0 < x \le 2 \text{ or } 3 \le x < 9\}$
    e. $A \cap C = \emptyset$ **f.** $B^c = \{x \in \mathbf{R} \mid x \le -1 \text{ or } x \ge 2\}$
    g. $A^c \cap B^c = \{x \in \mathbf{R} \mid x < -3 \text{ or } x \ge 2\}$ **h.** $A^c \cup B^c = \{x \in \mathbf{R} \mid x < 1 \text{ or } x > 2\}$
    i. $(A \cap B)^c = \{x \in \mathbf{R} \mid x < 1 \text{ or } x > 2\}$ **j.** $(A \cup B)^c = \{x \in \mathbf{R} \mid x \le 0 \text{ or } x \ge 4\}$

12. **a.** $A \cup B = \{x \in \mathbf{R} \mid -3 \le x < 2\}$ **b.** $A \cap B = \{x \in \mathbf{R} \mid -1 < x \le 0\}$
    c. $A^c = \{x \in \mathbf{R} \mid x < -3 \text{ or } x > 0\}$ **d.** $A \cup C = \{x \in \mathbf{R} \mid -3 \le x \le 0 \text{ or } 6 < x \le 8\}$
    e. $A \cap C = \emptyset$ **f.** $B^c = \{x \in \mathbf{R} \mid x \le -1 \text{ or } x \ge 2\}$
    g. $A^c \cap B^c = \{x \in \mathbf{R} \mid x < -3 \text{ or } x \ge 2\}$ **h.** $A^c \cup B^c = \{x \in \mathbf{R} \mid x \le -1 \text{ or } x > 0\}$
    i. $(A \cap B)^c = \{x \in \mathbf{R} \mid x \le -1 \text{ or } x > 0\}$ **j.** $(A \cup B)^c = \{x \in \mathbf{R} \mid x < -3 \text{ or } x \ge 2\}$

    Note that $(A \cap B)^c = A^c \cup B^c$ and that $(A \cup B)^c = A^c \cap B^c$.

13. **a.** $A \cap B = \{1111\}$

   c. $A - B = \{1110, 1000, 1001\}$

14. **a.**

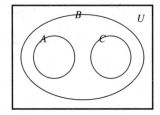

15. **a.**

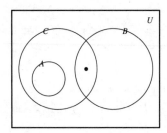

   b.

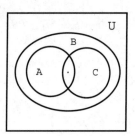

   c.

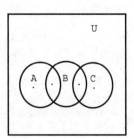

16. **a.** $A \cup (B \cap C) = \{a, b, c\}, (A \cup B) \cap C = \{b, c\},$ and

   $(A \cup B) \cap (A \cup C) = \{a, b, c, d\} \cap \{a, b, c, e\} = \{a, b, c\}.$

   Hence $A \cup (B \cap C) = (A \cup B) \cap (A \cup C).$

17. a

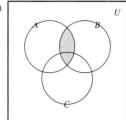

b

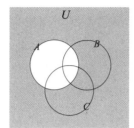

c

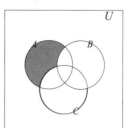

d

e

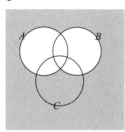

f

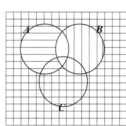

Region shaded ||| *is* A^c
Region shaded ≡ *is* B^c

Cross-hatched region is $A^c \cap B^c$

18. a. The number 0 is not in $\emptyset$ because $\emptyset$ has no elements.

   b. No. The left-hand set is the empty set; it does not have any elements. The right-hand set is a set with one element, namely $\emptyset$.

19. $A_1 - \{1, 1^2\} - \{1\}$, $A_2 = \{2, 2^2\} = \{2, 4\}$,
   $A_3 = \{3, 3^2\} = \{3, 9\}$, $A_4 = \{4, 4^2\} = \{4, 16\}$

   a. $A_1 \cup A_2 \cup A_3 \cup A_4 = \{1\} \cup \{2, 4\} \cup \{3, 9\} \cup \{4, 16\} = \{1, 2, 3, 4, 9, 16\}$

   b. $A_1 \cap A_2 \cap A_3 \cap A_4 = \{1\} \cap \{2, 4\} \cap \{3, 9\} \cap \{4, 16\} = \emptyset$

   c. A_1, A_2, A_3, and A_4 are not mutually disjoint, because $A_2 \cap A_4 = \{4\} \neq \emptyset$.

21. $C_0 = \{0, -0\} = \{0\}$, $C_1 = \{1, -1\}$, $C_2 = \{2, -2\}$, $C_3 = \{3, -3\}$, $C_4 = \{4, -4\}$

   a. $\bigcup_{i=0}^{4} C_i = \{0\} \cup \{1, -1\} \cup \{2, -2\} \cup \{3, -3\} \cup \{4, -4\} = \{-4, -3, -2, -1, 0, 1, 2, 3, 4\}$

   b. $\bigcup_{i=0}^{4} C_i = \{0\} \cap \{1, -1\} \cap \{2, -2\} \cap \{3, -3\} \cap \{4, -4\} = \emptyset$

   c. C_0, C_1, C_2, ... are mutually disjoint because no two of the sets have any elements in common.

   d. $\bigcup_{i=0}^{n} C_i = \{-n, -(n-1), \ldots, -2, -1, 0, 1, 2, \ldots, (n-1), n\}$

   e. $\bigcap_{i=0}^{n} C_i = \emptyset$

f. $\displaystyle\bigcup_{i=0}^{\infty} C_i = \mathbf{Z}$, the set of all integers

g. $\displaystyle\bigcap_{i=0}^{\infty} C_i = \emptyset$

22. $D_0 = [-0, 0] = \{0\}, D_1 = [-1, 1], D_2 = [-2, 2], D_3 = [-3, 3], D_4 = [-4, 4]$

a. $\displaystyle\bigcup_{i=0}^{4} D_i = \{0\} \cup [-1, 1] \cup [-2, 2] \cup [-3, 3] \cup [-4, 4] = [-4, 4]$

b. $\displaystyle\bigcap_{i=0}^{4} D_i = \{0\} \cap [-1, 1] \cap [-2, 2] \cap [-3, 3] \cap [-4, 4] = \{0\}$

c. $D_0, D_1, D_2, \ldots$, are not mutually disjoint. In fact, each $D_k \subseteq D_{k+1}$.

d. $\displaystyle\bigcup_{i=0}^{n} D_i = [-n, n]$

e. $\displaystyle\bigcap_{i=0}^{n} D_i = \{0\}$

f. $\displaystyle\bigcup_{i=0}^{\infty} D_i = \mathbf{R}$, the set of all real numbers

g. $\displaystyle\bigcap_{i=0}^{\infty} D_i = \{0\}$

24. $W_0 = (0, \infty), W_1 = (1, \infty), W_2 = (2, \infty), W_3 = (3, \infty), W_4 = (4, \infty)$

a. $\displaystyle\bigcup_{i=0}^{4} W_i = (0, \infty) \cup (1, \infty) \cup (2, \infty) \cup (3, \infty) \cup (4, \infty) = (0, \infty)$

b. $\displaystyle\bigcap_{i=0}^{4} W_i = (0, \infty) \cap (1, \infty) \cap (2, \infty) \cap (3, \infty) \cap (4, \infty) = (4, \infty)$

c. $W_0, W_1, W_2, \ldots$ are not mutually disjoint. In fact, $W_{k+1} \subseteq W_k$ for every integer $k \geq 0$.

d. $\displaystyle\bigcup_{i=0}^{n} W_i = (0, \infty)$

e. $\displaystyle\bigcap_{i=0}^{n} W_i = (n, \infty)$

f. $\displaystyle\bigcup_{i=0}^{\infty} W_i = (0, \infty)$

g. $\displaystyle\bigcap_{i=0}^{\infty} W_i = \emptyset$

27. **a.** No. The element d is in two of the sets.

b. Yes. Every element in $\{p, q, u, v, w, x, y, z\}$ is in one of the sets of the partition and no element is in more than one set of the partition.

c. No. The number 4 is in both sets $\{5, 4\}$ and $\{1, 3, 4\}$.

d. No. None of the sets contains 6.

e. Yes. Every element in $\{1, 2, 3, 4, 5, 6, 7, 8\}$ is in one of the sets of the partition and no element is in more than one set of the partition.

28. Yes. Every integer is either even or odd, and no integer is both even and odd.

30. Yes. By the quotient-remainder theorem, every integer can be represented in exactly one of the following forms: $4k$ or $4k+1$ or $4k+2$ or $4k+3$ for some integer k. Thus $\mathbf{Z} = A_0 \cup A_1 \cup A_2 \cup A_3$, $A_0 \cap A_1 = \emptyset, A_0 \cap A_2 = \emptyset, A_0 \cap A_3 = \emptyset, A_1 \cap A_2 = \emptyset, A_1 \cap A_3 = \emptyset$, and $A_2 \cap A_3 = \emptyset$.

31. **a.** $A \cap B = \{2\}$, so $\mathscr{P}(A \cap B) = \{\emptyset, \{2\}\}$.

  b. $A = \{1, 2\}$, so $\mathscr{P}(A) = \{\emptyset, \{1\}, \{2\}, \{1, 2\}\}$.

  c. $A \cup B = \{1, 2, 3\}$, so $\mathscr{P}(A \cup B) = \{\emptyset, \{1\}, \{2\}, \{3\}, \{1, 2\}, \{1, 3\}, \{2, 3\}, \{1, 2, 3\}\}$.

  d. $A \times B = \{(1, 2), (1, 3), (2, 2), (2, 3)\}$, so

  $$\mathscr{P}(A \times B) = \{\emptyset, \{(1,2)\}, \{(1,3)\}, \{(2,2)\}, \{(2,3)\}, \{(1,2),(1,3)\}, \{(1,2),(2,2)\},$$
  $$\{(1,2),(2,3)\}, \{(1,3),(2,2)\}, \{(1,3),(2,3)\}, \{(1,2),(2,3)\}, \{(1,3),(2,2)\},$$
  $$\{(1,3),(2,3)\}, \{(2,2),(2,3)\}, \{(1,2),(1,3),(2,2)\}, \{(1,2),(1,3),(2,3)\},$$
  $$\{(1,2),(2,2),(2,3)\}, \{(1,3),(2,2),(2,3)\}, \{(1,2),(1,3),(2,2),(2,3)\}\}.$$

32. **a.** $\mathscr{P}(A \times B) = \{\emptyset, \{(1, u)\}, \{(1, v)\}, \{(1, u), (1, v)\}\}$

33. **a.** $\mathscr{P}(\emptyset) = \{\emptyset\}$

  b. $\mathscr{P}(\mathscr{P}(\emptyset)) = \mathscr{P}(\{\emptyset\}) = \{\emptyset, \{\emptyset\}\}$

  c. $\mathscr{P}(\mathscr{P}(\mathscr{P}(\emptyset))) = \{\emptyset, \{\emptyset\}, \{\{\emptyset\}\}, \{\emptyset, \{\emptyset\}\}\}$

34. **a.** $A_1 \cup (A_2 \times A_3) = \{1\} \cup \{(u, m), (u, n), (v, m), (v, n)\} = \{1, (u, m), (u, n), (v, m), (v, n)\}$

35. **a.** $A \times (B \cup C) = \{a, b\} \times \{1, 2, 3\} = \{(a, 1), (a, 2), (a, 3), (b, 1), (b, 2), (b, 3)\}$

  b. $(A \times B) \cup (A \times C) = \{(a, 1), (a, 2), (b, 1), (b, 2), (a, 3), (b, 3)\}$
  $$= \{(a, 1), (a, 2), (b, 1), (b, 2), (a, 3), (b, 3)\}$$

36.

i	1				2				3			4
j		1	2	3	1	2	3	4	1		2	
found		no	yes		no		yes		no	yes		
answer	$A \subseteq B$											

Section 6.2

1. **a.** (1) A (2) $B \cup C$

  b. (1) $A \cap B$ (2) C

  d. (1) $(A \cup B) \cap C$ (2) $A \cup (B \cap C)$

2. **a.** (1) $A - B$ (2) A (3) A (4) B

3. **a.** by definition of subset (because $A \subseteq B$)

  b. by definition of subset (because $B \subseteq C$)

  c. by definition of subset

5. <u>Proof:</u> Suppose A and B are any sets.

  Proof that $B - A \subseteq B \cap A^c$: Suppose $x \in B - A$. By definition of set difference, $x \in B$ and $x \notin A$. It follows by definition of complement that $x \in B$ and $x \in A^c$, and so by definition of intersection, $x \in B \cap A^c$. *[Thus every element in $B - A$ is in $B \cap A^c$, and so $B - A \subseteq B \cap A^c$ by definition of subset.]*

  Proof that $B \cap A^c \subseteq B - A$: Suppose $x \in B \cap A^c$. By definition of intersection, $x \in B$ and $x \in A^c$. It follows by definition of complement that $x \in B$ and $x \notin A$, and so by definition of set difference, $x \in B - A$. *[Thus every element in $B \cap A^c$ is in $B - A$, and so $B \cap A^c \subseteq B - A$ by definition of subset.]*

  [Since both subset relations have been proved, $B - A = B \cap A^c$ by definition of set equality.]

6. (1) (a) $(A \cap B) \cup (A \cap C)$ (b) A (c) $x \in C$ (d) $x \in (A \cap B) \cup (A \cap C)$

   (2) (a) or   (b) and   (c) $x \in A \cap (B \cup C)$ (d) subset

   (3) (a) $A \cap (B \cup C) = (A \cap B) \cup (A \cap C)$

7. *Hint:* This is somewhat similar to the proof in Example 6.2.3.

8. Proof: Suppose A and B are any sets.

   Proof that $(A \cap B) \cup (A \cap B^c) \subseteq A$: Suppose $x \in (A \cap B) \cup (A \cap B^c)$. *[We must show that $x \in A$.]*

   By definition of union, $x \in A \cap B$ or $x \in A \cap B^c$.

   Case 1 ($x \in A \cap B$): In this case x is in both A and B, and so, in particular, $x \in A$.

   Case 2 ($x \in A \cap B^c$): In this case x is in A and x is not in B, and so, in particular, $x \in A$.

   Thus, in either case, $x \in A$ *[as was to be shown]*. So $(A \cap B) \cup (A \cap B^c) \subseteq A$ *[by definition of subset]*.

   Proof that $A \subseteq (A \cap B) \cup (A \cap B^c)$: Suppose $x \in A$. *[We must show that $x \in (A \cap B) \cup (A \cap B^c)$.]* Either $x \in B$ or $x \notin B$.

   Case 1 ($x \in B$): In this case since we also know that x is in A, by definition of intersection, $x \in A \cap B$.

   Case 2 ($x \notin B$): In this case, by definition of complement, $x \in B^c$, and, since we also know that x is in A, by definition of intersection, $x \in A \cap B^c$.

   Thus, in either case $x \in A \cap B$ or $x \in A \cap B^c$, and so, $x \in (A \cap B) \cup (A \cap B^c)$ by definition of union *[as was to be shown]*. So $A \subseteq (A \cap B) \cup (A \cap B^c)$ *[by definition of subset]*.

   Conclusion: Since both subset relations have been proved it follows by definition of set equality that $(A \cap B) \cup (A \cap B^c) = A$.

9. Proof: Suppose A, B, and C are any sets. *[To show that $(A - B) \cup (C - B) = (A \cup C) - B$, we must show that $(A - B) \cup (C - B) \subseteq (A \cup C) - B$ and that $(A \cup C) - B \subseteq (A - B) \cup (C - B)$.]*

   Proof that $(A - B) \cup (C - B) \subseteq (A \cap C) - B$: Suppose that x is any element in $(A - B) \cup (C - B)$. *[We must show that $x \in (A \cup C) - B$.]*

   By definition of union, $x \in A - B$ or $x \in C - B$.

   Case 1 ($x \in A - B$): By definition of set difference, $x \in A$ and $x \notin B$, and because $x \in A$, $x \in A \cup C$ by definition of union. So $x \in A \cup C$ and $x \notin B$, and thus, by definition of set difference, $x \in (A \cup C) - B$.

   Case 2 ($x \in C - B$): By definition of set difference, $x \in C$ and $x \notin B$, and because $x \in C$, $x \in A \cup C$ by definition of union. So $x \in A \cup C$ and $x \notin B$, and thus, by definition of set difference, $x \in (A \cup C) - B$.

   Hence, in both cases, $x \in (A \cup C) - B$ *[as was to be shown]*.

   So $(A - B) \cup (C - B) \subseteq (A \cup C) - B$ *[by definition of subset]*.

   Proof that $(A \cup C) - B \subseteq (A - B) \cup (C - B)$:

   Suppose that x is any element in $(A \cup C) - B$. *[We must show that $x \in (A - B) \cup (C - B)$.]*

   By definition of set difference, $x \in (A \cup C)$ and $x \notin B$.

   Thus, by definition of union, $x \in A$ or $x \in C$, and in both cases, $x \notin B$.

   Case 1 ($x \in A$ and $x \notin B$): By definition of set difference, $x \in A - B$, and so by definition of union, $x \in (A - B) \cup (C - B)$.

   Case 2 ($x \in C$ and $x \notin B$): By definition of set difference, $x \in C - B$, and so by definition of union, $x \in (A - B) \cup (C - B)$.

In both cases, $x \in (A - B) \cup (C - B)$ *[as was to be shown]*.

So $(A \cup C) - B \subseteq (A - B) \cup (C - B)$ *[by definition of subset]*.

Because both subset relations have been proved, we conclude that $(A - B) \cup (C - B) = (A \cup C) - B$.

10. <u>Proof:</u> Suppose A, B, and C are any sets.

We will show that $(A \cup B) \cap C \subseteq A \cup (B \cap C)$.

Suppose x is any element $(A \cup B) \cap C$.

By definition of intersection x is in $A \cup B$ and x is in C.

Then by definition of union x is in A or x is in B, and in both cases x is in C. It follows by definition of union that in case x is in A and x is in C, then x is in $A \cup (B \cap C)$ by virtue of being in A. And in case x is in $B \cap C$, then x is in $A \cup (B \cap C)$ by virtue of being in $B \cap C$. Thus in both cases x is in $A \cup (B \cap C)$, which proves that every element in $(A \cup B) \cap C$ is in $A \cup (B \cap C)$.

Hence $(A \cup B) \cap C \subseteq A \cup (B \cap C)$ by definition of subset.

11. <u>Proof:</u> Suppose A, B and C are any sets.

We will show that $A \cap (B - C) \subseteq (A \cap B) - (A \cap C)$.

Suppose $x \in A \cap (B - C)$. *[We must show that $x \in (A \cap B) - (A \cap C)$.]*

By definition of intersection, $x \in A$ and $x \in B - C$.

So, by the set difference law, $x \in A$ and both $x \in B$ and $x \notin C$.

Since $x \in A$ and $x \in B$, then $x \in A \cap B$ by definition of intersection.

And since $x \in A$ and $x \notin C$, then $x \notin A \cap C$ by definition of intersection.

(For if $x \in A \cap C$, then, by definition of intersection, x would be in C, which it is not.)

Thus $x \in A \cap B$ and $x \notin A \cap C$, and so, by the set difference law, $x \in (A \cap B) - (A \cap C)$.

This shows that every element in $A \cap (B - C)$ is in $(A \cap B) - (A \cap C)$,

and so, by definition of subset, $A \cap (B - C) \subseteq (A \cap B) - (A \cap C)$.

12. <u>Proof:</u>

Suppose A, B and C are any sets.

We will show that $(A \cup B) - C \subseteq (A - C) \cup (B - C)$.

Suppose $x \in (A \cup B) - C$. *[We must show that $x \in (A - C) \cup (B - C)$.]*

By the set difference law, $x \in A \cup B$ and $x \notin C$.

So, by definition of union, $x \in A$ or $x \in B$, and in both cases $x \notin C$.

In case 1, $x \in A$ and $x \notin C$, and thus, by the set difference law, $x \in A - C$.

In case 2, $x \in B$ and $x \notin C$, and thus, by the set difference law, $x \in B - C$.

Thus $x \in A - C$ or $x \in B - C$.

It follows that $x \in (A - C) \cup (B - C)$ by definition of union.

This shows that every element in $(A \cup B) - C$ is in $(A - C) \cup (B - C)$,

and so, by definition of subset, $(A \cup B) - C \subseteq (A - C) \cup (B - C)$.

13. <u>Proof:</u> Let A, B, and C be any sets.

Proof that $(A - B) \cap (C - B) \subseteq (A \cap C) - B$:

Suppose $x \in (A - B) \cap (C - B)$. *[We must show that $x \in (A \cap C) - B$.]*

By definition of intersection, $x \in A - B$ and $x \in C - B$.

By definition of set difference, $x \in A$ and $x \notin B$ and $x \in C$ and $x \notin B$

Thus $x \in A \cap C$ by definition of intersection.

In addition, $x \notin B$. Hence $x \in (A \cap C) - B$ by definition of set difference.

[Thus $(A - B) \cap (C - B) \subseteq (A \cap C) - B$ by definition of subset.]

Proof that $(A \cap C) - B \subseteq (A - B) \cap (C - B)$:

Suppose $x \in (A \cap C) - B$. *[We must show that $x \in (A - B) \cap (C - B)$.]*

By definition of set difference, $x \in A \cap C$ and $x \notin B$.

By definition of intersection, $x \in A$ and $x \in C$, and also $x \notin B$.

Hence both $x \in A$ and $x \notin B$ and also $x \in C$, and $x \notin B$.

So, by definition of set difference, $x \in A - B$ and $x \in C - B$.

Finally, by definition of intersection, $x \in (A - B) \cap (C - B)$.

[Thus $x \in (A \cap C) - B \subseteq (A - B) \cap (C - B)$ by definition of subset.]

[Since both subset relations have been proved, $(A - B) \cap (C - B) = (A \cap C) - B$ by definition of set equality.]

14. <u>Proof:</u> Let A and B be any sets.

    Proof that $A \cup (A \cap B) \subseteq A$:

    Suppose $x \in A \cup (A \cap B)$. *[We must show that $x \in A$.]*

    By definition of union, $x \in A$ or $x \in A \cap B$.

    In case $x \in A$, then clearly $x \in A$.

    In case $x \in A \cap B$, then, by definition of intersection, $x \in A$ and $x \in B$, and so, in particular, $x \in A$.

    Hence in either case $x \in A$.

    [Thus $A \cup (A \cap B) \subseteq A$ by definition of subset.]

    Proof that $A \subseteq A \cup (A \cap B)$:

    Suppose $x \in A$.

    Then by definition of union, $x \in A \cup (A \cap B)$.

    [Thus $A \subseteq A \cup (A \cap B)$ by definition of subset.]

    [Since both subset relations have been proved, $A \cup (A \cap B) = A$ by definition of set equality.]

15. <u>Proof:</u> Let A be any set. *[We must show that $A \cup \emptyset = A$.]*

    Proof that $A \cup \emptyset \subseteq A$: Suppose $x \in A \cup \emptyset$. Then $x \in A$ or $x \in \emptyset$ by definition of union. But $x \notin \emptyset$ since $\emptyset$ has no elements. Hence $x \in A$.

    Proof that $A \subseteq A \cup \emptyset$: Suppose $x \in A$. Then the statement "$x \in A$ or $x \in \emptyset$" is true. Hence $x \in A \cup \emptyset$ by definition of union. *[Alternatively, $A \subseteq A \cup \emptyset$ by the inclusion in union property.]*

    Since $A \cup \emptyset \subseteq A$ and $A \subseteq A \cup \emptyset$, then $A \cup \emptyset = A$ by definition of set equality.

16. <u>Proof:</u> Suppose $A, B,$ and C are any sets such that $A \subseteq B$. Let $x \in A \cap C$. By definition of intersection, $x \in A$ and $x \in C$. Now since $A \subseteq B$ and $x \in A$, then $x \in B$. Hence $x \in B$ and $x \in C$, and so, by definition of intersection, $x \in B \cap C$. *[Thus $A \cap C \subseteq B \cap C$ by definition of subset.]*

17. <u>Proof:</u>

    Suppose $A, B,$ and C are sets and $A \subseteq B$. Let $x \in A \cup C$. *[We must show that $x \in B \cup C$.]*

    Then by definition of union, $x \in A$ or $x \in C$.

In the first case, since, $A \subseteq B$, $x \in B$, and in the second case, $x \in C$.

Thus $x \in B$ or $x \in C$, and so, by definition of union, $x \in B \cup C$.

[Thus $A \cup C \subseteq B \cup C$ by definition of subset.]

18. <u>Proof:</u>

Suppose A and B are sets and $A \subseteq B$. Let $x \in B^c$. *[We must show that $x \in A^c$.]*

By definition of complement, $x \notin B$.

It follows that $x \notin A$ *[because if $x \in A$ then $x \in B$ (since $A \subseteq B$), and this would contradict the fact that $x \notin B$].*

Hence by definition of complement $x \in A^c$.

[Thus $B^c \subseteq A^c$ by definition of subset.]

19. <u>Proof:</u>

Suppose A, B, and C are sets and $A \subseteq B$ and $A \subseteq C$. Let $x \in A$. *[We must show that $x \in B \cap C$.]*

Since $x \in A$ and $A \subseteq B$, then $x \in B$ (by definition of subset).

Similarly, since $x \in A$ and $A \subseteq C$, then $x \in C$.

Hence $x \in B$ and $x \in C$, and so by definition of intersection, $x \in B \cap C$.

[By definition of subset, therefore, $A \subseteq B \cap C$.]

20. <u>Proof:</u>

Suppose A, B, and C are sets and $A \subseteq C$ and $B \subseteq C$. Let $x \in A \cup B$. *[We must show that $x \in C$.]*

By definition of union, $x \in A$ or $x \in B$.

Case 1 ($x \in A$): In this case, since $A \subseteq C$, by definition of subset, $x \in C$.

Case 1 ($x \in B$): In this case, since $B \subseteq C$, by definition of subset, $x \in C$.

In both cases, $x \in C$ *[as was to be shown]*.

21. <u>Proof:</u> Suppose A, B, and C are arbitrarily chosen sets.

Proof that $A \times (B \cup C) \subseteq (A \times B) \cup (A \times C)$: Suppose $(x, y) \in A \times (B \cup C)$. *[We must show that $(x, y) \in (A \times B) \cup (A \times C)$.]* Then $x \in A$ and $y \in B \cup C$. By definition of union, this means that $y \in B$ or $y \in C$.

Case 1 ($y \in B$): Then, since $x \in A$, $(x, y) \in A \times B$ by definition of Cartesian product. Hence $(x, y) \in (A \times B) \cup (A \times C)$ by definition of union.

Case 2 ($y \in C$): Then, since $x \in A$, $(x, y) \in A \times C$ by definition of Cartesian product. Hence $(x, y) \in (A \times B) \cup (A \times C)$ by definition of union.

Hence, in either case, $(x, y) \in (A \times B) \cup (A \times C)$ *[as was to be shown]*.

Thus $A \times (B \cup C) \subseteq (A \times B) \cup (A \times C)$ by definition of subset.

Proof that $(A \times B) \cup (A \times C) \subseteq A \times (B \cup C)$: Suppose $(x, y) \in (A \times B) \cup (A \times C)$. *[We must show that $(x, y) \in (A \times B) \cup (A \times C)$.]*

Then $(x, y) \in A \times B$ or $(x, y) \in A \times C$.

Case 1 ($(x, y) \in A \times B$: In this case, $x \in A$ and $y \in B$. Now since $y \in B$ then $y \in B \cup C$ by definition of union. Hence $x \in A$ and $y \in B \cup C$, and so, by definition of Cartesian product, $(x, y) \in A \times (B \cup C)$.

Case 2 ($(x, y) \in A \times C$: In this case $x \in A$ and $y \in C$. Now since $y \in C$, then $y \in B \cup C$ by definition of union. Hence $x \in A$ and $y \in B \cup C$, and so, by definition of Cartesian product, $(x, y) \in A \times (B \cup C)$.

Thus, in either case, $(x, y) \in A \times (B \cup C)$. *[Hence, by definition of subset, $(A \times B) \cup (A \times C) \subseteq A \times (B \cup C)$.]*

[Since both subset relations have been proved, we can conclude that $A \times (B \cup C) = (A \times B) \cup (A \times C)$ by definition of set equality.]

22. Proof:

Suppose A, B, and C are sets.

Proof that $A \times (B \cap C) \subseteq (A \times B) \cap (A \times C)$: Suppose $(x, y) \in A \times (B \cap C)$. *[We must show that $(x, y) \in (A \times B) \cap (A \times C)$.]*

By definition of Cartesian product, $x \in A$ and $y \in B \cap C$.

By definition of intersection, $y \in B$ and $y \in C$.

It follows that both statements "$x \in A$ and $y \in B$" and "$x \in A$ and $y \in C$" are true.

Hence by definition of Cartesian product, $(x, y) \in A \times B$ and $(x, y) \in A \times C$, and so by definition of intersection, $(x, y) \in (A \times B) \cap (A \times C)$.

[Thus $A \times (B \cap C) \subseteq (A \times B) \cap (A \times C)$ by definition of subset.]

Proof that $(A \times B) \cap (A \times C) \subseteq A \times (B \cap C)$: Suppose $(x, y) \in (A \times B) \cap (A \times C)$. *[We must show that $(x, y) \in A \times (B \cap C)$.]*

By definition of intersection, $(x, y) \in A \times B$ and $(x, y) \in A \times C$, and so by definition of Cartesian product $x \in A$ and $y \in B$ and also $x \in A$ and $y \in C$.

Consequently, the statement "$x \in A$ and both $y \in B$ and $y \in C$" is true.

It follows by definition of intersection that $x \in A$ and $y \in B \cap C$.

So, by definition of Cartesian product, $(x, y) \in A \times (B \cap C)$.

[Thus $(A \times B) \cap (A \times C) \subseteq A \times (B \cap C)$ by definition of subset.]

[Since both subset relations have been proved, $A \times (B \cap C) = (A \times B) \cap (A \times C)$ by definition of set equality.]

23. There is more than one error in the "proof." The most serious is the misuse of the definition of subset. To say that A is a subset of B means that for every x, **if** $x \in A$ **then** $x \in B$. It does not mean that there exists an element of A that is also an element of B. The second error in the proof occurs in the last sentence. Even if there is an element in A that is in B and an element in B that is in C, it does not follow that there is an element in A that is in C. For instance, suppose $A = \{1, 2\}$, $B = \{2, 3\}$, and $C = \{3, 4\}$. Then there is an element in A that is in B (namely 2) and there is an element in B that is in C (namely, 3), but there is no element in A that is in C.

24. The "proof" claims that because $x \notin A$ or $x \notin B$, it follows that $x \notin A \cup B$. But it is possible for "$x \notin A$ or $x \notin B$" to be true and "$x \notin A \cup B$" to be false. For example, let $A = \{1, 2\}$, $B = \{2, 3\}$, and $x = 3$. Then since $3 \notin \{1, 2\}$, the statement "$x \notin A$ or $x \notin B$" is true. But since $A \cup B = \{1, 2, 3\}$ and $3 \in \{1, 2, 3\}$, the statement "$x \notin A \cup B$" is false.

25. A correct proof of the given statement must show that if $x \in (A - B) \cup (A \cap B)$ then $x \in A$. This incorrect proof uses the assumption that $x \in A$ as a basis for concluding that $x \in A$. In other words, this incorrect proof assumes what is to be proved.

Another mistake is that the assertion "If $x \in A$ then $x \in A - B$" is not necessarily true. In fact, it is often false. For example, if $A = \{1, 2\}$ and $B = \{2\}$, then $A - B = \{1\}$, and so the statement "$2 \in A$" is true but the statement "$2 \in A - B$" is false.

26. **a.**

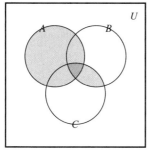

The shaded region is $A \cup (B \cap C)$.

b

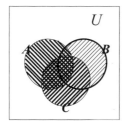

darkly shaded region is $A \cap (B \cup C)$

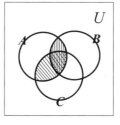

entire shaded region is $(A \cap B) \cup (A \cap C)$

c.

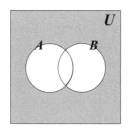

shaded region is $(A \cup B)^c$

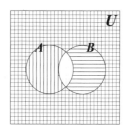

cross-hatched region is $A^c \cap B^c$

d.

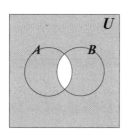

shaded region is $(A \cap B)^c$

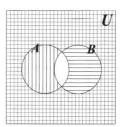

entire shaded region is $A^c \cup B^c$

27. **a.** $(A - B) \cap (B - A)$ **b.** intersection   **c.** $B - A$ **d.** B
    e. A **f.** A **g.** $(A - B) \cap (B - A) = \emptyset$

28. <u>Proof by contradiction:</u> Suppose not. That is, suppose there exist sets A and B such that $\overline{(A \cap B) \cap (A \cap B^c)} \neq \emptyset$. By definition of $\emptyset$, there is an element x in $(A \cap B) \cap (A \cap B^c)$. By definition of intersection, $x \in (A \cap B)$ and $x \in (A \cap B^c)$. Applying the definition of intersection again, we have that since $x \in (A \cap B)$, $x \in A$ and $x \in B$, and since $x \in (A \cap B^c)$, $x \in A$ and $x \notin B$. Thus, in particular, $x \in B$ and $x \notin B$, which is a contradiction. It follows that the supposition is false, and so $(A \cap B) \cap (A \cap B^c) = \emptyset$.

30. <u>Proof by contradiction:</u> Suppose not. That is, suppose there exists a set A such that $A \cap A^c \neq \emptyset$.

    [We must show that this supposition leads logically to a contradiction.]

By definition of $\emptyset$, there is an element x in $A \cap A^c$.

By definition of intersection, $x \in A$ and $x \in A^c$.

So by definition of complement, $x \in A$ and $x \notin A$, and this is a contradiction.

[Hence the supposition is false, and we conclude that $A \cap A^c = \emptyset$.]

32. Proof by contradiction: Let A be any set. Suppose $A \times \emptyset \neq \emptyset$. Then there would be an element (x, y) in $A \times \emptyset$. By definition of Cartesian product, $x \in A$ and $y \in \emptyset$. But there are no elements y such that $y \in \emptyset$. Hence there are no elements (x, y) in $A \times \emptyset$, which contradicts the supposition. *[Thus the supposition is false, and so $A \times \emptyset = \emptyset$.]*

33. Proof: Let A and B be sets such that $A \subseteq B$. *[We will show that $A \cap B^c = \emptyset$.]*

Suppose $A \cap B^c \neq \emptyset$; that is, suppose there were an element, say x, such that $x \in A \cap B^c$.

[We will use proof by contradiction to show that this is impossible.]

By definition of intersection, $x \in A$ and $x \in B^c$.

So $x \in A$ and $x \notin B$ by definition of complement.

But $A \subseteq B$ by hypothesis, and, since $x \in A$, then $x \in B$ by definition of subset.

Thus $x \notin B$ and also $x \in B$, which is a contradiction.

Hence the supposition that $A \cap B^c \neq \emptyset$ is false, and so $A \cap B^c = \emptyset$.

36. Proof: Let A, B, and C be any sets such that $C \subseteq B - A$. *[We will show that $A \cap C = \emptyset$.]*

Suppose $A \cap C \neq \emptyset$. Then there is an element, say x, such that $x \in A \cap C$.

By definition of intersection, $x \in A$ and $x \in C$.

Now since $x \in C$ and $C \subseteq B - A$, then $x \in B$ and $x \notin A$.

So $x \in A$ and $x \notin A$, which is a contradiction.

Hence the supposition is false, and thus $A \cap C = \emptyset$.

39. **a.** Proof:

Let A and B be any sets. *[We will show that $[(A - B) \cup (B - A) \cup (A \cap B] \subseteq A \cup B$ and that $A \cup B \subseteq (A - B) \cup (B - A) \cup (A \cap B)$.]*

Proof that $(A - B) \cup (B - A) \cup (A \cap B) \subseteq A \cup B$):

Suppose $x \in (A - B) \cup (B - A) \cup (A \cap B)$. *[We must show that $x \in A \cup B$.]*

By definition of union, $x \in A - B$ or $x \in B - A$ or $x \in A \cap B$.

(1) In case $x \in A - B$, by definition of set difference, $x \in A$ and $x \notin B$. So, in particular, $x \in A$, and thus, by definition of union, $x \in A \cup B$.

(2) In case $x \in B - A$, by definition of set difference, and $x \notin A$ and $x \in B$. So, in particular, $x \in B$, and thus, by definition of union, $x \in A \cup B$.

(3) In case $x \in A \cap B$, by definition of set difference, $x \in A$ and $x \in B$. Thus, by definition of union, $x \in A \cup B$.

Thus in all three possible cases, $x \in A \cup B$ *[as was to be shown]*.

Proof that $A \cup B \subseteq (A - B) \cup (B - A) \cup (A \cap B)$:

Suppose $x \in A \cup B$. *[We must show that $x \in (A - B) \cup (B - A) \cup (A \cap B)$.]*

By definition of union, $x \in A$ or $x \in B$.

(1) In case $x \in A$, either $x \in B$ or $x \notin B$. If $x \in B$, then, since x is also in A, $x \in A \cap B$ by definition of intersection. It follows by definition of union that $x \in (A - B) \cup (B - A) \cup (A \cap B)$. If $x \notin B$, then, since x is also in A, $x \in A - B$ by definition of set difference. It follows by definition of union that $x \in (A - B) \cup (B - A) \cup (A \cap B)$.

(2) In case $x \in B$, either $x \in A$ or $x \notin A$. If $x \in A$, then, since x is also in B, $x \in A \cap B$ by definition of intersection. It follows by definition of union that $x \in (A-B) \cup (B-A) \cup (A \cap B)$. If $x \notin A$, then, since x is also in B, $x \in B - A$ by definition of set difference. It follows by definition of union that $x \in (A - B) \cup (B - A) \cup (A \cap B)$.

Thus in both possible cases, $x \in (A - B) \cup (B - A) \cup (A \cap B)$ *[as was to be shown]*.

[Since both subset relations have been proved, $(A-B) \cup (B-A) \cup (A \cap B) = A \cup B$ by definition of set equality.]

b. To show that $(A - B), (B - A)$, and $(A \cap B)$ are mutually disjoint, we must show that the intersection of any two of them is the empty set. Now, by definition of set difference and set intersection, saying that $x \in A - B$ means that (1) $x \in A$ and $x \notin B$, saying that $x \in B - A$ means that (2) $x \in B$ and $x \notin A$, and saying that $x \in A \cap B$ means that (3) $x \in A$ and $x \in B$. Conditions (1)–(3) are mutually exclusive: no two of them can be satisfied at the same time. Thus no element can be in the intersection of any two of the sets, and, therefore, the intersection of any two of the sets is the empty set. Hence, $(A - B), (B - A)$, and $(A \cap B)$ are mutually disjoint.

40. Suppose that n is any positive integer and that A and $B_1, B_2, B_3, \ldots, B_n$ are any sets.

Proof that $A \cap \left(\bigcup_{i=1}^{n} B_i \right) \subseteq \bigcup_{i=1}^{n} (A \cap B_i)$:

Suppose x is any element in $A \cap \left(\bigcup_{i=1}^{n} B_i \right)$. *[We will show that $x \in \bigcup_{i=1}^{n} (A \cap B_i)$.]*

By definition of intersection, $x \in A$ and $x \in \bigcup_{i=1}^{n} B_i$. Since $x \in \bigcup_{i=1}^{n} B_i$, the definition of general union implies that $x \in B_i$ for some $i = 1, 2, \ldots, n$, and so, since $x \in A$, the definition of intersection implies that $x \in A \cap B_i$. Thus, by definition of general union, $x \in \bigcup_{i=1}^{n} (A \cap B_i)$ *[as was to be shown]*.

Proof that $\bigcup_{i=1}^{n} (A \cap B_i) \subseteq A \cap \left(\bigcup_{i=1}^{n} B_i \right)$:

Suppose x is any element in $\bigcup_{i=1}^{n} (A \cap B_i)$. *[We will show that $x \in A \cap \left(\bigcup_{i=1}^{n} B_i \right)$.]*

By definition of general union, $x \in A \cap B_i$ for some $i = 1, 2, \ldots, n$. Thus, by definition of intersection, $x \in A$ and $x \in B_i$. Since $x \in B_i$ for some $i = 1, 2, \ldots, n$, then by definition of general union, $x \in \bigcup_{i=1}^{n} B_i$. Hence $x \in A$ and $x \in \bigcup_{i=1}^{n} B_i$, and so, by definition of intersection, $x \in A \cap \left(\bigcup_{i=1}^{n} B_i \right)$ *[as was to be shown]*.

Conclusion: Since both subset relations have been proved, it follows by definition of set equality that

$$A \cap \left(\bigcup_{i=1}^{n} B_i \right) = \bigcup_{i=1}^{n} (A \cap B_i).$$

41. *Proof sketch:* If $x \in \bigcup_{i=1}^{n} (A_i - B)$, then $x \in A_i - B$ for some $i = 1, 2, \ldots, n$, and so, (1) for some $i = 1, 2, \ldots, n$, $x \in A_i$ (which implies that $x \in \left(\bigcup_{i=1}^{n} A_i \right)$) and (2) $x \notin B$. Conversely, if $x \in \left(\bigcup_{i=1}^{n} A_i \right) - B$, then $x \in \left(\bigcup_{i=1}^{n} A_i \right)$ and $x \notin B$, and so, by definition of general union,

$x \in A_i$ for some $i = 1, 2, \ldots, n$, and $x \notin B$. This implies that there is an integer i such that $x \in A_i - B$, and thus that $x \in \bigcup_{i=1}^{n} (A_i - B)$.

42. Proof:

Let $A_1, A_2, \ldots, A_n$ and B be any sets. *[We will show that* $\bigcap_{i=1}^{n}(A_i - B) = \left(\bigcap_{i=1}^{n}A_i\right) - B).]$

Proof that $\bigcap_{i=1}^{n}(A_i - B) \subseteq \left(\bigcap_{i=1}^{n}A_i\right) - B$:

Suppose $x \in \bigcap_{i=1}^{n}(A_i - B)$. *[We will show that* $x \in \left(\bigcap_{i=1}^{n}A_i\right) - B.]$
By definition of general intersection, $x \in A_i - B$ for every integer $i = 1, 2, \ldots, n$. And by definition of set difference, $x \in A_i$ and $x \notin B$ for every integer $i = 1, 2, \ldots, n$. It follows by definition of general intersection that $x \in \left(\bigcap_{i=1}^{n}A_i\right)$. In addition, since $x \notin B$, $x \in \left(\bigcap_{i=1}^{n}A_i\right) - B$ by definition of set difference *[as was to be shown]*.

Proof that $\left(\bigcap_{i=1}^{n}A_i\right) - B \subseteq \bigcap_{i=1}^{n}(A_i - B)$:

Suppose $x \in \left(\bigcap_{i=1}^{n}A_i\right) - B$. *[We will show that* $x \in \bigcap_{i=1}^{n}(A_i - B).]$

By definition of set difference, $x \in \left(\bigcap_{i=1}^{n}A_i\right)$ and $x \notin B$. By definition of general intersection, $x \in A_i$ for every integer $i = 1, 2, \ldots, n$. In addition, since $x \notin B$, $x \in (A_i - B)$ for every integer $i = 1, 2, \ldots, n$. So $x \in \bigcap_{i=1}^{n}(A_i - B)$ by definition of general intersection *[as was to be shown]*.

[Since both subset relations have been proved, $\bigcap_{i=1}^{n}(A_i - B) = \left(\bigcap_{i=1}^{n}A_i\right) - B$ *by definition of set equality.]*

Section 6.3

1. Counterexample: A, B, and C can be any sets where A has an element that is not in C. For instance, let $A = \{1, 2\}$, $B = \{2\}$, and $C = \{2\}$. Then

$$(A \cup B) \cap C = (\{1, 2\} \cup \{2\}) \cap \{2\} = \{1, 2\} \cap \{2\} = \{2\},$$

and

$$A \cup (B \cap C) = \{1, 2\} \cup (\{2\} \cap \{2\}) = \{1, 2\} \cup \{2\} = \{1, 2\}.$$

Thus $1 \in A \cup (B \cap C)$ but $1 \notin (A \cup B) \cap C$, and hence $(A \cup B) \cap C \neq A \cup (B \cap C)$ *[by definition of subset]*.

3. Counterexample: A, B, and C can be any sets where $A \subseteq C$ and B contains at least one element that is not in either A or C. For instance, let

$$A = \{1\}, \ B = \{2\}, \ \text{and} \ C = \{1, 3\}.$$

Then $A \nsubseteq B$ and $B \nsubseteq C$ but $A \subseteq C$.

5. False. <u>Counterexample</u>: A, B, and C can be any sets where all three sets have an element in common or where A and C have a common element that is not in B. For instance, let $A = \{1, 2, 3\}$, $B = \{2, 3\}$, and $C = \{3\}$. Then $B - C = \{2\}$, and so

$$A - (B - C) = \{1, 2, 3\} - \{2\} = \{1, 3\}.$$

On the other hand,

$$A - B = \{1, 2, 3\} - \{2, 3\} = \{1\}, \text{ and so } (A - B) - C = \{1\} - \{3\} = \{1\}.$$

Since $\{1, 3\} \neq \{1\}$, then $A - (B - C) \neq (A - B) - C$.

6. True. <u>Proof</u>: Let A and B be any sets.

   Proof that $A \cap (A \cup B) \subseteq A$: Suppose $x \in A \cap (A \cup B)$. By definition of intersection, $x \in A$ and $x \in A \cup B$. In particular, $x \in A$. Thus, by definition of subset, $A \cap (A \cup B) \subseteq A$.

   Proof that $A \subseteq A \cap (A \cup B)$: Suppose $x \in A$. Then by definition of union, $x \in A \cup B$. Hence $x \in A$ and $x \in A \cup B$, and so, by definition of intersection $x \in A \cap (A \cup B)$. Thus, by definition of subset, $A \subseteq A \cap (A \cup B)$.

   Because both $A \cap (A \cup B) \subseteq A$ and $A \subseteq A \cap (A \cup B)$ have been proved, we conclude that $A \cap (A \cup B) = A$.

9. True. <u>Proof</u>: Suppose A, B, and C are any sets such that $A \subseteq C$ and $B \subseteq C$.

   Suppose $x \in A \cup B$. By definition of union, $x \in A$ or $x \in B$.

   Now if $x \in A$ then $x \in C$ (because $A \subseteq C$), and if $x \in B$ then $x \in C$ (because $B \subseteq C$).

   Hence, in either case, $x \in C$. *[So, by definition of subset, $A \cup B \subseteq C$.]*

11. *Hint:* The statement is false. Consider sets U, A, B, and C as follows: $U = \{1, 2, 3, 4\}$, $A = \{1, 2\}$, $B = \{1, 2, 3\}$, and $C = \{2\}$.

12. True. <u>Proof</u>: Let A, B, and C be any sets. *[We must show that $A \cap (B - C) = (A \cap B) - (A \cap C)$.]*

    Proof that $A \cap (B - C) \subseteq (A \cap B) - (A \cap C)$:

    Suppose $x \in A \cap (B - C)$. *[We must show that $x \in (A \cap B) - (A \cap C)$.]*

    By definition of intersection, $x \in A$ and $x \in (B - C)$, and so

    $x \in A$ and, by definition of set difference, $x \in B$ and $x \notin C$.

    Now if x were in $A \cap C$, then x would be in C, which it is not.

    Thus $x \notin A \cap C$, and so $x \in A \cap B$ and $x \notin A \cap C$.

    Hence $x \in (A \cap B) - (A \cap C)$ by definition of set difference *[as was to be shown]*.

    [Therefore, $A \cap (B - C) \subseteq (A \cap B) - (A \cap C)$.]

    Proof that $(A \cap B) - (A \cap C) \subseteq A \cap (B - C)$:

    Suppose $x \in (A \cap B) - (A \cap C)$. *[We must show that $x \in A \cap (B - C)$.]*

    By definition of set difference, $x \in A \cap B$ and $x \notin A \cap C$, and so,

    by definition of intersection, $x \in A$ and $x \in B$, and also $x \notin A \cap C$.

    Now if x were in C then x would be in both A and C, and so x would be in $A \cap C$ which it is not.

    Thus $x \in A$ and $x \in B$ and $x \notin C$, and hence

    $x \in A$ and $x \in B - C$ by definition of set difference.

    Finally, by definition of intersection, $x \in A \cap (B - C)$ *[as was to be shown]*.

    [Therefore, $(A \cap B) - (A \cap C) \subseteq A \cap (B - C)$.]

    [Since both subset relations have been proved, $A \cap (B - C) = (A \cap B) - (A \cap C)$ by definition of set equality.]

14. *Hint:* The statement is true. *Sketch of part of proof:* Suppose $x \in A$. *[We must show that $x \in B$.]* Either $x \in C$ or $x \notin C$. In case $x \in C$, make use of the fact that $A \cap C \subseteq B \cap C$ to show that $x \in B$. In case $x \notin C$, make use of the fact that $A \cup C \subseteq B \cup C$ to show that $x \in B$.

15. Counterexample: A, B, and C can be any sets where B and C have an element that is not in A, or where A and B have an element that is not in C. For instance, let

$A = \{1\}$, $B = \{4\}$, and $C = \{1, 4\}$. Then

$A - (B \cup C) = (\{1\} \cup \{1, 4\}) = \{1, 4\}$ and

$A \cup (C - B) = \{1\} \cup \{1\} = \{1\}$. Thus

$4 \in A - (B \cup C)$ but $4 \notin A \cup (C - B)$, and hence

$A - (B \cup C) \nsubseteq A \cup (C - B)$ by definition of subset.

17. True. <u>Proof:</u> Suppose A and B are any sets with $A \subseteq B$. *[We must show that $\mathscr{P}(A) \subseteq \mathscr{P}(B)$.]* So suppose $X \in \mathscr{P}(A)$. Then $X \subseteq A$ by definition of power set. And because $A \subseteq B$, we also have that $X \subseteq B$ by the transitive property for subsets. Thus, by definition of power set, $X \in \mathscr{P}(B)$. This proves that for every X, if $X \in \mathscr{P}(A)$ then $X \in \mathscr{P}(B)$, and so $\mathscr{P}(A) \subseteq \mathscr{P}(B)$ *[as was to be shown]*.

18. False. <u>Counterexample:</u> For any sets A and B, the only sets that are in $\mathscr{P}(A) \cup \mathscr{P}(B)$ are subsets of either A or B. On the other hand, a set in $\mathscr{P}(A \cup B)$ can contain elements from both A and B. Thus, if at least one of A or B contains elements that are not in the other set, then $\mathscr{P}(A) \cup \mathscr{P}(B)$ and $\mathscr{P}(A \cup B)$ will not be equal. For instance, let $A = \{1\}$ and $B = \{2\}$. Then $\{1, 2\} \in \mathscr{P}(A \cup B)$ but $\{1, 2\} \notin \mathscr{P}(A) \cup \mathscr{P}(B)$.

19. *Hint:* The statement is true. To prove it, suppose A and B are any sets, and suppose $X \in \mathcal{P}(A) \cup \mathcal{P}(B)$. Note that if $X \subseteq A$, then $X \subseteq A \cup B$, and so $X \in \mathcal{P}(A \cup B)$.

20. True. <u>Proof:</u> Let A and B be any sets.

21. False. The elements of $\mathscr{P}(A \times B)$ are subsets of $A \times B$, whereas the elements of $\mathscr{P}(A) \times \mathscr{P}(B)$ are ordered pairs whose first element is a subset of A and whose second element is a subset of B.

<u>Counterexample:</u> Let $A = B = \{1\}$. Then $\mathscr{P}(A) = \{\emptyset, \{1\}\}$ and $\mathscr{P}(B) = \{\emptyset, \{1\}\}$, and so

$$\mathscr{P}(A) \times \mathscr{P}(B) = \{(\emptyset, \emptyset), (\emptyset, \{1\}), (\{1\}, \emptyset), (\{1\}, \{1\})\}.$$

On the other hand, $A \times B = \{(1, 1)\}$, and so $\mathscr{P}(A \times B) = \{\emptyset, \{(1, 1)\}\}$.

By inspection, it is clear that $\mathscr{P}(A) \times \mathscr{P}(B) \neq \mathscr{P}(A \times B)$.

22. **a.** *Statement:* $\forall$ set S, $\exists$ a set T such that $S \cap T = \emptyset$.

Negation: $\exists$ a set S such that $\forall$ set T, $S \cap T \neq \emptyset$. The statement is true. Given any set S, take $T = S^c$. Then $S \cap T = S \cap S^c = \emptyset$ by the complement law for $\cap$. Alternatively, T could be taken to be $\emptyset$.

23. *Hint:* $S_0 = \{\emptyset\}$, $S_1 = \{\{a\}, \{b\}, \{c\}\}$

24. **a.** $S_1 = \{\emptyset, \{t\}, \{u\}, \{v\}, \{t, u\}, \{t, v\}, \{u, v\}, \{t, u, v\}\}$

b. $S_2 = \{\{w\}, \{t, w\}, \{u, w\}, \{v, w\}, \{t, u, w\}, \{t, v, w\}, \{u, v, w\}, \{t, u, v, w\}\}$

c. Yes

d. S_1 and S_2 each have eight elements.

e. $S_1 \cup S_2$ has sixteen elements.

f. $S_1 \cup S_2 = \mathscr{P}(A)$.

25. *Hint:* The proof uses the same basic idea as the proof of Theorem 6.3.1. In this case let $P(n)$ be the sentence "If a set S has n elements, the number of subsets of S with an even number of elements equals the number of subsets of S with an odd number of elements."

26. *Hint:* Use mathematical induction. In the inductive step, you will consider the set of all nonempty subsets of $\{2, \ldots, k\}$ and the set of all nonempty subsets of $\{2, \ldots, k+1\}$. Any subset of $\{2, \ldots, k+1\}$ either contains $k+1$ or does not contain $k+1$. Thus

$$\begin{bmatrix} \text{the sum of all products} \\ \text{of elements of nonempty} \\ \text{subsets of } \{2, \ldots, k+1\} \end{bmatrix} = \begin{bmatrix} \text{the sum of all products} \\ \text{of elements of nonempty} \\ \text{subsets of } \{2, \ldots, k+1\} \\ \text{that do not contain } k+1 \end{bmatrix} + \begin{bmatrix} \text{the sum of all products} \\ \text{of elements of nonempty} \\ \text{subsets of } \{2, \ldots, k+1\} \\ \text{that contain } k+1 \end{bmatrix}$$

Now any subset of $\{2, \ldots, k+1\}$ that does not contain $k+1$ is a subset of $\{2, \ldots, k\}$. And any subset of $\{2, \ldots, k+1\}$ that contains $k+1$ is the union of a subset of $\{2, \ldots, k\}$ and $\{k+1\}$.

27. **a.** commutative law for $\cap$ **b.** distributive law   **c.** commutative law for $\cap$

28. Partial answer:

   a. set difference law   **b.** set difference law   **c.** commutative law for $\cap$ **d.** De Morgan's law

29. *Hint:* Remember to use the properties in Theorem 6.2.2 exactly as they are written. For example, the distributive law does not state that for all sets A, B, and C, $(A \cup B) \cap C = (A \cap C) \cup (B \cap C)$.

30. Proof: Let sets $A, B,$ and C be given. Then

$$\begin{aligned} (A \cap B) \cup C &= C \cup (A \cap B) && \text{by the commutative law for } \cup \\ &= (C \cup A) \cap (C \cup B) && \text{by the distributive law} \\ &= (A \cup C) \cap (B \cup C) && \text{by the commutative law for } \cup. \end{aligned}$$

31. Proof: Suppose A and B are sets. Then

$$\begin{aligned} A \cup (B - A) &= A \cup (B \cap A^c) && \text{by the set difference law} \\ &= (A \cup B) \cap (A \cup A^c) && \text{by the distributive law} \\ &= (A \cup B) \cap U && \text{by the complement law for } \cup \\ &= A \cup B && \text{by the identity law for } \cap. \end{aligned}$$

33. Proof: Let A and B be any sets. Then

$$\begin{aligned} (A - B) \cap (A \cap B) &= (A \cap B^c) \cap (A \cap B) && \text{by the set difference law} \\ &= A \cap [B^c \cap (A \cap B)] && \text{by the associative law for } \cap \\ &= A \cap [(A \cap B) \cap B^c] && \text{by the commutative law for } \cap \\ &= A \cap [A \cap (B \cap B^c)] && \text{by the associative law for } \cap \\ &= A \cap [A \cap \emptyset] && \text{by the complement law for } \cap \\ &= A \cap \emptyset && \text{by the identity law for } \cap \\ &= \emptyset && \text{by the identity law for } \cap. \end{aligned}$$

36. Proof: Let A and B be any sets. Then

$$\begin{aligned} ((A^c \cup B^c) - A)^c &= ((A^c \cup B^c) \cap A^c)^c && \text{by the set difference law} \\ &= (A^c \cup B^c)^c \cup (A^c)^c && \text{by De Morgan's law} \\ &= ((A^c)^c \cap (B^c)^c) \cup (A^c) && \text{by De Morgan's law} \\ &= (A \cap B) \cup A && \text{by the double complement law} \\ &= A \cup (A \cap B) && \text{by the commutative law for } \cup \\ &= A && \text{by the absorption law.} \end{aligned}$$

39. <u>Proof</u>: Let A and B be any sets. Then

$(A - B) \cup (B - A)$

$$
\begin{array}{lll}
= & (A \cap B^c) \cup (B \cap A^c) & \text{by the set difference law (used twice)} \\
= & [(A \cap B^c) \cup B] \cap [(A \cap B^c) \cup A^c)] & \text{by the distributive law} \\
= & [B \cup (A \cap B^c)] \cap [A^c \cup (A \cap B^c)] & \text{by the commutative law for } \cup \text{ (used twice)} \\
= & [(B \cup A) \cap (B \cup B^c)] \cap [(A^c \cup A) \cap (A^c \cup B^c)] & \text{by the distributive law (used twice)} \\
= & [(A \cup B) \cap (B \cup B^c)] \cap [(A \cup A^c) \cap (A^c \cup B^c)] & \text{by the commutative law for } \cup \text{ (used twice)} \\
= & [(A \cup B) \cap U] \cap [U \cap (A^c \cup B^c)] & \text{by the complement law for } \cup \text{ (used twice)} \\
= & [(A \cup B) \cap U] \cap [(A^c \cup B^c) \cap U] & \text{by the commutative law for } \cap \\
= & (A \cup B) \cap (A^c \cup B^c) & \text{by the identity law for } \cap \text{ (used twice)} \\
= & (A \cup B) \cap (A \cap B)^c & \text{by De Morgan's law} \\
= & (A \cup B) - (A \cap B) & \text{by the set difference law.}
\end{array}
$$

41. *Hint:* The answer is $\emptyset$.

42. Let A and B be any sets. Then

$(A - (A \cap B)) \cap (B - (A \cap B))$

$$
\begin{array}{lll}
= & (A \cap (A \cap B)^c) \cap (B \cap (A \cap B)^c) & \text{by the set difference law (used twice)} \\
= & A \cap ((A \cap B)^c \cap (B \cap (A \cap B)^c)) & \text{by the associative law for } \cap \\
= & A \cap (((A \cap B)^c \cap B) \cap (A \cap B)^c) & \text{by the associative law for } \cap \\
= & A \cap ((B \cap (A \cap B)^c) \cap (A \cap B)^c) & \text{by the commutative law for } \cap \\
= & A \cap (B \cap ((A \cap B)^c \cap (A \cap B)^c)) & \text{by the associative law for } \cap \\
= & A \cap (B \cap (A \cap B)^c) & \text{by the idempotent law for } \cap \\
= & (A \cap B) \cap (A \cap B)^c & \text{by the associative law for } \cap \\
= & \emptyset & \text{by the complement law for } \cap.
\end{array}
$$

44. **a.** <u>Proof (by contradiction)</u>: Suppose not. That is, suppose there exist sets A and B such that $A - B$ and B are not disjoint. Then $(A - B) \cap B \neq \emptyset$, which means there is an element x in $(A - B) \cap B$. By definition of intersection, $x \in A - B$ and $x \in B$, and by definition of set difference, $x \in A$ and $x \notin B$. Hence $x \in B$ and $x \notin B$, which is a contradiction. Thus the supposition is false, and so $A - B$ and B are disjoint.

b. Let A and B be any sets. Then

$$
\begin{array}{lll}
(A - B) \cap B & = & (A \cap B^c) \cap B & \text{by the set difference law} \\
& = & A \cap (B^c \cap B) & \text{by the associative law for } \cap \\
& = & A \cap (B \cap B^c) & \text{by the commutative law for } \cap \\
& = & A \cap \emptyset & \text{by the complement law for } \cap \\
& = & \emptyset & \text{by the universal bound law for } \cap.
\end{array}
$$

Thus $A - B$ and B are disjoint.

45. **a.** <u>Proof</u>: Let A, B, and C be any sets.

Proof that $(A - B) \cup (B - C) \subseteq (A \cup B) - (B \cap C)$: Suppose $x \in (A - B) \cup (B - C)$. By definition of union, $x \in A - B$ or $x \in B - C$.

Case 1 ($x \in A - B$): In this case, by definition of set difference, $x \in A$ and $x \notin B$. Then since $x \in A$, by definition of union, $x \in A \cup B$. Also, since $x \notin B$, then $x \notin B \cap C$ (for otherwise, by definition of intersection, x would be in B, which it is not). Thus $x \in A \cup B$ and $x \notin B \cap C$ $x \in A \cup B$ and $x \notin B \cap C$. and so, by definition of set difference, $x \in (A \cup B) - (B \cap C)$.

Case 2 ($x \in B - C$): In this case, by definition of set difference, $x \in B$ and $x \notin C$. Then since $x \in B$, by definition of union, $x \in A \cup B$. Also, since $x \notin C$, then $x \notin B \cap C$ (for otherwise, by definition of intersection, x would be in C, which it is not). Thus $x \in A \cup B$ and $x \notin B \cap C$. and so, by definition of set difference, $x \in (A \cup B) - (B \cap C)$.

Hence, in both cases, $x \in (A \cup B) - (B \cap C)$, and so, by definition of subset,

$$(A - B) \cup (B - C) \subseteq (A \cup B) - (B \cap C).$$

Proof that $(A \cup B) - (B \cap C) \subseteq (A - B) \cup (B - C)$: Suppose $x \in (A \cup B) - (B \cap C)$. By definition of set difference, $x \in A \cup B$ and $x \notin B \cap C$. Note that either $x \in B$ or $x \notin B$.

Case 1 $(x \in B)$: In this case $x \notin C$ because otherwise x would be in both B and C, which would contradict the fact that $x \notin B \cap C$. Thus, in this case, $x \in B$ and $x \notin C$, and so $x \in B - C$ by definition of set difference. Then $x \in (A - B) \cup (B - C)$ by definition of union.

Case 2 $(x \notin B)$: In this case, since $x \in A \cup B$, then $x \in A$. Hence $x \in A$ and $x \notin B$, and so $x \in A - B$ by definition of set difference. Then $x \in (A - B) \cup (B - C)$ by definition of union.

Hence, in both cases, $x \in (A - B) \cup (B - C)$, and so, by definition of subset,

$$(A \cup B) - (B \cap C) \subseteq (A - B) \cup (B - C).$$

Therefore, since both subset relations have been proved, we conclude that

$$(A - B) \cup (B - C) = (A \cup B) - (B \cap C)$$

by definition of set equality.

b. <u>Proof</u>: Let A, B, and C be any sets. Then

$(A - B) \cup (B - C)$

$=$	$(A \cap B^c) \cup (B \cap C^c)$	by the set difference law (used twice)
$=$	$((A \cap B^c) \cup B) \cap ((A \cap B^c) \cup C^c)$	by the distributive law
$=$	$(B \cup (A \cap B^c)) \cap ((A \cap B^c) \cup C^c)$	by the commutative law for $\cup$
$=$	$((B \cup A) \cap (B \cup B^c)) \cap ((A \cap B^c) \cup C^c)$	by the distributive law
$=$	$((B \cup A) \cap U) \cap ((A \cap B^c) \cup C^c)$	by the complement law for $\cup$
$=$	$(B \cup A) \cap ((A \cap B^c) \cup C^c)$	by the identity law for $\cap$
$=$	$(A \cup B) \cap ((A \cap B^c) \cup C^c)$	by the commutative law for $\cup$
$=$	$((A \cup B) \cap (A \cap B^c)) \cup ((A \cup B) \cap C^c)$	by the distributive law
$=$	$(((A \cup B) \cap A) \cap B^c) \cup ((A \cup B) \cap C^c)$	by the associative law for $\cap$
$=$	$((A \cap (A \cup B)) \cap B^c) \cup ((A \cup B) \cap C^c)$	by the commutative law for $\cap$
$=$	$(A \cap B^c) \cup ((A \cup B) \cap C^c)$	by the absorption law
$=$	$((A \cap B^c) \cup \emptyset) \cup ((A \cup B) \cap C^c)$	by the identity law for $\cup$
$=$	$((A \cap B^c) \cup (B \cap B^c)) \cup ((A \cup B) \cap C^c)$	by the complement law for $\cap$
$=$	$((B^c \cap A) \cup (B^c \cap B)) \cup ((A \cup B) \cap C^c)$	by the commutative law for $\cap$
$=$	$(B^c \cap (A \cup B)) \cup ((A \cup B) \cap C^c)$	by the distributive law
$=$	$((A \cup B) \cap B^c)) \cup ((A \cup B) \cap C^c)$	by the commutative law for $\cap$
$=$	$(A \cup B) \cap (B^c \cup C^c)$	by the distributive law
$=$	$(A \cup B) \cap (B \cap C)^c$	by De Morgan's law
$=$	$(A \cup B) - (B \cap C)$	by the set difference law.

c. Although writing down every detail of the element proof is somewhat tedious, its basic idea is not hard to see. In this case the element proof is probably easier than the algebraic proof.

46. **a.** $A \, \Delta \, B = (A - B) \cup (B - A) = \{1, 2\} \cup \{5, 6\} = \{1, 2, 5, 6\}$

47. <u>Proof</u>: Let A and B be any subsets of a universal set. By definition of Δ, showing that $A \, \Delta \, B = B \, \Delta \, A$ is equivalent to showing that $(A - B) \cup (B - A) = (B - A) \cup (A - B)$. This follows immediately from the commutative law for $\cup$.

48. <u>Proof</u>: Let A be any subset of a universal set. Then

$$
\begin{aligned}
A \triangle \emptyset &= (A - \emptyset) \cup (\emptyset - A) &&\text{by definition of } \triangle \\
&= (A \cap \emptyset^c) \cup (\emptyset \cap A^c) &&\text{by the set difference law} \\
&= (A \cap U) \cup (A^c \cap \emptyset^c) &&\text{by the complement of } \cup \text{ law and the commutative law for } \cap \\
&= A \cup \emptyset &&\text{by the identity law for } \cap \text{ and the universal bound law for } \cap \\
&= A &&\text{by the identity law for } \cup.
\end{aligned}
$$

51. <u>Lemma</u>: For any subsets A and B of a universal set U and for any element x,

   (1) $x \in A \triangle B \Leftrightarrow (x \in A \text{ and } x \notin B) \text{ or } (x \notin A \text{ and } x \in B)$

   (2) $x \notin A \triangle B \Leftrightarrow (x \notin A \text{ and } x \notin B) \text{ or } (x \in A \text{ and } x \in B)$.

   <u>Proof</u>:

   (1) Suppose A and B are any sets and x is any element. Then

$$
\begin{aligned}
x \in A \triangle B &\Leftrightarrow x \in (A - B) \cup (B - A) &&\text{by definition of } \triangle \\
&\Leftrightarrow x \in A - B \text{ or } x \in B - A &&\text{by definition of } \cup \\
&\Leftrightarrow (x \in A \text{ and } x \notin B) \text{ or } (x \in B \text{ and } x \notin A) &&\text{by definition of set difference.}
\end{aligned}
$$

   (2) Suppose A and B are any sets and x is any element.

   Observe that there are only four mutually exclusive possibilities for the relationship of x to A and B: $(x \in A \text{ and } x \notin B)$ or $(x \in B \text{ and } x \notin A)$ or $(x \in A \text{ and } x \in B)$ or $(x \notin A \text{ and } x \notin B)$.

   By part (1), the condition that $x \in A \triangle B$ is equivalent to the first two possibilities. So the condition that $x \notin A \triangle B$ is equivalent to the second two possibilities.

   In other words, $x \notin A \triangle B \Leftrightarrow (x \notin A \text{ and } x \notin B) \text{ or } (x \in A \text{ and } x \in B)$.

   <u>Theorem</u>: For all subsets A, B, and C of a universal set U, if $A \triangle C = B \triangle C$ then $A = B$.

   <u>Proof</u>: Let A, B, and C be any subsets of a universal set U, and suppose that $A \triangle C = B \triangle C$. *[We will show that $A = B$.]*

   Proof that $A \subseteq B$: Suppose $x \in A$. Either $x \in C$ or $x \notin C$. If $x \in C$, then $x \in A$ and $x \in C$ and so by the lemma, $x \notin A \triangle C$. But $A \triangle C = B \triangle C$. Thus $x \notin B \triangle C$ either. Hence, again by the lemma, since $x \in C$ and $x \notin B \triangle C$, then $x \in B$. On the other hand, if $x \notin C$, then by the lemma, since $x \in A$, $x \in A \triangle C$. But $A \triangle C = B \triangle C$. So, again by the lemma, since $x \notin C$ and $x \in B \triangle C$, then $x \in B$. Hence in either case, $x \in B$ *[as was to be shown]*.

   Proof that $B \subseteq A$: The proof is exactly the same as for $A \subseteq B$ with the letters A and B interchanged.

   Since $A \subseteq B$ and $B \subseteq A$, by definition of set equality $A = B$.

52. *Hint:* First show that for any sets A and B and for any element x,

   $x \in A \triangle B \Leftrightarrow (x \in A \text{ and } x \notin B) \text{ or } (x \in B \text{ and } x \notin A)$

   and

   $x \notin A \triangle B \Leftrightarrow (x \notin A \text{ and } x \notin B) \text{ or } (x \in B \text{ and } x \in A)$.

53. *Start of proof:* Suppose A and B are any subsets of a universal set U. By the universal bound law for union, $B \cup U = U$, and so, by the commutative law for union, $U \cup B = U$. Take the intersection of both sides of the equation with A.

54. <u>Proof</u>:

   Suppose A and B are any subsets of a universal set U.

   By the universal bound law for $\cap$, $B \cap \emptyset = \emptyset$, and so, by the commutative law for $\cap$, $\emptyset \cap B = \emptyset$.

   Take the union with A of both sides to obtain $A \cup (\emptyset \cap B) = A \cup \emptyset$.

The left-hand side of this equation is $A \cup (\emptyset \cap B) = (A \cup \emptyset) \cap (A \cup B) = A \cap (A \cup B)$ by the distributive law and the identity law for $\cup$.

And the right-hand side of the equation equals A by the by the identity law for $\cup$.

Hence $A \cap (A \cup B) = A$ *[as was to be shown]*.

Section 6.4

1. **a.** because 1 is an identity for $\cdot$ **b.** by the complement law for $+$
   c. by the distributive law for $+$ over $\cdot$ **d.** by the complement law for $\cdot$
   e. because 0 is an identity for $+$

3. **a.** by the commutative law for $\cdot$ **b.** by the distributive law for $\cdot$ over $+$
   c. by the identity law for $\cdot$ **d.** by the distributive law for $\cdot$ over $+$
   e. by the commutative law for $+$ **f.** by the identity law for $\cdot$

4. <u>Proof</u> (*Universal bound for* 0): For every a in B,

$$
\begin{aligned}
a \cdot 0 &= a \cdot (a \cdot \bar{a}) && \text{by the complement law for } \cdot \\
&= (a \cdot a) \cdot \bar{a} && \text{by the associative law for } \cdot \\
&= a \cdot \bar{a} && \text{by exercise 1} \\
&= 0 && \text{by the complement law for } \cdot.
\end{aligned}
$$

5. **a.** <u>Proof</u>: $0 \cdot 1 = 0$ because 1 is an identity for $\cdot$, and $0 + 1 = 1 + 0 = 1$ because $+$ is commutative and 0 is an identity for $+$. Thus, by the uniqueness of the complement law, $\bar{0} = 1$.

6. <u>Proof</u>: Suppose 0 and $0'$ are elements of B both of which are identities for $+$. Then both 0 and $0'$ satisfy the identity, complement, and universal bound laws. *[We will show that $0 = 0'$.]* By the identity law for $+$, for every $a \in B$,

$$
a + 0 = a \ (*) \quad \text{and} \quad a + 0' = a \ (**).
$$

It follows that

$$
\begin{aligned}
0' &= 0' + 0 && \text{by } (*) \text{ with } a = 0' \\
&= 0 + 0' && \text{by the commutative law for } + \\
&= 0 && \text{by } (**) \text{ with } a = 0.
\end{aligned}
$$

[This is what was to be shown.]

7. *Hint:* Suppose 1 and $1'$ are elements of B both of which are identities for $\cdot$. Then for every $a \in B$, by the identity law for $\cdot$, $a \cdot 1 = a$ and $a \cdot 1' = a$. It follows that $a \cdot 1 = a \cdot 1'$, and thus $\bar{a} + a \cdot 1 = \bar{a} + a \cdot 1'$, and so forth.

8. <u>Proof</u>: Suppose B is a Boolean algebra and a and b are any elements of B. We first prove that $(a \cdot b) + (\bar{a} + \bar{b}) = 1$.

$a \cdot b + (\bar{a} + \bar{b})$

$$
\begin{aligned}
&= ((a \cdot b) \cdot \bar{a}) + ((a \cdot b) \cdot \bar{b}) && \text{by the commutative law for } + \\
&= ((b \cdot a) \cdot a) \cdot (\bar{a} \cdot (\bar{b} \cdot b)) && \text{by the distributive law of } + \text{ over } \cdot \\
&= ((\bar{b} + \bar{a}) + a) \cdot (\bar{a} + (\bar{b} + b)) && \text{by the commutative and associative laws for } + \\
&= (\bar{b} + (\bar{a} + a)) \cdot (\bar{a} + (b + \bar{b})) && \text{by the associative and commutative laws for } + \\
&= (\bar{b} + (a + \bar{a})) \cdot (\bar{a} + 1) && \text{by the commutative and complement laws for } + \\
&= (\bar{b} + 1) \cdot 1 && \text{by the universal bound laws for } + \\
&= 1 \cdot 1 && \text{by the universal bound law for } + \\
&= 1 && \text{by the identity law for } \cdot.
\end{aligned}
$$

Next we prove that $(a \cdot b) \cdot (\bar{a} + \bar{b}) = 0$.

$(a \cdot b) \cdot (\bar{a} + \bar{b})$

$$
\begin{aligned}
&= ((a \cdot b) \cdot \bar{a}) + ((a \cdot b) \cdot \bar{b}) && \text{by the distributive law of } \cdot \text{ over } + \\
&= ((b \cdot a) \cdot \bar{a}) + (a \cdot (b \cdot \bar{b})) && \text{by the commutative and associative laws for } \cdot \\
&= (b \cdot (a \cdot \bar{a})) + (a \cdot 0) && \text{by the associative and complement laws for } \cdot \\
&= (b \cdot 0) + 0 && \text{by the complement and universal bound laws for } \cdot \\
&= 0 + 0 && \text{by the universal bound law for } \cdot \\
&= 0 && \text{by the identity law for } +.
\end{aligned}
$$

Because both $(a \cdot b) + (\bar{a} + \bar{b}) = 1$ and $(a \cdot b) \cdot (\bar{a} + \bar{b}) = 0$, it follows, by the uniqueness of the complement law, that $\overline{a \cdot b} = \bar{a} + \bar{b}$.

9. <u>Proof of De Morgan's law for $\cdot$</u>: By exercise 8, we know that for all x and y in B, $\overline{x \cdot y} = \bar{x} + \bar{y}$. So suppose a and b are any elements in B. Substitute $\bar{a}$ and $\bar{b}$ in place of x and y in this equation to obtain $\overline{\bar{a} \cdot \bar{b}} = \bar{\bar{a}} + \bar{\bar{b}}$, and since $\bar{\bar{a}} + \bar{\bar{b}} = a + b$ by the double complement law, we have $\overline{\bar{a} \cdot \bar{b}} = a + b$. Hence by the uniqueness of the complement law, the complement of $\bar{a} \cdot \bar{b}$ is $a + b$. It follows by definition of complement that

$$(\bar{a} \cdot \bar{b}) + (a + b) = 1 \qquad \text{and} \qquad (\bar{a} \cdot \bar{b}) \cdot (a + b) = 0.$$

By the commutative laws for $+$ and $\cdot$,

$$(a + b) + (\bar{a} \cdot \bar{b}) = 1 \qquad \text{and} \qquad (a + b) \cdot (\bar{a} \cdot \bar{b}) = 0,$$

and thus by the uniqueness of the complement law, the complement of $a + b$ is $\bar{a} \cdot \bar{b}$. In other words, $\overline{a + b} = \bar{a} \cdot \bar{b}$.

<u>Alternative Proof</u>: A different proof can be obtained by taking the proof for exercise 8 in Appendix B and changing every $+$ sign to a $\cdot$ sign and every $\cdot$ sign to a $+$ sign.

10. *Hint:* One way to prove the statement is to use the result of exercise 3. Some stages in the proof are the following:

$$
\begin{aligned}
y = (y + x) \cdot y &= \cdots = (x \cdot y) + (z \cdot y) = \cdots \\
&= z \cdot (x + y) = \cdots = z.
\end{aligned}
$$

11. **a.** (i) Because S has only two distinct elements, 0 and 1, we only need to check that $0 + 1 = 1 + 0$. This is true because both sums equal 1.

(v) *Partial answer:* Show that for all a, b, and c in B,

$$
\begin{aligned}
&a + (b \cdot c) = (a + b) \cdot (a + b). \\
&0 + (0 \cdot 0) = 0 + 0 = 0 \text{ and } (0 + 0) \cdot (0 + 0) \\
&\qquad\qquad = 0 \cdot 0 = 0 \qquad [a = b = c = 0] \\
&0 + (0 \cdot 1) = 0 + 0 = 0 \text{ and } (0 + 0) \cdot (0 + 1) \\
&\qquad\qquad = 0 \cdot 1 = 0 \qquad [a = b = 0, c = 1] \\
&0 + (1 \cdot 0) = 0 + 0 = 0 \text{ and } (0 + 1) \cdot (0 + 0) \\
&\qquad\qquad = 1 \cdot 0 = 0 \qquad [a = c = 0, b = 1] \\
&0 + (1 \cdot 1) = 0 + 1 = 1 \text{ and } (0 + 1) \cdot (0 + 1) \\
&\qquad\qquad = 1 \cdot 1 = 1 \qquad [a = 0, b = c = 1]
\end{aligned}
$$

b. *Hint:* Verify that $0 + x = x$ and that $1 \cdot x = x$ for every $x \in S$.

12. <u>Proof</u>: Suppose a is any element of a Boolean algebra B.

$$\begin{aligned}
a + 1 &= (a+1) \cdot 1 && \text{because 1 is an identity for } \cdot \\
&= (a+1) \cdot (a + \bar{a}) && \text{by the complement law for } + \\
&= a + 1 \cdot \bar{a} && \text{by the distributive law for } + \text{ over } \cdot \\
&= a + \bar{a} \cdot 1 && \text{by the commutative law for } \cdot \\
&= a + \bar{a} && \text{because 1 is an identity for } \cdot \\
&= 1 && \text{by the complement law for } + .
\end{aligned}$$

13. *Start of proof:* Suppose a and b are any elements of a Boolean algebra B.

$a \cdot b + a = a \cdot b + a \cdot 1$ because 1 is an identity for $\cdot$

15. <u>Proof of the associative law for +</u>: Suppose a, b, and c are any elements in a Boolean algebra B.

Part 1: We first prove that $(a + (b + c)) \cdot a = ((a + b) + c) \cdot a$.

$$(a + (b+c)) \cdot a \;=\; a \quad \text{by an absorption law (exercise 3).}$$

In addition:

$$\begin{aligned}
((a+b)+c) \cdot a &= a \cdot ((a+b)+c) && \text{by the commutative law for } \cdot \\
&= a \cdot (a+b) + a \cdot c && \text{by the distributive law for } \cdot \text{ over } + \\
&= (a+b) \cdot a + a \cdot c && \text{by the commutative law for } + \\
&= a + a \cdot c && \text{by the absorption law for } + \text{ over } \cdot \text{(exercise 3)} \\
&= a \cdot c + a && \text{by the commutative law for } + \\
&= a && \text{by the absorption law for } \cdot \text{ over } + \text{ (exercise 13).}
\end{aligned}$$

Since both quantities equal a, we obtain

$$(a + (b+c)) \cdot a = ((a+b)+c) \cdot a.$$

Part 2: We next prove that $(a + (b + c)) \cdot \bar{a} = ((a + b) + c) \cdot \bar{a}$.

$$\begin{aligned}
(a + (b+c)) \cdot \bar{a} &= \bar{a} \cdot (a + (b+c)) && \text{by the commutative law for } \cdot \\
&= \bar{a} \cdot a + \bar{a} \cdot (b+c) && \text{by the distributive law for } \cdot \text{ over } + \\
&= 0 + \bar{a} \cdot (b+c) && \text{by the complement law for } \cdot \\
&= \bar{a} \cdot (b+c) + 0 && \text{by the commutative law for } + \\
&= \bar{a} \cdot (b+c) && \text{because 0 is an identity for } + \text{ (exercise 4).}
\end{aligned}$$

In addition:

$$\begin{aligned}
((a+b)+c) \cdot \bar{a} &= \bar{a} \cdot ((a+b)+c) && \text{by the commutative law for } \cdot \\
&= \bar{a} \cdot (a+b) + \bar{a} \cdot c && \text{by the distributive law for } \cdot \text{ over } + \\
&= (\bar{a} \cdot a + \bar{a} \cdot b) + \bar{a} \cdot c && \text{by the distributive law for } \cdot \text{ over } + \\
&= (a \cdot \bar{a} + \bar{a} \cdot b) + \bar{a} \cdot c && \text{by the commutative law for } \cdot \\
&= (0 + \bar{a} \cdot b) + \bar{a} \cdot c && \text{by the complement law for } \cdot \\
&= (\bar{a} \cdot b + 0) + \bar{a} \cdot c && \text{by the commutative law for } + \\
&= \bar{a} \cdot b + \bar{a} \cdot c && \text{because 0 is an identity for } + \text{ (exercise 4)} \\
&= \bar{a} \cdot (b+c) && \text{by the distributive law for } \cdot \text{ over } + .
\end{aligned}$$

Since both quantities equal $\bar{a} \cdot (b + c)$, we obtain

$$(a + (b+c)) \cdot \bar{a} = ((a+b)+c) \cdot \bar{a}.$$

Part 3: Parts (1) and (2) show that

$$(a + (b + c)) \cdot a = ((a + b) + c) \cdot a \quad \text{and} \quad (a + (b + c)) \cdot \overline{a} = ((a + b) + c) \cdot \overline{a}.$$

By the test for equality law (exercise 14), we conclude that

$$a + (b + c) = (a + b) + c.$$

16. The sentence is not a statement because it is both true and false. If the sentence were true, then because it declares itself to be false, the sentence would be false. Therefore, the sentence is not true. On the other hand, if the sentence were false, then it would be false that "This sentence is false," and so the sentence would be true. Consequently, the sentence is false. It follows that the sentence is both true and false.

17. This sentence is a statement because it is true. Recall that the only way for an if-then statement to be false is for the hypothesis to be true and the conclusion false. In this case the hypothesis is not true. So regardless of what the conclusion states, the sentence is true. (This is an example of a statement that is vacuously true, or true by default.)

18. This statement contradicts itself. If it were true, then because it declares itself to be a lie, it would be false. Consequently, it is not true. On the other hand, if it were false, then it would be false that "the sentence in this box is a lie," and so the sentence would be true. Consequently, the sentence is not false. Thus the sentence is neither true nor false, which contradicts the definition of a statement. Hence the sentence is not a statement.

20. This sentence is not a statement because it is both true and false. If the sentence is true, then, by definition of an *or* statement, either the sentence is false or $1 + 1 = 3$. But $1 + 1 \neq 3$, and so the sentence is false. On the other hand, if the sentence is false, then (by DeMorgan's law) both of the following must be true: "This sentence is false" and "$1 + 1 = 3$." But it is not true that $1 + 1 = 3$. So it is impossible for the sentence to be false and hence the sentence is true. Consequently, the sentence is both true and false.

23. *Hint:* Suppose that apart from statement (ii), all of Nixon's other assertions about Watergate are evenly split between true and false.

24. No. Suppose there exists a computer program P that has as output a list of all computer programs that do not list themselves in their output. If P lists itself as output, then it would be on the output list of P, which consists of all computer programs that do not list themselves in their output. Hence P would not list itself as output. But if P does not list itself as output, then P would be a member of the list of all computer programs that do not list themselves in their output, and this list is exactly the output of P. Hence P would list itself as output. This analysis shows that the assumption of the existence of such a program P is contradictory, and so no such program exists.

27. Because the total number of strings consisting of 11 or fewer English words is finite, the number of such strings that describe integers must be also finite. Thus the number of integers described by such strings must be finite, and hence there is a largest such integer, say m. Let $n = m + 1$. Then n is "the smallest integer not describable in fewer than 12 English words." But this description of n contains only 11 words. So n is describable in fewer than 12 English words, which is a contradiction. (*Comment*: This contradiction results from the self-reference in the description of n.)

28. *Hint:* Show that any algorithm that solves the printing problem can be adapted to produce an algorithm that solves the halting problem.

Review Guide: Chapter 6

Definitions and Notation:

- How can you express the definition of subset formally as a universal conditional statement?
- What is a proper subset of a set?
- How are the definitions of subset and equality of sets related?
- What are Venn diagrams?
- What are the union, intersection, and difference of sets?
- What is the complement of a set?
- What is the relation between sets and interval notation?
- How are unions and intersections defined for indexed collections of sets?
- What does it mean for two sets to be disjoint?
- What does it mean for a collection of sets to be mutually disjoint?
- What is a partition of a set?
- What is the power set of a set?
- What is an ordered n-tuple?
- What is the Cartesian product of n sets, where $n \geq 2$?

Set Theory

- How do you use an element argument to prove that one set is a subset of another set?
- How are subsets used to show that two sets are equal?
- How are the procedural versions of set operations used to prove properties of sets?
- What are the commutative laws for sets?
- What are the associative laws for sets?
- What are the distributive laws for sets?
- What are De Morgan's laws for sets?
- What is the set difference law?
- Why is the empty set a subset of every set?
- How is the element method used to show that a set equals the empty set?
- What is the basic method for disproving a proposed set identity?
- How do you find the number of subsets of a set with a finite number of elements?
- What is an "algebraic method" for proving that one set equals another set?

Boolean Algebras, Russell's Paradox, and the Halting Problem

- What is a Boolean algebra?
- How are the following related: (1) a set of statement forms with operations $\vee$ and $\wedge$, (2) the set of all subsets of a universal set with operations $\cup$ and $\cap$, and (3) a Boolean algebra?
- How do you deduce additional properties of a Boolean algebra from the properties that define it?
- What is Russell's paradox?
- What is the Halting Problem?

Fill-in-the-Blank Review Questions

Section 6.1

1. The notation $A \subseteq B$ is read ____ and means that ____.
2. To use an element argument for proving that a set X is a subset of a set Y, you suppose that ____ and show that ____.
3. To disprove that a set X is a subset of a set Y, you show that there is ____.
4. An element x is in $A \cup B$ if, and only if, ____.
5. An element x is in $A \cap B$ if, and only if, ____.
6. An element x is in $B - A$ if, and only if, ____.
7. An element x is in A^c if, and only if, ____.
8. The empty set is a set with ____.
9. The power set of a set A is ____.
10. Sets A and B are disjoint if, and only if, ____.
11. A collection of nonempty sets $A_1, A_2, A_3, \ldots$ is a partition of a set A if, and only if, ____.

Section 6.2

1. To prove that a set X is a subset of a set $A \cap B$, you suppose that x is any element of X and you show that $x \in A$ ____ $x \in B$.
2. To prove that a set X is a subset of a set $A \cup B$, you suppose that x is any element of X and you show that $x \in A$ ____ $x \in B$.
3. To prove that a set $A \cup B$ is a subset of a set X, you start with any element x in $A \cup B$ and consider the two cases ____ and ____. You then show that in either case ____.
4. To prove that a set $A \cap B$ is a subset of a set X, you suppose that ____ and you show that ____.
5. To prove that a set X equals a set Y, you prove that ____ and that ____.
6. To prove that a set X does not equal a set Y, you need to find an element that is in ____ and not ____ or that is in ____ and not ____.

Section 6.3

1. Given a proposed set identity involving set variables A, B, and C, the most common way to show that the equation does not hold in general is to find concrete sets A, B, and C that, when substituted for the set variables in the equation, ____.
2. When using the algebraic method for proving a set identity, it is important to ____ for every step.
3. When citing a property from Theorem 6.2.2, it must be used ____ as it is stated.

Section 6.4

1. In the comparison between the structure of the set of statement forms and the set of subsets of a universal set, the *or* operation $\vee$ corresponds to ____, the *and* operation $\wedge$ corresponds to $\cap$, a tautology $\mathbf{t}$ corresponds to ____, a contradiction $\mathbf{c}$ corresponds to ____, and the negation operation $\sim$ corresponds to ____.
2. The operations of $+$ and $\cdot$ in a Boolean algebra are generalizations of the operations of ____ and ____ in the set of all statements forms in a given finite number of variables and the operations of ____ and ____ in the set of all subsets of a given set.

3. Russell showed that the following proposed "set definition" could not actually define a set:
_____.

Answers for Fill-in-the-Blank Review Questions

Section 6.1

1. the set A is a subset of the set B; for every x, if $x \in A$ then $x \in B$
(*Or:* every element of A is also an element of B)
2. x is any *[particular but arbitrarily chosen]* element of X; x is an element of Y
3. an element in X that is not in Y
4. x is in A or x is in B (*Or:* x is in at least one of the sets A and B)
5. x is in A and x is in B (*Or:* x is in both A and B)
6. x is in B and x is not in A
7. x is in the universal set and is not in A
8. no elements
9. the set of all subsets of A
10. $A \cap B = \emptyset$ (*Or:* A and B have no elements in common)
11. A is the union of all the sets $A_1, A_2, A_3, \ldots$ and $A_i \cap A_j = \emptyset$ for all $i, j = 1, 2, 3, \ldots$

Section 6.2

1. and
2. or
3. $x \in A$; $x \in B$; $x \in X$
4. $x \in A \cap B$ (*Or:* x is an element of both A and B); $x \in X$
5. $X \subseteq Y$; $Y \subseteq X$
6. X; in Y; Y; in X

Section 6.3

1. make the left-hand side unequal to the right-hand side.
Or: result in different values on the two sides of the equation.
2. cite one of the properties from Theorem 6.2.2
3. exactly

Section 6.4

1. the operation of union $\cup$; the operation of intersection $\cap$; a universal set U; the empty set $\emptyset$; the operation of complementation ^{c}
2. $\vee$; $\wedge$; $\cup$; $\cap$
3. the set of all sets that are not elements of themselves
Or: the set of all sets that are elements of themselves

Chapter 7: Functions

The aim of Section 7.1 is to promote a broad view of the function concept and to give you experience with the wide variety of functions that arise in discrete mathematics. Representation of functions by arrow diagrams is emphasized to prepare the way for the discussion of one-to-one and onto functions in Section 7.2.

Section 7.2 focuses on function properties. As you are learning about one-to-one and onto functions in this section, you may need to review the logical principles such as the negation of $\forall$, $\exists$, and if-then statements and the equivalence of a conditional statement and its contrapositive. These logical principles are needed to understand the equivalence of the two forms of the definition of one-to-one and what it means for a function not to be one-to-one or onto.

Sections 7.3 and 7.4 go together in the sense that the relations between one-to-one and onto functions and composition of functions developed in Section 7.3 are used to prove the fundamental theorem about cardinality in Section 7.4. The proofs that a composition of one-to-one functions is one-to-one or that a composition of onto functions is onto (and the related exercises) will test the degree to which you have learned to instantiate mathematical definitions in abstract contexts, apply the method of generalizing from the generic particular in a sophisticated setting, develop mental models of mathematical concepts that are both vivid and generic enough to reason with, and create moderately complex chains of deductions.

When you read Section 7.4, try to see the connections that link Russell's paradox, the halting problem, and the Cantor diagonalization argument.

Section 7.1

1. **a.** domain of $f = \{1, 3, 5\}$, co-domain of $f = \{s, t, u, v\}$

   b. $f(1) = v, f(3) = s, f(5) = v$

   c. range of $f = \{s, v\}$

   d. yes, no

   e. inverse image of $s = \{3\}$, inverse image of $u = \varnothing$, inverse image of $v = \{1, 5\}$

   f. $\{(1, v), (3, s), (5, v)\}$

3. **a.** True. The definition of function says that for any input there is one and only one output, so if two inputs are equal, their outputs must also be equal.

   b. False. The definition of function does not allow an element of the domain to be associated to two different elements of the co-domain, but it does allow an element of the co-domain to be the image of more than one element in the domain. For example, let $X = \{1, 2\}$ and $Y = \{a\}$ and define $f: X \to Y$ by specifying that $f(1) = f(2) = a$. Then f defines a function from X to Y for which a has two unequal preimages.

   c. True. The definition of function does not prohibit this occurrence.

   d. This statement is false. Each input to a function is related to only one output.

4. **a.** There are four functions from X to Y as shown below.

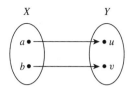 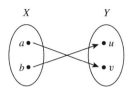

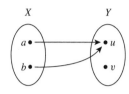 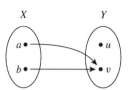

5. **a.** $I_Z(e) = e$

   b. $I_Z(b_i^{jk}) = b_i^{jk}$

6. **a.** The sequence can be given by the function $f : \mathbf{Z}^{nonneg} \to \mathbf{R}$ defined by the rule

   $$f(n) = \frac{(-1)^n}{2n+1} \quad \text{for each nonnegative integer } n.$$

   (This is one answer among many.)

   b. Define $F : \mathbf{Z}^{nonneg} \to \mathbf{R}$ as follows: for each nonnegative integer n, $F(n) = (-1)^n(2n)$.

7. **a.** $F(\{1, 3, 4\}) = 1$ [because $\{1, 3, 4\}$ has an odd number of elements]

8. **a.** $F(0) = (0^3 + 2 \cdot 0 + 4) \bmod 5 = 4 \bmod 5 = 4$

   b. $F(1) = (1^3 + 2 \cdot 1 + 4) \bmod 5 = 7 \bmod 5 = 2$

9. **a.** $S(1) = 1$

   b. $S(15) = 1 + 3 + 5 + 15 = 24$

   c. $S(17) = 1 + 17 = 18$

   d. $S(5) = 1 + 5 = 6$

   e. $S(18) = 1 + 2 + 3 + 6 + 9 + 18 = 39$

   f. $S(21) = 1 + 3 + 7 + 21 = 32$

10. **a.** $T(1) = \{1\}$

    b. $T(15) = \{1, 3, 5, 15\}$

    c. $T(17) = \{1, 17\}$

11. **a.** $F(4, 4) = (2 \cdot 4 + 1, 3 \cdot 4 - 2) = (9, 10)$

    b. $F(2, 1) = (2 \cdot 2 + 1, 3 \cdot 1 - 2) = (5, 1)$

12. **a.** $G(4, 4) = ((2 \cdot 4 + 1) \bmod 5, (3 \cdot 4 - 2) \bmod 5) = (9 \bmod 5, 10 \bmod 5) = (4, 0)$

    b. $G(2, 1) = ((2 \cdot 2 + 1) \bmod 5, (3 \cdot 1 - 2) \bmod 5) = (5 \bmod 5, 1 \bmod 5) = (0, 1)$

    c. $G(3, 2) = ((2 \cdot 3 + 1) \bmod 5, (3 \cdot 2 - 2) \bmod 5) = (7 \bmod 5, 4 \bmod 5) = (2, 4)$

    d. $G(1, 5) = ((2 \cdot 1 + 1) \bmod 5, 3 \cdot 5 - 2) \bmod 5) = (3 \bmod 5, 13 \bmod 5) = (3, 3)$

13.

x	$f(x)$	$g(x)$
0	$4^2 \bmod 5 = 1$	$(0^2 + 3 \cdot 0 + 1) \bmod 5 = 1$
1	$5^2 \bmod 5 = 0$	$(1^2 + 3 \cdot 1 + 1) \bmod 5 = 0$
2	$6^2 \bmod 5 = 1$	$(2^2 + 3 \cdot 2 + 1) \bmod 5 = 1$
3	$7^2 \bmod 5 = 4$	$(3^2 + 3 \cdot 3 + 1) \bmod 5 = 4$
4	$8^2 \bmod 5 = 4$	$(4^2 + 3 \cdot 4 + 1) \bmod 5 = 4$

The table shows that $f(x) = g(x)$ for every x in J_5. Thus by definition of equality of functions, $f = g$.

15. $F \cdot G$ and $G \cdot F$ are equal because for every real number x,

$$
\begin{aligned}
(F \cdot G)(x) &= F(x) \cdot G(x) \quad \text{by definition of } F \cdot G \\
&= G(x) \cdot F(x) \quad \text{by the commutative law} \\
&\qquad\qquad\qquad\quad \text{for multiplication of real numbers} \\
&= (G \cdot F)(x) \quad \text{by definition of } G \cdot F.
\end{aligned}
$$

16. No. For instance, let F and G be defined by the rules $F(x) = x$ and $G(x) = 0$ for each real numbers x. Then $(F - G)(2) = F(2) - G(2) = 2 - 0 = 2$, whereas $(G - F)(2) = G(2) - F(2) = 0 - 2 = -2$, and $2 \neq -2$. In fact, $G - F = F - G \Leftrightarrow F = G$.

17. **a.** $2^3 = 8$ **c.** $4^1 = 4$

18. **a.** $\log_3 81 = 4$ because $3^4 = 81$

b. $\log_2 1024 = 10$ because $2^{10} = 1024$

c. $\log_3 \left(\frac{1}{27}\right) = -3$ because $3^{-3} = \frac{1}{27}$

d. $\log_2 1 = 0$ because $2^0 = 1$

e. $\log_{10} \dfrac{1}{10} = -1$ because $10^{-1} = \dfrac{1}{10}$

f. $\log_3 3 = 1$ because $3^1 = 3$

g. $\log_2 2^k = k$ because the exponent to which 2 must be raised to obtain 2^k is k
Alternative answer: $\log_2 2^k = k$ because $2^k = 2^k$

19. Let b be any positive real number with $b \neq 1$. Since $b^1 = b$, then $\log_b b = 1$ by definition of logarithm.

21. <u>Proof:</u> Suppose b and u are any positive real numbers with $b \neq 1$. *[We must show that* $\log_b \left(\frac{1}{u}\right) = -\log_b(u)$.*]* Let $v = \log_b \left(\frac{1}{u}\right)$. By definition of logarithm, $b^v = \frac{1}{u}$. Multiplying both sides by u and dividing by b^v gives $u = b^{-v}$, and thus, by definition of logarithm, $-v = \log_b(u)$. When both sides of this equation are multiplied by -1, the result is $v = -\log_b(u)$. Therefore, $\log_b \left(\frac{1}{u}\right) = -\log_b(u)$ because both expressions equal v. *[This is what was to be shown.]*

22. *Hint:* Use a proof by contradiction. Suppose $\log_3 7$ is rational. Then $\log_3 7 = \frac{a}{b}$ for some integers a and b with $b \neq 0$.

Apply the definition of logarithm to rewrite $\log_3 7 = \frac{a}{b}$ in exponential form.

23. Suppose b and y are positive real numbers with $\log_b y = 3$.

By definition of logarithm, this implies that $b^3 = y$. Then

$$
y = b^3 = \frac{1}{\frac{1}{b^3}} = \frac{1}{\left(\frac{1}{b}\right)^3} = \left(\frac{1}{b}\right)^{-3}.
$$

24. Since $\log_b y = 2$, then $b^2 = y$. Now, by properties of exponents, $(b^2)^1 = y$, and so $\log_{b^2}(y) = 1$.

25. **a.** $p_1(2, y) = 2$, $p_1(5, x) = 5$, range of $p_1 = \{2, 3, 5\}$

26. **a.** $mod\,(67, 10) = 7$ and $div(67, 10) = 6$ since $67 = 10 \cdot 6 + 7$.

27. **a.** $f(aba) = 0$ *[because there are no b's to the left of the left-most a in aba]*
$f(bbab) = 2$ *[because there are two b's to the left of the left-most a in bbab]*
$f(b) = 0$ *[because the string contains no a's]*

b. $g(aba) = aba$, $g(bbab) = babb$, $g(b) = b$
The range of g is the set of all strings of a's and b's, which equals S.

28. **a.** $E(0110) = 000111111000$ and $D(111111000111) = 1101$

29. **a.** $H(10101, 00011) = 3$

30. **a.**

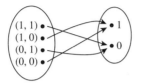

b.

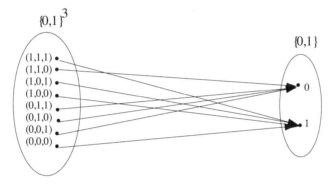

32. **a.** $f(1,1,1) = (4 \cdot 1 + 3 \cdot 1 + 2 \cdot 1) \, mod \, 2 = 9 \, mod \, 2 = 1$
$f(0,0,1) = (4 \cdot 0 + 3 \cdot 0 + 2 \cdot 1) \, mod \, 2 = 2 \, mod \, 2 = 0$

33. If g were well defined, then $g(1/2) = g(2/4)$ because $1/2 = 2/4$. However, $g(1/2) = 1 - 2 = -1$ and $g(2/4) = 2 - 4 = -2$. Since $-1 \neq -2$, $g(1/2) \neq g(2/4)$. Thus g is not well defined.

35. Student B is correct. If R were well defined, then $R(3)$ would have a uniquely determined value. However, on the one hand, $R(3) = 2$ because $(3 \cdot 2) \, mod \, 5 = 1$, and, on the other hand, $R(3) = 7$ because $(3 \cdot 7) \, mod \, 5 = 1$. Hence $R(3)$ does not have a uniquely determined value, and so R is not well defined.

36. Student D is correct. In order for S to be a function, $S(2)$ has to be that integer y in V such that $2y \, mod \, 4 = 1$. Because of the definition of V, the only possible values for y are 1, 2, and 3. But

$$2 \cdot 1 \, mod \, 4 = 2 \neq 1, \quad 2 \cdot 2 \, mod \, 4 = 4 \, mod \, 4 = 0 \neq 1, \quad and \quad 2 \cdot 3 \, mod \, 4 = 6 \, mod \, 4 = 2 \neq 1.$$

Thus there is no number y in V such that $S(2) = y$, and hence S does not satisfy property (1) of the definition of function.

38. **a.**

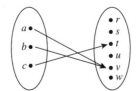

b. $f(A) = \{v\}$, $\quad f(X) = \{t, v\}$, $\quad f^{-1}(C) = \{c\}$, $\quad f^{-1}(D) = \{a, b\}$,
$f^{-1}(E) = \emptyset$, $\quad f^{-1}(Y) = \{a, b, c\} = X$

39. **a.**

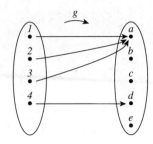

b. $g(A) = \{a\}$ $g(X) = \{a, d\}$ $g^{-1}(C) = \{1, 2, 3\}$ $g^{-1}(D) = \emptyset$
$g^{-1}(Y) = \{1, 2, 3, 4\} = X$

40. *Partial answer:* (i) $y \in F(A)$ or $y \in F(B)$ (ii) some   (iii) $A \cup B$ (iv) $F(A \cup B)$

41. The statement is true. <u>Proof:</u> Let F be a function from X to Y, and suppose $A \subseteq X, B \subseteq X$, and $A \subseteq B$. Let $y \in F(A)$. *[We must show that $y \in F(B)$.]* By definition of image of a set, $y = F(x)$ for some $x \in A$. Thus since $A \subseteq B$, $x \in B$, and so $y = F(x)$ for some $x \in B$. Hence $y \in F(B)$ *[as was to be shown]*.

42. The property is true.

    <u>Proof:</u> Suppose $y \in F(A \cap B)$. *[We must show that $y \in F(A) \cap F(B)$.]* By definition of image of a set, there exists an element x in $A \cap B$ such that $y = F(x)$. By definition of intersection, x is in A and x is in B., and so, by definition of image of an element, $F(x) \in F(A)$ and $F(x) \in F(B)$. Thus, by substitution, $y \in F(A)$ and $y \in F(B)$. It follows, by definition of intersection, that $y \in F(A) \cap F(B)$ *[as was to be shown]*.

43. The statement is false. <u>Counterexample:</u> Let $X = \{1, 2, 3\}$, let $Y = \{a, b\}$, and define a function $F: X \to Y$ by the arrow diagram shown below.

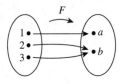

    Let $A = \{1, 2\}$ and $B = \{1, 3\}$. Then $F(A) = \{a, b\} = F(B)$, and so $F(A) \cap F(B) = \{a, b\}$. But $F(A \cap B) = F(\{1\}) = \{a\} \neq \{a, b\}$. And so $F(A) \cap F(B) \not\subseteq F(A \cap B)$.

    (This is just one of many possible counterexamples.)

45. The statement is true. <u>Proof:</u> Let F be a function from a set X to a set Y, and suppose $C \subseteq Y$, $D \subseteq Y$, and $C \subseteq D$. *[We must show that $F^{-1}(C) \subseteq F^{-1}(D)$.]* Suppose $x \in F^{-1}(C)$. Then $F(x) \in C$. Since $C \subseteq D$, $F(x) \in D$ also. Hence, by definition of inverse image, $x \in F^{-1}(D)$. *[So $F^{-1}(C) \subseteq F^{-1}(D)$.]*

46. *Hint:* $x \in F^{-1}(C \cup D) \Leftrightarrow F(x) \in C \cup D \Leftrightarrow F(x) \in C$ or $F(x) \in D$

48. The statement is true.

    <u>Proof 1:</u> Let $F: X \to Y$ be any function, and suppose that $C \subseteq Y$ and $D \subseteq Y$. *[We must show that $F^{-1}(C - D) = F^{-1}(C) - F^{-1}(D)$.]*

    Proof that $F^{-1}(C - D) \subseteq F^{-1}(C) - F^{-1}(D)$:

    Suppose x is any element in $F^{-1}(C - D)$. *[We must show that $x \in F^{-1}(C) - F^{-1}(D)$.]*

    By definition of inverse image, $F(x) \in C - D$, and so,

    by definition of set difference, $F(x) \in C$ and $F(x) \notin D$.

Then, by definition of inverse image, $x \in F^{-1}(C)$ and $x \notin F^{-1}(D)$, and so

by definition of set difference, $x \in F^{-1}(C) - F^{-1}(D)$ *[as was to be shown].*

Proof that $F^{-1}(C) - F^{-1}(D) \subseteq F^{-1}(C - D)$:

Suppose x is any element in $F^{-1}(C) - F^{-1}(D)$. *[We must show that $x \in F^{-1}(C - D)$.]*

By definition of set difference, $F^{-1}(C)$ and $x \notin F^{-1}(D)$, and so,

by definition of inverse image, $F(x) \in C$ and $F(x) \notin D$.

Then, by definition of set difference, $F(x) \in C - D$, and so,

by definition of inverse image $x \in F^{-1}(C - D)$ *[as was to be shown].*

Conclusion: Since both subset containments have been proved, we conclude that $F^{-1}(C - D) = F^{-1}(C) - F^{-1}(D)$ *[as was to be shown].*

Proof 2:*(This proof uses the logic of if-and-only-if statements.)*

Let $F: X \to Y$ be any function, and suppose that $C \subseteq Y$ and $D \subseteq Y$. *[We must show that $F^{-1}(C - D) = F^{-1}(C) - F^{-1}(D)$.]*

Suppose x is any element in X. Then, by definition of inverse image and set difference,

$$
\begin{aligned}
x \in F^{-1}(C - D) \quad &\Leftrightarrow \quad F(x) \in C - D \\
&\Leftrightarrow \quad F(x) \in C \text{ and } F(x) \notin D \\
&\Leftrightarrow \quad x \in F^{-1}(C) \text{ and } x \notin F^{-1}(D) \\
&\Leftrightarrow \quad x \in F^{-1}(C) - F^{-1}(D).
\end{aligned}
$$

The preceding steps show that if $x \in F^{-1}(C - D)$ then $x \in F^{-1}(C) - F^{-1}(D)$ (which implies that $F^{-1}(C - D) \subseteq F^{-1}(C) - F^{-1}(D)$), and if $x \in F^{-1}(C) - F^{-1}(D)$ then $x \in F^{-1}(C - D)$ (which implies that $F^{-1}(C) - F^{-1}(D) \subseteq F^{-1}(C - D)$).

Hence $F^{-1}(C - D) = F^{-1}(C) - F^{-1}(D)$.

51. **a.** $\phi(15) = 8$ *[because 1, 2, 4, 7, 8, 11, 13, and 14 have no common factors with 15 other than ± 1]*

    b. $\phi(2) = 1$ *[because the only positive integer less than or equal to 2 having no common factors with 2 other than ± 1 is 1]*

    c. $\phi(5) = 4$ *[because 1, 2, 3, and 4 have no common factors with 5 other than ± 1]*

    d. $\phi(12) = 4$ *[because 1, 5, 7, and 11 have no common factors with 12 other than ± 1]*

    e. $\phi(11) = 10$ *[because 1, 2, 3, 4, 5, 6, 7, 8, 9, and 10 have no common factors with 11 other than ± 1]*

    f. $\phi(1) = 1$ *[because 1 is the only positive integer which has no common factors with 1 other than ± 1]*

52. Proof: Let p be any prime number and n any integer with $n \geq 1$. There are p^{n-1} positive integers less than or equal to p^n that have a common factor other than ± 1 with p^n—namely, $p, 2p, 3p, \ldots, (p^{n-1})p$. Hence, there are $p^n - p^{n-1}$ positive integers less than or equal to p^n that do not have a common factor with p^n except for ± 1.

Section 7.2

1. The second statement is the contrapositive of the first.

2. **a.** most

3. There are many counterexamples to the given statement. This was given in Appendix B.

Counterexample: Consider the function defined by the following arrow diagram:

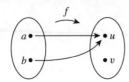

Observe that a is sent to exactly one element of Y, namely, u, and b is also sent to exactly one element of Y, namely, u also. So it is true that every element of X is sent to exactly one element of Y. But f is not one-to-one because $f(a) = f(b)$ whereas $a \neq b$. [Note that to say, "Every element of X is sent to exactly one element of Y" is just another way of saying that in the arrow diagram for the function there is only one arrow coming out of each element of X. But this statement is part of the definition of any function, not just of a one-to-one function.]

Another counterexample is the following:

Counterexample: Consider the function defined as the following relation: $f = \{(1,4),(2,4)\}$ with domain $\{1, 2\}$ and co-domain $\{4\}$. Each element of the domain is related to exactly one element of the co-domain because 1 is related to 4 (and not to anything else), and 2 is related to 4 (and not to anything else). But f is not one-to-one because $f(1) = f(2)$ and $1 \neq 2$.

4. *Hint:* The statement is true.

5. *Hint:* One of the incorrect ways is (b).

6. **a.** f is not one-to-one because $f(1) = 4 = f(9)$ and $1 \neq 9$. f is not onto because $f(x) \neq 3$ for any x in X.

   b. g is one-to-one because $g(1) \neq g(5), g(1) \neq g(9)$, and $g(5) \neq g(9)$. g is onto because each element of Y is the image of some element of X: $3 = g(5), 4 = g(9)$, and $7 = g(1)$.

7. **a.** F is not one-to-one because $F(c) = e = F(d)$ and $c \neq d$. F is onto because each element of Y is the image of some element of X: $e = F(c) = F(d), f = F(a)$, and $g = F(b)$.

9. For each part of this problem there are a number of correct answers.

   a.

   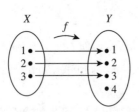

   b.   **c.**   **d.**

   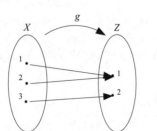   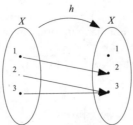   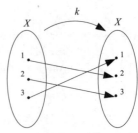

10. **a.** (i) **f is one-to-one.** Proof: Suppose $f(n_1) = f(n_2)$ for some integers n_1 and n_2. [We must show that $n_1 = n_2$] By definition of f, $2n_1 = 2n_2$, and dividing both sides by 2 gives $n_1 = n_2$ [as was to be shown].

(ii) *f is not onto*. <u>Counterexample</u>: Consider $1 \in \mathbf{Z}$. We claim that $1 \neq f(n)$, for any integer n. If there were an integer n such that $1 = f(n)$, then, by definition of f, $1 = 2n$. Dividing both sides by 2 would give $n = 1/2$. But $1/2$ is not an integer. Hence $1 \neq f(n)$ for any integer n, and so f is not onto.

b. *h is onto*. <u>Proof</u>: Suppose $m \in 2\mathbf{Z}$. *[We must show that there exists an integer such that h of that integer equals m.]* Since $m \in 2\mathbf{Z}$, $m = 2k$ for some integer k. Then $h(k) = 2k = m$. Hence there exists an integer (namely, k) such that $h(k) = m$ *[as to be shown]*.

11. *Hints:* **a.** (i) g is one-to-one (ii) g is not onto

b. *G is onto*. <u>Proof</u>: Suppose y is any element of $\mathbf{R}$. *[We must show that there is an element x in $\mathbf{R}$ such that $G(x) = y$. What would x be if it exists? Scratch work shows that x would have to equal $(y + 5)/4$. The proof must then show that x has the necessary properties.]* Let $x = (y+5)/4$. Then (1) $x \in \mathbf{R}$, and (2) $G(x) = G((y+5)/4) = 4[(y+5)/4] - 5 = (y+5) - 5 = y$ *[as was to be shown]*.

12. **a.** (i) *F is one-to-one*. <u>Proof</u>: Suppose n_1 and n_2 are in $\mathbf{Z}$ and $F(n_1) = F(n_2)$. *[We must show that $n_1 = n_2$.]* By definition of F, $2 - 3n_1 = 2 - 3n_2$. Subtracting 2 from both sides and dividing by -3 gives $n_1 = n_2$.

(ii). *F is not onto*. <u>Counterexample</u>: Let $m = 0$. Then m is in $\mathbf{Z}$ but $m \neq F(n)$ for any integer n. *[For if $m = F(n)$ then $0 = 2 - 3n$, and so $3n = 2$ and $n = 2/3$. But $2/3$ is not in $\mathbf{Z}$.]*

b. *G is onto*. <u>Proof</u>: Suppose y is any element of $\mathbf{R}$. *[We must show that there is an element x in $\mathbf{R}$ such that $G(x) = y$.]*

[Scratch work: If such an x exists, then, by definition of G, $y = 2 - 3x$ and so $3x = 2 - y$, or, equivalently, $x = (2 - y)/3$. Let's check to see if this works.]

Let $x = (2 - y)/3$. Then

$$(1)\ x \in \mathbf{R} \quad \text{and} \quad (2)\ G(x) = G\left(\frac{2-y}{3}\right) = 2 - 3\left(\frac{2-y}{3}\right) = 2 - (2 - y) = 2 - 2 + y = y.$$

[This is what was to be shown.]

13. **a.** (i) *H is not one-to-one*. <u>Counterexample</u>: $H(1) = 1 = H(-1)$ but $1 \neq -1$.

(ii) *H is not onto*. <u>Counterexample</u>: $H(x) \neq -1$ for any real number x because $H(x) = x^2$ and no real numbers have negative squares.

b. *K is onto*. <u>Proof</u>: Suppose y is any element of $\mathbf{R}^{nonneg}$. Let $x = \sqrt{y}$. Then (1) x is a real number because $y \geq 0$, and (2) by definition of K, $K(x) = K(\sqrt{y}) = (\sqrt{y})^2 = y$.

14. The "proof" claims that f is one-to-one because for each integer n there is only one possible value for $f(n)$. But to say that for each integer n there is only one possible value for $f(n)$ is just another way of saying that f satisfies one of the conditions necessary for it to be a function. To show that f is one-to-one, one must show that any integer n has a *different* function value from that of the integer m whenever $n \neq m$.

15. *f is one-to-one*. <u>Proof</u>: Suppose $f(x_1) = f(x_2)$ where x_1 and x_2 are nonzero real numbers. *[We must show that $x_1 = x_2$.]*

$$\text{By definition of } f, \quad \frac{x_1 + 1}{x_1} = \frac{x_2 + 1}{x_2}.$$

Cross-multiplying gives $\qquad x_1 x_2 + x_2 = x_1 x_2 + x_1,$

and so $x_1 = x_2$ by subtracting $x_1 x_2$ from both sides. *[This is what was to be shown.]*

16. ***f is not one-to-one***. <u>Counterexample</u>: Note that

$$\frac{x_1}{x_1^2 + 1} = \frac{x_2}{x_2^2 + 1} \Rightarrow x_1 x_2^2 + x_1 = x_2 x_1^2 + x_2$$

$$\Rightarrow x_1 x_2^2 - x_2 x_1^2 = x_2 - x_1$$

$$\Rightarrow x_1 x_2 (x_2 - x_1) = x_2 - x_1$$

$$\Rightarrow x_1 = x_2 \text{ or } x_1 x_2 = 1$$

Thus for a counterexample take any x_1 and x_2 with $x_1 \neq x_2$ but $x_1 x_2 = 1$. For instance, take $x_1 = 2$ and $x_2 = 1/2$. Then $f(x_1) = f(2) = 2/5$ and $f(x_2) = f(1/2) = 2/5$, but $2 \neq 1/2$.

18. ***f is one-to-one***.

<u>Proof 1</u>: Let x_1 and x_2 be any real numbers other than 1, and suppose that $f(x_1) = f(x_2)$. *[We must show that $x_1 = x_2$.]*

By definition of f, $\dfrac{x_1 + 1}{x_1 - 1} = \dfrac{x_2 + 1}{x_2 - 1}$.

Cross-multiplying gives

$(x_1 + 1)(x_2 - 1) = (x_2 + 1)(x_1 - 1)$ or, equivalently, $x_1 x_2 - x_1 + x_2 - 1 = x_1 x_2 - x_2 + x_1 - 1$.

Adding $1 - x_1 x_2$ to both sides gives

$$-x_1 + x_2 = -x_2 + x_1, \text{ or, equivalently, } 2x_1 = 2x_2.$$

Dividing both sides by 2 gives $x_1 = x_2$ *[as was to be shown]*.

<u>Proof 2</u>: Let x_1 and x_2 be any real numbers such that $f(x_1) = f(x_2)$.* *[We must show that $x_1 = x_2$.]*

By definition of f, $\dfrac{x_1 + 1}{x_1 - 1} = \dfrac{x_2 + 1}{x_2 - 1}$.

Then

	$(x_1 + 1)(x_2 - 1)$	$=$	$(x_2 + 1)(x_1 - 1)$
$\Rightarrow$	$x_1 x_2 - x_1 + x_2 - 1$	$=$	$x_1 x_2 - x_2 + x_1 - 1$
$\Rightarrow$	$-x_1 + x_2$	$=$	$-x_2 + x_1$
$\Rightarrow$	$2x_1$	$=$	$2x_2$
$\Rightarrow$	x_1	$=$	x_2

by cross-multiplying
by multiplying out
by adding $1 - x_1 x_2$ to both sides
by adding $x_1 + x_2$ to both sides
by dividing both sides by 2.

[This is what was to be shown.]

* When f is defined for x_1 and x_2, then neither x_1 nor x_2 can equal 1 because the denominator would equal zero.

19. **a.** Because $\frac{417302072}{7} \cong 59614581.7$ and $417302072 - 7 \cdot 59614581 = 5$, $H(417302072) = 417302072 \bmod 7 = 5$. But position 5 is already occupied, so the next position is checked. It is free, and thus the record is placed in position 6.

20. Recall that $\lfloor x \rfloor =$ that unique integer n such that $n \leq x$, $n + 1$.

a. Floor *is not one-to-one*. <u>Counterexample</u>: Floor$(0) = 0 = $ Floor$(1/2)$ but $0 \neq 1/2$.

b. Floor *is onto*. <u>Proof</u>: Suppose $m \in \mathbf{Z}$.
[We must show that there exists a real number y such that Floor $(y) = m$.]

Let $y = m$. Then Floor$(y) = $ Floor$(m) = m$ since m is an integer. Hence there exists a real number y such that Floor$(y) = m$ *[as was to be shown]*.

(In fact, Floor takes the value m for *all* real numbers in the interval $m \leq x$, $m + 1$.)

21. **a.** *L is not one-to-one*. <u>Counterexample</u>: $L(0) = L(1) = 1$ but $1 \neq 0$.

    b. *L is onto*. <u>Proof</u>: Suppose n is any nonnegative integer.
    [We must show that there exists a string s in S such that $L(s) = n$.]

    $$\text{Let } s = \begin{cases} \lambda(\text{the null string}) & \text{if } n = 0 \\ \underbrace{00\ldots0}_{n\ 0's} & \text{if } n > 0 \end{cases}$$

    Then $L(s) = $ the length of $s = n$ *[as was to be shown]*.

23. **a.** *F is not one-to-one*. <u>Counterexample</u>: Let $A = \{a\}$ and $B = \{b\}$. Then $F(A) = F(B)$ $= 1$ but $A \neq B$.

24. **a.** *N is not one-to-one*. <u>Counterexample</u>: Let $s_1 = a$ and $s_2 = ab$. Then $N(s_1) = N(s_2) = 1$ but $s_1 \neq s_2$.

    b. *N is not onto*. <u>Counterexample</u>: The number -1 is in **Z** but $N(s) \neq -1$ for any string s in S because no string has a negative number of a's.

26. **S is not one-to-one**. <u>Counterexample</u>: $S(6) = 1 + 2 + 3 + 6 = 12$ and $S(11) = 1 + 11 = 12$. So $S(6) = S(11)$ but $6 \neq 11$.

    S is not onto. <u>Counterexample</u>: In order for there to be a positive integer n such that $S(n)$ $= 5$, n would have to be less than 5. But $S(1) = 1$, $S(2) = 3$, $S(3) = 4$, and $S(4) = 7$. Hence there is no positive integer n such that $S(n) = 5$.

27. **a.** *T is one-to-one*. <u>Proof</u>: $T(n)$ is the set of all the positive divisors of n. Note that for each positive integer n, the largest element of $T(n)$ is n because n divides n and no integer larger than n divides n.

    So suppose n_1 and n_2 are positive integers and $T(n_1) = T(n_2)$. *[We must show that $n_1 = n_2$.]*

    Now $T(n_1)$ is the set of all the positive divisors of n_1 and $T(n_2)$ is the set of all the positive divisors of n_2.

    Since $T(n_1) = T(n_2)$, the largest element of $T(n_1)$, namely n_1, is the same as the largest element of $T(n_2)$, namely n_2.

    Hence $n_1 = n_2$ *[as was to be shown]*.

    b. *T is not onto*. <u>Counterexample</u>: The set $\{2\}$ is a finite subset of positive integers, but there is no positive integer n such that $T(n) = \{2\}$. The reason is that the number 1 divides every positive integer, and so 1 must be an element of $T(n)$ for each positive integer n. But $1 \notin \{2\}$. (There are many other examples that show T is not onto.)

28. **a.** *G is one-to-one*. <u>Proof</u>: Suppose (x_1, y_1) and (x_2, y_2) are any elements of **R** $\times$ **R** such that $G(x_1, y_1) = G(x_2, y_2)$. *[We must show that $(x_1, y_1) = (x_2, y_2)$.]*

    By definition of G, $(2y_1, -x_1) = (2y_2, -x_2)$, and, by definition of ordered pair,

    $$2y_1 = 2y_2 \quad \text{and} \quad -x_1 = -x_2.$$

    Dividing both sides of the equation on the left by 2 and both sides of the equation on the right by -1 gives that

    $$y_1 = y_2 \quad \text{and} \quad x_1 = x_2,$$

    and so, by definition of ordered pair, $(x_1, y_1) = (x_2, y_2)$ *[as was to be shown]*.

    b. *G is onto*. <u>Proof</u>: Suppose (u, v) is any element of **R** $\times$ **R**.
    *[We must show that there is an element (x, y) in **R** $\times$ **R** such that $G(x, y) = (u, v)$.]*
    Let $(x, y) = (-v, u/2)$. Then

    (1) $(x, y) \in$ **R** $\times$ **R** and (2) $G(x, y) = (2y, -x) = (2(u/2), -(-v)) = (u, v)$

    [as was to be shown].

30. **a.** ***J is one-to-one.*** <u>Proof</u>: Suppose (r_1, s_1) and (r_2, s_2) are in $\mathbf{Q} \times \mathbf{Q}$ and $J(r_1, s_1) = J(r_2, s_2)$. *[We must show that $(r_1, s_1) = (r_2, s_2)$.]* By definition of J,

$$r_1 + \sqrt{2}s_1 = r_2 + \sqrt{2}s_2 \quad \text{and thus} \quad r_1 - r_2 = \sqrt{2}(s_2 - s_1).$$

Note that both $r_1 - r_2$ and $s_2 - s_1$ are rational because they are differences of rational numbers (exercise 17 of Section 4.3).

Suppose $s_2 - s_1 \neq 0$. Then $\sqrt{2}(s_2 - s_1)$ is a product of a nonzero rational number and an irrational number ($\sqrt{2}$), and so it is irrational (exercise 13 of Section 4.7). As a consequence, the rational number $(r_1 - r_2)$ equals the irrational number ($\sqrt{2}(s_2 - s_1)$). Because this is impossible, the supposition that $s_2 - s_1 \neq 0$ must be false, and therefore $s_2 - s_1 = 0$.

Thus, by substitution,

$$r_1 - r_2 = \sqrt{2}(s_2 - s_1) = \sqrt{2} \cdot 0 = 0.$$

So $r_1 - r_2 = 0$ and $s_2 - s_1 = 0$ or, equivalently, $r_1 = r_2$ and $s_2 = s_1$.

Hence $(r_1, s_1) = (r_2, s_2)$ *[as was to be shown]*.

b. ***J is not onto.*** <u>Counterexample</u>: We show that J is not onto by giving an example of a real number that is not equal to $J(r, s)$ for any rational numbers r and s. For example, consider the number $\sqrt{3}$ and suppose there were rational numbers r and s such that

$$\sqrt{3} = r + \sqrt{2}s.$$

We will show that this supposition leads logically to a contradiction.]

Case 1 (s = 0): In this case, $\sqrt{3} = r$ where r is a rational number, which contradicts the fact that $\sqrt{3}$ is irrational (exercise 19, Section 4.8).

Case 2 (s ≠ 0): In this case,

$$\sqrt{3} - \sqrt{2}s = r$$
$$\Rightarrow \quad 3 + 2s^2 - 2s\sqrt{6} = r^2 \qquad \text{by squaring both sides}$$
$$\Rightarrow \quad -2s\sqrt{6} = r^2 - 3 - 2s^2 \qquad \text{by subtracting } 3 + 2s^2 \text{ from both sides}$$
$$\Rightarrow \quad \sqrt{6} = \frac{r^2 - 3 - 2s^2}{-2s} \qquad \text{by dividing both sides by } -2s.$$

Now both $r^2 - 3 - 2s^2$ and $-2s$ are rational numbers because products and differences of rational numbers are rational (exercises 15 and 17, Section 4.3), and $-2s$ is nonzero because it is a product of -2 and s, which are both nonzero numbers (zero product property). Thus $\sqrt{6}$ is a quotient of a rational number and a nonzero rational number, which is rational (by the result of exercise 16 in Section 4.3). But this contradicts the fact that $\sqrt{6}$ is irrational (by the result of exercise 25, Section 4.8).

Conclusion: Since a contradiction was obtained in both cases, we conclude that the supposition is false. That is, there are no rational numbers r and s such that $\sqrt{3} = r + \sqrt{2}s$. Therefore J is not onto.

31. **a.** ***F is one-to-one.*** <u>Proof</u>: Suppose $F(a, b) = F(c, d)$ for some ordered pairs (a, b) and (c, d) in $\mathbf{Z}^+ \times \mathbf{Z}^+$. *[We must show that $(a, b) = (c, d)$.]*

By definition of F, $3^a 5^b = 3^c 5^d$.

By the unique factorization of integers theorem (Theorem 4.4.5), the number of 3's on the left side of the equation must equal the number of 3's on the right side and the number of 5's on the left side of the equation must equal the number of 5's on the right side.

Hence $a = b$ and $c = d$, and thus $(a, b) = (c, d)$.

32. **a.** Let $x = \log_8 27$ and $y = \log_2 3$. *[The question is: Is $x = y$?]* By definition of logarithm, both of these equations can be written in exponential form as

$$8^x = 27 \quad \text{and} \quad 2^y = 3.$$

Now $8 = 2^3$. So

$$8^x = (2^3)^x = 2^{3x}.$$

Also $27 = 3^3$ and $3 = 2^y$. So

$$27 = 3^3 = (2^y)^3 = 2^{3y}.$$

Hence, since $8^x = 27$,

$$2^{3x} = 2^{3y}.$$

By (7.2.5.), then,

$$3x = 3y,$$

and so

$$x = y.$$

But $x = \log_8 27$ and $y = \log_2 3$, and so

$$\log_8 27 = y = \log_2 3$$

and the answer to the question is yes.

33. <u>Proof</u>: Suppose that b, x, and y are any positive real numbers such that $b \neq 1$.

$$\text{Let} \quad u = \log_b(x) \quad \text{and} \quad v = \log_b(y).$$

[We must show that $\log_b(\frac{x}{y}) = \log_b(x) - \log_b(y)$.]

By definition of logarithm, $b^u = x$ and $b^v = y$. It follows that

$$\frac{x}{y} = \frac{b^u}{b^v} = b^{u-v}$$

by substitution, property (7.2.3), and the fact that $b^{-v} = \dfrac{1}{b^v}$.

Translating $\dfrac{x}{y} = b^{u-v}$ into logarithmic form gives that

$$\log_b\left(\frac{x}{y}\right) = u - v,$$

and so, by substitution,

$$\log_b\left(\frac{x}{y}\right) = \log_b(x) - \log_b(y)$$

[as was to be shown].

35. *Start of Proof:* Suppose a, b, and y are any *[particular but arbitrarily chosen]* real numbers such that b and x are positive and $b \neq 1$. *[We must show that $\log_b(x^a) = a\log_b x$.]* Let $r = \log_b(x^a)$ and $s = \log_b(x)$.

36. No. <u>Counterexample</u>: Define $f: \mathbf{R} \to \mathbf{R}$ and $g: \mathbf{R} \to \mathbf{R}$ as follows: $f(x) = x$ and $g(x) = -x$ for every real number x. Then f and g are both one-to-one

[because for all real numbers x_1 and x_2, if $f(x_1) = f(x_2)$ then $x_1 = x_2$, and if $g(x_1) = g(x_2)$ then $-x_1 = -x_2$, and so $x_1 = x_2$ in this case as well].

On the other hand, $f + g$ is not one-to-one

[because $f + g$ satisfies the equation $(f + g)(x) = x + (-x) = 0$ for every real number x, and so, for instance, $(f + g)(1) = (f + g)(2)$ but $1 \neq 2$].

38. Yes. <u>Proof:</u> Let f be a one-to-one function from $\mathbf{R}$ to $\mathbf{R}$, and let c be any nonzero real number. Suppose $(c \cdot f)(x_1) = (c \cdot f)(x_2)$. *[We must show that $x_1 = x_2$.]* It follows by definition of $(c \cdot f)$ that $c \cdot (f(x_1)) = c \cdot f(x_2)$. Since $c \neq 0$, we may divide both sides of the equation by c to obtain $f(x_1) = f(x_2)$. And since f is one-to-one, this implies that $x_1 = x_2$ *[as was to be shown].*

39. If $f: \mathbf{R} \to \mathbf{R}$ is onto and c is any nonzero real number, then $(c \cdot f)$ is also onto.

    <u>Proof:</u> Suppose $f: \mathbf{R} \to \mathbf{R}$ is onto and c is any nonzero real number.

    Let y be any real number in $\mathbf{R}$. *[We must show that there exists a real number in $\mathbf{R}$ such that $(c \cdot f)(\text{that number}) = y.]*

    Since $c \neq 0$, y/c is a real number. So, since f is onto, there is a real number x in $\mathbf{R}$ with $f(x) = y/c$.

    Then $y = c \cdot (f(x)) = (c \cdot f)(x)$. Thus $(c \cdot f)$ is onto *[as was to be shown].*

40. **a.** *Hint:* The assumption that F is one-to-one is needed in the proof that $F^{-1}(F(A)) \subseteq A$. If $F(r) \in F(A)$, the definition of image of a set implies that there is an element x in A such that $F(r) = F(x)$.

    b. *Hint:* The assumption that F is one-to-one is needed in the proof that $F(A_1) \cap F(A_2) \subseteq F(A_1 \cap A_2)$. If $u \in F(A_1)$ and $u \in F(A_2)$, then the definition of image of a set implies that there are elements x_1 in A_1 and x_2 in A_2 such that $F(x_1) = u$ and $F(x_2) = u$ and, thus, that $F(x_1) = F(x_2)$.

42.

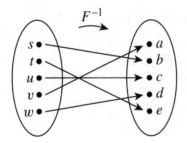

44. The function is not a one-to-one correspondence because it is not onto.

45. The function h is a one-to-one correspondence. The answer to exercise 10(b) shows that h is onto, so to finish showing that h is a one-to-one correspondence, we need to show that h is one-to-one. To prove this, suppose n_1 and n_2 are any integers such that $h(n_1) = h(n_2)$. By definition of h, this implies that $2n_1 = 2n_2$, and dividing both sides by 2 gives that $n_1 = n_2$. Hence h is one-to-one, and so h is a one-to-one correspondence.

    We claim that $h^{-1}(m) = \dfrac{m}{2}$ for every even integer m in $2\mathbf{Z}$. Given any $m \in 2\mathbf{Z}$,

    $$h^{-1}(m) = \frac{m}{2} \text{ if, and only if, } h(\frac{m}{2}) = m.$$

    But this is true because for every $m \in 2\mathbf{Z}$,

    $$h(\frac{m}{2}) = 2\left(\frac{m}{2}\right) = m.$$

46. The function g is not a one-to-one correspondence because it is not onto. For instance, if $m = 2$, it is impossible to find an integer n such that $g(n) = m$. (This is because if $g(n) = m$, then $4n - 5 = 2$, which implies that $n = 7/4$. Thus the only number n with the property that $g(n) = m$ is $7/4$. But $7/4$ is not an integer.)

47. The function G is a one-to-one correspondence. The answer to exercise 11b shows that G is onto, so to finish showing that G is a one-to-one correspondence, we need to show that G is one-to-one. To prove this, suppose $G(x_1) = G(x_2)$ for some real numbers x_1 and x_2 in **R**. *[We must show that $x_1 = x_2$.]* By definition of G, $4x_1 - 5 = 4x_2 - 5$. Add 5 to both sides of this equation and divide both sides by 4 to obtain $x_1 = x_2$. *[This is what was was to be shown.]* We claim that $G^{-1}(y) = \dfrac{y+5}{4}$ for each y in **R**. By definition of inverse function, this is true if, and only if, $G\left(\dfrac{y+5}{4}\right) = y$. Now

$$G\left(\frac{y+5}{4}\right) = 4\left(\frac{y+5}{4}\right) - 5 = (y+5) - 5 = y,$$

and so it is the case that $G^{-1}(y) = (y+5)/4$ for each y in **R**.

48. By the result of exercise 12a, F is not onto. Hence F is not a one-to-one correspondence.

50. The function L is not a one-to-one correspondence because by the solution to exercise 21, it is not one-to-one.

51. The function D is not a one-to-one correspondence because by the solution to exercise 22, it is not one-to-one.

52. The function f is a one-to-one correspondence. The answer to exercise 15 shows that f is one-to-one, and if the co-domain is taken to be the set of all real numbers not equal to 1, then f is also onto. The reason is that given any real number $y \neq 1$, if we take $x = \dfrac{1}{y-1}$, then x is a real number and

$$f(x) = f\left(\frac{1}{y-1}\right) = \frac{\frac{1}{y-1} + 1}{\frac{1}{y-1}} = \frac{1 + (y-1)}{1} = y.$$

Thus $f^{-1}(y) = \dfrac{1}{y-1}$ for each real number $y \neq 1$.

53. The function f is a one-to-one correspondence because the answer to exercise 16 in Appendix B shows that f is not one-to-one.

54. The function f is a one-to-one correspondence. By the result of exercise 17, f is one-to-one. f is also onto for the following reason. Given any real number y other than 3, let $x = \dfrac{1}{3-y}$. Then x is a real number (because $y \neq 3$) and

$$f(x) = f(\frac{1}{3-y}) = \frac{3\left(\frac{1}{3-y}\right) - 1}{\frac{1}{3-y}} = \frac{3\left(\frac{1}{3-y}\right) - 1}{\frac{1}{3-y}} \cdot \frac{(3-y)}{(3-y)} = \frac{3 - (3-y)}{1} = 3 - 3 + y = y.$$

This calculation also shows that $f^{-1}(y) = \dfrac{1}{3-y}$ for every real number $y \neq 3$.

57. **Algorithm 7.2.1 Checking Whether a Function is One-to-One**

[For a given function F with domain $X = \{a[1], a[2], \ldots, a[n]\}$, this algorithm discovers whether or not F is one-to-one. Initially, answer is set equal to "one-to-one". Then the values of $F(a[i])$ and $F(a[j])$ are systematically compared for indices i and j with $1 \leq i < j \leq n$. If at any point it is found that $F(a[i]) = F(a[j])$ and $a[i] \neq a[j]$, then F is not one-to-one, and so answer is set equal to "not one-to-one" and execution ceases. If after all possible values of i and j have been examined, the value of answer is still "one-to-one", then F is one-to-one.]

Input: *n [a positive integer]*, $a[1], a[2], \ldots, a[n]$ *[a one-dimensional array representing the set X]*, *F [a function with domain X]*

Algorithm Body:

    answer := *"one-to-one"*

    $i := 1$

    while ($i \leq n - 1$ and *answer* = *"one-to-one"*)

        $j := i + 1$

        while ($j \leq n$ and *answer* = *"one-to-one"*)

            if ($F(a[i]) = F(a[j])$ and $a[i] \neq a[j]$) **then** *answer* := *"not one-to-one"*

            $j := j + 1$

        end while

        $i := i + 1$

    end while

Output: *answer [a string]*

58. *Hint:* Let a function F be given and suppose the domain and co-domain of F are represented by the one-dimensional arrays $a[1], a[2], \ldots, a[n]$ and $b[1], b[2], \ldots, b[m]$, respectively. Introduce a variable *answer* whose initial value is "onto." For each $b[i]$ from $i = 1$ to m, make a search through $a[1], a[2], \ldots, a[n]$ to check whether $b[i] = F(a[j])$ for some $a[j]$. Introduce a Boolean variable to indicate whether a search has been successful. (Set the variable equal to 0 before the start of each search, and let it have the value 1 if the search is successful.) At the end of each search, check the value of the Boolean variable. If it is 0, then F is not onto. If all searches are successful, then F is onto.

Section 7.3

1. $g \circ f$ is defined as follows:

$$(g \circ f)(1) = g(f(1)) = g(5) = 1$$
$$(g \circ f)(3) = g(f(3)) = g(3) = 5$$
$$(g \circ f)(5) = g(f(5)) = g(1) = 3.$$

   $f \circ g$ is defined as follows:

$$(f \circ g)(1) = f(g(1)) = f(3) = 3$$
$$(f \circ g)(3) = f(g(3)) = f(5) = 1$$
$$(f \circ g)(5) = f(g(5)) = f(1) = 5.$$

   Then $g \circ f \neq f \circ g$ because, for example, $(g \circ f)(1) \neq (f \circ g)(1)$.

3. $(G \circ F)(x) = G(F(x)) = G(x^3) = x^3 - 1$ for every real number x.

   $(F \circ G)(x) = F(G(x)) = F(x - 1) = (x - 1)^3$ for every real number x.

   $G \circ F \neq F \circ G$; for instance, $(G \circ F)(2) = 2^3 - 1 = 7$, whereas $(F \circ G)(2) = (2 - 1)^3 = 1$.

6. $(G \circ F)(0) = G(F(0)) = G(7 \cdot 0) = G(0) = 0 \bmod 5 = 0$
   $(G \circ F)(1) = G(F(1)) = G(7 \cdot 1) = G(7) = 7 \bmod 5 = 2$
   $(G \circ F)(2) = G(F(2)) = G(7 \cdot 2) = G(14) = 14 \bmod 5 = 4$
   $(G \circ F)(3) = G(F(3)) = G(7 \cdot 3) = G(21) = 21 \bmod 5 = 1$
   $(G \circ F)(4) = G(F(4)) = G(7 \cdot 4) = G(28) = 28 \bmod 5 = 3$

7. **a.** *Partial answer:*
   $(L \circ M)(12) = L(M(12)) = L(12 \bmod 5) = L(2) = 2^2 = 4$
   $(M \circ L)(12) = M(L(12)) = M(12^2) = M(144) = 144 \bmod 5 = 4$

8. **a.** $(T \circ L)(abaa) = T(L(abaa)) = T(4) = 4 \bmod 3 = 1$

   b. $(T \circ L)(baaab) = T(L(baaab)) = T(5) = 5 \bmod 3 = 2$

   c. $(T \circ L)(aaa) = T(L(aaa)) = T(3) = 3 \bmod 3 = 0$

9. **a.** $(G \circ F)(2) = G(F(2)) = G(2^2/3) = G(4/3) = \lfloor 4/3 \rfloor = 1$

   b. $(G \circ F)(-3) = G(F(-3)) = G((-3)^2/3) = G(9/3) = \lfloor 3 \rfloor = 3$

   c. $(G \circ F)(5) = G(F(5)) = G(5^2/3) = G(25/3) = 8$

10. **a.** $(G \circ F)(8) = G(F(8)) = G(16) = \lfloor 16/2 \rfloor = 8$
    $(F \circ G)(8) = F(G(8)) = F(\lfloor 8/2 \rfloor) = F(4) = 8$
    $(G \circ F)(3) = G(F(3)) = G(6) = \lfloor 6/2 \rfloor = 3$
    $(F \circ G)(3) = F(G(3)) = F(\lfloor 3/2 \rfloor) = F(1) = 2$

    b. $G \circ F \neq F \circ G$ because $(G \circ F)(3) = (F \circ G)(3) = 2$ and $3 \neq 2$.

12. For every x in **R**, $(F^{-1} \circ F)(x) = F^{-1}(F(x)) = F^{-1}(3x+2) = \dfrac{(3x+2)-2}{3} = \dfrac{3x}{3} = x = I_{\mathbf{R}}(x).$

    Hence $F^{-1} \circ F = I_{\mathbf{R}}$ by definition of equality of functions.

    In addition, for every y in **R,**

    $$(F \circ F^{-1})(y) = F(F^{-1}(y)) = F\left(\frac{y-2}{3}\right) = 3\left(\frac{y-2}{3}\right) + 2 = (y-2) + 2 = y = I_{\mathbf{R}}(y).$$

    Hence $F \circ F^{-1} = I_{\mathbf{R}}$ by definition of equality of functions.

15. **a.** By definition of logarithm with base b, for each real number x, $\log_b(b^x)$ is the exponent to which b must be raised to obtain b^x. But this exponent is just x. So $\log_b(b^x) = x$.

    b. For all positive real numbers b and x, $\log_b x$ is the exponent to which b must be raised to obtain x. So if b is raised to this exponent, x is obtained. In other words, $b^{\log_b x} = x$.

16. *Hint:* Suppose f is any function from a set X to a set Y, and show that for every x in X, $(I_Y \circ f)(x) = f(x)$.

18. **a.** $s_k = s_m$ **b.** $z/2 = t/2$ **c.** $f(x_1) = f(x_2)$

19. The answer is no.

    <u>Counterexample:</u> Define f and g by the arrow diagrams

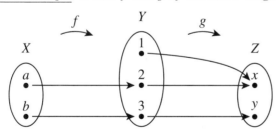

    Then $g \circ f$ is one-to-one but g is not one-to-one. (This is one counterexample among many. The exercise asks students to find a different one.)

21. For the given conditions, f must be one-to-one.

    <u>Proof:</u> Suppose $f\colon X \to Y$ and $g\colon Y \to Z$ are functions and $g \circ f\colon X \to Z$ is one-to-one.

    To show that f is one-to-one, suppose x_1 and x_2 are any elements in X with $f(x_1) = f(x_2)$. *[We must show that $x_1 = x_2$.]*

    By definition of function, $g(f(x_1)) = g(f(x_2))$, and so, by definition of function composition, $(g \circ f)(x_1) = (g \circ f)(x_2)$.

    Since $g \circ f$ is one-to-one, it follows that $x_1 = x_2$ *[as was to be shown]*.

22. *Hint :* Suppose $f\colon X \to Y$ and $g\colon Y \to Z$ are functions and $g \circ f$ is onto. Given $z \in Z$, there is an element x in X such that $(g \circ f)(x) = z$. (Why?) If $y = f(x)$, what can you deduce about $g(y)$?

24. The statement is true.

Proof: Suppose X is any set and f, g, and h are functions from X to X such that h is one-to-one and $h \circ f = h \circ g$. *[We must show that for every x in X, $f(x) = g(x)$.]*

Suppose x is any element in X.

Because $h \circ f = h \circ g$, we have that $(h \circ f)(x) = (h \circ g)(x)$ by definition of equality of functions.

It follows by definition of function composition that $h(f(x)) = h(g(x))$.

Since h is one-to-one, this implies that $f(x) = g(x)$ *[as was to be shown]*.

26.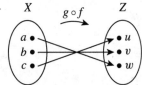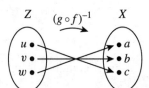

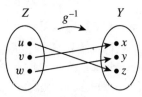 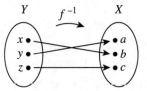

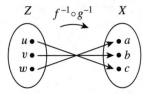

The functions $(g \circ f)^{-1}$ and $f^{-1} \circ g^{-1}$ are equal.

27. Let $\Leftrightarrow$ stand for the words "if, and only if". Then, by definition of inverse function, for all real numbers x, y, and z,

$$(g \circ f)^{-1}(z) = x \Leftrightarrow (g \circ f)(x) = z, \quad g^{-1}(z) = y \Leftrightarrow g(y) = z, \quad \text{and} \quad f^{-1}(y) = x \Leftrightarrow f(x) = y.$$

Now $g \circ f \colon \mathbf{R} \to \mathbf{R}$ is defined by

$$(g \circ f)(x) = g(f(x)) = g(x + 3) = -(x + 3) \text{ for each } x \in \mathbf{R}.$$

Then

$$(g \circ f)(x) = z \Leftrightarrow z = -x - 3 \Leftrightarrow x + z + 3 = 0 \Leftrightarrow x = -z - 3 \Leftrightarrow -z - 3 = (g \circ f)^{-1}(z).$$

Thus $(g \circ f)^{-1} \colon \mathbf{R} \to \mathbf{R}$ is defined by

$$(g \circ f)^{-1}(z) = -z - 3 \text{ for each real number } z.$$

Similarly, since $g \colon \mathbf{R} \to \mathbf{R}$ and $f \colon \mathbf{R} \to \mathbf{R}$ are defined by the equations

$$g(y) = -y \quad \text{and} \quad f(x) = x + 3,$$

then

$$g(y) = z \Leftrightarrow z = -y \Leftrightarrow y = -z \Leftrightarrow g^{-1}(z) = -z$$

and

$$f(x) = y \Leftrightarrow x + 3 = y \Leftrightarrow y = x - 3 \Leftrightarrow f^{-1}(y) = y - 3.$$

It follows that

$$(f^{-1} \circ g^{-1})(z) = f^{-1}(g^{-1}(z)) = f^{-1}(-z) = -z - 3 = -z - 3.$$

Thus $f^{-1} \circ g^{-1} \colon \mathbf{R} \to \mathbf{R}$ is defined by

$$(f^{-1} \circ g^{-1})(z) = -z - 3 \text{ for each real number } z.$$

Since both $(g \circ f)^{-1}$ and $f^{-1} \circ g^{-1}$ always take the same values, it follows by definition of equality of functions that $(g \circ f)^{-1} = f^{-1} \circ g^{-1}$.

29. *Hints:* (1) Theorems 7.3.3 and 7.3.4 taken together insure that $g \circ f$ is one-to-one and onto. (2) Use the inverse function property: $F^{-1}(b) = F^{-1}$.

30. The property is true.

<u>Proof 1</u>: Let X, Y, and Z be any sets, let $f : X \to Y$ and $g : Y \to Z$ be any functions, and let C be any subset of Z.

Proof that $((g \circ f)^{-1}(C) \subseteq f^{-1}(g^{-1}(C))$:

Suppose $x \in (g \circ f)^{-1}(C)$. *[We must show that $x \in f^{-1}(g^{-1}(C))$.]*

By definition of inverse image (for $g \circ f$), $(g \circ f)(x) \in C$, and so,

by definition of composition of functions, $g(f(x)) \in C$.

Then by definition of inverse image (for g), $f(x) \in g^{-1}(C)$, and

by definition of inverse image (for f), $x \in f^{-1}(g^{-1}(C))$.

So by definition of subset, $(g \circ f)^{-1}(C) \subseteq f^{-1}(g^{-1}(C))$.

Proof that $f^{-1}(g^{-1}(C)) \subseteq (g \circ f)^{-1}(C)$:

Suppose $x \in f^{-1}(g^{-1}(C))$. *[We must show that $x \in (g \circ f)^{-1}(C)$.]*

By definition of inverse image (for f), $f(x) \in g^{-1}(C)$, and so,

by definition of inverse image (for g), $g(f(x)) \in C$.

So by definition of composition of functions, $(g \circ f)(x) \in C$.

Then by definition of inverse image (for $g \circ f$), $x \in (g \circ f)^{-1}(C)$.

So by definition of subset, $f^{-1}(g^{-1}(C)) \subseteq (g \circ f)^{-1}(C)$.

Conclusion: Since each set is a subset of the other, the two sets are equal.

<u>Proof 2</u> *(using the logic of if-and-only-if statements)*

Let X, Y, and Z be any sets, let $f : X \to Y$ and $g : Y \to Z$ be any functions, and let C be any subset of Z.

Then $x \in (g \circ f)^{-1}(C) \Leftrightarrow (g \circ f)(x) \in C$ *[by definition of inverse image for $g \circ f$]*

$\Leftrightarrow g(f(x)) \in C$ *[by definition of composition of functions]*

$\Leftrightarrow f(x) \in g^{-1}(C)$ *[by definition of inverse image for g]*

$\Leftrightarrow x \in f^{-1}(g^{-1}(C))$ *[by definition of inverse image for f]*.

So both sets consist of the same elements, and thus, by definition of set equality, $(g \circ f)^{-1}(C) = f^{-1}(g^{-1}(C))$.

Section 7.4

1. The student should have replied that for A to have the same cardinality as B means that there is a function from A to B that is one-to-one and onto. A set cannot have the property of being onto or one-to-one another set; only a function can have these properties.

2. <u>Proof</u>: Define a function $f\colon \mathbf{Z}^+ \to S$ as follows: For every positive integer k, $f(k) = k^2$.

   f is one-to-one: [*We must show that for all k_1 and $k_2 \in \mathbf{Z}^+$, if $f(k_1) = f(k_2)$ then $k_1 = k_2$.*] Suppose k_1 and k_2 are positive integers and $f(k_1) = f(k_2)$. By definition of f, $(k_1)^2 = (k_2)^2$, so $k_1 = \pm k_2$. But k_1 and k_2 are *positive*. Hence $k_1 = k_2$.

   f is onto: [*We must show that for each $n \in S$, there exists $k \in \mathbf{Z}^+$ such that $n = f(k)$.*] Suppose $n \in S$. By definition of S, $n = k^2$ for some positive integer k. Then by definition of f, $n = f(k)$.

   Since there is a one-to-one, onto function (namely, f) from $\mathbf{Z}^+$ to S, the two sets have the same cardinality.

3. <u>Proof</u>: Define $f\colon \mathbf{Z} \to 3\mathbf{Z}$ by the rule $f(n) = 3n$ for each integer n.

   The function f is one-to-one because for any integers n_1 and n_2, if $f(n_1) = f(n_2)$ then $3n_1 = 3n_2$ and so $n_1 = n_2$.

   Also f is onto because if m is any element in $3\mathbf{Z}$, then $m = 3k$ for some integer k. Then $f(k) = 3k = m$ by definition of f.

   Thus, since there is a function $f\colon \mathbf{Z} \to 3\mathbf{Z}$ that is one-to-one and onto, $\mathbf{Z}$ has the same cardinality as $3\mathbf{Z}$.

4. <u>Proof</u>: Define a function $f\colon \mathbf{O} \to 2\mathbf{Z}$ as follows: $f(n) = n - 1$ for each odd integer n.

   Since n is odd, $n = 2k + 1$ for some integer k, and so $n - 1 = 2k$, which is even. Thus f is well-defined.

   Also f is one-to-one because for all odd integers n_1 and n_2, if $f(n_1) = f(n_2)$ then $n_1 - 1 = n_2 - 1$ and hence $n_1 = n_2$.

   Moreover f is onto because given any even integer m, then $m = 2k$ for some integer k, and so $m + 1 = 2k + 1$, which is odd, and $f(m + 1) = (m + 1) - 1 = m$ by definition of f.

   Thus, because there is a function $f\colon \mathbf{O} \to 2\mathbf{Z}$ that is one-to-one and onto, $\mathbf{O}$ has the same cardinality as $2\mathbf{Z}$.

6. There is a function from $2\mathbf{Z}$ to $\mathbf{Z}$ that is one-to-one but not onto.

   <u>Proof</u>: Define a function $I\colon 2\mathbf{Z} \to \mathbf{Z}$ as follows: $I(n) = n$ for each even integer n. I is clearly one-to-one because if $I(n_1) = I(n_2)$ then by definition of I, $n_1 = n_2$. But I is not onto because the range of I consists only of even integers. In other words, if m is any odd integer, then $I(n) \neq m$ for any even integer n.

   There is a function from $\mathbf{Z}$ to $2\mathbf{Z}$ that is onto but not one-to-one.

   <u>Proof</u>: Define a function $J\colon \mathbf{Z} \to 2\mathbf{Z}$ as follows $J(n) = 2\left\lfloor \dfrac{n}{2} \right\rfloor$ for each integer n. Then J is onto because for any even integer m, $m = 2k$ for some integer k. Let $n = 2k$. Then

   $$J(n) = J(2k) = 2\left\lfloor \frac{2k}{2} \right\rfloor = 2\lfloor k \rfloor = 2k = m.$$

   But J is not one-to-one because, for example,

   $$J(2) = 2\left\lfloor \frac{2}{2} \right\rfloor = 2 \cdot 1 = 2 \quad \text{and} \quad J(3) = 2\left\lfloor \frac{3}{2} \right\rfloor = 2 \cdot 1 = 2.$$

   So $J(2) = J(3)$ but $2 \neq 3$.

(More generally, given any integer k, if $m = 2k$, then

$$J(m) = 2 \left\lfloor \frac{m}{2} \right\rfloor = 2 \left\lfloor \frac{2k}{2} \right\rfloor = 2 \lfloor k \rfloor$$

and

$$J(m+1) = 2 \left\lfloor \frac{m+1}{2} \right\rfloor = 2 \left\lfloor \frac{2k+1}{2} \right\rfloor = 2 \left\lfloor k + \frac{1}{2} \right\rfloor = 2k.$$

So $J(m) = J(m+1)$ but $m \neq m+1$.)

7. **b.** For each positive integer n, $F(n) = (-1)^n \left\lfloor \dfrac{n}{2} \right\rfloor$.

8. It was shown in Example 7.4.2 that $\mathbf{Z}$ is countably infinite, which means that $\mathbf{Z}^+$ has the same cardinality as $\mathbf{Z}$. By exercise 3, $\mathbf{Z}$ has the same cardinality as $3\mathbf{Z}$. It follows by the transitive property of cardinality (Theorem 7.4.1 (c)) that $\mathbf{Z}^+$ has the same cardinality as $3\mathbf{Z}$, and so $3\mathbf{Z}$ is countably infinite *[by definition of countably infinite]*. Hence $3\mathbf{Z}$ is countable *[by definition of countable]*.

9. Proof: Define a function $f \colon \mathbf{Z}^+ \to \mathbf{Z}^{nonneg}$ as follows: $f(n) = n - 1$ for each positive integer n. *[We will show that $\mathbf{Z}^{nonneg}$ is countable.]*

   Observe that if $n \geq 1$ then $n - 1 \geq 0$, so f is well-defined.

   In addition, f is one-to-one because for all positive integers n_1 and n_2, if $f(n_1) = f(n_2)$ then $n_1 - 1 = n_2 - 1$ and hence $n_1 = n_2$.

   Moreover f is onto because if m is any nonnegative integer, then $m + 1$ is a positive integer and $f(m+1) = (m+1) - 1 = m$ by definition of f.

   Thus, because there is a function $f \colon \mathbf{Z}^+ \to \mathbf{Z}^{nonneg}$ that is one-to-one and onto, $\mathbf{Z}^+$ has the same cardinality as $\mathbf{Z}^{nonneg}$.

   It follows that $\mathbf{Z}^{nonneg}$ is countably infinite and hence countable.

10. Proof: Define $f \colon S \to U$ by the rule $f(x) = 2x$ for each real number x in S. Then f is one-to-one by the same argument as in exercise 10a of Section 7.2 with $\mathbf{R}$ in place of $\mathbf{Z}$. Furthermore, f is onto because if y is any element in U, then $0 < y < 2$ and so $0 < y/2 < 1$. Consequently, $y/2 \in S$ and $f(y/2) = 2(y/2) = y$. Hence f is a one-to-one correspondence, and so S and U have the same cardinality.

11. *Hint:* Define $h \colon S \to V$ as follows: $h(x) = 3x + 2$, for every $x \in S$.

12. Proof: Define $F \colon S \to W$ by the rule $F(x) = (b-a)x + a$ for each real number x in S.

    Then F is well-defined because if $0 < x < 1$, then $a < (b-a)x + a < b$.

    In addition, F is one-to-one because if x_1 and x_2 are in S and $F(x_1) = F(x_2)$, then

    $$(b-a)x_1 + a = (b-a)x_2 + a,$$

    and so *[by subtracting a and dividing by $b - a$]* $x_1 = x_2$.

    Furthermore, F is onto because if y is any element in W, then $a < y < b$. Subtracting a from the right-hand inequality and using the facts that $a < y$ and that $a < b$ gives that

    $$y - a < b - a \ (*) \quad \text{and} \quad y - a > 0 \ (**) \quad \text{and} \quad b - a > 0 \ (***).$$

    Dividing both sides of (*) by $b - a$ and dividing the left-hand side of (**) by the left-hand side of (***) gives that

    $$0 < \frac{y-a}{b-a} < 1.$$

Consequently, $\dfrac{y-a}{b-a} \in S$ and

$$h\left(\frac{y-a}{b-a}\right) = (b-a)\left(\frac{y-a}{b-a}\right) + a = y.$$

Hence F is a one-to-one correspondence, and so S and W have the same cardinality.

13.

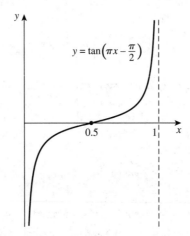

It is clear from the graph that f is one-to-one (since it is increasing) and that the image of f is all of $\mathbf{R}$ (since the lines $x = 0$ and $x = 1$ are vertical asymptotes). Thus S and $\mathbf{R}$ have the same cardinality.

15. <u>Proof</u>: Let B be the set of all bit strings (strings of 0's and 1's).

Define a function $F: \mathbf{Z}^+ \to B$ as follows: $F(1) = \lambda$, $F(2) = 0$, $F(3) = 1$, $F(4) = 00$, $F(5) = 01$, $F(6) = 10$, $F(7) = 11$, $F(8) = 000$, $F(9) = 001$, $F(10) = 010$, and so forth.

At each stage, all the strings of length k are counted before the strings of length $k+1$, and the strings of length k are counted in order of increasing magnitude when interpreted as binary representations of integers.

It is clear by inspection that F is both one-to-one and onto, which shows that the set of all bit strings is countably infinite and hence countable.

Note: A more formal definition for F is the following:

$$F(n) = \begin{cases} \lambda & \text{if } n = 1 \\ \text{the } k\text{-bit binary representation of } n - 2^k & \text{if } n > 1 \text{ and } \lfloor \log_2 n \rfloor = k. \end{cases}$$

For example, $F(7) = 11$ because $\lfloor \log_2 7 \rfloor = 2$ and the two-bit binary representation of $7 - 2^2$ ($= 3$) is 11.

16. In Example 7.4.4 it was shown that there is a one-to-one correspondence from $\mathbf{Z}^+$ to $\mathbf{Q}^+$. This implies that the positive rational numbers can be written as an infinite sequence: $r_1, r_2, r_3, r_4, \ldots$. Now the set $\mathbf{Q}$ of all rational numbers consists of the numbers in this sequence together with 0 and the negative rational numbers: $-r_1, -r_2, -r_3, -r_4, \ldots$.

Let $r_0 = 0$. Then the elements of the set of all rational numbers can be "counted" as follows:

$$r_0, r_1, -r_1, r_2, -r_2, r_3, -r_3, r_4, -r_4, \ldots$$

In other words, we can define a one-to-one correspondence as follows: for each integer $n \geq 1$,

$$G(n) = \begin{cases} r_{n/2} & \text{if } n \text{ is even} \\ -r_{(n-1)/2} & \text{if } n \text{ is odd} \end{cases}$$

This shows that $\mathbf{Q}$ is countably infinite and hence countable.

17. *Hint:* See the hints for exercises 18 and 19 in Section 4.3.

18. No. For instance, both $\sqrt{2}$ and $-\sqrt{2}$ are irrational (by Theorem 4.8.1 and exercise 23 in Section 4.7), and yet their average is $\dfrac{\sqrt{2} + (-\sqrt{2})}{2}$ which equals 0 and is rational.

More generally: If r is any rational number and x is any irrational number, then both $r + x$ and $r - x$ are irrational (by exercise 12 in Section 4.7 or by the combination of Theorem 4.7.3 and exercise 9 in Section 4.7). Yet the average of these numbers is $\dfrac{(r + x) + (r - x)}{2} \cdot \dfrac{2r}{2} = r$, which is rational.

19. *Hint:* Suppose r and s are real numbers with $s > r > 0$. Let n be an integer such that $n > \dfrac{\sqrt{2}}{s - r}$, and let $m = \left\lfloor \dfrac{nr}{\sqrt{2}} \right\rfloor + 1$. Show that $m > \dfrac{nr}{\sqrt{2}} \geq m - 1$, and use the fact that $s = r + (s - r)$ to conclude that $r < \dfrac{\sqrt{2}m}{n} < s$.

21. *Two examples of many:* Define $F \colon \mathbf{Z} \to \mathbf{Z}$ by the rule $F(n) = \begin{cases} n/2 & \text{if } n \text{ is even} \\ 0 & \text{if } n \text{ is odd} \end{cases}$.
Then F is onto because given any integer m, $m = F(2m)$. But F is not one-to-one because, for instance, $F(1) = F(3) = 0$.

Define $G \colon \mathbf{Z} \to \mathbf{Z}$ by the rule $G(n) = \lfloor n/2 \rfloor$ for each integer n. Then G is onto because given any integer m, $m = \lfloor m \rfloor = \lfloor (2m)/2 \rfloor = G(2m)$. But G is not one-to-one because, for instance, $G(2) = \lfloor 2/2 \rfloor = 1$ and $G(3) = \lfloor 3/2 \rfloor = 1$ and $2 \neq 3$.

22. *Hint:* Use the unique factorization of integers theorem (Theorem 4.4.5) and Theorem 7.4.3.

23. **a.** Define a function $G \colon \mathbf{Z}^{nonneg} \to \mathbf{Z}^{nonneg} \times \mathbf{Z}^{nonneg}$ as follows: Let $G(0) = (0, 0)$, and then follow the arrows in the diagram, letting each successive ordered pair of integers be the value of G for the next successive integer. Thus, for instance,

$$G(1) = (1, 0), \ G(2) = (0, 1), \ G(3) = (2, 0),$$

$$G(4) = (1, 1), \ G(5) = (0, 2), \ G(6) = (3, 0),$$

$$G(7) = (2, 1), \ G(8) = (1, 2), \text{ and so forth.}$$

b. *Hint:* Observe that if the top ordered pair of any given diagonal is $(k, 0)$, the entire diagonal (moving from top to bottom) consists of $(k, 0), (k-1, 1), (k-2, 2), \ldots, (2, k-2), (1, k-1), (0, k)$. Thus for every ordered pair (m, n) within any given diagonal, the value of $m + n$ is constant, and as you move down the ordered pairs in the diagonal, starting at the top, the value of the second element of the pair keeps increasing by 1.

24. The proof given below is adapted from one in *Foundations of Modern Analysis* by Jean Dieudonné, New York: Academic Press, 1969, page 14.

Proof: Suppose (a, b) and (c, d) are in $\mathbf{Z}^+ \times \mathbf{Z}^+$ and $(a, b) \neq (c, d)$.

Case 1, $a + b \neq c + d$: By interchanging (a, b) and (c, d) if necessary, we may assume that $a + b < c + d$. Then

$$H(a,b) = b + \frac{(a+b)(a+b+1)}{2} \qquad \text{by definition of } H$$

$$\leq a + b + \frac{(a+b)(a+b+1)}{2} \qquad \text{because } a \geq 0$$

$$< (a+b+1) + \frac{(a+b)(a+b+1)}{2} \qquad \text{because } a+b < a+b+1$$

$$< \frac{2(a+b+1)}{2} + \frac{(a+b)(a+b+1)}{2} \qquad \text{by algebra}$$

$$< \frac{(a+b+1)(a+b+2)}{2} \qquad \text{by factoring out } (a+b+1)$$

$$< \frac{(c+d)(c+d+1)}{2} \qquad \begin{array}{l}\text{since } a+b < c+d \text{ and } a,\, b,\, c, \\ \text{and } d \text{ are integers, } a+b+1 \leq c+d\end{array}$$

$$< d + \frac{(c+d)(c+d+1)}{2} \qquad \text{because } d \geq 0$$

$$< H(c,d) \qquad \text{by definition of } H.$$

Therefore, $H(a,b) \neq H(c,d)$.

Case 2, $a + b = c + d$: First observe that in this case $b \neq d$. For if $b = d$, then subtracting b from both sides of $a + b = c + d$ gives $a = c$, and so $(a,b) = (c,d)$, which contradicts our assumption that $(a,b) \neq (c,d)$. Hence,

$$H(a,b) = b + \frac{(a+b)(a+b+1)}{2} = b + \frac{(c+d)(c+d+1)}{2} \neq d + \frac{(c+d)(c+d+1)}{2} = H(c,d),$$

and so $H(a,b) \neq H(c,d)$.

Thus both in case 1 and in case 2, $H(a,b) \neq H(c,d)$, and hence H is one-to-one.

25. *Hint:* There are at least two different approaches to this problem. One is to use the method discussed in Section 4.3. Another is to suppose that $1.999999\ldots < 2$ and derive a contradiction. (Show that the difference between 2 and $1.999999\ldots$ can be made smaller than any given positive number.)

26. <u>Proof</u>: Let A be an infinite set. Construct a countably infinite subset $a_1, a_2, a_3, \ldots$ of A, by letting a_1 be any element of A, letting a_2 be any element of A other than a_1, letting a_3 be any element of A other than a_1 or a_2, and so forth. This process never stops (and hence $a_1, a_2, a_3, \ldots$ is an infinite sequence) because A is an infinite set. More formally,

1. Let a_1 be any element of A.

2. For each integer $n \neq 2$, let a_n be any element of $A - \{a_1, a_2, a_3, \ldots, a_{n-l}\}$. Such an element exists, for otherwise $A - \{a_1, a_2, a_3, \ldots, a_{n-1}\}$ would be empty and A would be finite.

27. <u>Proof</u>: Suppose A is any countably infinite set, B is any set, and $g: A \to B$ is onto. Since A is countably infinite, there is a one-to-one correspondence $f: \mathbf{Z}^+ \to A$. Then, in particular, f is onto, and so by Theorem 7.3.4, $g \circ f$ is an onto function from $\mathbf{Z}^+$ to B.

Define a function $h: B \to \mathbf{Z}^+$ as follows: Suppose x is any element of B. Since $g \circ f$ is onto, $\{m \in \mathbf{Z}^+ \mid (g \circ f)(m) = x\} \neq \emptyset$. Thus, by the well-ordering principle for the integers, this set has a least element. In other words, there is a least positive integer n with $(g \circ f)(n) = x$. Let $h(x)$ be this integer.

We claim that h is a one-to-one. Suppose $h(x_1) = h(x_2) = n$. By definition of h, n is the least positive integer with $(g \circ f)(n) = x_1$. Moreover, by definition of h, n is the least positive integer with $(g \circ f)(n) = x_2$. Hence $x_1 = (g \circ f)(n) = x_2$.

Thus h is a one-to-one correspondence between B and a subset S of positive integers (the range of h). Since any subset of a countable set is countable (Theorem 7.4.3), S is countable, and so there is a one-to-one correspondence between B and a countable set. It follows from the transitive property of cardinality that B is countable.

29. *Hint:* Suppose A and B are any two countably infinite sets. Then there are one-to-one correspondences $f\colon \mathbf{Z}^+ \to A$ and $g\colon \mathbf{Z}^+ \to B$.

   Case 1 ($A \cap B = \emptyset$): In this case define $h\colon \mathbf{Z}^+ \to A \cup B$ as follows: $h\colon \mathbf{Z}^+ \to A \cup B$ as follows: For all integers $n \geq 1$,

   $$h(n) = \begin{cases} f(n/2) & \text{if } n \text{ is even} \\ g((n+1)/2) & \text{if } n \text{ is odd} \end{cases}$$

   Show that h is one-to-one and onto.

   Case 2 ($A \cap B \neq \emptyset$): In this case let $C = B - A$. Then $A \cup B = A \cup C$ and $A \cap C = \emptyset$.

   If C is countably infinite, use the result of case 1 to complete the proof. If C is finite, use the result of exercise 28 to complete the proof.

30. <u>Proof by contradiction:</u> Suppose not. That is, suppose the set of all irrational numbers were countable.

   Then the set of all real numbers could be written as a union of two countably infinite sets: the set of all rational numbers and the set of all irrational numbers.

   By exercise 29 this union is countably infinite, and so the set of all real numbers would be countably infinite and hence countable.

   But this contradicts the fact that the set of all real numbers is uncountable (which follows immediately from Theorems 7.4.2 and 7.4.3 or Corollary 7.4.4).

   Hence the set of all irrational number is uncountable.

31. *Hint:* Consider the following cases:(1) A and B are both finite, (2) at least one of A or B is infinite and $A \cap B = \emptyset$, (3) at least one of A or B is infinite and $A \cap B \neq \emptyset$. In case 3 use the facts that $A \cup B = (A - B) \cup (B - A) \cup (A \cap B)$ and the sets $A - B$, $B - A$, and $A \cap B$ are mutually disjoint.

32. *Hint:* Use the one-to-one correspondence $F\colon \mathbf{Z}^+ \to \mathbf{Z}$ of Example 7.4.2 to define a function $G\colon \mathbf{Z}^+ \times \mathbf{Z}^+ \to \mathbf{Z} \times \mathbf{Z}$ by the formula $G(m, n) = (F(m), F(n))$. Show that G is a one-to-one correspondence, and use the result of exercise 22 and the transitive property of cardinality.

33. <u>Proof:</u> First note that there are as many equations of the form $x^2 + bx + c = 0$ as there are pairs (b, c) where b and c are in $\mathbf{Z}$.

   By exercise 32, the set of all such pairs is countably infinite, and so the set of equations of the form $x^2 + bx + c = 0$ is countably infinite.

   Next observe that, by the quadratic formula, each equation $x^2 + bx + c = 0$ has at most two solutions (which may be complex numbers):

   $$x = \frac{-b + \sqrt{b^2 - 4c}}{2} \quad \text{and} \quad x = \frac{-b - \sqrt{b^2 - 4c}}{2}.$$

   Let

   $$R_1 = \left\{ x \mid x = \frac{-b + \sqrt{b^2 - 4c}}{2} \quad \text{for some integers } b \text{ and } c \right\},$$

   $$R_2 = \left\{ x \mid x = \frac{-b - \sqrt{b^2 - 4c}}{2} \quad \text{for some integers } b \text{ and } c \right\},$$

   and $R = R_1 \cup R_2$. Then R is the set of all solutions of equations of the form $x^2 + bx + c = 0$ where b and c are integers.

Define functions F_1 and F_2 from the set of equations of the form $x^2 + bx + c = 0$ to the sets R_1 and R_2 as follows:

$$F_1(x^2 + bx + c = 0) = \frac{-b + \sqrt{b^2 - 4c}}{2} \quad \text{and} \quad F_2(x^2 + bx + c = 0) = \frac{-b - \sqrt{b^2 - 4c}}{2}.$$

Then F_1 and F_2 are onto functions defined on countably infinite sets, and so, by exercise 27, R_1 and R_2 are countable. Since any union of two countable sets is countable (exercise 31), $R = R_1 \cup R_2$ is countable.

34. *Hint for Solution 1:* Define a function $f \colon \mathscr{P}(S) \to T$ as follows: For each subset A of S, let $f(A) = \chi_A(x)$, the *characteristic function* of A, where

$\chi_A \colon S \to \{0, 1\}$ is defined by the rule

$$\chi_A(x) = \begin{cases} 1 & \text{if } x \in A \\ 0 & \text{if } x \notin A \end{cases}.$$

Show that f is one-to-one (for all subsets A_1 and A_2 in S, if $\chi_{A_2} = \chi_{A_2}$ then $A_1 = A_2$) and that f is onto (given any function $g \colon S \to \{0, 1\}$, there is a subset A of S such that $g = \chi_A$).

Hint for Solution 2: Define $H \colon T \to \mathscr{P}(S)$ by letting $H(f) = \{\, x \in S \mid f(x) = 1 \,\}$. Show that H is a one-to-one correspondence.

35. Partial proof (by contradiction): Suppose not. Suppose there is a one-to-one, onto function $f \colon S \to \mathscr{P}(S)$. Let $A = \{x \in S \mid x \notin f(x)\}$.

Then $A \in \mathscr{P}(S)$, and since f is onto, there is a $z \in S$ such that $A = f(z)$. *[Now derive a contradiction!]*

36. Proof: Let B be the set of all functions from $\mathbf{Z}^+$ to $\{0, 1\}$ and let D be the set of all functions from $\mathbf{Z}^+$ to $\{0, 1, 2, 3, 4, 5, 6, 7, 8, 9\}$.

Elements of B can be represented as infinite sequences of 0's and 1's (for instance, 01101010110...) and elements of D can be represented as infinite sequences of digits from 0 to 9 inclusive (for instance, 20775931124...).

We define a function $H \colon B \to D$ as follows: For each function f in B, consider the representation of f as an infinite sequence of 0's and 1's.

Such a sequence is also an infinite sequence of digits chosen from 0 to 9 inclusive (one formed without using 2,3,...,9), which represents a function in D. We define this function to be $H(f)$.

More formally, for each $f \in B$, let $H(f)$ be the function in D defined by the rule $H(f)(n) = f(n)$ for all $n \in \mathbf{Z}^+$.

It is clear from the definition that H is one-to-one.

We define a function $K \colon D \to B$ as follows: For each function g in D, consider the representation of g as a sequence of digits from 0 to 9 inclusive.

Replace each of these digits by its 4-bit binary representation adding leading 0's if necessary to make a full four bits. (For instance, 2 would be replaced by 0010.)

The result is an infinite sequence of 0's and 1's, which represents a function in B. This function is defined to be $K(g)$.

Note that K is one-to-one because if $g_1 \neq g_2$ then the sequences representing g_1 and g_2 must have different digits in some position m, and so the corresponding sequences of 0's and 1's will differ in at least one of the positions $4m - 3, 4m - 2, 4m - 1$, or $4m$, which are the locations of the 4-bit binary representations of the digits in position m.

It can be shown that whenever there are one-to-one functions from one set to a second and from the second set back to the first, then the two sets have the same cardinality. This fact is known as the Schröder-Bernstein theorem after its two discoverers. For a proof see, for example, *Set Theory and Metric Spaces* by Irving Kaplansky, *A Survey of Modern Algebra*, Third Edition, by Garrett Birkhoff and Saunders MacLane, *Naive Set Theory* by Paul Halmos, or *Topology* by James R. Munkres. The above discussion shows that there are one-to-one functions from B to D and from D to B, and hence by the Schröder-Bernstein theorem the two sets have the same cardinality.

37. *Hint:* Since A and B are countable, their elements can be listed as $A: a_1, a_2, a_3, \ldots$ and $B: b_1, b_2, b_3, \ldots$. Represent the elements of $A \times B$ in a grid:

$$
\begin{array}{ccc}
(a_1, b_1) & (a_1, b_2) & (a_1, b_3) \ldots \\
(a_2, b_1) & (a_2, b_2) & (a_2, b_3) \ldots \\
(a_3, b_1) & (a_3, b_2) & (a_3, b_3) \ldots \\
\vdots & \vdots & \vdots
\end{array}
$$

Now use a counting method similar to that of Example 7.4.4.

Review Guide: Chapter 7

Definitions: How are the following terms defined?

- function f from a set X to a set Y
- Let f be a function from a set X to a set Y.
  - the domain, co-domain, and range of f
  - the value of f at x, where x is in X
  - the image of x under f, where x is in X
  - the output of f for the input x, where x is in X
  - the image of X under f
  - an inverse image of y, where y is in Y
  - the identity function on a set
  - the image of A, where $A \subseteq X$
  - the inverse image of B, where $B \subseteq Y$
- logarithm with base b of a positive number x and the logarithmic function with base b
- Hamming distance function
- Boolean function
- one-to-one function
- onto function
- exponential function with base b
- one-to-one correspondence
- inverse function
- composition of functions
- cardinality
- countable and uncountable sets.

Functions

- How do you draw an arrow diagram for a function defined on a finite set?
- Given a function defined by an arrow diagram or by a formula, how do you find values of the function, the range of the function, and the inverse image of an element in its co-domain?
- How do you show that two functions are equal?
- What is the relation between a sequence and a function?
- Can you give an example of a function defined on a power set? a function defined on a Cartesian product?
- What is an example of an encoding function? a decoding function?
- If the claim is made that a given formula defines a function from a set X to a set Y, how do you determine that the "function" is not well-defined?

One-to-one and Onto

- How do you show that a function is not one-to-one?
- How do you show that a function defined on an infinite set is one-to-one?
- How do you show that a function is not onto?
- How do you show that a function defined on an infinite set is onto?
- How do you determine if a given function has an inverse function?
- How do you find an inverse function if it exists?

Exponents and Logarithms

- What are the four laws of exponents?
- What are the properties of logarithms that correspond to the laws of exponents?
- How can you use the laws of exponents to derive properties of logarithms?
- How are the logarithmic function with base b and the exponential function with base b related?

Composition of Functions

- How do you compute the composition of two functions?
- What is the composition of a function with its inverse?
- Why is a composition of one-to-one functions one-to-one?
- Why is a composition of onto functions onto?

Applications of Functions

- What is a Hash function?
- How do you show that one set has the same cardinality as another?
- How do you show that a given set is countably infinite? countable?
- How do you show that the set of all positive rational numbers is countable?
- How is the Cantor diagonalization process used to show that the set of real numbers between 0 and 1 is uncountable?
- How do you show that the set of all computer programs in a given computer language is countable?

Fill-in-the-Blank Review Questions

Section 7.1

1. Given a function f from a set X to a set Y, $f(x)$ is _____.
2. Given a function f from a set X to a set Y, if $f(x) = y$, then y is called _____ or _____ or _____ or _____.
3. Given a function f from a set X to a set Y, the range of f (or the image of X under f) is _____.
4. Given a function f from a set X to a set Y, if $f(x) = y$, then x is called _____ or _____.
5. Given a function f from a set X to a set Y, if $y \in Y$, then $f^{-1}(y) = $ _____ and is called _____.
6. Given functions f and g from a set X to a set Y, $f = g$ if, and only if, _____.
7. Given positive real numbers x and b with $b \neq 1$, $\log_b x = $ _____.
8. Given a function f from a set X to a set Y and a subset A of X, $f(A) = $ _____.
9. Given a function f from a set X to a set Y and a subset C of Y, $f^{-1}(C) = $ _____.

Section 7.2

1. If F is a function from a set X to a set Y, then F is one-to-one if, and only if, _____.
2. If F is a function from a set X to a set Y, then F is not one-to-one if, and only if, _____.
3. If F is a function from a set X to a set Y, then F is onto if, and only if, _____.
4. If F is a function from a set X to a set Y, then F is not onto if, and only if, _____.

5. The following two statements are _____:

   $\forall\, u, v \in U$, if $H(u) = H(v)$ then $u = v$.

   $\forall\, u, v \in U$, if $u \neq v$ then $H(u) \neq H(v)$.

6. Given a function $F\colon X \to Y$ and an infinite set X, to prove that F is one-to-one, you suppose that _____ and then you show that _____.

7. Given a function $F\colon X \to Y$ and an infinite set X, to prove that F is onto, you suppose that _____ and then you show that _____.

8. Given a function $F\colon X \to Y$, to prove that F is not one-to-one, you _____.

9. Given a function $F\colon X \to Y$, to prove that F is not onto, you _____.

10. A one-to-one correspondence from a set X to a set Y is a _____ that is _____.

11. If F is a one-to-one correspondence from a set X to a set Y and y is in Y, then $F^{-1}(y)$ is _____.

Section 7.3

1. If f is a function from X to Y', g is a function from Y to Z, and $Y' \subseteq Y$, then $g \circ f$ is a function from _____ to _____, and $(g \circ f)(x)$ _____ for every x in X.

2. If f is a function from X to Y and I_X and I_Y are the identity functions from X to X and Y to Y, respectively, then $f \circ I_X = $ _____ and $I_Y \circ f = $ _____.

3. If f is a one-to-one correspondence from X to Y, then $f^{-1} \circ f = $ _____ and $f \circ f^{-1} = $ _____.

4. If f is a one-to-one function from X to Y and g is a one-to-one function from Y to Z, you prove that $g \circ f$ is one-to-one by supposing that _____ and then showing that _____.

5. If f is an onto function from X to Y and g is an onto function from Y to Z, you prove that $g \circ f$ is onto by supposing that _____ and then showing that _____.

Section 7.4

1. A set is finite if, and only if, _____.

2. To prove that a set A has the same cardinality as a set B you must _____.

3. The reflexive property of cardinality says that given any set A, _____.

4. The symmetric property of cardinality says that given any sets A and B, _____.

5. The transitive property of cardinality says that given any sets A, B, and C, _____.

6. A set is called countably infinite if, and only if, _____.

7. A set is called countable if, and only if, _____.

8. In each of the following, fill in the blank with the word countable or the word uncountable.

   (a) The set of all integers is _____.

   (b) The set of all rational numbers is _____.

   (c) The set of all real numbers between 0 and 1 is _____.

   (d) The set of all real numbers is _____.

9. The Cantor diagonalization process is used to prove that _____.

Answers for Fill-in-the-Blank Review Questions

Section 7.1

1. the unique output element y in Y that is related to x by f

2. the value of f at x; the image of x under f; the output of f for the input x

3. the set of all y in Y such that $f(x) = y$

4. an inverse image of y under f; a preimage of y

5. $\{x \in X \mid f(x) = y\}$; the inverse image of y

6. $f(x) = g(x)$ for every $x \in X$

7. the exponent to which b must be raised to obtain x
   Or: $\log_b y = x \Leftrightarrow b^y = x$

8. $\{y \in Y \mid y = f(x) \text{ for some } x \in A\}$ (*Or:* $\{f(x) \mid x \in A\}$)

9. $\{x \in X \mid f(x) \in C\}$

Section 7.2

1. for all x_1 and x_2 in X, if $F(x_1) = F(x_2)$ then $x_1 = x_2$

2. there exist elements x_1 and x_2 in X such that $F(x_1) = F(x_2)$ and $x_1 \neq x_2$

3. for all y in Y, there exists at least one element x in X such that $f(x) = y$

4. there exists an element y in Y such that for all elements x in X, $f(x) \neq y$

5. logically equivalent ways of expression what it means for H to be a one-to-one function (The second is the contrapositive of the first.)

6. x_1 and x_2 are any *[particular but arbitrarily chosen]* elements in X with the property that $F(x_1) = F(x_2)$; $x_1 = x_2$

7. y is any *[particular but arbitrarily chosen]* element in Y; there exists at least one element x in X such that $F(x) = y$

8. show that there are concrete elements x_1 and x_2 in X with the property that $F(x_1) = F(x_2)$ and $x_1 \neq x_2$

9. show that there is a concrete element y in Y with the property that $F(x) \neq y$ for any element x in X

10. function from X to Y; both one-to-one and onto

11. the unique element x in X such that $F(x) = y$ (in other words, $F^{-1}(y)$ is the unique preimage of y in X)

Section 7.3

1. X; Z; $g(f(x))$

2. f; f

3. I_X; I_Y

4. x_1 and x_2 are any *[particular but arbitrarily chosen]* elements in X with the property that $(g \circ f)(x_1) = (g \circ f)(x_2)$; $x_1 = x_2$

5. z is any *[particular but arbitrarily chosen]* element in Z; there exists at least one element x in X such that $(g \circ f)(x) = z$

Section 7.4

1. it is the empty set or there is a one-to-one correspondence from $\{1, 2, \ldots, n\}$ to it, where n is a positive integer

2. show that there exists a function from A to B that is one-to-one and onto;
   Or: show that there exists a one-to-one correspondence from A to B

3. A has the same cardinality as A

4. if A has the same cardinality as B, then B has the same cardinality as A

5. if A has the same cardinality as B and B has the same cardinality as C, then A has the same cardinality as C

6. it has the same cardinality as the set of all positive integers

7. it is finite or countably infinite

8. countable; countable; uncountable; uncountable

9. the set of all real numbers between 0 and 1 is uncountable

Chapter 8: Relations

The first section of this chapter focuses on understanding equivalent ways to specify and represent relations, both finite and infinite. In Section 8.2 the reflexivity, symmetry, and transitivity properties of relations are introduced and explored, and in Section 8.3 equivalence relations are discussed. As you work on these sections, you will frequently use the fact that the same proof outlines are used to prove and disprove universal conditional statements no matter what their mathematical context.

Section 8.4 deepens and extends the discussion of congruence relations in Sections 8.2 and 8.3 through applications to modular arithmetic and cryptography. The section is designed to make it possible to give you meaningful practice with RSA cryptography without having to spend several weeks on the topic. After a brief introduction to the idea of cryptography, the first part of the section is devoted to helping develop the facility with modular arithmetic that is needed to perform the computations for RSA cryptography, especially finding least positive residues of integers raised to large positive powers and using the Euclidean algorithm to compute positive inverses modulo a number. Proofs of the underlying mathematical theory are left to the end of the section.

Partial order relations, discussed in Section 8.5, are reflexive, antisymmetric, and transitive. Projects, such as building a product in a factory, or constructing a compiler for a computer, or analyzing the dataflow in a complex system, involve a variety of tasks, some of which can be performed in parallel. The use of partial order relations helps optimize the efficiency of such projects. They are also important in the study of advanced topics in logic and theoretical computer science.

Section 8.1

1. **a.** $0\ E\ 0$ because $0 - 0 = 0 = 2 \cdot 0$, so $2 \mid (0 - 0)$.

   $5\ \cancel{E}\ 2$ because $5 - 2 = 3$ and $3 \neq 2k$ for any integer k, so $2 \nmid (5 - 2)$.

   $(6, 6) \in E$ because $6 - 6 = 0 = 2 \cdot 0$, so $2 \mid (6 - 6)$.

   $(-1, 7) \in E$ because $-1 - 7 = -8 = 2 \cdot (-4)$, so $2 \mid (-1 - 7)$.

2. *Hint:* To prove a statement of the form $p \leftrightarrow (q \vee r)$, you need to prove both (1) $p \to (q \vee r)$ and (2) $(q \vee r) \to p$. The easiest way to prove $p \to (q \vee r)$ is to prove the logically equivalent statement form $(p \wedge \sim q) \to r$. And the easiest way to prove $(q \vee r) \to p$ is to prove the logically equivalent statement form $(q \to p) \wedge (r \to p)$. In this case, suppose m and n are any integers, and let p be "$m - n$ is even," let q be "both m and n are even," and let r be "both m and n are odd."

3. **a.** $10\ T\ 1$ because $10 - 1 = 9 = 3 \cdot 3$, and so $3 \mid (10 - 1)$.

   $1\ T\ 10$ because $1 - 10 = -9 = 3 \cdot (-3)$, and so $3 \mid (1 - 10)$.

   $2\ T\ 2$ because $2 - 2 = 0 = 3 \cdot 0$, and so $3 \mid (2 - 2)$.

   $8\ \cancel{T}\ 1$ because $8 - 1 = 7 \neq 3k$, for any integer k. So $3 \nmid (8 - 1)$.

   b. *One possible answer:* $3, 6, 9, -3, -6$

   c. *One possible answer:* $4, 7, 10, -2, -5$

   d. *One possible answer:* $5, 8, 11, -1, -4$

   e. <u>Theorem</u>: 1. All integers of the form $3k$ are related by T to 0.

           2. All integers of the form $3k + 1$ are related by T to 1.

           3. All integers of the form $3k + 2$ are related by T to 2.

   <u>Proof for part (2)</u>: Let n be any integer of the form $n = 3k + 1$ for some integer k. By substitution, $n - 1 = (3k + 1) - 1 = 3k$, and so by definition of divisibility, $3 \mid (n - 1)$. Hence by definition of T, $n\ T\ 1$.

The proofs for parts (1) and (3) are identical to the proof for part (2) with 0 and 2, respectively, substituted in place of 1.

4. **a.** Yes, because 15 and 25 are both divisible by 5, which is prime.

   b. No, because the prime factors of 22 are 2 and 11 and the only prime factor of 27 is 3, and so 22 and 27 have no common prime factor.

5. **a.** Yes, because both $\{a, b\}$ and $\{b, c\}$ have two elements.

6. **a.** No, because $\{a\} \cap \{c\} = \emptyset$.

   b. Yes, because $\{a, b\} \cap \{b, c\} = \{b\} \neq \emptyset$.

   c. Yes, because $\{a, b\} \cap \{a, b, c\} = \{a, b\} \neq \emptyset$.

7. **a.** Yes. $1 \, R \, (-9) \Leftrightarrow 5 \mid (1^2 - (-9)^2)$. But $1^2 - (-9)^2 = 1 - 81 = -80$, and $5 \mid (-80)$ because $-80 = 5 \cdot (-16)$.

8. **a.** Yes, because both $abaa$ and $abba$ have the same first two characters ab.

   b. No, because the first two characters of $aabb$ are different from the first two characters of $bbaa$.

9. **a.** Yes, because the sum of the characters in 0121 is 4 and the sum of the characters in 2200 is also 4.

   b. No, because the sum of the characters in 1011 is 3, whereas the sum of the characters in 2101 is 4.

   c. No, because the sum of the characters in 2212 is 7, whereas the sum of the characters in 2121 is 6.

   d. Yes, because the sum of the characters in 1220 is 5 and the sum of the characters in 2111 is also 5.

10. $R = \{(3, 4), (3, 5), (3, 6), (4, 5), (4, 6), (5, 6)\}$ $R^{-1} = \{(4, 3), (5, 3), (6, 3), (5, 4), (6, 4), (6, 5)\}$

12. **a.** No. If $F: X \to Y$ is not onto, then F^{-1} fails to be defined on all of Y. In other words, there is an element y in Y such that $(y, x) \notin F^{-1}$ for any $x \notin X$. Consequently, F^{-1} does not satisfy property (1) of the definition of function.

   b. No. If $F: X \to Y$ is not one-to-one, then there exist x_1 and x_2 in X and y in Y such that $(x_1, y) \in F$ and $(x_2, y) \in F$ and $x_1 \neq x_2$. But this implies that there exist x_1 and x_2 in X and y in Y such that $(y, x_1) \in F^{-1}$ and $(y, x_2) \in F^{-1}$ and $x_1 \neq x_2$. Consequently, F^{-1} does not satisfy property (2) of the definition of function.

13.

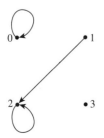

15.

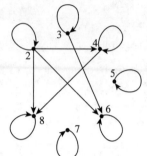

16. *Hint:* See Example 8.1.6.

18.

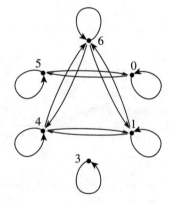

19. $A \times B = \{(2,6), (2,8), (2,10), (4,6), (4,8), (4,10)\}$
   $R = \{(2,6), (2,8), (2,10), (4,8)\}$
   $S = \{(2,6), (4,8)\}$
   $R \cup S = R, \quad R \cap S = S$

21.

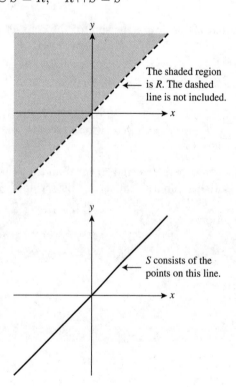

The shaded region is R. The dashed line is not included.

S consists of the points on this line.

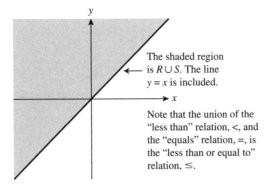

The shaded region is $R \cup S$. The line $y = x$ is included.

Note that the union of the "less than" relation, $<$, and the "equals" relation, $=$, is the "less than or equal to" relation, $\leq$.

The graph of the intersection of R and S is obtained by finding the set of all points common to both graphs. But there are no points for which both $x < y$ and $x = y$. Hence $R \cap S = \emptyset$ and the graph consists of no points at all.

24. **a.** 574329   Tak Kurosawa
    011985   John Schmidt

   b. 466581   Mary Lazars
    778400   Jamal Baskers

Section 8.2

1. **a.** R_1 :

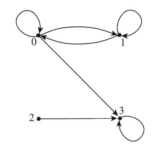

   b. R_1 is not reflexive: $2 \not{R_1} 2$.

   c. R_1 is not symmetric: $2 R_1 3$ but $3 \not{R_1} 2$.

   d. R_1 is not transitive: $1 R_1 0$ and $0 R_1 3$ but $1 \not{R_1} 3$.

3. **a.** R_3 : $0 \bullet$ $\bullet 1$

   b. R_3 is not reflexive: $(0, 0) \notin R_3$.

   c. R_3 is symmetric. (If R_3 were not symmetric, there would be elements x and y in $A = \{0, 1, 2, 3\}$ such that $(x, y) \in R_3$ but $(y, x) \in R_3$. It is clear by inspection that no such elements exist.)

   d. R_3 is not transitive: $(2, 3) \in R_3$ and $(3, 2) \in R_3$ but $(2, 2) \notin R_3$.

6. **a.** R_6: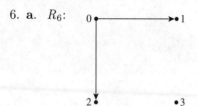

b. R_6 is not reflexive: $(0, 0) \notin R_6$.

c. R_6 is not symmetric: $(0, 1) \in R_6$ but $(1, 0) \notin R_6$.

d. R_6 is transitive by default. (If R_6 were not transitive, there would be elements x, y, and z in $\{0, 1, 2, 3\}$ such that $(x, y) \in R_6$ and $(y, z) \notin R_6$ and $(x, z) \notin R_6$. It is clear by inspection that no such elements exist.)

7. **a.** R_7: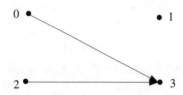

b. R_7 is not reflexive because, for example, $(0, 0) \notin R_7$.

c. R_7 is not symmetric because, for example, $(0, 3) \in R_7$ but $(3, 0) \notin R_7$.

d. R_7 is transitive (by default).

8. **a.** R_8:

b. R_8 is not reflexive because, for example, $(2, 2) \notin R_8$.

c. R_8 is symmetric.

d. R_8 is transitive.

9. ***R is reflexive:*** R is reflexive $\Leftrightarrow$ for every real number x, $x \, R \, x$. By definition of R, this means that for every real number x, $x \geq x$. In other words, for every real number x, $x > x$ or $x = x$, which is true.

R is not symmetric: R is symmetric $\Leftrightarrow$ for all real numbers x and y, if $x \, R \, y$ then $y \, R \, x$. By definition of R, this means that for all real numbers x and y, if $x \geq y$ then $y \geq x$. The following counterexample shows that this is false. $x = 1$ and $y = 0$. Then $x \geq y$, but $y \not\geq x$ because $1 \geq 0$ and $0 \not\geq 1$.

R is transitive: R is transitive $\Leftrightarrow$ for all real numbers x, y, and z, if $x \, R \, y$ and $y \, R \, z$ then $x \, R \, z$. By definition of R, this means that for all real numbers x, y, and z, if $x \geq y$ and $y \geq z$ then $x \geq z$. This is true by definition of $\geq$ and the transitive property of order for the real numbers. (See Appendix A, T18.)

11. ***D is reflexive:*** For D to be reflexive means that for every real number x, $x \, D \, x$. By definition of D, this means that for every real number x, $xx = x^2 \geq 0$, which is true.

D is symmetric: For D to be symmetric means that for all real numbers x and y, if $x\,D\,y$ then $y\,D\,x$. By definition of D, this means that for all real numbers x and y, if $xy \geq 0$ then $yx \geq 0$, which is true by the commutative law of multiplication.

D is not transitive: For D to be transitive means that for all real numbers x, y, and z, if $x\,D\,y$ and $y\,D\,z$ then $x\,D\,z$. By definition of D, this means that for all real numbers x, y, and z, if $xy \geq 0$ and $yz \geq 0$ then $xz \geq 0$. This is false because there exist real numbers x, y, and z such that $xy \geq 0$ and $yz \geq 0$ but $xz \not\geq 0$. As a counterexample, let $x = 1$, $y = 0$, and $z = -1$. Then $x\,D\,y$ and $y\,D\,z$ because $1 \cdot 0 \geq 0$ and $0 \cdot (-1) \geq 0$. But $x\,\not{D}\,z$ because $1 \cdot (-1) \not\geq 0$.

12. *E is reflexive:* *[We must show that for every integer m, $m\,E\,m$.]* Suppose m is any integer. Since $m - m = 0$ and $4 \mid 0$, we have that $4 \mid (m - m)$. Consequently, $m\,E\,m$ by definition of E.

   E is symmetric: *[We must show that for all integers m and n, if $m\,E\,n$ then $n\,E\,m$.]* Suppose m and n are any integers such that $m\,E\,n$. By definition of E, this means that $4 \mid (m - n)$, and so, by definition of divisibility, $m - n = 4r$ for some integer r.

   Now $\quad n - m = -(m - n) \quad$ and hence, by substitution, $\quad n - m = -(4r) = 4(-r)$.

   It follows that $4 \mid (n - m)$ by definition of divisibility (since $-r$ is an integer), and thus $n\,E\,m$ by definition of E.

   E is transitive: *[We must show that for all integers m, n, and p if $m\,E\,n$ and $n\,E\,p$ then $m\,E\,p$.]* Suppose m, n, and p are any integers such that $m\,E\,n$ and $n\,E\,p$. By definition of E this means that $4 \mid (m - n)$ and $4 \mid (n - p)$, and so, by definition of divisibility, $m - n = 4r$ for some integer r and $n - p = 4s$ for some integer s. Now

   $$m - p = (m - n) + (n - p) \quad \text{and hence, by substitution,} \quad m - p = 4r + 4s = 4(r + s).$$

   It follows that $4 \mid (m - p)$ by definition of divisibility (since $r + s$ is an integer), and thus $m\,E\,p$ by definition of E.

15. *D is reflexive:* *[We must show that for every positive integer m, $m\,D\,m$.]* Suppose m is any positive integer. Since $m = m \cdot 1$, by definition of divisibility $m \mid m$. Hence $m\,D\,m$ by definition of D.

   D is not symmetric: For D to be symmetric would mean that for all positive integers m and n, if $m\,D\,n$ then $n\,D\,m$. By definition of divisibility, this would mean that for all positive integers m and n, if $m \mid n$ then $n \mid m$. A counterexample shows that this is false. Let $m = 2$ and $n = 4$. Then $m \mid n$ because $2 \mid 4$ but $n \nmid m$ because $4 \nmid 2$.

   D is transitive: To prove transitivity of D, we must show that for all positive integers m, n, and p, if $m\,D\,n$ and $n\,D\,p$ then $m\,D\,p$. By definition of D, this means that for all positive integers m, n, and p, if $m \mid n$ and $n \mid p$ then $m \mid p$. This is true by Theorem 4.4.3 (the transitivity of divisibility).

18. *Hint:* Q is reflexive, symmetric, and transitive.

20. *E is reflexive:* **E** is reflexive $\Leftrightarrow$ for every subset A of X, $A\,\mathbf{E}\,A$. By definition of **E**, this means that for every subset A of X, A has the same number of elements as A, which is true.

   E is symmetric: **E** is symmetric $\Leftrightarrow$ for all subsets A and B of X, if $A\,\mathbf{E}\,B$ then $B\,\mathbf{E}\,A$. By definition of **E**, this means that if A has the same number of elements as B, then B has the same number of elements as A, which is true.

   E is transitive: **E** is transitive $\Leftrightarrow$ for all subsets A, B, and C of X, if $A\,\mathbf{E}\,B$ and $B\,\mathbf{E}\,C$ then $A\,\mathbf{E}\,C$. By definition of **E**, this means that for all subsets, A, B, and C of X, if A has the same number of elements as B and B has the number of elements as C, then A has the same number of elements as C, which is true.

21. For each set X, let $N(X)$ be the number of elements in X.

    L *is not reflexive*: **L** is reflexive $\Leftrightarrow$ for every set $A \in \mathscr{P}(X)$, A **L** A. By definition of **L** this means that for every set A in $\mathscr{P}(X)$, $N(A) < N(A)$. In fact, this is false for every set in $\mathscr{P}(X)$ because, by the trichotomy law for the real numbers (Appendix A, T17), no real number is less than itself. As a concrete example, let $A = \emptyset$. Then $N(A) = 0$, and 0 is not less than 0.

    L *is not symmetric*: For **L** to be symmetric would mean that for all sets A and B in $\mathscr{P}(X)$, if A **L** B then B **L** A. By definition of **L**, this would mean that for all sets A and B in $\mathscr{P}(X)$, if $N(A) < N(B)$, then $N(B) < N(A)$. This is false for all sets A and B in $\mathscr{P}(X)$ because, by the trichotomy law for the real numbers (Appendix A, T17), if one real number is less than a second, the second is not less than the first. As a concrete example, take $A = \emptyset$ and $B = \{a\}$. Then $N(A) = 0$ and $N(B) = 1$. Then A is related to B by **L** (since $0 < 1$), but B is not related to A by **L** (since $1 \not< 0$).

    L *is transitive*: To prove transitivity of **L**, we must show that for all sets A, B, and C in $\mathscr{P}(X)$, if A **L** B and B **L** C then A **L** C. By definition of **L** this means that for all sets A, B, and C in $\mathscr{P}(X)$, if $N(A) < N(B)$ and $N(B) < N(C)$, then $N(A) < N(C)$. This is true by the transitivity property of order (Appendix A, T18).

23. **S *is reflexive:*** **S** is reflexive $\Leftrightarrow$ for every subset A of X, A **S** A. By definition of **S**, this means that for every subset A of X, $A \subseteq A$. This is true because every set is a subset of itself.

    S *is symmetric:* **S** $\Leftrightarrow$ for all subsets A and B of X, if A **S** B then B **S** A. By definition of **S**, this means that for all subsets A and B of X, if $A \subseteq B$ then $B \subseteq A$. This is false because it is assumed that $X \neq \emptyset$ and so there is an element, say a, in X. As a counterexample, take $A = \emptyset$ and $B = \{a\}$. Then $A \subset B$ but $B \not\subseteq A$.

    S *is transitive:* **S** is transitive $\Leftrightarrow$ for all subsets A, B, and C of X, if A **S** B and B **S** C, then A **S** C. By definition of **S**, this means that for all subsets A, B, and C of X, if $A \subseteq B$ and $B \subseteq C$ then $A \subseteq C$, which is true by the transitive property of subsets (Theorem 6.2.1 (3)).

24. **U *is not reflexive***: **U** is reflexive $\Leftrightarrow$ for every set A in $\mathscr{P}(X)$, A **U** A. By definition of **U** this means that for every set A in $\mathscr{P}(X)$, $A \neq A$. This is false for every set in $\mathscr{P}(X)$. For instance, let $A = \emptyset$. It is not true that $\emptyset \neq \emptyset$.

    U *is symmetric*: **U** is symmetric $\Leftrightarrow$ for all sets A and B in $\mathscr{P}(X)$, if A **U** B then B **U** A. By definition of **U**, this means that for all sets A and B in $\mathscr{P}(X)$, if $A \neq B$, then $B \neq A$. But this is true.

    U *is not transitive*: **U** is transitive $\Leftrightarrow$ for all sets A, B, and C in $\mathscr{P}(X)$, if A **U** B and B **U** C then A **U** C. By definition of **U** this means that for all sets A, B, and C in $\mathscr{P}(X)$, if $A \neq B$ and $B \neq Z$, then $A \neq C$. But this is false as the following counterexample shows. Since $X \neq \emptyset$, there exists an element x in X. Let $A = \{x\}$, $B = \emptyset$, and $C = \{x\}$. Then $A \neq B$ and $B \neq Z$, but $A = C$.

25. **R *is reflexive:*** Suppose s is any string in A. Then s R s because s has the same first two characters as s.

    R *is symmetric:* Suppose s and t are any strings in A such that s R t. By definition of R, s has the same first two characters as t. It follows that t has the same first two characters as s, and so t R s.

    R *is transitive:* Suppose s, t, and u are any strings in A such that s R t and t R u. By definition of R, s has the same first two characters as t and t has the same first two characters as u. It follows that s has the same two characters as u, and so s R u.

27. **I *is reflexive:*** *[We must show that for every statement p, p **I** p.]* Let p be any statement, and consider the statement $p \to p$. The only way a conditional statement can be false is for its hypothesis to be true and its conclusion false. Both the hypothesis and the conclusion in

$p \to p$ have the same truth value. Thus it is impossible for $p \to p$ to be false, and so it must be true.

I *is not symmetric:* **I** is symmetric $\Leftrightarrow$ for all statements p and q, if p **I** q then q **I** p. By definition of **I**, this means that for all statements p and q, if $p \to q$ then $q \to p$. This is false: there are statements p and q such that $p \to q$ is true and $q \to p$ is false. For instance, let p be "10 is divisible by 4" and let q be "10 is divisible by 2" Then $p \to q$ is "If 10 is divisible by 4, then 10 is divisible by 2," which is true because its hypothesis, p, is false. On the other hand, $q \to p$ is "If 10 is divisible by 2, then 10 is divisible by 4." This is false because its hypothesis, q, is true and its conclusion, p, is false.

I *is transitive:* [We must show that for all statements p, q, and r, if p **I** q and q **I** r then p **I** r.] Suppose p, q, and r are statements such that p **I** q and q **I** r. By definition of **I**, this means that $p \to q$ and $q \to r$ are both true. By transitivity of if-then (Example 2.3.6 and exercise 20 of Section 2.3), we can conclude that $p \to r$ is true. Hence, by definition of **I**, p **I** r.

28. **F** *is reflexive:* **F** is reflexive $\Leftrightarrow$ for every element (x, y) in **R** $\times$ **R**, (x, y) **F** (x, y). By definition of **F**, this means that for every element (x, y) in **R** $\times$ **R**, $x = x$, which is true.

F *is symmetric:* [We must show that for all elements (x_1, y_1) and (x_2, y_2) in **R** $\times$ **R**, if (x_1, y_1) **F** (x_2, y_2) then (x_2, y_2) **F** (x_1, y_1).] Suppose (x_1, y_1) and (x_2, y_2) are elements of **R** $\times$ **R** such that (x_1, y_1) **F** (x_2, y_2). By definition of **F**, this means that $x_1 = x_2$. By symmetry of equality, $x_2 = x_1$. Thus, by definition of **F**, (x_2, y_2) **F** (x_1, y_1).

F *is transitive:* [We must show that for all elements (x_1, y_1), (x_2, y_2), and (x_3, y_3) in **R** $\times$ **R**, if (x_1, y_1) **F** (x_2, y_2) and (x_2, y_2) **F** (x_3, y_3) then (x_1, y_1) **F** (x_3, y_3).] Suppose (x_1, y_1), (x_2, y_2), and (x_3, y_3) are elements of **R** $\times$ **R** such that (x_1, y_1) **F** (x_2, y_2) and (x_2, y_2) **F** (x_3, y_3). By definition of **F**, this means that $x_1 = x_2$ and $x_2 = x_3$. By transitivity of equality, $x_1 = x_3$. Hence, by definition of **F**, (x_1, y_1) **F** (x_3, y_3).

30. **R** *is reflexive*: R is reflexive $\Leftrightarrow$ for every point p in A, $p\,R\,p$. By definition of R this means that for every point p in A, p and p both lie on the same half line emanating from the origin. This is true.

R *is symmetric*:: [We must show that for all points p_1 and p_2 in A, if $p_1 R\, p_2$ then $p_2 R\, p_1$.] Suppose p_1 and p_2 are points in A such that $p_1 R\, p_2$. By definition of R this means that p_1 and p_2 lie on the same half line emanating from the origin. But this implies that p_2 and p_1 lie on the same half line emanating from the origin. So by definition of R, $p_2 R\, p_1$.

R *is transitive*: [We must show that for all points p_1, p_2 and p_3 in A, if $p_1 R\, p_2$ and $p_2 R\, p_3$ then $p_1 R\, p_3$.] Suppose p_1, p_2, and p_3 are points in A such that $p_1 R\, p_2$ and $p_2 R\, p_3$. By definition of R, this means that p_1 and p_2 lie on the same half line emanating from the origin and p_2 and p_3 lie on the same half line emanating from the origin.

Since two points determine a line, it follows that both p_1 and p_3 lie on the same half line determined by the origin and p_2. Thus p_1 and p_3 lie on the same half line emanating from the origin. So by definition of R, $p_1 R\, p_3$.

31. **R** *is reflexive:* R is reflexive $\Leftrightarrow$ for every person p in A, $p\,R\,p$. By definition of R, this means that for every person p living in the world today, p lives within 100 miles of p, which is true.

R *is symmetric:* [We must show that for all people p and q in A, if $p\,R\,q$ then $q\,R\,p$.] Suppose p and q are people in A such that $p\,R\,q$. By definition of R, this means that p lives within 100 miles of q. Since this implies that q lives within 100 miles of p, by definition of R, $q\,R\,p$.

R *is not transitive:* R is transitive $\Leftrightarrow$ for all people p, q, and r, if $p\,R\,q$ and $q\,R\,r$ then $p\,R\,r$. This is false. As a counterexample, take p to be an inhabitant of Chicago, Illinois, q an inhabitant of Kankakee, Illinois, and r an inhabitant of Champaign, Illinois. Then $p\,R\,q$

because Chicago is less than 100 miles from Kankakee, and $q\,R\,r$ because Kankakee is less than 100 miles from Champaign, but $p\,\cancel{R}\,r$ because Chicago is not less than 100 miles from Champaign.

33. **R is not reflexive**: R is reflexive $\Leftrightarrow$ for every line l in A, $l\,R\,l$. By definition of R this means that for every line l in the plane, l is perpendicular to itself. This is false for every line in the plane since no line is perpendicular to itself.

    R is symmetric: *[We must show that for all lines l_1 and l_2 in A, if $l_1R\,l_2$ then $l_2R\,l_1$.]* Suppose l_1 and l_2 are lines in A such that $l_1R\,l_2$. By definition of R this means that l_1 is perpendicular to l_2. In this case, l_2 is perpendicular to l_1, and so, by definition of R, $l_2R\,l_1$.

    R is not transitive: R is transitive $\Leftrightarrow$ for all lines l_1, l_2, and l_3 in A, if $l_1R\,l_2$ and $l_2R\,l_3$ then $l_1R\,l_3$. A counterexample shows that this is false. For instance, let l_1 and l_3 be the horizontal axis and let l_2 be the vertical axis. Then $l_1R\,l_2$ and $l_2R\,l_3$ because the horizontal axis is perpendicular to the vertical axis and the vertical axis is perpendicular to the horizontal axis. But $l_1\,\cancel{R}\,l_3$ because the horizontal axis is not perpendicular to itself.

34. Proof: Suppose R is any reflexive relation on a set A.

    [We must show that R^{-1} is reflexive. To show this, we must show that for every x in A, $x\,R^{-1}$ x.] Given any element x in A, since R is reflexive, $x\,R\,x$, and by definition of relation, this means that $(x,\,x)\in R$. It follows, by definition of the inverse of a relation, that $(x,\,x)\in R^{-1}$, and so, by definition of relation, $x\,R^{-1}\,x$ *[as was to be shown]*.

36. Proof: Suppose R is a transitive relation on a set A. To show that R^{-1} is transitive, we suppose that x, y, and z are any elements of A such that $x\,R^{-1}\,y$ and $y\,R^{-1}\,z$. *[We must show that $x\,R^{-1}\,z$.]* By definition of R^{-1},

    $$y\,R\,x \text{ and } z\,R\,y, \quad \text{or, equivalently,} \quad z\,R\,y \text{ and } y\,R\,x,$$

    and, since R is transitive, $z\,R\,x$. Thus, by definition of R^{-1}, $z\,R^{-1}\,x$ *[as was to be shown]*.

37. **$R\cap S$ is reflexive:** Suppose R and S are reflexive. *[To show that $R\cap S$ is reflexive, we must show that $\forall x\in A$, $(x,x)\in R\cap S$.]* So suppose $x\in A$. Since R is reflexive, $(x,x)\in R$, and since S is reflexive, $(x,x)\in S$. Thus, by definition of intersection, $(x,x)\in R\cap S$ *[as was to be shown]*.

38. *Hint:* The answer is yes.

39. **$R\cap S$ is transitive**: Suppose R and S are transitive. *[To show that $R\cap S$ is transitive, we must show that $\forall x,y,z\in A$, if $(x,y)\in R\cap S$ and $(y,z)\in R\cap S$ then $(x,z)\in R\cap S$.]* So suppose x, y, and z are elements of A such that $(x,y)\in R\cap S$ and $(y,z)\in R\cap S$. By definition of intersection, $(x,y)\in R$, $(x,y)\in S$, $(y,z)\in R$, and $(y,z)\in S$. It follows that $(x,z)\in R$ because R is transitive and $(x,y)\in R$ and $(y,z)\in R$. Also $(x,z)\in S$ because S is transitive and $(x,y)\in S$ and $(y,z)\in S$. Thus by definition of intersection $(x,z)\in R\cap S$.

41. **$R\cup S$ is symmetric**: To show that $R\cup S$ is symmetric, we must show that for all x and y in A, if $(x,y)\in R\cup S$ then $(y,x)\in R\cup S$. So suppose (x,y) is a particular but arbitrarily chosen element in $R\cup S$. *[We must show that $(y,x)\in R\cup S$.]* By definition of union, $(x,y)\in R$ or $(x,y)\in S$. In case $(x,y)\in R$, then $(y,x)\in R$ because R is symmetric, and hence $(y,x)\in R\cup S$ by definition of union. In case $(x,y)\in S$ then $(y,x)\in S$ because S is symmetric, and hence $(y,x)\in R\cup S$ by definition of union. Thus, in both cases, $(y,x)\in R\cup S$ *[as was to be shown]*.

42. **$R\cup S$ is not necessarily transitive**: As a counterexample, let $R=\{(0,1)\}$ and $S=\{(1,2)\}$. Then both R and S are transitive (by default), but $R\cup S=\{(0,1),(1,2)\}$ is not transitive because $(0,1)\in R\cup S$ and $(1,2)\in R\cup S$ but $(0,2)\notin R\cup S$. As another counterexample, let $R=\{(x,y)\in \mathbf{R}\times\mathbf{R}\mid x<y\}$ and let $S=\{(x,y)\in \mathbf{R}\times\mathbf{R}\mid x>y\}$.

Then both R and S are transitive because of the transitivity of order for the real numbers. But $R \cup S = \{(x, y) \in \mathbf{R} \times \mathbf{R} \mid x \neq y\}$ is not transitive because, for instance, $(1, 2) \in R \cup S$ and $(2, 1) \in R \cup S$ but $(1, 1) \notin R \cup S$.

43. R_1 is not irreflexive because $(0, 0) \in R_1$.

   R_1 is not asymmetric because $(0, 1) \in R_1$ and $(1, 0) \in R_1$.

   R_1 is not intransitive because $(0, 1) \in R_1$ and $(1, 0) \in R_1$ and $(0, 0) \in R_1$.

45. R_3 is irreflexive because no element of A is related by R_3 to itself.

   R_3 is not asymmetric because $(2, 3) \in R_3$ and $(3, 2) \in R_3$.

   R_3 is intransitive. To see why, observe that R_3 consists only of $(2, 3)$ and $(3, 2)$. Now $(2, 3) \in R_3$ and $(3, 2) \in R_3$ but $(2, 2) \notin R_3$. Also $(3, 2) \in R_3$ and $(2, 3) \in R_3$ but $(3, 3) \notin R_3$.

48. R_6 is irreflexive because no element of A is related by R_6 to itself.

   R_6 is asymmetric because R_6 consists only of $(0, 1)$ and $(0, 2)$ and neither $(1, 0)$ nor $(2, 0)$ is in R_6 is not intransitive. Let $x = y = z = 0$. Then $(x, y) \in R_6$ and $(y, z) \in R_6$ and $(x, z) \in R_6$.

51. $R^t = R \cup \{(0, 0), (0, 3), (1, 0), (3, 1), (3, 2), (3, 3), (0, 2), (1, 2)\}$

   $= \{(0, 0), (0, 1), (0, 2), (0, 3), (1, 0), (1, 1), (1, 2), (1, 3), (2, 2), (3, 0), (3, 1), (3, 2), (3, 3)\}$

54. ***Algorithm—Test for Reflexivity***

   [The input for this algorithm is a binary relation R defined on a set A, which is represented as the one-dimensional array $a[1], a[2], \ldots, a[n]$. To test whether R is reflexive, a variable called answer is initially set equal to "yes," and each element $a[i]$ of A is examined in turn to see whether it is related by R to itself. If any element is not related to itself by R, then answer is set equal to "no," the while loop is not repeated, and processing terminates.]

   Input: n *[a positive integer]*, $a[1], a[2], \ldots, a[n]$ *[a one-dimensional array representing a set A]*, R *[a subset of $A \times A$]*

   Algorithm Body:

   $i := 1$, *answer* := "yes"

   while (*answer* = "yes" and $i \leq n$)

      if $(a[i], a[i]) \notin R$ **then** *answer* := "no"

      $i := i + 1$

   end while

   Output: *answer [a string]*

Section 8.3

1. **a.** $c \, R \, c$ **b.** $b \, R \, a$, $c \, R \, b$, $e \, R \, d$ **c.** $a \, R \, c$ **d.** $c \, R \, c$, $b \, R \, a$, $c \, R \, b$, $e \, R \, d$, $a \, R \, c$, $c \, R \, a$

2. **a.** $R = \{(0, 0), (0, 2), (2, 0), (2, 2), (1, 1), (3, 3), (3, 4), (4, 3), (4, 4)\}$

3. $[0] = \{0, 4\}$, $[1] = \{1, 3\}$, $[2] = \{2\}$, $[3] = \{1, 3\}$

   There are three distinct equivalence classes: $[0] = \{0, 4\} = [4]$, $[1] = \{1, 3\} = [3]$, $[2] = \{2\}$

5. $[1] = \{1, 5, 9, 13, 17\}$, $[2] = \{2, 6, 10, 14, 18\}$, $[3] = \{3, 7, 11, 15, 19\}$, $[4] = \{4, 8, 12, 16, 20\}$,

   $[5] = \{5, 9, 13, 17, 1\} = [1]$

   There are four distinct equivalence classes: $[1]$, $[2]$, $[3]$, $[4]$

6. $[0] = \{0, 3, -3\}$, $[1] = \{1, 4, -2\}$, $[2] = \{2, 5, -1, -4\}$, $[3] = \{3, 0, -3\} = [3]$

   There are three distinct equivalence classes: $[0] = \{0, 3, -3\} = [3]$, $[1] = \{1, 4, -2\}$, $[2] = \{2, 5, -1, -4\} = [5]$.

7. distinct equivalence classes: $\{(1, 3), (3, 9)\}$, $\{(2, 4), (-4, -8), (3, 6)\}$, $\{(1, 5)\}$

8. distinct equivalence classes: $\{\emptyset\}$, $\{\{a\}, \{b\}, \{c\}\}$, $\{\{a, b\}, \{a, c\}, \{b, c\}\}$, $\{\{a, b, c\}\}$

9. distinct equivalence classes: $\{\emptyset\}$, $\{\{0\}, \{1, -1\}, \{-1, 0, 1\}\}$, $\{\{1\}, \{0, 1\}\}$, $\{\{-1\}, \{0, -1\}\}$

11. distinct equivalence classes:

   $[0] = \{x \in A \mid 4 \text{ divides } (x^2 - 0)\} = \{x \in A \mid 4 \text{ divides } x^2\} = \{-4, -2, 0, 2, 4\}$

   $[1] = \{x \in A \mid 4 \text{ divides } (x^2 - 1^2)\} = \{x \in A \mid 4 \text{ divides } (x^2 - 1)\} = \{-3, -1, 1, 3\}$

12. distinct equivalence classes:

   $[0] = \{x \in A \mid 5 \text{ divides } (x^2 - 0)\} = \{0\}$

   $[1] = \{x \in A \mid 5 \text{ divides } (x^2 - 1)\} = \{x \in A \mid 5 \text{ divides } (x - 1)(x + 1)\} = \{1, -1, 4, -4\}$

   $[2] = \{x \in A \mid 5 \text{ divides } (x^2 - 2^2)\} = \{x \in A \mid 5 \text{ divides } (x - 2)(x + 2)\} = \{2, -2, 3, -3\}$

13. distinct equivalence classes: $\{aaaa, aaab, aaba, aabb\}$, $\{abaa, abab, abba, abbb\}$, $\{baaa, baab, baba, babb\}$, $\{bbaa, bbab, bbba, bbbb\}$

15. **a.** True. $17 - 2 = 15$ and $5 \mid 15$.   **b.** False. $4 - (-5) = 9$ and $7 \nmid 9$

   c. True. $-2 - (-8) = -6$ and $3 \mid (-6)$.   **d.** True. $-6 - 22 = -28$ and $2 \mid (-28)$.

16. **a.** $[7] = [4] = [19]$, $[-4] = [17]$, $[-6] = [27]$

17. **a.** <u>Proof</u>: Suppose that m and n are integers such that $m \equiv n(\text{mod } 3)$. *[We must show that $m \bmod 3 = n \bmod 3$]*

   By definition of congruence, $3 \mid (m - n)$, and so, by definition of divisibility, $m - n = 3a$ for some integer a.

   Let $r = m \bmod 3$. Then $m = 3b + r$ for some integer b.

   Since $m - n = 3a$, it follows by substitution that $m - n = (3b + r) - n = 3a$. Then $(3b + r) = 3a + n$, which implies that $3b - 3a + r = n$, or, equivalently, $n = 3(b - a) + r$.

   Now $b - a$ is an integer and $0 \le r < 3$. So, by definition of *mod*, $n \bmod 3 = r$, which equals $m \bmod 3$.

   Suppose that m and n are integers such that $m \bmod 3 = n \bmod 3$. *[We must show that $m \equiv n(\text{mod } 3)$.]*

   Let $r = m \bmod 3 = n \bmod 3$. Then, by definition of *mod*, $m = 3p + r$ and $n = 3q + r$ for some integers p and q.

   By substitution, $m - n = (3p + r) - (3q + r) = 3(p - q)$.

   Since $p - q$ is an integer, it follows that $3 \mid (m - n)$, and so, by definition of congruence, $m \equiv n \ (\text{mod } 3)$.

18. **a.** *One possible answer:* Let $A = \{1, 2\}$ and $B = \{2, 3\}$. Then $A \ne B$, so A and B are distinct. But A and B are not disjoint since $2 \in A \cap B$.

   b. Let $A_1 = \{1, 2\}$, $A_2 = \{2, 3\}$, $x = 1$, $y = 2$, and $z = 3$. Then both x and y are in A_1 and both y and z are in A_2, but x and z are not both in either A_1 or A_2.

19. **b. (1)** <u>Proof</u>:

S is reflexive because for each student x at a college, x has the same age as x.

S is symmetric because for all students x and y at a college, if x is the same age as y then y is the same age as x.

S is transitive because for all students x, y, and z at a college, if x is the same age as y and y is the same age as z then x is the same age as z.

S is an equivalence relation because it is reflexive, symmetric, and transitive.

(2) There is one equivalence class for each age that is represented in the student body of your college. Each equivalence class consists of all the students of a given age.

20. **(1)** The solution to exercise 12 in Section 8.2 proved that E is reflexive, symmetric, and transitive. Thus E is an equivalence relation.

(2) Observe that for any integer a, the equivalence class of a is

$$\begin{aligned} [a] &= \{x \in Z \mid x\,E\,a\} && \text{by definition of equivalence class} \\ &= \{x \in Z \mid x - a \text{ is divisible by 4}\} && \text{by definition of } E \\ &= \{x \in Z \mid x - a = 4k \text{ for some integer } k\} && \text{by definition of divisibility} \\ &= \{x \in Z \mid x = 4k + a \text{ for some integer } k\} && \text{by algebra.} \end{aligned}$$

Now when any integer a is divided by 4, the only possible remainders are 0, 1, 2, and 3 and no integer has two distinct remainders when it is divided by 4. Thus every integer is contained in exactly one of the following four equivalence classes:

$$\begin{aligned} &\{x \in Z \mid x = 4k \text{ for some integer } k\} \\ &\{x \in Z \mid x = 4k + 1 \text{ for some integer } k\} \\ &\{x \in Z \mid x = 4k + 2 \text{ for some integer } k\} \\ &\{x \in Z \mid x = 4k + 3 \text{ for some integer } k\} \end{aligned}$$

21. **(1)** We first show that for all integers m and n, $m\,R\,n$ if, and only if, m and n are both even or m and n are both odd, or, in other words, if, and only if, m and n have the same parity.

Case 1 (m and n are both even): In this case $7m$ and $5n$ are both even because each is a product of an even integer and an integer (exercise 23, Section 4.2). Thus $7m - 5n$ is even because it is a difference of two even integers (exercise 25, Section 4.2). Hence m R n.

Case 2 (m and n are both odd): In this case $7m$ and $5n$ are both odd because each is a product of two odd integers (exercise 20, Section 4.2). Thus $7m - 5n$ is even because it is a difference of two even integers (exercise 25, Section 4.2). Hence m R n.

Case 3 (m is even and n is odd): In this case $7m$ is even because it is a product of an even integer and an integer, and $5n$ is odd because it is a product of two odd integers (exercise 20, Section 4.2). Thus $7m - 5n$ is odd because it is a difference of an even integer minus an odd integer (Theorem 4.2.2, Section 4.2). Hence m is not related to n by R.

Case 3 (m is odd and n is even): In this case $7m$ is odd because it is a product of two odd integers (exercise 20, Section 4.2), and $5n$ is even because it is a product of an even integer and an integer (exercise 23, Section 4.2). Thus $7m - 5n$ is odd because it is a difference of an odd integer minus an even integer (Theorem 4.2.1, Section 4.2). Hence m is not related to n by R.

R is reflexive because any integer m has the same parity as itself, and so m R m.

R is symmetric because for all integers m and n, m R n is true if, and only if, both m and n have the same parity, which is true if, and only if, both n and m have the same parity, which implies that n R m.

R is transitive: Suppose that m, n, and p are any integers such that m R n and n R p. *[We must show that m R p.]* Then m and n have the same parity and n and p have the same parity. But because n has the same parity as both m and p then m and p have the same parity as each other. Thus $m\ R\ p$ *[as was to be shown]*.

(2) There are two distinct equivalence classes for R: the set of all even integers and the set of all odd integers.

24. **(1)** Proof:

R is reflexive because for each identifier x in A, x has the same memory location as x.

R is symmetric because for all identifiers x and y in A, if x has the same memory location as y then y has the same memory location as x.

R is transitive because for all identifiers x, y, and z in A, if x has the same memory location as y and y has the same memory location as z then x has the same memory location as z.

R is an equivalence relation because it is reflexive, symmetric, and transitive.

(2) There are as many distinct equivalence classes as there are distinct memory locations that are used to store variables during execution of the program. Each equivalence class consists of all variables that are stored in the same location.

25. **(1)** Proof: A is reflexive because each real number has the same absolute value as itself.

A is symmetric because for all real numbers x and y, if $|x| = |y|$ then $|y| = |x|$.

A is transitive because for all real numbers x, y, and z, if $|x| = |y|$ and $|y| = |z|$ then $|x| = |z|$.

A is an equivalence relation because it is reflexive, symmetric, and transitive.

(2) The distinct classes are all sets of the form $\{x, -x\}$, where x is a real number.

26. *Hints*: (1) D is reflexive, symmetric, and transitive. The proofs are very similar to the proofs in exercise 17.

(2) There are two distinct equivalence classes. Note that $m^2 - n^2 = (m - n)(m + n)$ for all integers m and n. In addition, $3 \mid (m - n)$ or $3 \mid (m + n) \Leftrightarrow$ either $m - n = 3r$ or $m + n = 3r$, for some integer r

27. **(1)** Proof:

R *is reflexive*: Suppose m is any integer. Since $m^2 - m^2 = 0$ and $4 \mid 0$, we have that $4 \mid (m^2 - m^2)$. Consequently, $m\ R\ m$ by definition of R.

R *is symmetric*: Suppose m and n are any integers such that $m\ R\ n$. By definition of R this means that $4 \mid (m^2 - n^2)$, and so, by definition of divisibility, $m^2 - n^2 = 4k$ for some integer k. Now $n^2 - m^2 = -(m^2 - n^2)$. Hence by substitution, $n^2 - m^2 = -(4k) = 4 \cdot (-k)$. It follows that $4 \mid (n^2 - m^2)$ by definition of divisibility (since $-k$ is an integer), and thus $n\ R\ m$ by definition of R.

R *is transitive*: Suppose m, n and p are any integers such that $m\ R\ n$ and $n\ R\ p$. By definition of R, this means that $4 \mid (m^2 - n^2)$ and $4 \mid (n^2 - p^2)$, and so, by definition of divisibility, $m^2 - n^2 = 4k$ for some integer k, and $n^2 - p^2 = 4l$ for some integer l. Now $m^2 - p^2 = (m^2 - n^2) + (n^2 - p^2)$. Hence by substitution, $m^2 - p^2 = 4k + 4l = 4(k + l)$. It follows that $4 \mid (m^2 - p^2)$ by definition of divisibility (since $k + l$ is an integer), and thus $m\ R\ p$ by definition of R.

R is an equivalence relation because it is reflexive, symmetric, and transitive.

(2) If m is even, then $m = 2a$ for some integer a, and so $m^2 - 0^2 = (2a)^2 = 4a^2$, which is divisible by 4. Hence $m \in [0]$.

If m is odd, then $m = 2a + 1$ for some integer a, and so $m^2 - 1^2 = (2a+1)^2 - 1 = 4a^2 + 4a + 1 - 1 = 4a^2 + 4a$, which is divisible by 4. Hence $m \in [1]$.

Thus there are two distinct equivalence classes:

$[0] = \{m \in \mathbf{Z} \mid m \text{ is even}\}$ and $[1] = \{m \in \mathbf{Z} \mid m \text{ is odd}\}$.

28. **(1)** <u>Proof</u>: I is reflexive because the difference between each real number and itself is 0, which is an integer.

    I is symmetric because for all real numbers x and y, if $x - y$ is an integer, then $y - x = (-1)(x - y)$, which is also an integer.

    I is transitive because for all real numbers x, y, and z, if $x - y$ is an integer and $y - z$ is an integer, then $x - z = (x - y) + (y - z)$ is the sum of two integers and thus is an integer.

    I is an equivalence relation because it is reflexive, symmetric, and transitive.

    (2) There is one class for each real number x with $0 \leq x < 1$. The distinct classes are all sets of the form $\{y \in \mathbf{R} \mid y = n + x,$ for some integer $n\}$, where x is a real number such that $0 \leq x < 1$.

29. **(1)** <u>Proof</u>: P is reflexive because each ordered pair of real numbers has the same first element as itself.

    P is symmetric for the following reason: Suppose (w, x) and (y, z) are ordered pairs of real numbers such that $(w, x) \, P(y, z)$. Then, by definition of P, $w = y$. Now by the symmetric property of equality, this implies that $y = w$, and so, by definition of P, $(y, z) \, P(w, x)$.

    P is transitive for the following reason: Suppose (u, v), (w, x), and (y, z) are ordered pairs of real numbers such that $(u, v) \, P(w, x)$ and $(w, x) \, P(y, z)$. Then, by definition of P, $u = w$ and $w = y$. It follows from the transitive property of equality that $u = y$. Hence, by definition of P, $(u, v) \, P \, (y, z)$.

    P is an equivalence relation because it is reflexive, symmetric, and transitive.

    (2) There is one equivalence class for each real number. The distinct equivalence classes are all sets of ordered pairs $\{(x, y) \in \mathbf{R} \times \mathbf{R} \mid x = a\}$, for each real number a. Equivalently, the equivalence classes consist of all vertical lines in the Cartesian plane.

30. **(1)** <u>Proof</u>:

    Q is reflexive because each ordered pair has the same second element as itself.

    Q is symmetric for the following reason: Suppose (w, x) and (y, z) are ordered pairs of real numbers such that $(w, x) \, Q \, (y, z)$. Then, by definition of Q, $x = z$. By the symmetric property of equality, this implies that $z = x$, and so, by definition of Q, $(y, z) \, Q \, (w, x)$.

    Q is transitive for the following reason: Suppose (u, v), (w, x), and (y, z) are ordered pairs of real numbers such that $(u, v) \, Q \, (w, x)$ and $(w, x) \, Q \, (y, z)$. Then, by definition of Q, $v = x$ and $x = z$. By the transitive property of equality, this implies that $v = z$, and so, by definition of Q, $(u, v)(y, z) \, Q \, (y, z)$.

    Q is an equivalence relation because it is reflexive, symmetric, and transitive.

    (2) There is one equivalence class for each real number. The distinct equivalence classes are all the sets of the form $\{(x, y) \in \mathbf{R} \times \mathbf{R} \mid y = b\}$ where b is a real number. Equivalently, the distinct equivalence classes are all the vertical lines in the Cartesian plane.

32. *Solution:* There is one equivalence class for each real number t such that $0 \leq t < \pi$. One line in each class goes through the origin, and that line makes an angle of t with the positive horizontal axis.

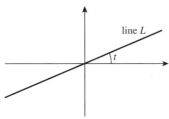

    Alternatively, there is one equivalence class for every possible slope: all real numbers plus "undefined."

33. The distinct equivalence classes can be identified with the points on a geometric figure, called a *torus*, that has the shape of the surface of a doughnut.

Each point in the interior of the rectangle $\{(x, y) \mid 0 < x < 1 \text{ and } 0 < y < 1\}$ is only equivalent to itself.

Each point on the top edge of the rectangle is in the same equivalence class as the point vertically below it on the bottom edge of the rectangle (so we can imagine identifying these points by gluing them together — this gives us a cylinder).

In addition, each point on the left edge of the rectangle is in the same equivalence class as the point horizontally across from it on the right edge of the rectangle (so we can also imagine identifying these points by gluing them together — this brings the two ends of the cylinder together to produce a torus).

34. The programmer's equals method is not an equivalence relation. If points p, q, and r all lie on a straight line with q in the middle, and if p is c units from q and q is c units from r, then p is more than c units from r. Thus the relation is not transitive.

36. <u>Proof</u>: Suppose R is an equivalence relation on a set A and $a \in A$. Because R is an equivalence relation, R is reflexive, and because R is reflexive, each element of A is related to itself by R. In particular, $a \, R \, a$. Hence, by definition of equivalence class, $a \in [a]$.

38. <u>Proof</u>: Suppose R is an equivalence relation on a set A and a, b, and c are elements of A with $b \, R \, c$ and $c \in [a]$. Since $c \in [a]$, then $c \, R \, a$ by definition of equivalence class. Now R is transitive because R is an equivalence relation. Thus, since $b \, R \, c$ and $c \, R \, a$, then $b \, R \, a$. It follows that $b \in [a]$ by definition of equivalence class.

39. <u>Proof</u>: Suppose R is an equivalence relation on a set A, a and b are in A, and $[a] = [b]$.

Since R is reflexive, $a \, R \, a$, and so by definition of class, $a \in [a]$. *[Alternatively, one could reference exercise 36 here.]*

Since $[a] = [b]$, we have that $a \in [b]$ by definition of set equality.

But then by definition of equivalence class, $a \, R \, b$.

40. <u>Proof</u>: Suppose a, b, and x are in A, $a \, R \, b$, and $x \in [a]$. By definition of equivalence class, $x \, R \, a$. So $x \, R \, a$ and $a \, R \, b$, and thus, by transitivity, $x \, R \, b$. Hence $x \in [b]$.

41. *Hint:* To show that $[a] = [b]$, show that $[a] \subseteq [b]$ and $[a] \subseteq [b]$. To show that $[a] = [b]$, show that for every x in A, if $x \in [a]$ then $x \in [b]$.

42. **a.** Suppose $(a, b) \in A$. By commutativity of multiplication for the real numbers, $ab = ba$. But then by definition of R, $(a, b)R(a, b)$, and so R is reflexive.

b. Suppose $(a, b), (c, d) \in A$ and $(a, b)R(c, d)$. By definition of R, $ad = bc$, and so by commutativity of multiplication for the real numbers and symmetry of equality, $cb = da$. But then by definition of R, $(c, d)R(a, b)$, and so R is symmetric.

c. *One possible answer:* $(2, 6), (-2, -6), (3, 9), (-3, -9)$ are all in $[(2, 6)] = [(1, 3)]$.

d. *One possible answer:* $(2, 5), (4, 10), (-2, -5)$, and $(6, 15)$ are all in $[(2, 5)]$.

43. **a.** <u>Proof</u>: Suppose that $(a, b), (a', b'), (c, d)$, and (c', d') are any elements of A such that

$$[(a, b)] = [(a', b')] \text{ and } [(c, d)] = [(c', d')].$$

By definition of R,

$$ab' = ba'(*) \text{ and } cd' = dc'(**).$$

We must show that

$$[(a, b)] + [(c, d)] = [(a', b')] + [(c', d')].$$

By definition of the addition on A, this equation is true if, and only if,

$$[(ad + bc, bd)] = [(a'd' + b'c', b'd')].$$

And, by definition of the relation, this equation is true if, and only if,

$$(ad + bc)b'd' = bd(a'd' + b'c').$$

After multiplying out, this becomes

$$adb'd' + bcb'd' = bda'd' + bdb'c',$$

and regrouping, turns it into

$$(ab')(dd') + (cd')(bb') = (ba')(dd') + (dc')(bb').$$

Substituting the values from (*) and (**) shows that this last equation is true.

c. Suppose that (a, b) is any element of A. We must show that

$$[(a, b)] + [(0, 1)] = [(a, b)].$$

By definition of the addition on A, this equation is true if, and only if,

$$[(a \cdot 1 + b \cdot 0, b \cdot 1)] = [(a, b)].$$

And this last equation is true because $a \cdot 1 + b \cdot 0 = a$ and $b \cdot 1 = b$.

e. Suppose that (a, b) is any element of A. We must show that

$$[(a, b)] + [(-a, b)] = [(-a, b)] + [(a, b)] = [(0, 1)].$$

By definition of the addition on A, this equation is true if, and only if,

$$[(ab + b(-a), bb)] = [(0, 1)],$$

or, equivalently,

$$[(0, bb)] = [(0, 1)].$$

By definition of the relation, this last equation is true if, and only if, $0 \cdot 1 = bb \cdot 0$, which is true.

44. **a.** Let (a, b) be any element of $\mathbf{Z}^+ \times \mathbf{Z}^+$. We must show that $(a, b) \ R \ (a, b)$. By definition of R, this relationship holds if, and only if, $a + b = b + a$. But this equation is true by the commutative law of addition for real numbers. Hence R is reflexive.

c. *Hint:* You will need to show that for any positive integers a, b, c, and d, if $a + d = c + b$ and $c + f = d + e$, then $a + f = b + e$.

d. *One possible answer:* (1, 1), (2, 2), (3, 3), (4, 4), (5, 5)

g. Observe that for any positive integers a and b, the equivalence class of (a, b) consists of all ordered pairs in $\mathbf{Z}^+ \times \mathbf{Z}^+$ for which the difference between the first and second coordinates equals $a - b$. Thus there is one equivalence class for each integer: positive, negative, and zero. Each positive integer n corresponds to the class of $(n + 1, 1)$; each negative integer $-n$ corresponds to the class of $(1, n + 1)$; and zero corresponds to the class $(1, 1)$.

45. The given argument assumes that from the fact that the statement "$\forall x$ in A, if $x \ R \ y$ then $y \ R \ x$" is true, it follows that given any element x in R, there must exist an element y in R such that $x \ R \ y$ and $y \ R \ x$. This is false. For instance, consider the following relation R defined on $A = \{1, 2\}$: $R = \{(1, 1)\}$. This relation is symmetric and transitive, but it is not reflexive. Given $2 \in A$, there is no element y in A such that $(2, y) \in R$. Thus we cannot go on to use symmetry to say that $(y, 2) \in R$ and transitivity to conclude that $(2, 2) \in R$.

47. **c.** Ways and Means .

Section 8.4

1. **a.** ZKHUH VKDOO ZH PHHW

   b. IN THE CAFETERIA

3. **a.** The relation $3 \mid (25 - 19)$ is true because $25 - 19 = 6$ and $3 \mid 6$ (since $6 = 3 \cdot 2$).

   b. By definition of congruence modulo n, to show that $25 \equiv 19 \pmod 3$, one must show that $3/(25 - 19)$. This was verified in part (a).

   c. To show that $25 = 19 + 3k$ for some integer k, one solves the equation for k and checks that the result is an integer. In this case, $k = (25 - 19)/3 = 2$, which is an integer. Thus $25 = 19 + 2 \cdot 3$.

   d. When 25 is divided by 3, the remainder is 1 because $25 = 3 \cdot 8 + 1$. When 19 is divided by 3, the remainder is also 1 because $19 = 3 \cdot 6 + 1$. Thus 25 and 19 have the same remainder when divided by 3.

   e. By definition, 25 *mod* 3 is the remainder obtained when 25 is divided by 3, and 19 *mod* 3 is the remainder obtained when 19 is divided by 3. In part (d) these two numbers were shown to be equal.

6. <u>Proof:</u> Given any integer $n > 1$ and any integer a with $0 \le a < n$, the notation $[a]$ denotes the equivalence class of a for the relation of congruence modulo n.

   We first show that given any integer m, m is in one of the classes $[0], [1], [2], \ldots, [n-1]$.

   The reason is that, by the quotient-remainder theorem, $m = nk + a$, where k and a are integers and $0 \le u < n$, and so, by Theorem 8.4.1, $m \equiv a \pmod n$. It follows by Lemma 8.3.2 that $[m] = [a]$.

   Next we use an argument by contradiction to show that all the equivalence classes $[0], [1], [2], \ldots, [n-1]$ are distinct.

   Suppose this is not the case. That is, suppose a and b are integers with $0 \le a < n$ and $0 \le b < n$, $a \ne b$, and $[a] = [b]$. Without loss of generality, we may assume that $a > b \ge 0$, which implies that $-a < -b \le 0$. Adding a to all parts of the inequality gives $0 < a - b \le a$. By Exercise 8.3.39, $[a] = [b]$ implies that $a \equiv b \pmod n$. Hence, by Theorem 8.4.1, $n \mid (a-b)$, and so, by the quotient-remainder theorem (Theorem 4.4.1), $n \le a - b$. But $a < n$. Thus $n \le a - b \le a < n$, which is contradictory. Therefore the supposition is false, and we conclude that all the equivalence classes $[0], [1], [2], \ldots, [n-1]$ are distinct.

7. **a.** $128 \equiv 2 \pmod 7$ because $128 - 2 = 126 = 7 \cdot 18$, and $61 \equiv 5 \pmod 7$ because $61 - 5 = 56 = 7 \cdot$

   b. $128 + 61 \equiv (2 + 5) \pmod 7$ because $128 + 61 = 189, 2 + 5 = 7$, and $189 - 7 = 182 = 7 \cdot 26$

   c. $128 - 61 \equiv (2 - 5) \pmod 7$ because $128 - 61 = 67, 2 - 5 = -3$, and $67 - (-3) = 70 = 7 \cdot 10$

   d. $128 \cdot 61 \equiv (2 \cdot 5) \pmod 7$ because $128 \cdot 61 = 7808, 2 \cdot 5 = 10$, and $7808 - (10) = 7798 = 7 \cdot 1114$

   e. $128^2 \equiv 2^2 \pmod 7$ because $128^2 = 16384, 2^2 = 4$, and $16384 - 4 = 16380 = 7 \cdot 2340$.

9. **a.** <u>Proof:</u> Suppose $a, b, c, d,$ and n are integers with $n > 1$, $a \equiv c \pmod n$, and $b \equiv d \pmod n$. By Theorem 8.4.1, $a - c = nr$ and $b - d = ns$ for some integers r and s. Then

$$(a + b) - (c + d) = (a - c) + (b - d) = nr + ns = n(r + s).$$

   Now $r + s$ is an integer, and so, by Theorem 8.4.1, $a + b \equiv (c + d) \pmod n$.

   By definition, $a - c = nr$ and $b - d = ns$ for some integers r and s. Then

$$(a - b) - (c - d) = (a - c) - (b - d) = nr - ns = n(r - s).$$

   But $r - s$ is an integer, and so, by definition, $a - b \equiv (c - d) \pmod n$.

12. **a.** <u>Proof (by mathematical induction)</u>: Let the property $P(n)$ be the congruence

$$10^n \equiv 1 \pmod 9. \quad \leftarrow \ P(n)$$

Show that $P(0)$ is true: The left-hand side of $P(0)$ is $10^0 = 1$ and the right-hand side is also 1.

Show that for every integer $k \geq 0$, if $P(k)$ is true then $P(k+1)$ is true: Let k be any integer with $k \geq 0$, and suppose $P(k)$ is true. That is, suppose

$$10^k \equiv 1 \pmod 9.(\ast) \qquad \leftarrow \ \begin{array}{l} P(k) \\ \text{inductive hypothesis} \end{array}$$

We must show that

$$10^{k+1} \equiv 1 \pmod 9. \quad \leftarrow \ P(k+1)$$

By Theorem 8.4.1,
$$10 \ \equiv 1 \pmod 9 (\ast\ast)$$

because $10 - 1 = 9 = 9 \cdot 1$. And by Theorem 8.4.3, we can multiply the left- and right-hand sides of $(\ast)$ and $(\ast\ast)$ to obtain

$$10^k \cdot 10 \equiv 1 \cdot 1 \pmod 9 \text{ or, equivalently, } 10^{k+1} \equiv 1 \pmod 9.$$

Hence $P(k+1)$ is true *[as was to be shown]*.

<u>Alternative Proof</u>: Note that $10^k \equiv 1 \pmod 9$ because $10 - 1 = 9$ and $9 \mid 9$. Thus by Theorem 8.4.3(4), $10^n \equiv 1^n \equiv 1 \pmod 9$.

b. <u>Proof</u>: Suppose a is a positive integer. Then $a = \sum_{k=0}^{n} d_k 10^k$, for some nonnegative integer n and integers d_k where $0 \leq d_k < 10$ for every $k = 1, 2, \ldots, n$. By Theorem 8.4.3,

$$a = \sum_{k=0}^{n} d_k 10^k \equiv \sum_{k=0}^{n} d_k \cdot 1 \ \equiv \sum_{k=0}^{n} d_k \pmod 9$$

because, by part (a), each $10^k \equiv 1 \pmod 9$. Hence, by Theorem 8.4.1, both a and $\sum_{k=0}^{n} d_k$ have the same remainder upon division by 9, and thus if either one is divisible by 9, so is the other.

14. $14^1 \ mod \ 55 = 14$
$14^2 \ mod \ 55 = 196 \ mod \ 55 = 31$
$14^4 \ mod \ 55 = (14^2 \ mod \ 55)^2 \ mod \ 55 = 31^2 \ mod \ 55 = 26$
$14^8 \ mod \ 55 = (14^4 \ mod \ 55)^2 \ mod \ 55 = 26^2 \ mod \ 55 = 16$
$14^{16} \ mod \ 55 = (14^8 \ mod \ 55)^2 \ mod \ 55 = 16^2 \ mod \ 55 = 36$

15. $4^{27} \ mod \ 55 = 14^{16+8+2+1} \ mod \ 55$
$\quad = \{(14^{16} \ mod \ 55)(14^8 \ mod \ 55)(14^2 \ mod \ 55)(14^1 \ mod \ 55)\} \ mod \ 55$
$\quad = (36 \cdot 16 \cdot 31 \cdot 14) \ mod \ 55 = 249984 \ mod \ 55 = 9$

16. Note that $307 = 256 + 32 + 16 + 2 + 1$.
$675^1 \ mod \ 713 = 675$
$675^2 \ mod \ 713 = 18$
$675^4 \ mod \ 713 = 18^2 \ mod \ 713 = 324$
$675^8 \ mod \ 713 = 324^2 \ mod \ 713 = 165$
$675^{16} \ mod \ 713 = 165^2 \ mod \ 713 = 131$
$675^{32} \ mod \ 713 = 131^2 \ mod \ 713 = 49$

$675^{64} \ mod \ 713 = 49^2 \ mod \ 713 = 262$

$675^{128} \ mod \ 713 = 262^2 \ mod \ 713 = 196$

$675^{256} \ mod \ 713 = 196^2 \ mod \ 713 = 627$

Thus

$675^{307} \ mod \ 713 = 675^{256+32+16+2+1} \ mod \ 713$

$= (675^{256} \cdot 675^{32} \cdot 675^{16} \cdot 675^2 \cdot 675^1) \ mod \ 713$

$= (627 \cdot 49 \cdot 131 \cdot 18 \cdot 675) \ mod \ 713 = 3.$

18. Note that $307 = 256 + 32 + 16 + 2 + 1$

$48^1 \ mod \ 713 = 48$

$48^2 \ mod \ 713 = 165$

$48^4 \ mod \ 713 = 165^2 \ mod \ 713 = 131$

$48^8 \ mod \ 713 = 131^2 \ mod \ 713 = 49$

$48^{16} \ mod \ 713 = 49^2 \ mod \ 713 = 262$

$48^{32} \ mod \ 713 = 262^2 \ mod \ 713 = 196$

$48^{64} \ mod \ 713 = 196^2 \ mod \ 713 = 627$

$48^{128} \ mod \ 713 = 627^2 \ mod \ 713 = 266$

$48^{256} \ mod \ 713 = 266^2 \ mod \ 713 = 169$

Thus

$48^{307} \ mod \ 713 = 48^{256+32+16+2+1} \ mod \ 713$

$= (48^{256} 48^{32} 48^{16} 48^2 48^1) \ mod \ 713$

$= (169 \cdot 196 \cdot 262 \cdot 165 \cdot 48) \ mod \ 713 = 12.$

19. The letters in HELLO translate numerically into 08, 05, 12, 12, and 15. By Example 8.4.9, the H is encrypted as 17. To encrypt E, we compute $5^3 \ mod \ 55 = 15$. To encrypt L, we compute $12^3 \ mod \ 55 = 23$. And to encrypt O, we compute $15^3 \ mod \ 55 = 20$. Thus the ciphertext is 17 15 23 23 20. (In practice, individual letters of the alphabet are grouped together in blocks during encryption so that deciphering cannot be accomplished through knowledge of frequency patterns of letters or words.)

21. The letters in EXCELLENT translate numerically into 05, 24, 03, 05,12, 12, 05, 14, 20. The solutions for exercises 19 (in Appendix B) and 20 (above) show that E, L, and C are encrypted as 15, 23, and 27, respectively. To encrypt X, we compute $24^3 \ mod \ 55 = 19$, to encrypt N, we compute $14^3 \ mod \ 55 = 49$, and to encrypt T, we compute $20^3 \ mod \ 55 = 25$. So the ciphertext is 15 19 27 15 23 23 15 49 25.

22. By Example 8.4.10, the decryption key is 27. Thus the residues modulo 55 for 13^{27}, 20^{27}, and 9^{27} must be found and then translated into letters of the alphabet.

Because $27 = 16 + 8 + 2 + 1$, we first perform the following computations:

$13^1 \equiv 13 \ (\text{mod } 55)$	$20^1 \equiv 20 \ (\text{mod } 55)$	$9^1 \equiv 9 \ (\text{mod } 55)$
$13^2 \equiv 4 \ (\text{mod } 55)$	$20^2 \equiv 15 \ (\text{mod } 55)$	$9^2 \equiv 26 \ (\text{mod } 55)$
$13^4 \equiv 4^2 \equiv 16 \ (\text{mod } 55)$	$20^4 \equiv 15^2 \equiv 5 \ (\text{mod } 55)$	$9^4 \equiv 26^2 \equiv 16 \ (\text{mod } 55)$
$13^8 \equiv 16^2 \equiv 36 \ (\text{mod } 55)$	$20^8 \equiv 25^2 \equiv 5 \ (\text{mod } 55)$	$9^8 \equiv 16^2 \equiv 36 \ (\text{mod } 55)$
$13^{16} \equiv 36^2 \equiv 31 \ (\text{mod } 55)$	$20^{16} \equiv 25^2 \equiv 20 \ (\text{mod } 55)$	$9^{16} \equiv 36^2 \equiv 31 \ (\text{mod } 55)$

Then we compute

$13^{27} \ mod \ 55 = (31 \cdot 36 \cdot 4 \cdot 13) \ mod \ 55 = 7,$

$20^{27} \ mod \ 55 = (20 \cdot 25 \cdot 15 \cdot 20) \ mod \ 55 = 15,$

$9^{27} \ mod \ 55 = (31 \cdot 36 \cdot 26 \cdot 9) \ mod \ 55 = 4.$

Finally, because 7, 15, and 4 translate into letters as G, O, and D, we see that the message is GOOD.

24. By Example 8.4.10, the decryption key is 27. Thus the residues modulo 55 for 51^{27}, 14^{27}, 49^{27}, and 15^{27} must be found and then translated into letters of the alphabet. Because $27 = 16 + 8 + 2 + 1$, we first perform the following computations:

$$51^1 \equiv 51 \ (\mathrm{mod} \ 55) \qquad 14^1 \equiv 14 \ (\mathrm{mod} \ 55) \qquad 49^1 \equiv 49 \ (\mathrm{mod} \ 55)$$
$$51^2 \equiv 16 \ (\mathrm{mod} \ 55) \qquad 14^2 \equiv 31 \ (\mathrm{mod} \ 55) \qquad 49^2 \equiv 36 \ (\mathrm{mod} \ 55)$$
$$51^4 \equiv 16^2 \equiv 36 \ (\mathrm{mod} \ 55) \qquad 14^4 \equiv 31^2 \equiv 26 \ (\mathrm{mod} \ 55) \qquad 49^4 \equiv 36^2 \equiv 31 \ (\mathrm{mod} \ 55)$$
$$51^8 \equiv 36^2 \equiv 31 \ (\mathrm{mod} \ 55) \qquad 14^8 \equiv 26^2 \equiv 16 \ (\mathrm{mod} \ 55) \qquad 49^8 \equiv 31^2 \equiv 26 \ (\mathrm{mod} \ 55)$$
$$51^{16} \equiv 31^2 \equiv 26 \ (\mathrm{mod} \ 55) \qquad 14^{16} \equiv 16^2 \equiv 36 \ (\mathrm{mod} \ 55) \qquad 49^{16} \equiv 26^2 \equiv 16 \ (\mathrm{mod} \ 55)$$

Then we compute

$51^{27} \ mod \ 55 = (26 \cdot 31 \cdot 16 \cdot 51) \ mod \ 55 = 6,$

$14^{27} \ mod \ 55 = (36 \cdot 16 \cdot 31 \cdot 14) \ mod \ 55 = 9,$

$49^{27} \ mod \ 55 = (16 \cdot 26 \cdot 36 \cdot 49) \ mod \ 55 = 14.$

In addition, we know from the solution to exercise 23 above that $15^{27} \ mod \ 55 = 5$. And 6, 9, 14, and 5 translate into letters as F, I, N, and E. So the message is FINE.

25. *Hint:* By Theorem 5.2.2, using a in place of r and $n - 1$ in place of n, we have

$$1 + a + a^2 + \cdots + a^{n-1} = \frac{a^n - 1}{a - 1}.$$

Multiplying both sides by $a - 1$ gives

$$a^n - 1 = (a - 1)(1 + a + a^2 + \cdots + a^{n-1}).$$

Step 1: $6664 = 765 \cdot 8 + 544$, and so $544 = 6664 - 765 \cdot 8$

Step 2: $765 = 544 \cdot 1 + 221$, and so $221 = 765 - 544$

Step 3: $544 = 221 \cdot 2 + 102$, and so $102 = 544 - 221 \cdot 2$

Step 4: $221 = 102 \cdot 2 + 17$, and so $17 = 221 - 102 \cdot 2$

Step 5: $102 = 17 \cdot 6 + 0$

Thus $\gcd(6664, 765) = 17$ (which is the remainder obtained just before the final division). Substitute back through steps 4–1 to express 17 as a linear combination of 6664 and 765:

$$\begin{aligned}
17 &= 221 - 102 \cdot 2 \\
&= 221 - (544 - 221 \cdot 2) = 221 \cdot 5 - 544 \cdot 2 \\
&= (765 - 544) \cdot 5 - 544 \cdot 2 = 765 \cdot 5 - 544 \cdot 7 \\
&= 765 \cdot 5 - (6664 - 765 \cdot 8) \cdot 7 = (-7) \cdot 6664 + 61 \cdot 765.
\end{aligned}$$

(When you have finished this final step, it is wise to verify that you have not made a mistake by checking that the final expression really does equal the greatest common divisor.)

27. **Step 1:** $4158 = 1568 \cdot 2 + 1022$, and so $1022 = 4158 - 1568 \cdot 2$

Step 2: $1568 = 1022 \cdot 1 + 546$, and so $546 = 1568 - 1022$

Step 3: $1022 = 546 \cdot 1 + 476$, and so $476 = 1022 - 546$

Step 4: $546 = 476 \cdot 1 + 70$, and so $70 = 546 - 476$

Step 5: $476 = 70 \cdot 6 + 56$, and so $56 = 476 - 70 \cdot 6$

Step 6: $70 = 56 \cdot 1 + 14$, and so $14 = 70 - 56$

Step 7: $56 = 14 \cdot 4 + 0$, and so $\gcd(4158, 1568) = 14$,

which is the remainder obtained just before the final division.

Substitute back through steps 6–1:

$$14 = 70 - 56 = 70 - (476 - 70 \cdot 6) = 70 \cdot 7 - 476$$
$$= (546 - 476) \cdot 7 - 476 = 7 \cdot 546 - 8 \cdot 476$$
$$= 7 \cdot 546 - 8 \cdot (1022 - 546) = 15 \cdot 546 - 8 \cdot 1022$$
$$= 15 \cdot (1568 - 1022) - 8 \cdot 1022 = 15 \cdot 1568 - 23 \cdot 1022$$
$$= 15 \cdot 1568 - 23 \cdot (4158 - 1568 \cdot 2) = 61 \cdot 1568 - 23 \cdot 4158$$

(It is always a good idea to verify that no mistake has been made by verifying that the final expression really does equal the greatest common divisor. In this case, a computation shows that the answer is correct.)

28.

a	330	156	18	12	6
b	156	18	12	6	0
r		18	12	6	0
q		2	8	1	2
s	1	0	1	-8	9
t	0	1	-2	17	-19
u	0	1	-8	9	-26
v	1	-2	17	-19	55
$newu$		1	-8	9	-26
$newv$		-2	17	-19	55
$sa + tb$	330	18	-6	6	6

30. <u>Proof:</u> Suppose a and b are positive integers, $S = \{x \mid x \text{ is a positive integer and } x = as + bt$ for some integers s and $t\}$, and c is the least element of S. We will show that $c \mid b$.

By the quotient-remainder theorem, $b = cq + r$ (*) for some integers q and r with $0 \le r < c$.

Now because c is in S, $c = as + bt$ for some integers s and t. Thus, by substitution into equation (*),

$$r = b - cq = b - (as + bt)q = a(-sq) + b(1 - tq).$$

Hence, by definition of S, either $r = 0$ or $r \in S$.

But if $r \in S$, then $r \ge c$ because c is the least element of S, and thus both $r < c$ and $r \ge c$ would be true, which would be a contradiction.

Therefore, $r \notin S$, and thus by elimination, we conclude that $r = 0$.

It follows that $b - cq = 0$, or, equivalently, $b = cq$, and so $c \mid b$ *[as was to be shown]*.

31. **a.** ***Step 1:*** $210 = 13 \cdot 16 + 2$, and so $2 = 210 - 16 \cdot 13$

    Step 2: $13 = 2 \cdot 6 + 1$, and so $1 = 13 - 2 \cdot 6$

    Step 3: $6 = 1 \cdot 6 + 0$, and so $\gcd(210, 13) = 1$

Substitute back through steps 2 and 1:

$$1 = 13 - 2 \cdot 6$$
$$= 13 - (210 - 16 \cdot 13) \cdot 6 = (-6) \cdot 210 + 97 \cdot 13$$

Thus $210 \cdot (-6) \equiv 1 \pmod{13}$, and so -6 is an inverse for 210 modulo 13.

b. Compute $13 - 6 = 7$. Note that $7 \equiv -6 \pmod{13}$ because $7 - (-6) = 13 = 13 \cdot 1$. Thus, by Theorem 8.4.3(3), $210 \cdot 7 \equiv 210 \cdot (-6) \pmod{13}$. By part (a), -6 is an inverse for 210 modulo 13, and so $210 \cdot (-6) \equiv 1 \pmod{13}$. It follows, by the symmetric and transitive properties of congruence, that $210 \cdot 7 \equiv 1 \pmod{13}$, and so 7 is a positive inverse for 210 modulo 13.

c. This problem can be solved using either the result of part (a) or that of part (b). By part (b) $210 \cdot 7 \equiv 1 \pmod{13}$. Multiply both sides by 8 and apply Theorem 8.4.3(3) to obtain $210 \cdot 56 \equiv 8 \pmod{13}$. Thus a positive solution for $210x \equiv 8 \pmod{13}$ is $x = 56$. Note that the least positive residue corresponding to this solution is also a solution. By Theorem 8.4.1, $56 \equiv 4 \pmod{13}$ because $56 = 13 \cdot 4 + 4$, and so, by Theorem 8.4.3(3), $210 \cdot 56 \equiv 210 \cdot 4 \equiv 9 \pmod{13}$. This shows that 4 is also a solution for the congruence, and because $0 \le 4 < 13$, 4 is the least positive solution for the congruence.

33. <u>Proof:</u> Suppose a, b, and c are integers such that $\gcd(a, b) = 1$, $a \mid c$, and $b \mid c$. We will show that $ab \mid c$.

    By Corollary 8.4.6 (or by Theorem 8.4.5), there exist integers s and t such that $as + bt = 1$.

    Also, by definition of divisibility, $c = au = bv$, for some integers u and v. Hence, by substitution,

    $$c = asc + btc = as(bv) + bt(au) = ab(sv + tu).$$

    But $sv + tu$ is an integer, and so, by definition of divisibility, $ab \mid c$ *[as was to be shown]*.

35. <u>Proof:</u> Let a be any integer and let n be any positive integer, and suppose s and t are any inverses for a modulo n. Thus $as \equiv 1 \pmod{n}$ and $at \equiv 1 \pmod{n}$. Note that $ast = (as) \cdot t = (at) \cdot s$ By Theorem 8.4.3(3), $(as) \cdot t \equiv t \pmod{n}$ and $(at) \cdot s \equiv s \pmod{n}$. Thus, by symmetry and transitivity of congruence modulo n, $s \equiv t \pmod{n}$. Because s and t were chosen arbitrarily, we conclude that any two inverses for a are congruent modulo n.

36. The numeric equivalents of H, E, L, and P are 08, 05, 12, and 16. To encrypt these letters, the following quantities must be computed: $8^{43} \bmod 713$, $5^{43} \bmod 713$, $12^{43} \bmod 713$, and $16^{43} \bmod 713$. We use the fact that $43 = 32 + 8 + 2 + 1$.

    H: $8 \equiv 8 \pmod{713}$
    $8^2 \equiv 64 \pmod{713}$
    $8^4 \equiv 64^2 \equiv 531 \pmod{713}$
    $8^8 \equiv 531^2 \equiv 326 \pmod{713}$
    $8^{16} \equiv 326^2 \equiv 39 \pmod{713}$
    $8^{32} \equiv 39^2 \equiv 95 \pmod{713}$
    Thus the ciphertext is $8^{43} \bmod 713$
    $= (95 \cdot 326 \cdot 64 \cdot 8) \bmod 713 = 233.$

    E: $5 \equiv 5 \pmod{713}$
    $5^2 \equiv 25 \pmod{713}$
    $5^4 \equiv 625 \pmod{713}$
    $5^8 \equiv 625^2 \equiv 614 \pmod{713}$
    $5^{16} \equiv 614^2 \equiv 532 \pmod{713}$
    $5^{32} \equiv 532^2 \equiv 676 \pmod{713}$
    Thus the ciphertext is $5^{43} \bmod 713$
    $= (676 \cdot 614 \cdot 25 \cdot 5) \bmod 713 = 129.$

    L: $12 \equiv 12 \pmod{713}$
    $12^2 \equiv 144 \pmod{713}$
    $12^4 \equiv 144^2 \equiv 59 \pmod{713}$
    $12^8 \equiv 59^2 \equiv 629 \pmod{713}$
    $12^{16} \equiv 629^2 \equiv 639 \pmod{713}$
    $12^{32} \equiv 639^2 \equiv 485 \pmod{713}$
    Thus the ciphertext is $12^{43} \bmod 713$
    $= (485 \cdot 629 \cdot 144 \cdot 12) \bmod 713 = 48.$

    P: $16 \equiv 16 \pmod{713}$
    $16^2 \equiv 256 \pmod{713}$
    $16^4 \equiv 256^2 \equiv 653 \pmod{713}$
    $16^8 \equiv 653^2 \equiv 35 \pmod{713}$
    $16^{16} \equiv 35^2 \equiv 512 \pmod{713}$
    $16^{32} \equiv 512^2 \equiv 473 \pmod{713}$
    Thus the ciphertext is $16^{43} \bmod 713$
    $= (473 \cdot 35 \cdot 256 \cdot 16) \bmod 713 = 128.$

    Therefore, the encrypted message is 233 129 048 128. (Again, note that in practice, individual letters of the alphabet are grouped together in blocks during encryption so that deciphering cannot be accomplished through knowledge of frequency patterns of letters or words. We kept them separate so that the numbers in the computations would be smaller and easier to work with.)

37. The numeric equivalents of C, O, M, and E are 03, 15, 13, and 05. To encrypt these letters, the following quantities must be computed: $3^{43} \bmod 713$, $15^{43} \bmod 713$, $13^{43} \bmod 713$, and $5^{43} \bmod 713$. Note that $43 = 32 + 8 + 2 + 1$.

C: $3 \equiv 3 \pmod{713}$
$3^2 \equiv 9 \pmod{713}$
$3^4 \equiv 9^2 \equiv 81 \pmod{713}$
$3^8 \equiv 81^2 \equiv 144 \pmod{713}$
$3^{16} \equiv 144^2 \equiv 59 \pmod{713}$
$3^{32} \equiv 59^2 \equiv 629 \pmod{713}$
Thus the ciphertext is
$3^{43} \bmod 713$
$= (629 \cdot 144 \cdot 9 \cdot 3) \bmod 713 = 675$.

O: $15 \equiv 15 \pmod{713}$
$15^2 \equiv 225 \pmod{713}$
$15^4 \equiv 225^2 \equiv 2 \pmod{713}$
$15^8 \equiv 2^2 \equiv 4 \pmod{713}$
$15^{16} \equiv 4^2 \equiv 16 \pmod{713}$
$15^{32} \equiv 16^2 \equiv 256 \pmod{713}$
Thus the ciphertext is
$15^{43} \bmod 713$
$= (256 \cdot 4 \cdot 225 \cdot 15) \bmod 713 = 89$.

M: $13 \equiv 13 \pmod{713}$
$13^2 \equiv 169 \pmod{713}$
$13^4 \equiv 169^2 \equiv 41 \pmod{713}$
$13^8 \equiv 41^2 \equiv 255 \pmod{713}$
$13^{16} \equiv 255^2 \equiv 142 \pmod{713}$
$13^{32} \equiv 142^2 \equiv 200 \pmod{713}$
Thus the ciphertext is
$13^{43} \bmod 713$
$= (200 \cdot 255 \cdot 169 \cdot 13) \bmod 713 = 476$.

E: $5 \equiv 5 \pmod{713}$
$5^2 \equiv 25 \pmod{713}$
$5^4 \equiv 625 \pmod{713}$
$5^8 \equiv 625^2 \equiv 614 \pmod{713}$
$5^{16} \equiv 614^2 \equiv 532 \pmod{713}$
$5^{32} \equiv 532^2 \equiv 676 \pmod{713}$
Thus the ciphertext is
$5^{43} \bmod 713$
$= (676 \cdot 614 \cdot 25 \cdot 5) \bmod 713 = 129$.

Therefore, the encrypted message is 675 089 476 129.]

(Again, note that, in practice, individual symbols are grouped together in blocks during encryption so that deciphering cannot be accomplished through knowledge of frequency patterns of letters or words. We kept them separate so that the numbers in the computations would be smaller and easier to work with.)

39. By exercise 38, the decryption key, d, is 307. Hence, to decrypt the message, the following quantities must be computed: $675^{307} \bmod 713$, $89^{307} \bmod 713$, and $48^{307} \bmod 713$. We use the fact that $307 = 256 + 32 + 16 + 2 + 1$.

$675 \equiv 675 \pmod{713}$
$675^2 \equiv 18 \pmod{713}$
$675^4 \equiv 18^2 \equiv 324 \pmod{713}$
$675^8 \equiv 324^2 \equiv 165 \pmod{713}$
$675^{16} \equiv 165^2 \equiv 131 \pmod{713}$
$675^{32} \equiv 131^2 \equiv 49 \pmod{713}$
$675^{64} \equiv 49^2 \equiv 262 \pmod{713}$
$675^{128} \equiv 262^2 \equiv 196 \pmod{713}$
$675^{256} \equiv 196^2 \equiv 627 \pmod{713}$

$89 \equiv 89 \pmod{713}$
$89^2 \equiv 78 \pmod{713}$
$89^4 \equiv 78^2 \equiv 380 \pmod{713}$
$89^8 \equiv 380^2 \equiv 374 \pmod{713}$
$89^{16} \equiv 374^2 \equiv 128 \pmod{713}$
$89^{32} \equiv 128^2 \equiv 698 \pmod{713}$
$89^{64} \equiv 698^2 \equiv 225 \pmod{713}$
$89^{128} \equiv 225^2 \equiv 2 \pmod{713}$
$89^{256} \equiv 2^2 \equiv 4 \pmod{713}$

$48 \equiv 48 \pmod{713}$
$48^2 \equiv 165 \pmod{713}$
$48^4 \equiv 131 \pmod{713}$
$48^8 \equiv 49 \pmod{713}$
$48^{16} \equiv 262 \pmod{713}$
$48^{32} \equiv 196 \pmod{713}$
$48^{64} \equiv 627 \pmod{713}$
$48^{128} \equiv 627^2 \equiv 266 \pmod{713}$
$48^{256} \equiv 266^2 \equiv 169 \pmod{713}$

Thus the decryption for 675 is

$$675^{307} \bmod 713 = (675^{256+32+16+2+1}) \bmod 713$$
$$= (627 \cdot 49 \cdot 131 \cdot 18 \cdot 675) \bmod 713 = 3,$$

which corresponds to the letter C.

The decryption for 89 is

$$89^{307} \bmod 713 = (89^{256+32+16+2+1}) \bmod 713$$
$$= (4 \cdot 698 \cdot 128 \cdot 78 \cdot 89) \bmod 713 = 15,$$

which corresponds to the letter O.

The decryption for 48 is

$$48^{307} \bmod 713 = (48^{256+32+16+2+1}) \bmod 713$$
$$= (169 \cdot 196 \cdot 262 \cdot 165 \cdot 48) \bmod 713 = 12,$$

which corresponds to the letter L.

Thus the decrypted message is COOL.

41. **a.** *Hint:* For the inductive step, assume $p \mid q_1 q_2 \cdots q_{s+1}$ and let $a = q_1 q_2 \cdots q_s$. Then $p \mid a q_{s+1}$, and either $p = q_{s+1}$ or Euclid's lemma and the inductive hypothesis can be applied.

42. **a.** When $a = 15$ and $p = 7$,

$$a^{p-1} = 15^6 = 11390625 \equiv 1 \ (\text{mod } 7) \text{ because } 11390625 - 1 = 7 \cdot 1627232.$$

b. When $a = 8$ and $p = 11$,

$$a^{p-1} = 8^{10} = 1073741824 \equiv 1 (\text{mod } 11) \text{ because } 1073741824 - 1 = 11 \cdot 97612893.$$

Section 8.5

1. **a.**

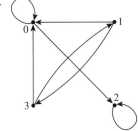

R_1 is not antisymmetric: $1 \ R_1 \ 3$ and $3 \ R_1 \ 1$ and $1 \neq 3$.

b.

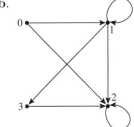

R_2 is antisymmetric: There are no cases where $a \ R \ b$ and $b \ R \ a$ and $a \neq b$.

2. R is not antisymmetric. Let x and y be any two distinct people of the same age. Then $x \ R \ y$ and $y \ R \ x$ but $x \neq y$.

3. R is not antisymmetric.

Counterexample: Let $s = 0$ and $t = 1$. Then $s \ R \ t$ and $t \ R \ s$ because $L(s) \leq L(t)$ and $L(t) \leq L(s)$, since both $L(s)$ and $L(t)$ equal 1, but $s \neq t$.

5. R is a partial order relation.

Proof:

R is reflexive: Suppose $(a, b) \in \mathbf{R} \times \mathbf{R}$. Then $(a, b) \ R \ (a, b)$ because $a = a$ and $b \leq b$.

R is antisymmetric: Suppose (a, b) and (c, d) are ordered pairs of real numbers such that $(a, b) \ R \ (c, d)$ and $(c, d) \ R \ (a, b)$. Then

$$\text{either } a < c \quad \text{or} \quad \text{both } a = c \text{ and } b \leq d$$

and

$$\text{either } c < a \quad \text{or} \quad \text{both } c = a \text{ and } d \leq b.$$

Thus

$$a \leq c \text{ and } c \leq a, \text{ and so } a = c.$$

Consequently,

$$b \le d \quad \text{and} \quad d \le b, \text{ and so } b = d.$$

Hence $(a, b) = (c, d)$.

R is transitive: Suppose (a, b), (c, d), and (e, f) are ordered pairs of real numbers such that $(a, b) \, R \, (c, d)$ and $(c, d) \, R \, (e, f)$. Then

$$\text{either } a < c \quad \text{or} \quad \text{both } a = c \text{ and } b \le d$$

and

$$\text{either } c < e \quad \text{or} \quad \text{both } c = e \text{ and } d \le f.$$

It follows that one of the following cases must occur.

Case 1 (a < c and c < e): Then by transitivity of $<$, $a < e$, and so $(a, b) \, R \, (e, f)$ by definition of R.

Case 2 (a < c and c = e): Then by substitution, $a < e$, and so $(a, b) \, R \, (e, f)$ by definition of R.

Case 3 (a = c and c < e): Then by substitution, $a < e$, and so $(a, b) \, R \, (e, f)$ by definition of R.

Case 4 (a = c and c = e): Then by definition of R, $b \le d$ and $d \le f$, and so by transitivity of $\le$, $b \le f$. Hence $a = e$ and $b \le f$, and so $(a, b) \, R \, (e, f)$ by definition of R.

In each case, $(a, b) \, R \, (e, f)$. Therefore, R is transitive. Since R is reflexive, antisymmetric, and transitive, R is a partial order relation.

6. R is a partial order relation.

   Proof:

   R is reflexive: Suppose $r \in P$. Then $r = r$, and so by definition of R, $r \, R \, r$.

   R is antisymmetric: Suppose $r, s \in P$ and $r \, R \, s$ and $s \, R \, r$. *[We must show that $r = s$.]*

   By definition of R, either r is an ancestor of s or $r = s$ and either s is an ancestor of r or $s = r$.

   Now it is impossible for both r to be an ancestor of s and s to be an ancestor of r. Hence one of these conditions must be false, and so $r = s$ *[as was to be shown]*.

   R is transitive: Suppose $r, s, t \in P$ and $r \, R \, s$ and $s \, R \, t$. *[We must show that $r \, R \, t$.]*

   By definition of R, either r is an ancestor of s or $r = s$ and either s is an ancestor of t or $s = t$.

   In case r is an ancestor of s and s is an ancestor of t, then r is an ancestor of t, and so $r \, R \, t$.

   In case r is an ancestor of s and $s = t$, then r is an ancestor of t, and so $r \, R \, t$.

   In case $r = s$ and s is an ancestor of t, then r is an ancestor of t, and so $r \, R \, t$.

   In case $r = s$ and $s = t$, then $r = t$, and so $r \, R \, t$. Thus in all four possible cases, $r \, R \, t$ *[as was to be shown]*.

   Conclusion: Since R is reflexive, antisymmetric, and transitive, R is a partial order relation.

8. R is not a partial order relation because R is not antisymmetric.

   Counterexample: $1 \, R \, 3$ (because $1 + 3$ is even) and $3 \, R \, 1$ (because $3 + 1$ is even) but $1 \ne 3$.

9. R is not a partial order relation because R is not antisymmetric.

   Counterexample: Let $x = 2$ and $y = -2$. Then $x \, R \, y$ because $(-2)^2 \le 2^2$, and $y \, R \, x$ because $2^2 \le (-2)^2$. But $x \ne y$ because $2 \ne -2$.

10. No. Counterexample: Define relations R and S on the set $\{1, 2\}$ as follows: $R = \{(1, 2)\}$ and $S = \{(2, 1)\}$. Then both R and S are antisymmetric, but $R \cup S = \{(1, 2), (2, 1)\}$ is not antisymmetric because $(1, 2) \in R \cup S$ and $(2, 1) \in R \cup S$ but $1 \ne 2$.

11. **a.** True, by (1).   **b.** False. By (1), $bba \preceq bbab$.

12. <u>Proof:</u>

$\preceq$ *is reflexive*: Suppose s is in S. If $s = \lambda$, then $s \preceq s$ by (3). If $s \neq \lambda$, then $s \preceq s$ by (1). Hence in either case, $s \preceq s$.

$\preceq$ *is antisymmetric*: Suppose s and t are in S and $s \preceq t$ and $t \preceq s$. *[We must show that $s = t$.]*

By definition of S, either $s = \lambda$ or $s = a_1 a_2 \ldots a_m$ and either $t = \lambda$ or $t = b_1 b_2 \ldots b_n$ for some positive integers m and n and elements $a_1, a_2, \ldots, a_m$ and $b_1, b_2, \ldots, b_n$ in A.

If $s \preceq t$ by virtue of condition (2), then it is false that $t \preceq s$. *[For suppose $s \preceq t$ by virtue of condition (2). Then for some integer k with $k \leq m$, $k \leq n$, and $k \geq 1$, $a_i = b_i$ for every $i = 1, 2, \ldots, k - 1$, and both $a_k R b_k$ and $a_k \neq b_k$. In this situation, it is impossible to have $t \preceq s$ by virtue of either condition (1) or (3). But it is also impossible for $t \preceq s$ by virtue of condition (2) because it would imply that $b_k R a_k$. However, since R is a partial order relation and $a_k R b_k$ it would follow that $a_k = b_k$, which would contradict the fact that $a_k \neq b_k$. Hence the assumption that $s \preceq t$ by virtue of condition (2) leads to the conclusion that $t \not\preceq s$, which contradicts the supposition.]*

An almost identical argument shows that it is impossible to have $t \preceq s$ by virtue of condition (2).

Hence $s \preceq t$ and $t \preceq s$ by virtue either of condition (1) or of condition (3).

In case $s \preceq t$ by virtue of condition (1), then neither s nor t is the null string and so $t \preceq s$ by virtue of condition (1). Then by (1) $m \leq n$ and $a_i = b_i$ for every $i = 1, 2, \ldots, m$ and $n \leq m$ and $b_i = a_i$ for every $i = 1, 2, \ldots, m$, and so $s = t$.

In case $s \preceq t$ by virtue of condition (3), then $s = \lambda$, and so since $t \preceq s$, then $t \preceq \lambda$. But the only situation with this result is condition (3) with $t = \lambda$, which implies that $s = t = \lambda$.

Thus in all possible cases, if $s \preceq t$ and $t \preceq s$, then $s = t$ *[as was to be shown]*.

$\preceq$ *is transitive*: Suppose s and t are any elements of S such that $s \preceq t$ and $t \preceq u$. *[We must show that $s \preceq u$.]*

By definition of S, (1) either $s = \lambda$ or $s = a_1 a_2 \ldots a_m$, (2) either $t = \lambda$ or $t = b_1 b_2 \ldots b_n$, and (3) either $u = \lambda$ or $u = c_1 c_2 \ldots c_p$ for some positive integers m, n, and p and elements $u_1, u_2, \ldots, a_m$, $b_1, b_2, \ldots, b_n$, and $c_1, c_2, \ldots, c_p$ in A.

Case 1 ($s = \lambda$): In this case, $s \preceq u$ by (3).

Case 2 ($s \neq \lambda$): In this case, since $s \preceq t$, $t \neq \lambda$, and since $t \preceq u$, $u \neq \lambda$.

Subcase a ($s \preceq t$ by condition (1) and $t \preceq u$ by condition (1)): Then $m \leq n$, and $n \leq p$, and $a_i = b_i$ for every $i = 1, 2, \ldots, m$, and $b_j = c_j$ for every $j = 1, 2, \ldots, n$. It follows that $a_i = c_i$ for every $i = 1, 2, \ldots, m$, and so by (1), $s \preceq u$.

Subcase b ($s \preceq t$ by condition (1) and $t \preceq u$ by condition (2)): Then $m \leq n$ and $a_i = b_i$ for every $i = 1, 2, \ldots, m$. In addition, there is an integer k, with $k \leq n$ and $k \leq p$, such that $k \geq 1$, and $b_j = c_j$ for every $j = 1, 2, \ldots, k - 1$, and also $b_k R c_k$, and $b_k \neq c_k$.

If $k \leq m$, then s and u satisfy condition (2) *[because $a_i = b_i$ for every $i = 1, 2, \ldots, m$ and so $k \leq m$, $k \leq p$, $k \geq 1$, and $a_i = b_i = c_i$ for every $i = 1, 2, \ldots, k - 1$, and also $a_k R c_k$, and $a_k \neq c_k$]*.

If $k > m$, then s and u satisfy condition (1) *[because $a_i = b_i = c_i$ for every $i = 1, 2, \ldots, m$]*.

Thus in either case $s \preceq u$.

Subcase c ($s \preceq t$ by condition (2) and $t \preceq u$ by condition (1)): Then there is an integer k with $k \leq m$, $k \leq n$, and $k \geq 1$, such that $a_i = b_i$ for every $i = 1, 2, \ldots, k - 1$, and also $a_k R b_k$, and $a_k \neq b_k$. In addition, $n \leq p$ and $b_j = c_j$ for every $j = 1, 2, \ldots, n$. Then s and u satisfy condition (2) *[because $k \leq n$, $k \leq p$ (since $k \leq n$ and $n \leq p$), $k \geq 1$, $a_i = b_i = c_i$ for* every $i = 1, 2, \ldots, k-1$ *(since $k - 1 < n$)*, $a_k \, R \, c_k$ *(since $b_k = c_k$ because $k \leq n$)*, and $a_k \neq c_k$ *(since $b_k = c_k$ and $a_k \neq b_k$)]*. Thus $s \preceq u$.

Subcase d ($s \preceq t$ by condition (2) and $t \preceq u$ by condition (2)): Then there is an integer k with $k \leq m$, $k \leq n$, and $k \geq 1$, such that $a_i = b_i$ for every $i = 1, 2, \ldots, k-1$, and also $a_k \, R \, b_k$, and $a_k \neq b_k$. In addition, there is an integer l with $l \leq n$ and $l \leq p$, such that $l \geq 1$ and $b_j = c_j$ for every $j = 1, 2, \ldots, l-1$, and also $b_l \, R \, c_l$, and $b_l \neq c_l$.

If $k < l$, then $a_i = b_i = c_i$ for every $i = 1, 2, \ldots, k-1$, $a_k \, R \, b_k$, $b_k = c_k$ (in which case $a_k \, R \, c_k$), and $a_k \neq c_k$ (since $a_k \neq b_k$). Thus if, $k < l$, then $s \preceq u$ by condition (2).

If $k = l$, then $b_k \, R \, c_k$ (in which case $a_k \, R \, c_k$ by transitivity of R) and $b_k \neq c_k$. It follows that $a_k \neq c_k$ *[for if $a_k = c_k$, then $a_k \, R \, b_k$ and $b_k \, R \, a_k$, which implies that $a_k = b_k$ (since R is a partial order) and contradicts the fact that $a_k \neq b_k$]*. Thus if $k = l$, then $s \preceq u$ by condition (2).

If $k > l$, then $a_i = b_i = c_i$ for every $i = 1, 2, \ldots, l-1$, $a_l \, R \, c_l$ (because $b_l \, R \, c_l$ and $a_l = b_l$), and $a_l \neq c_l$ (because $b_l \neq c_l$ and $a_l = b_l$). Thus if $k > l$, then $s \preceq u$ by condition (2).

Hence, regardless of whether $k < l$, $k = l$, or $k > l$, we conclude that $s \preceq u$.

The above arguments show that in all possible situations, if $s \preceq t$ and $t \preceq u$ then $s \preceq u$ *[as was to be shown]*. Hence $\preceq$ *is* transitive.

Conclusion: Since $\preceq$ *is* reflexive, antisymmetric, and transitive, $\preceq$ *is* a partial order relation.

13. $R_1 = \{(a, a), (b, b)\}, \quad R_2 = \{(a, a), (b, b), (a, b)\}, \quad R_3 = \{(a, a), (b, b), (b, a)\}$

14. **a.**

$R_1 = \{(a, a), (b, b), (c, c)\}$ $R_2 = \{(a, a), (b, b), (c, c), (b, a)\}$
$R_3 = \{(a, a), (b, b), (c, c), (c, a)\}$ $R_4 = \{(a, a), (b, b), (c, c), (b, a), (c, a)\}$
$R_5 = \{(a, a), (b, b), (c, c), (c, b), (c, a)\}$ $R_6 = \{(a, a), (b, b), (c, c), (b, c), (b, a)\}$
$R_7 = \{(a, a), (b, b), (c, c), (c, b), (b, a), (c, a)\}$ $R_8 = \{(a, a), (b, b), (c, c), (b, c), (b, a), (c, a)\}$
$R_9 = \{(a, a), (b, b), (c, c), (b, c)\}$ $R_{10} = \{(a, a), (b, b), (c, c), (c, b)\}$

15. <u>Proof:</u>

Suppose R is a relation on a set A and R is reflexive, symmetric, transitive, and anti-symmetric. We will show that R is the identity relation on A.

First note that for all x and y in A, if $x \, R \, y$ then, because R is symmetric, $y \, R \, x$. But then, because R is also anti-symmetric $x = y$. Thus for all x and y in A, if $x \, R \, y$ then $x = y$.

This argument, however, does not prove that R is the identity relation on A because the conclusion would also follow from the hypothesis (by default) in the case where $A \neq \emptyset$ and $R = \emptyset$.

But when $A \neq \emptyset$, it is impossible for R to equal $\emptyset$ because R is reflexive, which means that $x \, R \, x$ for every x in A.

Thus every element in A is related by R to itself, and no element in A is related to anything other than itself. It follows that R is the identity relation on A.

16. **a.**

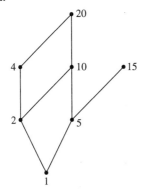

17. **a.**

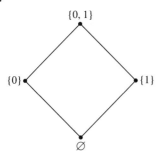

18.

21. **a.** <u>Proof</u>: *[We must show that for all a and b in A, a | b or b | a.]* Let a and b be particular but arbitrarily chosen elements of A. By definition of A, there are nonnegative integers r and s such that $a = 2^r$ and $b = 2^s$. Now either $r \leq s$ or $s < r$. If $r \leq s$, then

$$b = 2^s = 2^r \cdot 2^{s-r} = a \cdot 2^{s-r},$$

where $s - r \geq 0$. It follows, by definition of divisibility, that $a \mid b$. By a similar argument, if $s < r$, then $b \mid a$. Hence either $a \mid b$ or $b \mid a$ *[as was to be shown]*.

b.

```
•————•————•————•————•
1    2    2²   2³   2⁴
```
$$1 \quad 2 \quad 2^2 \quad 2^3 \quad 2^4$$

22. greatest element: none; least element: 1; maximal elements: 15, 20; minimal element: 1

24. greatest element: $\{0, 1\}$; least element: $\emptyset$; maximal element: $\{0, 1\}$; minimal element: $\emptyset$

26. greatest element: $(1, 1)$; least element: $(0, 0)$; maximal element: $(1, 1)$; minimal element: $(0, 0)$

27. greatest element: $(1,1)$ least element: $(0,0)$

maximal elements: $(1,1)$ minimal elements: $(0,0)$

30. **a.** No greatest element and no least element

    b. Least element is 0, greatest element is 1

    c. No greatest element and no least element

    d. greatest element: 9   least element: 1

31. R is a total order relation because it is reflexive, antisymmetric, and transitive (so it is a partial order) and because $[b, a, c, d]$ is a chain that contains every element of A: $b\ R\ c$, $c\ R\ a$, and $a\ R\ d$.

33. A is not totally ordered by the given relation because $9 \nmid 12$ and $12 \nmid 9$.

34. *Hint:* Let R' be the restriction of R to B and show that R' is reflexive, antisymmetric, and transitive. In each case, this follows almost immediately from the fact that R is reflexive, antisymmetric, and transitive.

35. *One possible solution:* $\emptyset \subseteq \{w\} \subseteq \{w, x\} \subseteq \{w, x, y\} \subseteq \{w, x, y, z\}$

36. $\{2, 4, 12, 24\}$ or $\{3, 6, 12, 24\}$

38. <u>Proof</u>: Suppose A is a partially ordered set with respect to a relation $\preceq$. By definition of total order, A is totally ordered if, and only if, any two elements of A are comparable. By definition of chain, this is true if, and only if, A is a chain.

39. <u>Proof (by mathematical induction)</u>: Let A be a set that is totally ordered with respect to a relation $\preceq$, and let the property $P(n)$ be the sentence

$$\text{Any subset of } A \text{ with } n \text{ elements has both} \quad \leftarrow\ P(n)$$
$$\text{a least element and a greatest element.}$$

Show that $P(1)$ is true:

If $A = \emptyset$, then $P(1)$ is true by default. So assume that A has at least one element, and suppose $S = \{a_1\}$ is a subset of A with one element. Because $\preceq$ is reflexive, $a_1 \preceq a_1$. So, by definition of least element and greatest element, a_1 is both a least element and a greatest element of S, and thus the property is true for $n = 1$.

Show that for every integer $k \geq 1$, if $P(k)$ is true then $P(k+1)$ is true: Let k be any integer with $k \geq 1$ and suppose that

$$\text{Any subset of } A \text{ with } k \text{ elements has both} \quad \leftarrow\ \begin{array}{l} P(k) \\ \text{inductive hypothesis} \end{array}$$
$$\text{a least element and a greatest element.}$$

We must show that

$$\text{Any subset of } A \text{ with } k+1 \text{ elements has both} \quad \leftarrow\ P(k+1)$$
$$\text{a least element and a greatest element.}$$

If A has fewer than $k+1$ elements, then the statement is true by default. So assume that A has at least $k+1$ elements and that $S = \{a_1, a_2, \ldots, a_{k+1}\}$ is a subset of A with $k+1$ elements. By inductive hypothesis, $S - \{a_{k+1}\}$ has both a least element s and a greatest element t. Now because A is totally ordered, a_{k+1} and s are comparable. If $a_{k+1} \preceq s$, then, by transitivity of $\preceq$, a_{k+1} is the least element of S; otherwise, s remains the least element of S. And if $t \preceq a_{k+1}$, then, by transitivity of $\preceq$, a_{k+1} is the greatest element of S; otherwise, t remains the greatest element of S. Thus S has both a greatest element and a least element *[as was to be shown]*.

40. **a.** <u>Proof by contradiction</u>: Suppose not. Suppose A is a nonempty, finite set that is partially ordered with respect to a relation $\preceq$, and suppose no element of A is minimal. Construct a sequence of elements $x_1, x_2, x_3, \ldots$ of A as follows:

1. Pick any element of A and call it x_1.

2. For each $i = 2, 3, 4, \ldots$, pick x_i to be an element of A for which $x_i \preceq x_{i-1}$ and $x_i \neq x_{i-1}$. *[Such an element must exist because otherwise x_{i-1} would be minimal, and we are supposing that no element of A is minimal.]* Now $x_i \neq x_j$ for any $i \neq j$. *[For if $x_i = x_j$, where $i < j$, then on the one hand, $x_j \preceq x_{j-1} \preceq \ldots \preceq x_{i+1} \preceq x_i$ and so $x_j \preceq x_{i+1}$. On the other hand, since $x_i = x_j$ and $x_i \succeq x_{i+1}$, then $x_j \succeq x_{i+1}$. Hence by antisymmetry, $x_j = x_{i+1}$, and so $x_i = x_{i+1}$ because $x_i = x_j$. But this contradicts the definition of the sequence $x_1, x_2, x_3, \ldots$]* Thus $x_1, x_2, x_3, \ldots$ is an infinite sequence of distinct elements, and consequently $\{x_1, x_2, x_3, \ldots\}$ is an infinite subset of the finite set A, which is impossible. Hence the supposition is false and we conclude that any partially ordered subset of a finite set has a minimal element.

42.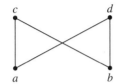

44. One such total order is 1, 5, 2, 15, 10, 4, 20.

45. One such total order is 3, 9, 2, 6, 18, 4, 12, 8.

46. One such total order is (0, 0), (1, 0), (0, 1), (1, 1).

48. One such total order is $\emptyset, \{a\}, \{b\}, \{c\}, \{d\}, \{a, b\}, \{a, c\}, \{a, d\}, \{b, c\}, \{b, d\}, \{c, d\},$
$\{a, b, c\}, \{a, b, d\}, \{a, c, d\}, \{b, c, d\}, \{a, b, c, d\}.$

50. **a.** *One possible answer:* 1, 6, 10, 9, 5, 7, 2, 4, 8, 3

51. **a.** 33 hours **b.** Critical path: 1, 2, 5, 8, 9

Review Guide: Chapter 8

Definitions: How are the following terms defined?

- congruence modulo 2 relation
- inverse of a relation from a set A to a set B
- relation on a set
- directed graph of a relation on a set
- n-ary relation (and binary, ternary, quaternary relations)
- reflexive, symmetric, and transitive properties of a relation on a set
- congruence modulo 3 relation
- transitive closure of a relation on a set
- equivalence relation on a set
- equivalence class
- congruence modulo n relation
- representative of an equivalence class
- m is congruent to n modulo d
- plaintext and cyphertext
- residue of a modulo n
- complete set of residues modulo n
- d is a linear combination of a and b
- a and b are relatively prime; $a_1, a_2, \ldots, a_n$ are pairwise relatively prime
- an inverse of a modulo n
- antisymmetric relation
- partial order relation
- lexicographic order
- Hasse diagram
- a and b are comparable
- poset
- total order relation
- chain, length of a chain
- maximal element, greatest element, minimal element, least element
- topological sorting
- compatible partial order relations
- PERT and CPM
- critical path

Properties of Relations on Sets and Equivalence Relations

- How do you show that a relation on a finite set is reflexive? symmetric? transitive?
- How do you show that a relation on an infinite set is reflexive? symmetric? transitive?
- How do you show that a relation on a set is not reflexive? not symmetric? not transitive?
- How do you find the transitive closure of a relation?
- What is the relation induced by a partition of a set?
- Given an equivalence relation on a set A, what is the relationship between the distinct equivalence classes of the relation and subsets of the set A?
- Given an equivalence relation on a set A and an element a in A, how do you find the equivalence class of a?
- In what way are rational numbers equivalence classes?

Cryptography

- How does the Caesar cipher work?
- If a, b, and n are integers with $n > 1$, what are some different ways of expressing the fact that $n \mid (a - b)$?
- How do you reduce a number modulo n?
- If n is an integer with $n > 1$, is congruence modulo n an equivalence relation on the set of all integers?
- How do you add, subtract, and multiply integers modulo an integer $n > 1$?
- What is an efficient way to compute a^k where a is an integer with $a > 1$ and k is a large integer?
- How do you express the greatest common divisor of two integers as a linear combination of the integers?
- When can you find an inverse modulo n for a positive integer a, and how do you find it?
- How do you encrypt and decrypt messages using RSA cryptography?
- What is Euclid's lemma? How is it proved?
- What is Fermat's little theorem? How is it proved?
- Why does the RSA cipher work?

Partial Order Relations

- How do you show that a relation on a set is or is not antisymmetric?
- If A is a set with a partial order relation R, S is a set of strings over A, and a and b are in S, how do you show that $a \preceq b$, where $\preceq$ denotes the lexicographic ordering of S?
- How do you construct the Hasse diagram for a partial order relation?
- How do you find a chain in a partially ordered set?
- Given a set with a partial order, how do you construct a topological sorting for the elements of the set?
- Given a job scheduling problem consisting of a number of tasks, some of which must be completed before others can be begun, how can you use a partial order relation to determine the minimum time needed to complete the job?

Fill-in-the-Blank Review Questions

Section 8.1

1. If R is a relation from A to B, $x \in A$, and $y \in B$, the notation $x \, R \, y$ means that _____.
2. If R is a relation from A to B, $x \in A$, and $y \in B$, the notation $x \, \cancel{R} \, y$ means that _____.
3. If R is a relation from A to B, $x \in A$, and $y \in B$, then $(y, x) \in R^{-1}$ if, and only if, _____.
4. A relation on a set A is a relation from _____ to _____.
5. If R is a relation on a set A, the directed graph of R has an arrow from x to y if, and only if, _____.

Section 8.2

1. For a relation R on a set A to be reflexive means that _____.
2. For a relation R on a set A to be symmetric means that _____.
3. For a relation R on a set A to be transitive means that _____.
4. To show that a relation R on an infinite set A is reflexive, you suppose that _____ and you show that _____.

5. To show that a relation R on an infinite set A is symmetric, you suppose that ____ and you show that ____.

6. To show that a relation R on an infinite set A is transitive, you suppose that ____ and you show that ____.

7. To show that a relation R on a set A is not reflexive, you ____.

8. To show that a relation R on a set A is not symmetric, you ____.

9. To show that a relation R on a set A is not transitive, you ____.

10. Given a relation R on a set A, the transitive closure of R is the relation R^t on A that satisfies the following three properties: ____, ____ and ____.

Section 8.3

1. For a relation on a set to be an equivalence relation, it must be ____.

2. The notation $m \equiv n \pmod{d}$ is read ____ and means that ____.

3. Given an equivalence relation R on a set A and given an element a in A, the equivalence class of a is denoted ____ and is defined to be ____.

4. If A is a set, R is an equivalence relation on A, and a and b are elements of A, then either $[a] = [b]$ or ____.

5. If A is a set and R is an equivalence relation on A, then the distinct equivalence classes of R form ____.

6. Let $A = \mathbf{Z} \times (\mathbf{Z} - \{0\})$, and define a binary relation R on A by specifying that for all (a, b) and (c, d) in A, $(a, b)R(c, d)$ if, and only if, $ad = bc$. Then there is exactly one equivalence class of R for each ____.

Section 8.4

1. When letters of the alphabet are encrypted using the Caesar cipher, the encrypted version of a letter is ____.

2. If a, b, and n are integers with $n > 1$, the following are all different ways of expressing the fact that $n \mid (a - b)$: ____, ____, ____, ____.

3. If a, b, c, d, m and n are integers with $n > 1$ and if $a \equiv c \pmod{n}$ and $b \equiv d \pmod{n}$, then $a + b \equiv$ ____, $a - b \equiv$ ____, $ab \equiv$ ____, and $a^m \equiv$ ____.

4. If a, n, and k are positive integers with $n > 1$, an efficient way to compute $a^k \pmod{n}$ is to write k as a ____ and use the facts about computing products and powers modulo n.

5. To express a greatest common divisor of two integers as a linear combination of the integers, you use the extended ____ algorithm.

6. To find an inverse for a positive integer a modulo an integer n with $n > 1$, you express the number 1 as ____.

7. To encrypt a message M using RSA cryptography with public key pq and e, you use the formula ____, and to decrypt a message C, you use the formula ____, where ____.

8. Euclid's lemma says that for all integers a, b, and c if $\gcd(a, c) = 1$ and $a \mid bc$, then ____.

9. Fermat's little theorem says that if p is any prime number and a is any integer such that $p \nmid a$ then ____.

10. The crux of the proof that the RSA cipher works is that if (1) p and q are distinct large prime numbers, (2) $M < pq$, (3) M is relatively prime to pq, (4) e is relative ly prime to $(p-1)(q-1)$, and (5) d is a positive inverse for e modulo $(p-1)(q-1)$, then $M = $ ____.

Section 8.5

1. For a binary relation R on a set A to be antisymmetric means that ____.

2. To show that a binary relation R on an infinite set A is antisymmetric, you suppose that _____ and you show that _____.

3. To show that a binary relation R on a set A is not antisymmetric, you _____.

4. To construct a Hasse diagram for a partial order relation, you start with a directed graph of the relation in which all arrows point upward and you eliminate _____, _____, and _____.

5. If A is a set that is partially ordered with respect to a relation $\preceq$ and if a and b are elements of A, we say that a and b are comparable if, and only if, _____ or _____.

6. A relation $\preceq$ on a set A is a total order if, and only if, _____.

7. If A is a set that is partially ordered with respect to a relation $\preceq$, and if B is a subset of A, then B is a chain if, and only if, for all a and b in B, _____.

8. Let A be a set that is partially ordered with respect to a relation $\preceq$, and let a be an element of A.

   (a) a is maximal if, and only if, _____.
   (b) a is a greatest element of A if, and only if, _____.
   (c) a is called minimal if, and only if, _____.
   (d) a is called a least element of A if, and only if, _____.

9. Given a set A that is partially ordered with respect to a relation $\preceq$, the relation $\preceq'$ is a topological sorting for $\preceq$, if, and only if, $\preceq'$ is a _____ and for all a and b in A if $a \preceq b$ then _____.

10. PERT and CPM are used to produce efficient _____.

Answers for Fill-in-the-Blank Review Questions

Section 8.1

1. x is related to y by R
2. x is not related to y by R
3. $(x, y) \in R$
4. A to A
5. x is related to y by R

Section 8.2

1. for every x in A; $x \, R \, x$
2. for all x and y in A, if $x \, R \, y$ then $y \, R \, x$
3. for all x, y, and z in A, if $x \, R \, y$ and $y \, R \, z$ then $x \, R \, z$
4. x is any element of A; $x \, R \, x$
5. x and y are any elements of A such that $x \, R \, y$; $y \, R \, x$
6. x, y, and z are any elements of A such that $x \, R \, y$ and $y \, R \, z$; $x \, R \, z$
7. show the existence of an element x in A such that $x \, \not\!R \, x$
8. show the existence of elements x and y in A such that $x \, R \, y$ but $y \, \not\!R \, x$
9. show the existence of elements x, y, and z in A such that $x \, R \, y$ and $y \, R \, z$ but $x \, \not\!R \, z$
10. R^t is transitive; $R \subseteq R^t$; if S is any other transitive relation that contains R, then $R^t \subseteq S$

Section 8.3

1. reflexive, symmetric, and transitive

2. m is congruent to n modulo d; d divides $m - n$

3. $[a]$; the set of all x in A such that $x \, R \, a$

4. $[a] \cap [b] = \emptyset$

5. a partition of A

6. rational number

Section 8.4

1. three places in the alphabet to the right of the letter, with X wrapped around to A, Y to B, and Z to C

2. $a \equiv b \pmod{n}$;
   $a = b + kn$ for some integer k;
   a and b have the same nonnegative remainder when divided by n;
   $a \bmod n = b \bmod n$

3. $(c + d) \pmod{n}$; $(c - d) \pmod{n}$; $(cd) \pmod{n}$; $c^m \pmod{n}$

4. sum of powers of 2

5. Euclidean

6. a linear combination of a and n

7. $C = M^e \bmod pq$; $M = C^d \bmod pq$; d is a positive inverse for e modulo $(p - 1)(q - 1)$

8. $a \mid b$

9. $a^{p-1} \equiv 1 \pmod{p}$

10. $M^{ed} \pmod{pq}$

Section 8.5

1. for all a and b in A, if $a \, R \, b$ and $b \, R \, a$ then $a = b$

2. a and b are any elements of A with $a \, R \, b$ and $b \, R \, a$; $a = b$

3. show the existence of elements a and b in A such that $a \, R \, b$ and $b \, R \, a$ and $a \neq b$

4. all loops; all arrows whose existence is implied by the transitive property; the direction indicators on the arrows

5. $a \preceq b$; $b \preceq a$

6. for any two elements a and b in A; either $a \preceq b$ or $b \preceq a$

7. a and b are comparable

8. (a) for every b in A either $b \preceq a$ or b and a are not comparable
   (b) for every b in A, $b \preceq a$
   (c) for every b in A either $a \preceq b$ or b and a are not comparable
   (d) for every b in A, $a \preceq b$

9. total order; $a \preceq b$

10. scheduling of tasks

Chapter 9: Counting and Probability

The primary aim of this chapter is to foster intuitive understanding for fundamental principles of counting and probability and an ability to apply them in a wide variety of situations. It is helpful to get into the habit of beginning a counting problem by listing (or at least imagining) some of the objects you are trying to count. If you see that all the objects to be counted can be matched up with the integers from m to n inclusive, then the total is $n - m + 1$ (Section 9.1). If you see that all the objects can be produced by a multi-step process, then the total can be found by counting the distinct paths from root to leaves in a possibility tree that shows the outcomes of each successive step (Section 9.2). And if each step of the process can be performed in a fixed number of ways (regardless of how the previous steps were performed), then the total can be calculated by applying the multiplication rule (Section 9.2).

If the objects to be counted can be separated into disjoint categories, then the total is just the sum of the subtotals for each category (Section 9.3). And if the categories are not disjoint, the total can be counted using the inclusion/exclusion rule (Section 9.3). If the objects to be counted can be represented as all the subsets of size r of a set with n elements, then the total is $\binom{n}{r}$ for which there is a computational formula (Section 9.5). Finally if the objects can be represented as all the multisets of size r of a set with n elements, then the total is $\binom{n+r-1}{r}$ (Section 9.6).

Section 9.4 introduces the pigeonhole principle, which provides a way to answer questions about how many of a certain object are needed to guarantee certain results and is used to show that certain results are guaranteed if a certain number of objects are present. The section includes the reasoning for why every rational number has a decimal expansion that either terminates or repeats.

Pascal's formula and the binomial theorem are discussed in Section 9.7. Each is proved both algebraically and combinatorially. Pascal's formula and the binomial theorem are discussed in Section 9.7. Each is proved both algebraically and combinatorially. Exercise 28 of Section 9.7 is intended to help you see how Pascal's formula is applied in the algebraic proof of the binomial theorem.

Sections 9.8 and 9.9 develop the axiomatic theory of probability through the concepts of expected value, conditional probability, independence, and Bayes' theorem. Exercise 20 of Section 9.1 can be solved directly by reasoning about the sample space, but it can also be solved using conditional probability, which is discussed in Section 9.9.

Section 9.1

2. 3/4, 1/2, 1/2

3. $\{1\spadesuit, 2\spadesuit, 3\spadesuit, 4\spadesuit, 5\spadesuit, 6\spadesuit, 7\spadesuit, 8\spadesuit, 9\spadesuit, 10\spadesuit, 1\heartsuit, 2\heartsuit, 3\heartsuit, 4\heartsuit, 5\heartsuit, 6\heartsuit, 7\heartsuit, 8\heartsuit, 9\heartsuit, 10\heartsuit\}$,
   Probability $= 20/52 \cong 38.5\%$

5. $\{10\clubsuit, J\clubsuit, Q\clubsuit, K\clubsuit, A\clubsuit, 10\spadesuit, J\spadesuit, Q\spadesuit, K\spadesuit, A\spadesuit, 10\heartsuit, J\heartsuit, Q\heartsuit, K\heartsuit, A\heartsuit, 10\spadesuit, J\spadesuit, Q\spadesuit, K\spadesuit, A\spadesuit\}$,
   Probability $= 20/52 = 5/13 \cong 38.5\%$

6. $\{2\clubsuit, 3\clubsuit, 4\clubsuit, 2\diamondsuit, 3\diamondsuit, 4\diamondsuit, 2\heartsuit, 3\heartsuit, 4\heartsuit, 2\spadesuit, 3\spadesuit, 4\spadesuit\}$, Probability $= 12/52 = 3/13 \cong 23.1\%$

7. $\{26, 35, 44, 53, 62\}$, Probability $= 3/8 \cong 37.5\%$

9. $\{11, 12, 13, 14, 15, 21, 22, 23, 24, 31, 32, 33, 41, 42, 51\}$, Probability $= 15/36 = 41\frac{2}{3}\%$

11. **a.** $\{HHH, HHT, HTH, HTT, THH, THT, TTH, TTT\}$
    b. (i) $\{HTT, THT, TTH\}$, probability $= 3/8 \cong 37.5\%$

12. **a.** $\{BBB, BBG, BGB, BGG, GBB, GBG, GGB, GGG\}$
    b. (i) $\{GBB, BGB, BBG\}$, probability $= 3/8 = 37.5\%$
       (ii) $\{GGB, GBG, BGG, GGG\}$ Probability $= 4/8 = 1/2 = 50\%$
       (iii) $\{BBB\}$ Probability $= 1/8 = 12.5\%$

13. **a.** {*CCC, CCW, CWC, CWW, WCC, WCW, WWC, WWW*}

    b. (i) {*CWW, WCW, WWC*}, probability = 3/8 = 37.5%

14. **a.** Probability = 3/8 = 37.5%

15. The methods used to compute the probabilities in exercises 12, 13, and 14 are exactly the same as those in exercise 11. The only difference in the solutions are the symbols used to denote the outcomes; the probabilities are identical. These exercises illustrate the fact that computing various probabilities that arise in connection with tossing a coin is mathematically identical to computing probabilities in other, more realistic situations. So if the coin tossing model is completely understood, many other probabilities can be computed without difficulty.

16. **a.** {*RRR, RRB, RRY, RBR, RBB, RBY, RYR, RYB, RYY, BRR, BRB, BRY, BBR, BBB, BBY, BYR, BYB, BYY, YRR, YRB, YRY, YBR, YBB, YBY, YYR, YYB, YYY*}

    b. {*RBY, RYB, YBR, BRY, BYR, YRB*}, Probability = 6/27 = 2/9 ≅ 22.2%

    c. {*RRB, RBR, BRR, RRY, RYR, YRR, BBR, BRB, RBB, BBY, BYB, YBB, YYR, YRY, RYY, YYB, YBY, BYY*}, Probability = 18/27 = 2/3 = $66\frac{2}{3}$%

18. **a.** {$B_1B_1, B_1B_2, B_1W, B_2B_1, B_2B_2, B_2W, WB_1, WB_2, WW$}

    b. {$B_1B_1, B_1B_2, B_2B_1, B_2B_2$}, probability = 4/9 ≅ 44.4%

    c. {B_1W, B_2W, WB_1, WB_2}, probability = 4/9 ≅ 44.4%

21. **a.**   10 11   12   13 14   15   16 17   18 ...   96   97 98   99

    $$\updownarrow \qquad\qquad \updownarrow \qquad\qquad \updownarrow \qquad \updownarrow \qquad\qquad\qquad \updownarrow$$
    $$3\cdot4 \qquad\quad 3\cdot5 \qquad\quad 3\cdot6 \quad\; 3\cdot32 \qquad\quad 3\cdot33$$

    The above diagram shows that there are as many positive two-digit integers that are multiples of 3 as there are integers from 4 to 33 inclusive. By Theorem 9.1.1, there are $33 - 4 + 1$, or 30, such integers.

    b. There are $99 - 10 + 1 = 90$ positive two-digit integers in all, and by part (a), 30 of these are multiples of 3. So the probability that a randomly chosen positive two-digit integer is a multiple of 3 is $30/90 = 1/3 = 33\frac{1}{3}$%.

    c. Of the integers from 10 through 99 that are multiples of 4, the smallest is 12 (=4 · 3) and the largest is 96 (= 4 · 24). Thus there are $24 - 3 + 1 = 22$ two-digit integers that are multiples of 4. Hence the probability that a randomly chosen two-digit integer is a multiple of 4 is $22/90 = 36\frac{2}{3}$%.

23. **c.** Because $\left\lfloor \dfrac{39}{2} \right\rfloor = 19$, the probability is $\dfrac{39 - 19 + 1}{39} = \dfrac{21}{39}$.

    d. Probability $= \dfrac{m - 3 + 1}{n} = \dfrac{m - 2}{n}$

24. **a.** (i) If n is even, there are $\left\lfloor \frac{n}{2} \right\rfloor = \frac{n}{2}$ elements in the sub-array.

    (ii) If n is odd, there are $\left\lfloor \frac{n}{2} \right\rfloor = \frac{n-1}{2}$ elements in the sub-array.

    b. There are n elements in the array, so

    (i) The probability that an element is in the given sub-array when n is even is $\dfrac{\frac{n}{2}}{n} = \frac{1}{2}$,

    (ii) The probability that an element is in the given sub-array when n is odd is $\dfrac{\frac{n-1}{2}}{n} = \frac{n-1}{2n}$.

26. Let k be the 27th element in the array. By Theorem 9.1.1, $k - 42 + 1 = 27$, and so $k = 42 + 27 - 1 = 68$. Thus the 27th element in the array is $A[68]$.

27. Let k be the 62nd element in the array. By Theorem 9.1.1, $k - 29 + 1 = 62$, so $k = 62 + 29 - 1 = 90$. Thus the 62nd element in the array is $B[90]$.

28. Let m be the smallest of the integers. By Theorem 9.1.1, $279 - m + 1 = 56$, and so $m = 279 - 56 + 1 = 224$. Thus the smallest of the integers is 224.

30.
$$
\begin{array}{cccccccccc}
1 & 2 & 3 & 4 & 5 & 6 & \ldots & 998 & 999 & 1000 & 1001 \\
& \updownarrow & & \updownarrow & & \updownarrow & & \updownarrow & & \updownarrow & \\
& 2\cdot 1 & & 2\cdot 2 & & 2\cdot 3 & & 2\cdot 499 & & 2\cdot 500 &
\end{array}
$$

The diagram above shows that there are as many even integers between 1 and 1001 as there are integers from 1 to 500 inclusive. There are 500 such integers.

31.
$$
\begin{array}{ccccccccccc}
1 & 2 & 3 & 4 & 5 & 6 & 7 & 8 & 9 & \ldots & 999 & 1000 & 1001 \\
& & \updownarrow & & & \updownarrow & & & \updownarrow & & \updownarrow & & \\
& & 3\cdot 1 & & & 3\cdot 2 & & & 3\cdot 3 & & 3\cdot 333 & &
\end{array}
$$

Thus there are 333 multiples of 3 between 1 and 1001.

32. **a.** M Tu W Th F Sa Su M Tu W Th F Sa Su ... F Sa Su M
    1 2 3 4 5 6 7 8 9 10 11 12 13 14   362 363 364 365
            $\updownarrow$ $\updownarrow$ $\updownarrow$
            $7\cdot 1$ $7\cdot 2$ $7\cdot 52$

Sundays occur on the 7th day of the year, the 14th day of the year, and in fact on all days that are multiples of 7. There are 52 multiples of 7 between 1 and 365, and so there are 52 Sundays in the year.

33. <u>Proof (by mathematical induction)</u>: Let the property $P(n)$ be the sentence

> The number of integers from m to n inclusive is $n - m + 1$. $\leftarrow P(n)$

We will prove by mathematical induction that the property is true for every integer $n \geq m$.

Show that $P(m)$ is true: $P(m)$ is true because there is just one integer, namely m, from m to m inclusive. Substituting m in place of n in the formula $n - m + 1$ gives $m - m + 1 = 1$, which is correct.

Show that for every integer $k \geq m$, if $P(k)$ is true then $P(k + 1)$ is true: Let k be any integer with $k \geq m$ and suppose that

> The number of integers from m to k inclusive is $k - m + 1$. $\leftarrow$ $\begin{array}{c} P(k) \\ \text{inductive hypothesis} \end{array}$

We must show that

> The number of integers from m to $k + 1$ inclusive is $(k + 1) - m + 1$. $\leftarrow P(k + 1)$

Consider the sequence of integers from m to $k + 1$ inclusive:

$$\underbrace{m, \quad m + 1, \quad m + 2, \quad \ldots, \quad k,}_{k - m + 1} \quad (k + 1).$$

By inductive hypothesis there are $k - m + 1$ integers from m to k inclusive. So there are $(k - m + 1) + 1$ integers from m to $k + 1$ inclusive. But $(k - m + 1) + 1 = (k + 1) - m + 1$. So there are $(k + 1) - m + 1$ integers from m to $k + 1$ inclusive *[as was to be shown]*.

Section 9.2

1.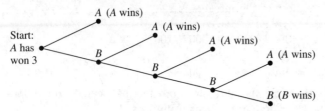

Game 4   Game 5   Game 6   Game 7

There are five ways to complete the series: A, B–A, B–B–A, B–B–B–A, and B–B–B–B.

3. Four ways: A–A–A–A, B–A–A–A–A, B–B–A–A–A–A, and B–B–B–A–A–A–A

4. Two ways: A–B–A–B–A–B–A and B–A–B–A–B–A–B

6. **a.**

Step 1:   Step 2:   Step 3:
Choose urn. Choose ball 1. Choose ball 2.

b. There are 12 equally likely outcomes of the experiment.

c. $2/12 = 1/6 = 16\frac{2}{3}\%$ **d.** $8/12 = 2/3 = 66\frac{2}{3}\%$

8. By the multiplication rule, the answer is $3 \cdot 2 \cdot 2 = 12$.

9. **a.** In going from city A to city B, one may take any of the 3 roads. In going from city B to city C, one may take any of the 5 roads. So, by the multiplication rule, there are $3 \cdot 5 = 15$ ways to travel from city A to city C via city B.

b. A round-trip journey can be thought of as a four-step operation:

Step 1: Go from A to B.
Step 2: Go from B to C.
Step 3: Go from C to B.
Step 4: Go from B to A.

Since there are 3 ways to perform step 1, 5 ways to perform step 2, 5 ways to perform step 3, and 3 ways to perform step 4, by the multiplication rule, there are $3 \cdot 5 \cdot 5 \cdot 3 = 225$ round-trip routes.

c. In this case the steps for making a round-trip journey are the same as in part (b), but since no route segment may be repeated, there are only 4 ways to perform step 3 and only 2 ways to perform step 4. So, by the multiplication rule, there are $3 \cdot 5 \cdot 4 \cdot 2 = 120$ round-trip routes in which no road is traversed twice.

11. **a.** Imagine constructing a bit string of length 8 as an eight-step process:

Step 1: Choose either a 0 or a 1 for the left-most position,
Step 2: Choose either a 0 or a 1 for the next position to the right.

$\vdots$

Step 8: Choose either a 0 or a 1 for the right-most position.

Since there are 2 ways to perform each step, the total number of ways to accomplish the entire operation, which is the number of different bit strings of length 8, is $2\cdot2\cdot2\cdot2\cdot2\cdot2\cdot2\cdot2 = 2^8 = 256$.

b. Imagine that there are three 0's in the three left-most positions, and imagine filling in the remaining 5 positions as a five-step process, where step i is to fill in the $(i+3)$rd position. Since there are 2 ways to perform each of the 5 steps, there are 2^5 ways to perform the entire operation. So there are 2^5, or 32, 8-bit strings that begin with three 0's.

c. If a bit string of length 8 begins and ends with a 1, then the six middle positions can be filled with any bit string of length 6. Hence there are $2^6 = 64$ such strings.

12. **a.** Think of creating a hexadecimal number that satisfies the given requirements as a five-step process.

Step 1:	Choose the left-most hexadecimal digits. It can be any of the 9 hexadecimal digits from 3 through B.
Steps 2–4:	Choose the three hexadecimal digits for the middle three positions. Each can be any of the 16 hexadecimal digits.
Step 5:	Choose the right-most hexadecimal digit. It can be any of the 11 hexadecimal digits from 5 through F.

There are 9 ways to perform step 1, 16 ways to perform each of steps 2 through 4, and 11 ways to perform step 5. Thus, the total number of specified hexadecimal numbers is $9 \cdot 16 \cdot 16 \cdot 16 \cdot 11 = 405{,}504$.

b. Think of creating a string of hexadecimal digits that satisfies the given requirements as a 6-step process.

Step 1: Choose the first hexadecimal digit. It can be any hexadecimal digit from 4 through D (which equals 13). There are $13 - 4 + 1 = 10$ of these.

Steps 2–5: Choose the second through the fifth hexadecimal digits. Each can be any one of the 16 hexadecimal digits.

Step 6: Choose the last hexadecimal digit. It can be any hexadecimal digit from 2 through E (which equals 14). There are $14 - 2 + 1 = 13$ of these.

So the total number of the specified hexadecimal numbers is $10 \cdot 16 \cdot 16 \cdot 16 \cdot 16 \cdot 13 = 8{,}519{,}680$.

13. **a.** In each of the four tosses there are two possible results: Either a head (H) or a tail (T) is obtained. Thus, by the multiplication rule, the number of outcomes is $2 \cdot 2 \cdot 2 \cdot 2 = 2^4 = 16$.

b. There are six outcomes with two heads: *HHTT, HTHT, HTTH, THHT, THTH, TTHH*. Thus the probability of obtaining exactly two heads is $6/16 = 3/8$.

14. **a.** Think of creating license plates that satisfy the given conditions as the following seven-step process: In steps 1–4 choose the letters to put in positions 1–4, and in steps 5–7, choose the digits to put in positions 5–7. Since there are 26 letters and 10 digits and since repetition is allowed, there are 26 ways to perform each of steps 1–4 and 10 ways to perform each of steps 5–7. Thus the number of license plates is

$$26 \cdot 26 \cdot 26 \cdot 26 \cdot 10 \cdot 10 \cdot 10 = 456{,}976{,}000.$$

b. In this case there is only one way to perform step 1 (because the first letter must be an A) and only one way to perform step 7 (because the last digit must be a 0). Therefore, the number of license plates is $26 \cdot 26 \cdot 26 \cdot 10 \cdot 10 = 1,757,600$.

d. In this case there are 26 ways to perform step 1, 25 ways to perform step 2, 24 ways to perform step 5, and 8 ways to perform step 6, so the number of license plates is $26 \cdot 25 \cdot 24 \cdot 23 \cdot 10 \cdot 9 \cdot 8 = 258,336,000$.

15. Think of creating combinations that satisfy the given requirements as multi-step processes in which each of steps 1-3 is to choose a number from 1 to 30, inclusive.

   a. Because there are 30 choices of numbers in each of steps 1–3, there are $30^3 = 27,000$ possible combinations for the lock.

   b. In this case we are given that no number may be repeated. So there are 30 choices for step 1, 29 for step 2, and 28 for step 3. Thus there are $30 \cdot 29 \cdot 28 = 24,360$ possible combinations for the lock.

16. **a.** Two solutions:

   (i) By the multiplication rule, the number of integers from 10 through 99

$$= \begin{bmatrix} \text{the number of ways to} \\ \text{pick the first digit} \end{bmatrix} \begin{bmatrix} \text{the number of ways to} \\ \text{pick the second digit} \end{bmatrix}$$

$$= 9 \cdot 10 = 90$$

   (ii) By Theorem 9.1.1, the number of integers from 10 through $99 = 90 - 10 + 1 = 90$.

   b. Because odd integers end in 1, 3, 5, 7, or 9, the number of odd integers from 10 through 99

$$= \begin{bmatrix} \text{the number of ways to} \\ \text{pick the first digit} \end{bmatrix} \begin{bmatrix} \text{the number of ways to} \\ \text{pick the second digit} \end{bmatrix}$$

$$= 9 \cdot 5 = 45.$$

   An alternative solution for part (b) uses the listing method shown in Example 9.1.4.

   c.
$$\begin{bmatrix} \text{the number of integers} \\ \text{with distinct digits} \end{bmatrix} = \begin{bmatrix} \text{the number of} \\ \text{ways to pick} \\ \text{the first digit} \end{bmatrix} \begin{bmatrix} \text{the number of} \\ \text{ways to pick} \\ \text{the sec ond digit} \end{bmatrix}$$
$$= 9 \cdot 9 = 81$$

   d.
$$\begin{bmatrix} \text{the number of odd integers} \\ \text{with distinct digits} \end{bmatrix} = \begin{bmatrix} \text{the number of} \\ \text{ways to pick} \\ \text{the second digit} \end{bmatrix} \begin{bmatrix} \text{the number of} \\ \text{ways to pick} \\ \text{the first digit} \end{bmatrix}$$
$$= 5 \cdot 8 = 40$$

   because the first digit can't equal 0, nor can it equal the second digit

   e. $81/90 = 9/10, 40/90 = 4/9$

18. **a.** Let step 1 be to choose either the number 2 or one of the letters corresponding to the number 2 on the keypad, let step 2 be to choose either the number 1 or one of the letters corresponding to the number 1 on the keypad, and let steps 3 and 4 be to choose either the number 3 or one of the letters corresponding to the number 3 on the keypad. There are 4 ways to perform step 1, 3 ways to perform step 2, and 4 ways to perform each of steps 3 and 4. So by the multiplication rule, there are $4 \cdot 3 \cdot 4 \cdot 4 = 192$ ways to perform the entire operation. Thus there are 192 different PINs that are keyed the same as 2133. Note that on a computer keyboard, these PINs would not be keyed the same way.

   b. Constructing a PIN that is obtainable by the same keystroke sequence as 5031 can be thought of as the following four-step process:

Step 1: Choose either the digit 5 or one of the three letters on the same key as the digit 5.

Step 2: Choose the digit 0.

Step 3: Choose the digit 3 or one of the three letters on the same key as the digit 3.

Step 4: Choose either the digit 1 or one of the two letters on the same key as the digit 1.

There are four ways to perform steps 1 and 3, one way to perform step 2, and three ways to perform step 4. So by the multiplication rule there are $4 \cdot 1 \cdot 4 \cdot 3 = 48$ different PINs that are keyed the same as 5031.

c. Constructing a numeric PIN with no repeated digit can be thought of as the following four-step process. Steps 1–4 are to choose the digits in position 1–4 (counting from the left). Because no digit may be repeated, there are 10 ways to perform step one, 9 ways to perform step two, 8 ways to perform step three, and 7 ways to perform step four. Thus the number of numeric PINs with no repeated digit is $10 \cdot 9 \cdot 8 \cdot 7 = 5040$.

19.

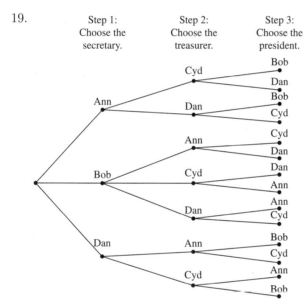

There are 14 different paths from "root" to "leaf" of this possibility tree, and so there are 14 ways the officers can be chosen. Because $14 = 2 \cdot 7$, reordering the steps will not make it possible to use the multiplication rule alone to solve this problem.

20. **a.** The number of ways to perform step 4 is not constant; it depends on how the previous steps were performed. For instance, if 3 digits had been chosen in steps 1–3, then there would be $10 - 3 = 7$ ways to perform step 4, but if 3 letters had been chosen in steps 1–3, then there would be 10 ways to perform step 4.

21. **a.** There are 2^{mn} relations from A to B because a relation from A to B is any subset of $A \times B$, $A \times B$ is a set with mn elements (since A has m elements and B has n elements), and the number of subsets of a set with mn elements is 2^{mn} (by Theorem 6.3.1).

b. In order to define a function from A to B we must specify exactly one image in B for each of the m elements in A. So we can think of constructing a function from A to B as an m-step process, where step i is to choose an image for the ith element of A (for $i = 1, 2, \ldots, m$). Because there are n choices of image for each of the m elements, by the multiplication rule, the total number of functions is $\underbrace{n \cdot n \cdot n \cdots n}_{m \text{ factors}} = n^m$.

c. The fraction of relations from A to B that are functions is $\dfrac{n^m}{2^{nm}} = \left(\dfrac{n}{2^n}\right)^m$

22. **a.** The answer is $4 \cdot 4 \cdot 4 = 4^3 = 64$. Imagine creating a function from a 3-element set to a 4-element set as a three-step process: Step 1 is to send the first element of the 3-element set to an element of the 4-element set (there are four ways to perform this step); step 2 is to send the second element of the 3-element set to an element of the 4-element set (there are also four ways to perform this step); and step 3 is to send the third element of the 3-element set to an element of the 4-element set (there are four ways to perform this step). Thus the entire process can be performed in $4 \cdot 4 \cdot 4$ different ways.

24. The outer loop is iterated 30 times, and during each iteration of the outer loop there are 15 iterations of the inner loop. Hence, by the multiplication rule, the total number of iterations of the inner loop is $30 \cdot 15 = 450$.

27. The outer loop is iterated $50 - 5 + 1 = 46$ times, and during each iteration of the outer loop there are $20 - 10 + 1 = 11$ iterations of the inner loop. Hence, by the multiplication rule, the total number of iterations of the inner loop is $46 \cdot 11 = 506$.

29. *Hint:* An efficient solution is to add leading zeros as needed to make each number five digits long. For instance, write 1 as 00001. Then, instead of choosing digits for the positions, choose positions for the digits. The answer is 720.

30. **a.** Call one of the integers r and the other s. Since r and s have no common factors, if p_i is a factor of r, then p_i is not a factor of s.

    So for each $i = 1, 2, \ldots, m$, either $p_i{}^{k_i}$ is a factor of r or $p_i{}^{k_i}$ is a factor of s.

    Thus, constructing r can be thought of as an m-step process in which step i is to decide whether $p_i{}^{k_i}$ is a factor of r or not.

    There are two ways to perform each step, and so the number of different possible r's is 2^m.

    Observe that once r is specified, s is completely determined because $s = n/r$.

    Hence the number of ways n can be written as a product of two positive integers rs which have no common factors is 2^m. Note that this analysis assumes that order matters because, for instance, $r = 1$ and $s = n$ will be counted separately from $r = n$ and $s = 1$.

    b. Each time that we can write n as rs, where r and s have no common factors, we can also write $n = sr$. So if order matters, there are twice as many ways to write n as a product of two integers with no common factors as there are if order does not matter. Thus if order does not matter, there are $2^m/2 = 2^{m-1}$ ways to write n as a product of two integers with no common factors.

31. **a.** There are $a + 1$ divisors: $1, p, p^2, \ldots, p^a$.

    b. A divisor is a product of any one of the $a + 1$ numbers listed in part (a) times any one of the $b + 1$ numbers $1, q, q^2, \ldots, q^b$. So, by the multiplication rule, there are $(a + 1)(b + 1)$ divisors in all.

32. **a.** Since the nine letters of the word *ALGORITHM* are all distinct, there are as many arrangements of these letters in a row as there are permutations of a set with nine elements: $9! = 362,880$.

    b. In this case there are effectively eight symbols to be permuted (because $\boxed{AL}$ may be regarded as a single symbol). So the number of arrangements is $8! = 40,320$.

33. **a.** The number of ways the 6 people can be seated equals the number or permutations of a set of 6 elements, namely, $6! = 720$.

    b. Assuming that the row is bounded by two aisles, arranging the people in the row can be regarded as the following 2-step process:

    Step 1: Choose the aisle seat for the doctor. *[There are 2 ways to do this.]*

Step 2: Choose an ordering for the remaining people. *[There are 5! ways to do this.]*

Thus, by the multiplication rule, the answer is $2 \cdot 5! = 240$.

(If it is assumed that one end of the row is against a wall, then there is only one aisle seat and the answer is $5! = 120$.)

c. Each married couple can be regarded as a single item, so the number of ways to order the 3 couples is $3! = 6$.

34. The same reasoning as in Example 9.2.9 gives an answer of $4! = 24$.

35. *WX, WY, WZ, XW, XY, XZ, YW, YX, YZ, ZW, ZX, ZY*

36. *stu, stv, sut, suv, svt, svu, tsu, tsv, tus, tuv, tvs, tvu, ust, usv, uts, utv, uvs, uvt, vst, vsu, vts, vtu, vus, vut*

39. **a.** $P(9,3) = \dfrac{9 \cdot 8 \cdot 7 \cdot 6!}{6!} = 504$

$\quad$ **b.** $P(9,6) = 9!/(9-6)! = 9!/3! = 9 \cdot 8 \cdot 7 \cdot 6 \cdot 5 \cdot 4 = 60,480$

$\quad$ **c.** $P(8,5) = \dfrac{8 \cdot 7 \cdot 6 \cdot 5 \cdot 4 \cdot 3!}{3!} = 6,720$

$\quad$ **d.** $P(7,4) = 7!/(7-4)! = 7!/3! = 7 \cdot 6 \cdot 5 \cdot 4 = 840$

41. <u>Proof</u>: Let n be an integer and $n \geq 2$. Then

$$
\begin{aligned}
P(n+1,2) - P(n,2) &= \frac{(n+1)!}{[(n+1)-2]!} - \frac{n!}{(n-2)!} = \frac{(n+1)!}{(n-1)!} - \frac{n!}{(n-2)!} \\
&= \frac{(n+1) \cdot n \cdot (n-1)!}{(n-1)!} - \frac{n \cdot (n-1) \cdot (n-2)!}{(n-2)!} \\
&= n^2 + n - (n^2 - n) = 2n = 2 \cdot \frac{n \cdot (n-1)!}{(n-1)!} \\
&= 2 \cdot \frac{n!}{(n-1)!} = 2P(n,1).
\end{aligned}
$$

42. <u>Proof 1</u>: Let n be any integer such that $n \geq 3$. By the first version of the formula in Theorem 9.2.3,

$$
\begin{aligned}
P(n+1,3) - P(n,3) &= (n+1)n(n-1) - n(n-1)(n-2) \\
&= n(n-1)[(n+1) - (n-2)] \\
&= n(n-1)(n+1-n+2) \\
&= 3n(n-1) \\
&= 3P(n,2).
\end{aligned}
$$

Proof 2: Let n be any integer such that $n \geq 3$. By the second version of the formula in Theorem 9.2.3,

$$
\begin{aligned}
P(n+1, 3) - P(n, 3) &= \frac{(n+1)!}{((n+1) - 3)!} - \frac{n!}{(n-3)!} \\
&= \frac{(n+1)!}{(n-2)!} - \frac{n!}{(n-3)!} \\
&= \frac{(n+1) \cdot n!}{(n-2)!} - \frac{(n-2) \cdot n!}{(n-2) \cdot (n-3)!} \\
&= \frac{n!((n+1) - (n-2))}{(n-2)!} \\
&= \frac{n!}{(n-2)!} \cdot 3 \\
&= 3P(n, 2).
\end{aligned}
$$

45. Proof (by mathematical induction): Let the property $P(n)$ be the sentence

The number of permutations of a set with n elements is $n!$. $\leftarrow$ $P(n)$

We will prove by mathematical induction that the property is true for every integer $n \geq 1$.

Show that $P(1)$ is true: $P(1)$ is true because if a set consists of one element there is just one way to order it, and $1! = 1$.

Show that for every integer $k \geq 1$, if $P(k)$ is true then $P(k+1)$ is true: Let k be any integer with $k \geq 1$ and suppose that

The number of permutations of a set with k elements is $k!$. $\leftarrow$ $P(k)$
inductive hypothesis

We must show that

number of permutations of a set with k elements is $(k+1)!$. $\leftarrow$ $P(k+1)$

Let X be a set with $k+1$ elements. The process of forming a permutation of the elements of X can be considered a two-step operation as follows:

Step 1: Choose the element to write first.

Step 2: Write the remaining elements of X in some order.

Since X has $k+1$ elements, there are $k+1$ ways to perform step 1, and by inductive hypothesis there are $k!$ ways to perform step 2. Hence by the multiplication rule there are $(k+1)k! = (k+1)!$ ways to form a permutation of the elements of X. But this means that there are $(k+1)!$ permutations of X *[as was to be shown]*.

47. **a.**

1 2 3	1 2 3	1 2 3	1 2 3	1 2 3	1 2 3
↓ ↓ ↓	↓ ↓ ↓	↓ ↓ ↓	↓ ↓ ↓	↓ ↓ ↓	↓ ↓ ↓
1 2 3	2 1 3	3 2 1	1 3 2	2 3 1	3 1 2

b.

1 2 3 4	1 2 3 4
↓ ↓ ↓ ↓	↓ ↓ ↓ ↓
1 2 3 4	3 2 1 4

Section 9.3

1. **a.** Think of creating a bit string with n bits as an n-step process where a general step k is to place either a 0 or a 1 in the kth position. Since there are two ways to do this for each position, by the multiplication rule, the number of bit strings of length k is 2^k. Now the set of all bit strings consisting of from 1 through 4 bits can be broken into four disjoint subsets:

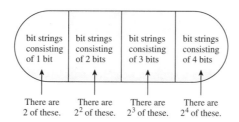

Applying the addition rule to the figure shows that there are $2 + 2^2 + 2^3 + 2^4 = 30$ bit strings consisting of from one through four bits.

b. By reasoning similar to that of part (a), there are $2^5 + 2^6 + 2^7 + 2^8 = 480$ bit strings of from five through eight bits.

3. **a.** The set of integers from 1 through 999 with no repeated digit can be broken into three disjoint subsets: those from 1 through 9, those from 1 through 99, and those from 100 through 999. Now constructing an integer from 100 through 999 with no repeated digit can be thought of as a three-step process.

Step 1: Choose a digit for the left-most position (where there are 9 choices because 0 cannot be chosen).
Step 2: Choose a digit for the middle position (where there are also 9 choices because the digit in the left-most position cannot be reused but 0 can be used).
Step 3: Choose a digit for the right-most position (where there are 8 choices because neither of the other two digits can be reused).

Thus there are $9 \cdot 9 \cdot 8$ integers from 100 through 999 with no repeated digit. Similar reasoning shows that there are $9 \cdot 9$ integers from 10 through 99 with no repeated digit. Finally, there are clearly 9 integers from 1 through 9 with no repeated digit. Hence, by the addition rule, the number of integers from 1 through 999 with no repeated digit is $9 + 9 \cdot 9 + 9 \cdot 9 \cdot 8 = 738$.

b.

$$\left[\begin{array}{l} \text{number of integers from 1 through 999} \\ \text{with at least one repeated digit} \end{array} \right]$$

$$= \left[\begin{array}{l} \text{total number of} \\ \text{integers from} \\ \text{1 through 999} \end{array} \right] - \left[\begin{array}{l} \text{number of integers} \\ \text{from 1 through 999} \\ \text{with no repeated digits} \end{array} \right]$$

$$= 999 - 738 = 261$$

c. The probability that an integer chosen at random has at least one repeated digit is $261/999 \cong 26.1\%$.

4. Use the multiplication rule to count the elements in each of the three sets containing 1, 2, and 3 letters, respectively. Then, because these sets are disjoint, use the addition rule to compute the total number of elements in the three sets taken together.

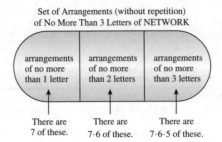

Set of Arrangements (without repetition)
of No More Than 3 Letters of NETWORK

arrangements of no more than 1 letter	arrangements of no more than 2 letters	arrangements of no more than 3 letters

There are 7 of these.   There are 7·6 of these.   There are 7·6·5 of these.

Applying the addition rule to the figure above shows that there are $7 + 7 \cdot 6 + 7 \cdot 6 \cdot 5 = 259$ arrangements of three letters of the word *NETWORK* if repetition of letters is not permitted.

6. In this exercise the 26 letters in the alphabet plus the 10 digits give a total of 36 symbols that can be used on a license plate.

   a. Imagine constructing a license plate with 4 symbols as a four-step process: step 1 is to fill in the first symbol, step 2 is to fill in the second symbol, step 3 is to fill in the third symbol, and step 4 is to fill in the fourth symbol. Because any one of the 36 symbols can be used in each step, by the multiplication rule, the number of license plates that use four symbols is 36^4. Similarly, the number that use 5 symbols is 36^5, and the number that use six symbols is 36^6. Thus because license plates can have anywhere from 4 to 6 symbols, the total number of plates with repeated symbols allowed is

$$36^4 + 36^5 + 36^6 = 2{,}238{,}928{,}128.$$

   b. When repetition is not allowed, the number of license plates that use four symbols is $36 \cdot 35 \cdot 34 \cdot 33$. The reason is that in the second step the symbol used in the first step cannot be used, so there are only 35 choices for the second step. In the third step, neither of the symbols used in the first two steps can be used, and so there are only 34 choices for the third step. And in the fourth step, none of the symbols used in the first three steps can be used, and so there are only 33 choices for the fourth step. Similarly, the number of license plates that use 5 symbols is $36 \cdot 35 \cdot 34 \cdot 33 \cdot 32$, and the number that use six symbols is $36 \cdot 35 \cdot 34 \cdot 33 \cdot 32 \cdot 31$. Thus the total number of license plates is

$$36 \cdot 35 \cdot 34 \cdot 33 + 36 \cdot 35 \cdot 34 \cdot 33 \cdot 32 + 36 \cdot 35 \cdot 34 \cdot 33 \cdot 32 \cdot 31 = 1{,}449{,}063{,}000.$$

   c. Consider two sets: the set of plates with repetition not allowed and the set of plates that have a repeated symbol. Note that these two sets have no elements in common, and that since every license plate either has a repeated symbol or does not have a repeated symbol, every license plate considered in part (a) is in one of the two sets. In other words, the set of all license plates with repetition allowed is composed of two disjoint subsets: the set of plates with repetition not allowed and the set of plates that have a repeated symbol. Thus, by the difference rule, the number of license plates with a repeated symbol is the difference between the number of plates with repetition allowed minus the number of plates with repetition not allowed:

$$2{,}238{,}928{,}128 - 1{,}449{,}063{,}000 = 789{,}865{,}128.$$

   d. The probability that a license plate chosen at random has at least one repeated symbol is $\dfrac{789{,}865{,}128}{2{,}238{,}928{,}128} \cong 35.3\%$.

7. **a.** The 26 letters in the alphabet plus the 10 digits plus the 14 special characters give a total of 50 symbols that can be used. By the multiplication rule, the number of passwords with 3, 4, and 5 symbols is 50^3, 50^4, and 50^5. Since the sets consisting of these passwords are disjoint, by the addition rule, the number of passwords is

$$50^3 + 50^4 + 50^5 = 318{,}875{,}000.$$

8. **a.** *Hint:* One approach is to divide the license plates into four groups depending on the number of digits and letters they contain. Another approach is to consider creating a license plate as a two-step process: *step 1*: either choose one digit or do not choose a digit; and *step 2*: choose 4 or 5 letters.

9. **a.** Each column of the table below corresponds to a pair of values of i and j for which the inner loop will be iterated.

i	1	2	→	3		→	4			→
j	1	1	2	1	2	3	1	2	3	4
	1	2		3			4			

Since there are $1 + 2 + 3 + 4 = 10$ columns, the inner loop will be iterated ten times.

b. On the ith iteration of the outer loop, there are i iterations of the inner loop, and this is true for each $i = 1, 2, \ldots, n$. Therefore, the total number of iterations of the inner loop is $1 + 2 + 3 + \cdots + n = n(n + 1)/2$.

11. **a.** The answer is the number of permutations of the five letters in *QUICK*, which equals $5! = 120$.

b. Because *QU* (in order) is to be considered as a single unit, the answer is the number of permutations of the four symbols $\boxed{QU}$, *I, C, K*. This is $4! = 24$.

c. By part (b), there are $4!$ arrangements of $\boxed{QU}$, *I, C, K*. Similarly, there are $4!$ arrangements of $\boxed{UQ}$, *I, C, K*. Therefore, by the addition rule, there are $4! + 4! = 48$ arrangements in all.

12. **a.** The number of ways to arrange the 6 letters of the word *THEORY* in a row is $6! = 720$

b. When the *TH* in the word *THEORY* are treated as an ordered unit, there are only 5 items to arrange, *TH, E, O, R,* and *Y.* and so there are $5!$ orderings. Similarly, there are $5!$ orderings for the symbols *HT, E, O, R,* and *Y*. Thus, by the addition rule, the total number of orderings is $5! + 5! = 120 + 120 = 240$.

13. **a.**
$$\begin{bmatrix} \text{the number of ways to place eight people} \\ \text{in a row keeping } A \text{ and } B \text{ together} \end{bmatrix}$$
$$= \begin{bmatrix} \text{the number of ways to arrange} \\ \boxed{AB} \, C\,D\,E\,F\,G\,H \end{bmatrix} + \begin{bmatrix} \text{the number of ways to arrange} \\ \boxed{BA} \, C\,D\,E\,F\,G\,H \end{bmatrix}$$
$$= 7! + 7! = 5{,}040 + 5{,}040 = 10{,}080$$

b.
$$\begin{bmatrix} \text{the number of ways to arrange the eight} \\ \text{people in a row keeping } A \text{ and } B \text{ apart} \end{bmatrix}$$
$$= \begin{bmatrix} \text{the total number of ways to} \\ \text{place the eight people in a row} \end{bmatrix} - \begin{bmatrix} \text{the number of ways to place the eight} \\ \text{people in a row keeping } A \text{ and } B \text{ together} \end{bmatrix}$$
$$= 8! - 10{,}080 = 40{,}320 - 10{,}080 = 30{,}240$$

14. the number of variable names
$$= \begin{bmatrix} \text{the number of numeric} \\ \text{variable names} \end{bmatrix} + \begin{bmatrix} \text{the number of string} \\ \text{variable names} \end{bmatrix}$$
$$= (26 + 26 \cdot 36) + (26 + 26 \cdot 36) = 1{,}924$$

15. The set of all possible identifiers may be divided into 30 non-overlapping subsets depending on the number of characters in the identifier. Constructing one of the identifiers in the kth subset can be regarded as a k-step process, where each step consists in choosing a symbol for

one of the characters (say, going from left to right). Because the first character must be a letter, there are 26 choices for step 1, and because subsequent letters can be letters or digits or underscores there are 37 choices for each subsequent step. By the addition rule, we add up the number of identifiers in each subset to obtain a total. But because 82 of the resulting strings cannot be used as identifiers, by the difference rule, we subtract 82 from the total to obtain the final answer. Thus we have

$$(26 + 26 \cdot 37 + 26 \cdot 37^2 + \cdots + 26 \cdot 37^{29}) - 82 = 26(1 + 37 + 37^2 + \cdots + 37^{29}) - 82$$

$$= 26 \cdot \sum_{k=0}^{29} 37^k - 82 = 26 \left(\frac{37^{30} - 1}{37 - 1} \right) - 82 \cong 8.030 \times 10^{46}.$$

16. **a.** $10 \cdot 9 \cdot 8 \cdot 7 \cdot 6 \cdot 5 \cdot 4 = 604{,}800$

b.
$$\left[\begin{array}{c} \text{the number of phone numbers} \\ \text{with at least one repeated digit} \end{array} \right]$$

$$= \left[\begin{array}{c} \text{the total number} \\ \text{of phone numbers} \end{array} \right] - \left[\begin{array}{c} \text{the number of phone numbers} \\ \text{with no repeated digits} \end{array} \right]$$

$$= 10^7 - 604{,}800 = 9{,}395{,}200$$

c. $\dfrac{9{,}395{,}200}{10^7} \cong 93.95\%$

18. **a.** <u>Proof</u>: Let A and B be mutually disjoint events in a sample space S. By the addition rule, $N(A \cup B) = N(A) + N(B)$. Therefore, by the equally likely probability formula,

$$P(A \cup B) = \frac{N(A \cup B)}{N(S)} = \frac{N(A) + N(B)}{N(S)}$$

$$= \frac{N(A)}{N(S)} + \frac{N(B)}{N(S)} = P(A) + P(B).$$

b. <u>Proof</u>: Let A and B be events in a sample space S with equally likely outcomes, and assume that A and B are mutually disjoint. By the inclusion/exclusion rule (Theorem 9.3.3), $N(A \cup B) = N(A) + N(B) - N(A \cap B)$. So by the equally likely probability formula,

$$P(A \cup B) = \frac{N(A \cup B)}{N(S)} = \frac{N(A) + N(B) - N(A \cap B)}{N(S)} = \frac{N(A)}{N(S)} + \frac{N(B)}{N(S)} - \frac{N(A \cap B)}{N(S)}$$

$$= P(A) + P(B) - P(A \cap B).$$

19. *Hint:* Justify the following answer: $39 \cdot 38 \cdot 38$.

20. **a.** Use strings of five digits to represent integers from 1 to 100,000 that contain the digit 6 exactly once. For example, use 00306 to represent 306. Strings of six digits are not needed because 100,000 does not contain a 6. Imagine constructing a five-digit string that contains exactly one 6 as a five-step operation to fill in five positions with five digits: $\underset{1}{\underline{}} \, \underset{2}{\underline{}} \, \underset{3}{\underline{}} \, \underset{4}{\underline{}} \, \underset{5}{\underline{}}$.

Step 1: Choose one of the five positions for the 6.
Step 2: Choose a digit for the left-most remaining position.
Step 3: Choose a digit for the next remaining position to the right.
Step 4: Choose a digit for the next remaining position to the right.
Step 5: Choose a digit for the right-most position.

Since there are 5 choices for step 1 (any one of the five positions) and 9 choices for each of steps 2–5 (any digit except 6), by the multiplication rule, the number of ways to perform this operation is $5 \cdot 9 \cdot 9 \cdot 9 \cdot 9 = 32{,}805$. Hence there are 32,805 integers from 1 to 100,000 that contain the digit 6 exactly once.

21. Call the employees U, V, W, X, Y, and Z, and suppose that U and V are the married couple. Let A be the event that U and V have adjacent desks. Since the desks of U and V can be adjacent either in the order UV or in the order VU, the number of desk assignments with U and V adjacent is the same as the sum of the number of permutations of the symbols $\boxed{UV}$, W, X, Y, Z plus the number of permutations of the symbols $\boxed{VU}$, W, X, Y, Z. By the multiplication rule each of these is 5!, and so by the addition rule the sum is $2 \cdot 5!$. Since the total number of permutations of U, V, W, X, Y, Z is 6!,

$$P(A) = 2 \cdot \frac{5!}{6!} = \frac{2}{6} = \frac{1}{3}.$$

Hence by the formula for the probability of the complement of an event,

$$P(A^c) = 1 - P(A) = 1 - \frac{1}{3} = \frac{2}{3}.$$

So the probability that the married couple have nonadjacent desks is 2/3.

23. **a.** Let A = the set of integers that are multiples of 4 and B = the set of integers that are multiples of 7. Then $A \cap B$ = the set of integers that are multiples of 28.

Now $N(A) = 250$ since 1 2 3   4   5 6 7   8   ... 999   1000,

$$\updownarrow \qquad\qquad \updownarrow \qquad\qquad \updownarrow$$
$$4 \cdot 1 \qquad 4 \cdot 2 \ \ldots \qquad 4 \cdot 250$$

or, equivalently, since $1{,}000 = 4 \cdot 250$.

Also $N(B) = 142$ since 1 2 3 4 5 6   7   ...   14   ...   994   995 ... 1000

$$\updownarrow \qquad \updownarrow \qquad \updownarrow$$
$$7 \cdot 1 \qquad 7 \cdot 2 \ \ldots \ 7 \cdot 142$$

or, equivalently, since $1{,}000 = 7 \cdot 142 + 6$.

And $N(A \cap B) = 35$ since 1 2 3 ...   28   ...   56   ...   980   ... 1000,

$$\updownarrow \qquad \updownarrow \qquad \updownarrow$$
$$28 \cdot 1 \qquad 28 \cdot 2 \ \ldots \ 28 \cdot 35$$

or, equivalently, since $1{,}000 = 28 \cdot 35 + 20$.

So $N(A \cup B) = 250 + 142 - 35 = 357$.

24. **a.** Let A and B be the sets of all integers from 1 through 1,000 that are multiples of 2 and 9 respectively. Then $N(A) = 500$ and $N(B) = 111$ (because $9 = 9 \cdot 1$ is the smallest integer in B and $999 = 9 \cdot 111$ is the largest). Also $A \cap B$ is the set of all integers from 1 through 1,000 that are multiples of 18, and $N(A \cap B) = 55$ (because $18 = 18 \cdot 1$ is the smallest integer in $A \cap B$ and $990 = 18 \cdot 55$ is the largest). It follows from the inclusion/exclusion rule that the number of integers from 1 through 1,000 that are multiples of 2 or 9 equals

$$N(A \cup B) = N(A) + N(B) - N(A \cap B) = 500 + 111 - 55 = 556.$$

b. The probability is $556/1000 = 55.6\%$.

c. $1000 - 556 = 444$

25. **a.** Length 0: λ

Length 1: 0, 1

Length 2: 00, 01, 10, 11

Length 3: 000, 001, 010, 011, 100, 101, 110

Length 4: 0000, 0001, 0010, 0011, 0100, 0101, 0110, 1000, 1001, 1010, 1011, 1100, 1101

b. By part (a), $d_0 = 1$, $d_1 = 2$, $d_2 = 4$, $d_3 = 7$, and $d_4 = 13$.

c. Let k be an integer with $k \geq 3$. Any string of length k that does not contain the bit pattern 111 starts either with a 0 or with a 1. If it starts with a 0, this can be followed by any string of $k - 1$ bits that does not contain the pattern 111. There are d_{k-1} of these. If the string starts with a 1, then the first two bits are 10 or 11. If the first two bits are 10, then these can be followed by any string of $k - 2$ bits that does not contain the pattern 111. There are d_{k-2} of these. If the string starts with a 11, then the third bit must be 0 (because the string does not contain 111), and these three bits can be followed by any string of $k - 3$ bits that does not contain the pattern 111. There are d_{k-3} of these. Therefore, for every integer $k \geq 3, d_k = d_{k-1} + d_{k-3}$.

d. By parts (b) and (c), $d_5 = d_4 + d_3 + d_2 = 13 + 7 + 4 = 24$. This is the number of bit strings of length five that do not contain the pattern 111.

26. **a.** *Hint:* $s_k = 2s_{k-1} + 2s_{k-2}$

b. *Hint:* For every integer $n \geq 0$,

$$s_n = \frac{\sqrt{3}+2}{2\sqrt{3}}(1 + \sqrt{3})^n + \frac{\sqrt{3}-2}{2\sqrt{3}}(1 - \sqrt{3})^n.$$

27. **a.** Let k be an integer with $k \geq 3$. The set of bit strings of length k that do not contain the pattern 101 can be partitioned into $k + 1$ subsets: the subset of strings that start with 0 and continue with any bit string of length $k - 1$ not containing 101 *[there are a_{k-1} of these]*, the subset of strings that start with 100 and continue with any bit string of length $k - 3$ not containing 101 *[there are a_{k-3} of these]*, the subset of strings that start with 1100 and continue with any bit string of length $k - 4$ not containing 101 *[there are a_{k-4} of these]*, the subset of strings that start with 11100 and continue with any bit string of length $k - 5$ not containing 101 *[there are a_{k-5} of these]*, until the following subset of strings is obtained: $\{\underbrace{11\ldots1}_{k-3\ 1\text{'s}}001, \underbrace{11\ldots1}_{k-3\ 1\text{'s}}000\}$ *[there are 2 of these and a_1 equals 2]*. In addition, the three single-element sets $\{\underbrace{11\ldots1}_{k-2\ 1\text{'s}}00\}$, $\{\underbrace{11\ldots1}_{k-1\ 1\text{'s}}0\}$, and $\{\underbrace{11\ldots1}_{k-1\ 1\text{'s}}1\}$ are in the partition, and since $a_0 = 1$ (because the only bit string of length zero that satisfies the condition is λ), $3 = a_0 + 2$. Thus by the addition rule,

$$a_k = a_{k-1} + a_{k-3} + a_{k-4} + \cdots + a_1 + a_0 + 2.$$

b. By part (a), if $k \geq 4$,

$$
\begin{aligned}
a_k &= a_{k-1} + a_{k-3} + a_{k-4} + \cdots + a_1 + a_0 + 2 \\
a_{k-1} &= a_{k-2} + a_{k-4} + a_{k-5} + \cdots + a_1 + a_0 + 2.
\end{aligned}
$$

Subtracting the second equation from the first gives

$$
\begin{aligned}
a_k - a_{k-1} &= a_{k-1} + a_{k-3} - a_{k-2} \\
\Rightarrow \qquad a_k &= 2a_{k-1} + a_{k-3} - a_{k-2}. \text{ (Call this equation (*).)}
\end{aligned}
$$

Note that $a_2 = 4$ (because all four bit strings of length 2 satisfy the condition) and $a_3 = 7$ (because all eight bit strings of length 3 satisfy the condition except 101). Thus equation (*) is also satisfied when $k = 3$ because in that case the right-hand side of the equation becomes $2a_2 + a_0 - a_1 = 2 \cdot 4 + 1 - 2 = 7$, which equals the left-hand side of the equation.

28. **a.** $a_3 = 3$ (The three permutations that do not move more than one place from their "natural" positions are 213, 132, and 123.)

29. **a.** $11001010_2 = 2 + 2^3 + 2^6 + 2^7 = 202$

$00111000_2 = 2^3 + 2^4 + 2^5 = 56$

$01101011_2 = 1 + 2 + 2^3 + 2^5 + 2^6 = 107$

$11101110_2 = 2 + 2^2 + 2^3 + 2^5 + 2^6 + 2^7 = 238$

So the answer is 202.56.107.238.

b. The network ID for a Class A network consists of 8 bits and begins with 0. If all possible combinations of eight 0's and 1's that start with a 0 were allowed, there would be 2 choices (0 or 1) for each of the 7 positions from the second through the eighth. This would give $2^7 = 128$ possible ID's. But because neither 00000000 nor 01111111 is allowed, the total is reduced by 2, so there are 126 possible Class A networks.

c. Let $w.x.y.z$ be the dotted decimal form of the IP address for a computer in a Class A network. Because the network IDs for a Class A network go from 00000001 $(= 1)$ through 01111110 $(= 126)$, w can be any integer from 1 through 126. In addition, each of x, y, and z can be any integer from 0 $(= 00000000)$ through 255 $(= 11111111)$, except that x, y, and z cannot all be 0 simultaneously and cannot all be 255 simultaneously.

d. Twenty-four positions are allocated for the host ID in a Class A network. If each could be either 0 or 1, there would be $2^{24} = 16,777,216$ possible host IDs. But neither all 0's nor all 1's is allowed, which reduces the total by 2. Thus there are 16,777,214 possible host IDs in a Class A network.

i. Observe that $140 = 128 + 8 + 4 = 10001100_2$, which begins with 10. Thus the IP address comes from a Class B network. An alternative solution uses the result of Example 9.3.5: Network IDs for Class B networks range from 128 through 191.

Thus, since $128 \leq 140 \leq 191$, the given IP address is from a Class B network.

30. To get a sense of the problem, we compute s_4 directly. If there are four seats in the row, there can be a single student in any one of the four seats or there can be a pair of students in seats 1&3, 1&4, or 2&4. No other arrangements are possible because with more than two students, two would have to sit next to each other. Thus $s_4 = 4 + 3 = 7$. In general, if there are k chairs in a row, then

$$
\begin{aligned}
s_k \;=\; & s_{k-1} \quad \text{(the number of ways a nonempty set of students can sit} \\
& \qquad \text{in the row with no two students adjacent and chair } k \text{ empty)} \\
& +s_{k-2} \quad \text{(the number of ways students can sit in the row with chair } k \\
& \qquad \text{occupied, chair } k \;\; 1 \text{ empty, and chairs 1 through} \\
& \qquad k-2 \text{ occupied by a nonempty set of students in such a} \\
& \qquad \text{way that no two students are adjacent)} \\
& +1 \quad \text{(for the seating in which chair } k \text{ is occupied} \\
& \qquad \text{and all the other chairs are empty} \\
\;=\; & s_{k-1} + s_{k-2} + 1 \text{ for all integers } k \geq 3.
\end{aligned}
$$

31. **a.** There are 12 possible birth months for A, 12 for B, 12 for C, and 12 for D, so the total is $12^4 = 20,736$.

b. If no two people share the same birth month, there are 12 possible birth months for A, 11 for B, 10 for C, and 9 for D. Thus the total is $12 \cdot 11 \cdot 10 \cdot 9 = 11,880$.

c. If at least two people share the same birth month, the total number of ways birth months could be associated with A, B, C, and D is $20,736 - 11,880 = 8,856$.

d. The probability that at least two of the four people share the same birth month is $\frac{8,856}{20,736} \cong 42.7\%$.

e. When there are five people, the probability that at least two share the same birth month is $\frac{12^5 - 12 \cdot 11 \cdot 10 \cdot 9 \cdot 8}{12^5} \cong 61.8\%$, and when there are more than five people, the probability is even greater. Thus, since the probability for four people is less than 50%, the group must contain five or more people for the probability to be at least 50% that two or more share the same birth month.

32. *Hint:* Analyze the solution to exercise 31.

33. **a.** The number of students who checked at least one of the statements is $N(H) + N(C) + N(D) - N(H \cap C) - N(H \cap D) - N(C \cap D) + N(H \cap C \cap D) = 28 + 26 + 14 - 8 - 4 - 3 + 2 = 55$.

b. By the difference rule, the number of students who checked none of the statements is the total number of students minus the number who checked at least one statement. This is $100 - 55 = 45$.

c.

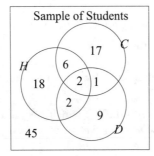

Sample of Students

d. The number of students who checked #1 and #2 but not #3 is $N(H \cap D) - N(H \cap C \cap D) = 8 - 2 = 6$.

e. 1   **f.** 17

35. Let

M = the set of married people in the sample,

Y = the set of people between 20 and 30 in the sample, and

F = the set of females in the sample.

Then the number of people in the set $M \cup Y \cup F$ is less than or equal to the size of the sample. And so

$$
\begin{aligned}
1{,}200 \; &\geq \; N(M \cup Y \cup F) \\
&= \; N(M) + N(Y) + N(F) - N(M \cap Y) - N(M \cap F) - N(Y \cap F) + N(M \cap Y \cap F) \\
&= \; 675 + 682 + 684 - 195 - 467 - 318 + 165 \\
&= \; 1{,}226.
\end{aligned}
$$

This is impossible since $1{,}200 < 1{,}226$, so the polltaker's figures are inconsistent. They could not have occurred as a result of an actual sample survey.

36. **a.** by the double complement law and the difference rule   **b.** by De Morgan's law

c. by the inclusion/exclusion rule

37. Let A be the set of all positive integers less than 1,000 that are not multiples of 2, and let B be the set of all positive integers less than 1,000 that are not multiples of 5. Since the only prime factors of 1,000 are 2 and 5, the number of positive integers that have no common factors with 1,000 is $N(A \cap B)$. Let the universe U be the set of all positive integers less than 1,000. Then A^c is the set of positive integers less than 1,000 that are multiples of 2, B^c is the set of positive integers less than 1,000 that are multiples of 5, and $A^c \cap B^c$ is the set of positive integers less than 1,000 that are multiples of 10. By one of the procedures discussed in Section 9.1 or 9.2, it is easily found that $N(A^c) = 499$, $N(B^c) = 199$, and $N(A^c \cap B) = 99$. Thus, by the inclusion/exclusion rule,

$$N(A^c \cup B^c) = N(A^c) + N(B^c) - N(A^c \cap B^c) = 499 + 199 - 99 = 599.$$

But by De Morgan's law, $N(A^c \cup B^c) = N((A \cap B)^c)$, and so

$$N((A \cap B)^c) = 599. \qquad (*)$$

Now since $(A \cap B)^c = U - (A \cap B)$, by the difference rule we have

$$N((A \cap B)^c) = N(U) - N(A \cap B). \qquad (**)$$

Equating the right-hand sides of (*) and (**) gives $N(U) - N(A \cap B) = 599$. And because $N(U) = 999$, we conclude that $999 - N(A \cap B) = 599$, or, equivalently, $N(A \cap B) = 999 - 599 = 400$. So there are 400 positive integers less than 1,000 that have no common factor with 1,000.

39. Imagine each integer from 1 through 999,999 as a string of six digits with leading 0's included. For each $i = 1, 2, 3$, let A_i be the set of all integers from 1 through 999,999 that do not contain the digit i. We want to compute $N(A_1^c \cap A_2^c \cap A_3^c)$. By De Morgan's law,

$$A_1^c \cap A_2^c \cap A_3^c = (A_1 \cup A_2)^c \cap A_3^c = (A_1 \cup A_2 \cup A_3)^c = U - (A_1 \cup A_2 \cup A_3),$$

and so, by the difference rule,

$$N(A_1^c \cap A_2^c \cap A_3^c) = N(U) - N(A_1 \cup A_2 \cup A_3).$$

By the inclusion/exclusion rule,

$$N(A_1 \cup A_2 \cup A_3) = N(A_1) + N(A_2) + N(A_3) - N(A_1 \cap A_2) - N(A_1 \cap A_3) - N(A_2 \cap A_3) + N(A_1 \cap A_2 \cap A_3).$$

Now $N(A_1) = N(A_2) = N(A_3) = 9^6$ because in each case any of nine digits may be chosen for each character in the string (for A_i these are all the ten digits except i). Also each $N(A_i \cap A_j) = 8^6$ because in each case any of eight digits may be chosen for each character of the string (for $A_i \cap A_j$ these are all the ten digits except i and j). Similarly, $N(A_1 \cap A_2 \cap A_3) = 7^6$ because any digit except 1, 2, and 3 may be chosen for each character in the string. Thus

$$N(A_1 \cup A_2 \cup A_3) = 3 \cdot 9^6 - 3 \cdot 8^6 + 7^6,$$

and so, by the difference rule,

$$N(A_1^c \cap A_2^c \cap A_3^c) = N(U) - N(A_1 \cup A_2 \cup A_3) = 10^6 - (3 \cdot 9^6 - 3 \cdot 8^6 + 7^6) = 74,460.$$

40. *Hint:* Let A and B be the sets of all positive integers less than or equal to n that are divisible by p and q, respectively. Then $\phi(n) = n - (N(A \cup B))$.

42. **a.** $g_3 = 1$, $g_4 = 1$, $g_5 = 2$ (*LWLLL* and *WWLLL*)

b. $g_6 = 4$ (*WWWLLL, WLWLLL, LWWLLL, LLWLLL*)

c. If $k \geq 6$, then any sequence of k games must begin with exactly one of the possibilities: W, LW, or LLW. The number of sequences of k games that begin with W is g_{k-1} because the succeeding $k - 1$ games can consist of any sequence of wins and losses except that the first sequence of three consecutive losses occurs at the end. Similarly, the number of sequences of k games that begin with LW is g_{k-2} and the number of sequences of k games that begin with LLW is g_{k-3}. Therefore, $g_k = g_{k-1} + g_{k-2} + g_{k-3}$ for every integer $k \geq 6$.

43. **c.** *Hint:* Divide the set of all derangements into two subsets: one subset consists of all derangements in which the number 1 changes places with another number, and the other subset consists of all derangements in which the number 1 goes to position $i \neq 1$ but i does not go to position 1. The answer is $d_k = (k-1)d_{k-1} + (k-1)d_{k-2}$. Can you justify it?

45. <u>Proof (by mathematical induction)</u>: Let the property $P(k)$ be the sentence

> If a finite set A equals the union of k distinct mutually
> disjoint subsets subsets $A_1, A_2, \ldots, A_k$, then $\leftarrow \ P(k)$
> $N(A) = N(A_1) + N(A_2) + \cdots + N(A_k)$.

We will prove by mathematical induction that $P(k)$ is true for every integer $k \geq 1$.

Show that $P(1)$ is true: $P(1)$ is true because if a finite set A equals the "union" of one subset A_1, then $A = A_1$, and so $N(A) = N(A_1)$.

Show that for every integer $i \geq 1$, if $P(i)$ is true then $P(i+1)$ is true: Let i be any integer with $i \geq 1$ and suppose that

> If a finite set A equals the union of i distinct mutually $P(i)$
> disjoint subsets subsets $A_1, A_2, \ldots, A_i$, then $\leftarrow$
> $N(A) = N(A_1) + N(A_2) + \cdots + N(A_i)$.   inductive hypothesis

We must show that

> If a finite set A equals the union of $i+1$ distinct mutually
> disjoint subsets subsets $A_1, A_2, \ldots, A_{i+1}$, then $\leftarrow \ P(i+1)$
> $N(A) = N(A_1) + N(A_2) + \cdots + N(A_{i+1})$.

Let A be a finite set that equals the union of $i+1$ distinct mutually disjoint subsets $A_1, A_2, \ldots, A_{i+1}$. Then $A = A_1 \cup A_2 \cup \cdots \cup A_{i+1}$ and $A_i \cap A_j = \emptyset$ for every integer i and j with $i \neq j$.

Let B be the set $A_1 \cup A_2 \cup \cdots \cup A_i$. Then $A = B \cup A_{i+1}$ and $B \cap A_{i+1} = \emptyset$.

[For if $x \in B \cap A_{i+1}$, then $x \in A_1 \cup A_2 \cup \cdots \cup A_i$ and $x \in A_{i+1}$, which implies that $x \in A_j$, for some j with $1 \leq j \leq i$, and $x \in A_{i+1}$. But A_j and A_i are disjoint. Thus no such x exists.]

Hence A is the union of the two mutually disjoint sets B and A_{i+1}. Since B and A_{i+1} have no elements in common, the total number of elements in $B \cup A_{i+1}$ can be obtained by first counting the elements in B, next counting the elements in A_{i+1}, and then adding the two numbers together.

It follows that $N(B \cup A_{i+1}) = N(B) + N(A_{i+1})$ which equals $N(A_1) + N(A_2) + \cdots + N(A_i) + N(A_{i+1})$ by inductive hypothesis. Hence $P(i+1)$ is true *[as was to be shown]*.

48. <u>Proof (by mathematical induction)</u>: Let the property $P(n)$ be the general inclusion/exclusion rule. We will prove by mathematical induction that $P(n)$ is true for every integer $n \geq 2$.

Show that $P(2)$ is true: $P(2)$ was proved in one way in the text preceding Theorem 9.3.3 and in another way in the solution to exercise 46.

Show that for every integer $r \geq 2$, if $P(r)$ is true then $P(r+1)$ is true: Let r be any integer with $r \geq 2$ and suppose that the general inclusion/exclusion rule holds for any collection of r finite sets. (This is the inductive hypothesis.) Let $A_1, A_2, \ldots, A_{r+1}$ be finite sets. Then

$N(A_1 \cup A_2 \cup \cdots \cup A_{r+1})$

$= N(A_1 \cup (A_2 \cup A_3 \cup \cdots \cup A_{r+1}))$ by the associative law for $\cup$

$= N(A_1) + N(A_2 \cup A_3 \cup \cdots \cup A_{r+1}) - N(A_1 \cap (A_2 \cup A_3 \cup \cdots \cup A_{r+1}))$

$\qquad\qquad\qquad\qquad\qquad\qquad$ by the inclusion/exclusion rule for two sets

$= N(A_1) + N(A_2 \cup A_3 \cup \cdots \cup A_{r+1}) - N((A_1 \cap A_2) \cup (A_1 \cap A_3) \cup \cdots \cup (A_1 \cap A_{r+1}))$

$\qquad\qquad\qquad\qquad\qquad\qquad$ by the generalized distributive law for sets
$\qquad\qquad\qquad\qquad\qquad\qquad$ (exercise 37, Section 6.2)

$= N(A_1) + \left(\sum_{2 \leq i \leq r+1} N(A_i) - \sum_{2 \leq i < j \leq r+1} N(A_i \cap A_j) \right)$

$$+ \sum_{2 \leq i < j < k \leq r+1} N(A_i \cap A_j \cap A_k) - \cdots + (-1)^{r+1} N(A_2 \cap A_3 \cap \cdots \cap A_{r+1}) \Big)$$

$$- \Big(\sum_{2 \leq i \leq r+1} N(A_1 \cap A_i) - \sum_{2 \leq i < j \leq r+1} N((A_1 \cap A_i) \cap (A_1 \cap A_j)) + \cdots$$

$$+ (-1)^{r+1} N((A_1 \cap A_2) \cap (A_1 \cap A_3) \cap \cdots \cap (A_1 \cap A_{r+1})) \Big)$$

by inductive hypothesis

$$= N(A_1) + \Big(\sum_{2 \leq i \leq r+1} N(A_i) - \sum_{2 \leq i < j \leq r+1} N(A_i \cap A_j)$$

$$+ \sum_{2 \leq i < j < k \leq r+1} N(A_i \cap A_j \cap A_k) - \cdots + (-1)^{r+1} N(A_2 \cap A_3 \cap \cdots \cap A_{r+1}) \Big)$$

$$- \Big(\sum_{2 \leq i \leq r+1} N(A_1 \cap A_i) - \sum_{2 \leq i < j \leq r+1} N(A_1 \cap A_i \cap A_j) + \cdots$$

$$+ (-1)^{r+1} N(A_1 \cap A_2 \cap A_3 \cap \cdots \cap A_{r+1}) \Big)$$

$$= \sum_{1 \leq i \leq r+1} N(A_i) - \sum_{1 \leq i < j \leq r+1} N(A_i \cap A_j) + \sum_{1 \leq i < j < k \leq r+1} N(A_i \cap A_j \cap A_k)$$

$$- \cdots + (-1)^{r+2} N(A_1 \cap A_3 \cap \cdots \cap A_{r+1}).$$

[This is what was to be proved.]

49. *Hint:* Use the solution method described in Section 5.8. The answer is $s_k = 2s_{k-1} + 3s_{k-2}$ for every integer $k \geq 4$.

Section 9.4

1. **a.** No. For instance, the aces of the four different suits could be selected.

   b. Yes. Let x_1, x_2, x_3, x_4, x_5 be the five cards. Consider the function S that sends each card to its suit.

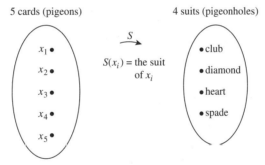

5 cards (pigeons)   4 suits (pigeonholes)

$S(x_i) =$ the suit of x_i

• club
• diamond
• heart
• spade

   By the pigeonhole principle, S is not one-to-one: $S(x_i) = S(x_j)$ for some two cards x_i and x_j. Hence at least two cards have the same suit.

3. Yes. Denote the residents by $x_1, x_2, \ldots, x_{500}$. Consider the function B from residents to birthdays that sends each resident to his or her birthday:

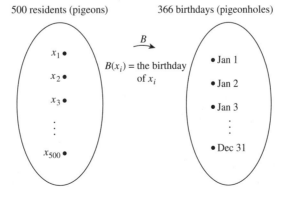

500 residents (pigeons)   366 birthdays (pigeonholes)

$B(x_i) =$ the birthday of x_i

• Jan 1
• Jan 2
• Jan 3
⋮
• Dec 31

By the pigeonhole principle, B is not one-to-one: $B(x_i) = B(x_j)$ for some two residents x_i and x_j. Hence at least two residents have the same birthday.

5. **a.** Yes. There are only three possible remainders that can be obtained when an integer is divided by 3: 0, 1, and 2. Thus, by the pigeonhole principle, if four integers are each divided by 3, then at least two of them must have the same remainder.

More formally, call the integers n_1, n_2, n_3, and n_4, and consider the function R that sends each integer to the remainder obtained when that integer is divided by 3:

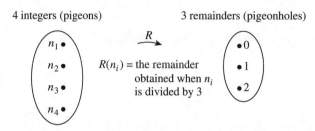

By the pigeonhole principle, R is not one-to-one: $R(n_i) = R(n_j)$ for some two integers n_i and n_j. Hence at least two integers must have the same remainder.

b. No. For instance, $\{0, 1, 2\}$ is a set of three integers no two of which have the same remainder when divided by 3.

6. **a.** Yes.

Solution 1: There are 6 possible remainders that can be obtained when an integer is divided by 6, namely 0, 1, 2, 3, 4, 5. Apply the pigeonhole principle, thinking of the 7 integers as the pigeons and the possible remainders as the pigeonholes. Each pigeon flies into the pigeonhole that is the remainder obtained when it is divided by 6. Since $7 > 6$, the pigeonhole principle says that at least two pigeons must fly into the same pigeonhole. So at least two of the numbers must have the same remainder when divided by 6.

Solution 2: Let X be the set of seven integers and Y the set of all possible remainders obtained through division by 6, and consider the function R from X (the pigeons) to Y (the pigeonholes) defined by the rule: $R(n) = n \bmod 6$ ($=$ the remainder obtained by the integer division of n by 6). Now X has 7 elements and Y has 6 elements (0, 1, 2, 3, 4, and 5). Hence by the pigeonhole principle, R is not one-to-one: $R(n_1) = R(n_2)$ for some integers n_1 and n_2 with $n_1 \neq n_2$. But this means that n_1 and n_2 have the same remainder when divided by 6.

b. No. Consider the set $\{1, 2, 3, 4, 5, 6, 7\}$. This set has seven elements no two of which have the same remainder when divided by 8.

7. *Hint:* Look at Example 9.4.3.

9. **a.** Yes.

Solution 1: Only six of the numbers from 1 to 12 are even (namely, 2, 4, 6, 8, 10, 12), so at most six even numbers can be chosen from between 1 and 12 inclusive. Hence if seven numbers are chosen, at least one must be odd.

Solution 2: Partition the set of all integers from 1 through 12 into six subsets (the pigeonholes), each consisting of an odd and an even number: $\{1, 2\}$, $\{3, 4\}$, $\{5, 6\}$, $\{7, 8\}$, $\{9, 10\}$, $\{11, 12\}$. If seven integers (the pigeons) are chosen from among 1 through 12, then, by the pigeonhole principle, at least two must be from the same subset. But each subset contains one odd and one even number. Hence at least one of the seven numbers is odd.

Solution 3: (a formal version of Solution 2): Let $S = \{x_1, x_2, x_3, x_4, x_5, x_6, x_7\}$ be a set of seven numbers chosen from the set $T = \{1, 2, 3, 4, 5, 6, 7, 8, 9, 10, 11, 12\}$, and let P be the following partition of T: $\{1, 2\}$, $\{3, 4\}$, $\{5, 6\}$, $\{7, 8\}$, $\{9, 10\}$, and $\{11, 12\}$. Since each element of S lies in exactly one subset of the partition, we can define a function F from S to P by letting $F(x_i)$ be the subset that contains x_i.

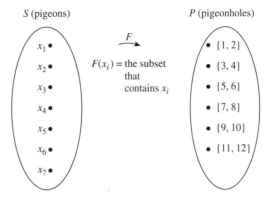

Since S has 7 elements and P has 6 elements, by the pigeonhole principle, F is not one-to-one. Thus two distinct numbers of the seven are sent to the same subset, which implies that these two numbers are the two distinct elements of the subset. Therefore, since each pair consists of one odd and one even integer, one of the seven numbers is odd.

b. No. For instance, none of the 10 numbers 1, 3, 5, 7, 9, 11, 13, 15, 17, 19 is even.

10. Yes. There are n even integers in the set $\{1, 2, 3, \ldots, 2n\}$, namely, $2(= 2 \cdot \underline{1})$, $4(= 2 \cdot \underline{2})$, $6(= 2 \cdot \underline{3}), \ldots, 2\underline{n}(= 2 \cdot n)$. So the maximum number of even integers that can be chosen is n. Thus if $n + 1$ integers are chosen, at least one of them must be odd.

12. The answer is 27. There are only 26 black cards in a standard 52-card deck, so at most 26 black cards can be chosen. Hence if 27 are taken, at least one must be red.

14. There are 61 integers from 0 through 60. Of these, 31 are even ($0 = 2 \cdot \underline{0}, 2 = 2 \cdot \underline{1}, 4 = 2 \cdot \underline{2}, \ldots, 60 = 2 \cdot \underline{30}$) and so 30 are odd. Hence if 32 integers are chosen, at least one must be odd, and if 31 integers are chosen, at least one must be even.

15. There are $n + 1$ even integers from 0 to $2n$ inclusive:

$$0\,(= 2 \cdot \underline{0}), \quad 2\,(= 2 \cdot \underline{1}), \quad 4\,(= 2 \cdot \underline{2}), \ldots, \quad 2n\,(= 2 \cdot \underline{n}).$$

So a maximum of $n + 1$ even integers can be chosen. Thus if at least $n + 2$ integers are chosen, one is sure to be odd. Similarly, there are n odd integers from 0 to $2n$ inclusive, namely

$$1\,(= 2 \cdot \underline{1} - 1), \quad 3\,(= 2 \cdot \underline{2} - 1), \ldots, \quad 2n - 1\,(= 2 \cdot \underline{n} - 1).$$

It follows that if at least $n + 1$ integers are chosen, one is sure to be even.

(An alternative way to reach the second conclusion is to note that there are $2n + 1$ integers from 0 to $2n$ inclusive. Because $n + 1$ of them are even, the number of odd integers is $(2n + 1) - (n + 1) = n$.)

17. The answer is 8. (There are only seven possible remainders for division by 7: 0, 1, 2, 3, 4, 5, 6. Hence if 8 are chosen, at least two must be the same.)

18. There are 15 distinct remainders that can be obtained through integer division by 15 (0, 1, 2, $\ldots$, 14). Hence at least 16 integers must be chosen in order to be sure that at least two have the same remainder when divided by 15.

20. **a.** The answer is 20,483 because the possible remainders are 0, 1, 2, $\ldots$, 20482.

b. The length of the repeating section of the decimal representation of 5/20483 is less than or equal to 20,482. The reason is that 20,482 is the number of nonzero remainders that can be obtained when a number is divided by 20,483. Thus, in the long-division process of dividing 5 by

20,483, either some remainder is 0 and the decimal expansion terminates, or only nonzero remainders are obtained and at some point within the first 20,482 successive divisions, a nonzero remainder is repeated. At that point the digits in the developing decimal expansion begin to repeat because the sequence of successive remainders repeats those previously obtained.

21. The length of the repeating section of the decimal representation of 683/1493 is less than or equal to 1,492. The reason is that there are 1,492 nonzero remainders that can be obtained when a number is divided by 1,493. Thus, in the long-division process of dividing $683.0000\ldots$ by 1,493, either some remainder is 0 and the decimal expansion terminates (in which case the length of the repeating section is 0) or, only nonzero remainders are obtained and at some point within the first 1,492 successive divisions, a nonzero remainder is repeated. At that point the digits in the developing decimal expansion begin to repeat because the sequence of successive remainders repeats those previously obtained.

22. This number is irrational because the decimal expansion neither terminates nor repeats.

24. Let A be the set of the thirteen chosen numbers, and let B be the set of all prime numbers between 1 and 40. Note that $B = \{2, 3, 5, 7, 11, 13, 17, 19, 23, 29, 31, 37\}$. For each x in A, let $F(x)$ be the smallest prime number that divides x. Since A has 13 elements and B has 12 elements, by the pigeonhole principle F is not one-to-one. Thus $F(x_1) = F(x_2)$ for some $x_1 \neq x_2$ in A. By definition of F, this means that the smallest prime number that divides x_1 equals the smallest prime number that divides x_2. Therefore, two numbers in A – namely, x_1 and x_2 – have a common divisor greater than 1. *[Strictly speaking, only integers less than or equal to 20 can divide integers less than or equal to 40. So we could have made the set B even smaller.]*

25. Yes. This follows from the generalized pigeonhole principle with 30 pigeons, 12 pigeonholes, and $k = 2$, using the fact that $30 > 2 \cdot 12$.

26. No. For instance, the birthdays of the 30 people could be distributed as follows: three birthdays in each of the six months January through June and two birthdays in each of the six months July through December.

27. Yes. Let X be the set of 2,000 people (the pigeons) and Y the set of all 366 possible birthdays (the pigeonholes). Define a function $B\colon X \to Y$ by specifying that $B(x) = x$'s birthday. Now $2000 > 4 \cdot 366 = 1464$, and so by the generalized pigeonhole principle, there must be some birthday y such that $B^{-1}(y)$ has at least $4 + 1 = 5$ elements. Hence at least 5 people must share the same birthday.

29. The answer is $x = 3$. There are 18 years from 17 through 34. Now $40 > 18 \cdot 2$, so by the generalized pigeonhole principle, you can be sure that there are at least $x = 3$ students of the same age. However, since $18 \cdot 3 > 40$, you cannot be sure of having more than three students with the same age. (For instance, three students could be each of the ages 17 through 20, and two could be each of the ages from 21 through 34.) So x cannot be taken to be greater than 3.

30. Consider the maximum number of pennies that can be chosen without getting at least five from the same year. This maximum, which is 12, is obtained when four pennies are chosen from each of the three years. Hence at least thirteen pennies must be chosen to be sure of getting at least five from the same year.

31. *Hint:* Use the same type of reasoning as in Example 9.4.6.

32. *Hints:* (1) The number of subsets of the six integers is $2^6 = 64$. (2) Since each integer is less than 13, the largest possible sum is 57. (Why? How is this sum obtained?)

33. Proof: Suppose A is a set of six positive integers each of which is less than 15. By Theorem 6.3.1, $\mathscr{P}(A)$, the power set of A, has $2^6 = 64$ elements, and so A has 63 nonempty subsets. Let k be the smallest number in the set A.

Given any nonempty subset of A, the sum of all the elements in the subset lies in the range from k through $k + 10 + 11 + 12 + 13 + 14 = k + 60$, and, by Theorem 9.1.1, there are $(k + 60) - k + 1 = 61$ integers in this range. Let S be the set of all possible sums of the elements that are in a nonempty subset of A. Then S has at most 61 elements.

Define a function F from the set of nonempty subsets of A to S as follows: For each nonempty subset X in A, let $F(X)$ be the sum of the elements of X. Because A has 63 nonempty subsets and S has 61 elements, the pigeonhole principle guarantees that F is not one-to-one. Thus there exist distinct nonempty subsets A_1 and A_2 of A such that $F(A_1) = F(A_2)$, which implies that the elements of A_1 add up to the same sum as the elements of A_2.

Note: In fact, it can be shown that it is always possible to find disjoint subsets of A with the same sum. To see why this is true, consider again the sets A_1 and A_2 found in the preceding proof. Then $A_1 \neq A_2$ and $F(A_1) = F(A_2)$. By definition of F, $F(A_1 - A_2) + F(A_1 \cap A_2) = $ the sum of the elements in $A_1 - A_2$ plus the sum of the elements in $A_1 \cap A_2$. But $A_1 - A_2$ and $A_1 \cap A_2$ are disjoint and their union is A_1. So $F(A_1 - A_2) + F(A_1 \cap A_2) = F(A_1)$. By the same reasoning, $F(A_2 - A_1) + F(A_1 \cap A_2) = F(A_2)$. Since $F(A_1) = F(A_2)$, we have that $F(A_1 - A_2) = F(A_1) - F(A_1 \cap A_2) = F(A_2) - F(A_1 \cap A_2) = F(A_2 - A_1)$. Hence the elements in $A_1 - A_2$ add up to the same sum as the elements in $A_2 - A_1$. But $A_1 - A_2$ and $A_2 - A_1$ are disjoint because $A_1 - A_2$ contains no elements of A_2 and $A_2 - A_1$ contains no elements of A_1.

35, *Hint:* Let X be the set consisting of the given 52 positive integers, and let Y be the set containing the following elements:

$$\{00\}, \{50\}, \{01, 99\}, \{02, 98\}, \{03, 97\}, \ldots, \{48, 52\}, \{49, 51\}.$$

Define a function F from X to Y by the rule $F(x) = $ the set containing the last two digits of x. Use the pigeonhole principle to argue that F is not one-to-one, and show how the desired conclusion follows.

36. <u>Proof:</u> Suppose that 101 integers are chosen from 1 to 200 inclusive. Call them $x_1, x_2, \ldots, x_{101}$. Represent each of these integers in the form $x_i = 2^{k_i} \cdot a_i$ where a_i is the uniquely determined odd integer obtained by dividing x_i by the highest possible power of 2. Because each x_i satisfies the condition $1 \leq x_i \leq 200$, each a_i satisfies the condition $1 \leq a_i \leq 199$. Define a function F from $X = \{x_1, x_2, \ldots, x_{101}\}$ to the set Y of all odd integers from 1 to 199 inclusive by the rule $F(x_i) = $ that odd integer a_i such that x_i equals $2^{k_i} \cdot a_i$. Now X has 101 elements and Y has 100 elements, namely

$$1 = 2 \cdot \underline{1} - 1, \ 3 = 2 \cdot \underline{2} - 1, \ 5 = 2 \cdot \underline{3} - 1, \ldots, \ 199 = 2 \cdot \underline{100} - 1.$$

Hence by the pigeonhole principle, F is not one-to-one: there exist integers x_i and x_j such that $F(x_i) = F(x_j)$ and $x_i \neq x_j$.

But $x_i = 2^{k_i} \cdot a_i$ and $x_j = 2^{k_j} \cdot a_j$ and $F(x_i) = a_i$ and $F(x_j) = a_j$. Thus $x_i = 2^{k_i} \cdot a_i$ and $x_j = 2^{k_j} \cdot a_i$. If $k_j > k_i$, then

$$x_j = 2^{k_j} \cdot a_i = 2^{k_j - k_i} \cdot 2^{k_i} \cdot a_i = 2^{k_j - k_i} \cdot x_i,$$

and so x_j is divisible by x_i. Similarly, if $k_j < k_i$, x_i is divisible by x_j. Hence, in either case, one of the numbers is divisible by another.

37. *Hint:* For each $k = 1, 2, \ldots, n$, let $a_k = x_1 + x_2 + \cdots + x_k$. If some a_k is divisible by n, then the problem is solved: the consecutive subsequence is $x_1, x_2, \ldots, x_k$. If no a_k is divisible by n, then $a_1, a_2, a_3, \ldots, a_n$ satisfies the hypothesis of part (a). Hence $a_j - a_i$ is divisible by n for some integers i and j with $j > i$. Write $a_j - a_i$ in terms of the x_i's to derive the given conclusion.

38. *Hint:* Let $a_1, a_2, \ldots, a_{n^2+1}$ be any sequence of $n^2 + 1$ distinct real numbers, and suppose that this sequence contains neither a strictly increasing subsequence of length $n + 1$ nor a strictly decreasing subsequence of length $n + 1$. Let S be the set of all ordered pairs of integers (i, d), where $1 \leq i \leq n$ and $1 \leq d \leq n$. For each term a_k in the sequence, let $F(a_k) = (i_k, d_k)$, where i_k is the length of the longest increasing sequence starting at a_k, and d_k is the length of the longest decreasing sequence starting at a_k. Suppose that F is one-to-one and derive a contradiction.

39. Let S be any set consisting entirely of integers from 1 through 100, and suppose that no integer in S divides any other integer in S. Factor out the highest power of 2 to write each integer in S as $2^k \cdot m$, where m is an odd integer.

    Now consider any two such integers in S, say $2^r \cdot a$ and $2^s \cdot b$. Observe that $a \neq b$. The reason is that if $a = b$, then whichever integer contains the fewer number of factors of 2 divides the other integer. (For example, $2^2 \cdot 3 \mid 2^4 \cdot 3$.)

    Thus there can be no more integers in S than there are distinct odd integers from 1 through 100, namely 50.

    Furthermore, it is possible to find a set T of 50 integers from 1 through 100 no one of which divides any other. For instance, $T = 51, 52, 53, \ldots, 99, 100$.

    Hence the largest number of elements that a set of integers from 1 through 100 can have so that no one element in the set is divisible by any other is 50.

Section 9.5

1. **a.** 2-combinations: $\{x_1, x_2\}$, $\{x_1, x_3\}$, $\{x_2, x_3\}$. Hence, $\dbinom{3}{2} = 3$.

   b. Unordered selections: $\{a, b, c, d\}$, $\{a, b, c, e\}$, $\{a, b, d, e\}$, $\{a, c, d, e\}$, $\{b, c, d, e\}$. Hence, $\dbinom{5}{4} = 5$.

3. $P(7, 2) = \dbinom{7}{2} \cdot 2!$

5. **a.** $\dbinom{6}{0} = \dfrac{6!}{0!(6-0)!} = \dfrac{6!}{1 \cdot 6!} = 1$ **b.** $\dbinom{6}{1} = \dfrac{6!}{0!(6-1)!} = \dfrac{6 \cdot 5!}{1 \cdot 5!} = 6$

6. **a.** the number of committees of $6 = \dbinom{15}{6} = \dfrac{15!}{(15-6)!6!} = \dfrac{15 \cdot 14 \cdot 13 \cdot 12 \cdot 11 \cdot \overset{7}{10} \cdot \overset{5}{9!}}{9! \cdot 6 \cdot 5 \cdot 4 \cdot 3 \cdot 2} = 5{,}005$

   b. $\begin{bmatrix} \text{the number of committees that} \\ \text{don't contain } A \text{ and } B \text{ together} \end{bmatrix}$

   $= \begin{bmatrix} \text{the number of committees} \\ \text{with } A \text{ and five others} - \\ \text{none of them } B \end{bmatrix} + \begin{bmatrix} \text{the number of committees} \\ \text{with } B \text{ and five others} - \\ \text{none of them } A \end{bmatrix} + \begin{bmatrix} \text{the number of} \\ \text{committees with} \\ \text{neither } A \text{ nor } B \end{bmatrix}$

   $= \dbinom{13}{5} + \dbinom{13}{5} + \dbinom{13}{6} = 1{,}287 + 1{,}287 + 1{,}716 = 4{,}290$

   Alternative solution:

   $\begin{bmatrix} \text{the number of committees that} \\ \text{don't contain } A \text{ and } B \text{ together} \end{bmatrix} = \begin{bmatrix} \text{the total number} \\ \text{of committees} \end{bmatrix} - \begin{bmatrix} \text{the number of committees} \\ \text{that contain both } A \text{ and } B \end{bmatrix}$

   $= \dbinom{15}{6} - \dbinom{13}{4} = 5{,}005 - 715 = 4{,}290$

c. $\left[\begin{array}{c}\text{the number of committees}\\\text{with both }A\text{ and }B\end{array}\right] + \left[\begin{array}{c}\text{the number of committees}\\\text{with neither }A\text{ nor }B\end{array}\right]$

$$= \binom{13}{4} + \binom{13}{6} = 715 + 1{,}716 = 2{,}431$$

d. (i) $\left[\begin{array}{c}\text{the number of subsets of}\\\text{three men chosen from eight}\end{array}\right] + \left[\begin{array}{c}\text{the number of subsets of}\\\text{three women chosen from seven}\end{array}\right]$

$$= \binom{8}{3} + \binom{7}{3} = 56 \cdot 35 = 1{,}960$$

(ii) $\left[\begin{array}{c}\text{the number of committees}\\\text{with at least one woman}\end{array}\right] = \left[\begin{array}{c}\text{the total number}\\\text{of committees}\end{array}\right] - \left[\begin{array}{c}\text{the number of}\\\text{all-male committees}\end{array}\right]$

$$= \binom{15}{6} - \binom{8}{6} = 5{,}005 - 28 = 4{,}977$$

e. $\left[\begin{array}{c}\text{the number of}\\\text{ways to choose}\\\text{two freshmen}\end{array}\right] \cdot \left[\begin{array}{c}\text{the number of}\\\text{ways to choose}\\\text{two sophomores}\end{array}\right] \cdot \left[\begin{array}{c}\text{the number of}\\\text{ways to choose}\\\text{two juniors}\end{array}\right] \cdot \left[\begin{array}{c}\text{the number of}\\\text{ways to choose}\\\text{two seniors}\end{array}\right]$

$$= \binom{3}{2}\binom{4}{2}\binom{3}{2}\binom{5}{2} = 540$$

8. *Hint:* The answers are **a.** 1001, **b.** (i) 420, (ii) all 1001 require proof, (iii) 175, **c.** 506, **d.** 561

9. **a.** The number of committees of six that can be formed from the 40 members of the club is

$$\binom{40}{6} = 3{,}838{,}380.$$

b. A committee with at least three people who favored the bylaw change either contains 3 in favor and 3 against, 4 in favor and 2 against, 5 in favor and 1 against, or 6 in favor and none against. Thus the number of committees of six that contain at least three club members who favored the bylaw change equals

$$\binom{24}{3}\binom{16}{3} + \binom{24}{4}\binom{16}{2} + \binom{24}{5}\binom{16}{1} + \binom{24}{6}\binom{16}{0} - 3{,}223{,}220.$$

11. **a.** (1) 4 [*because there are as many royal flushes as there are suits*]

(2) $\dfrac{4}{\binom{52}{5}} = \dfrac{4}{2{,}598{,}960} \cong 0.0000015$

b. (1) The number of hands with a straight flush is $4 \cdot 9 = 36$ [*the number of suits, namely 4, times the number of lowest denominated card in the straight, namely 9, because aces can be low*]

(2) probability $= \dfrac{36}{\binom{52}{5}} = \dfrac{36}{2{,}598{,}960} \cong 0.0000139$

c. (1) $13 \cdot \binom{48}{1} = 624$ [*because one can first choose the denomination of the four-of-a-kind and then choose one additional card from the 48 remaining*]

$$\dfrac{624}{\binom{52}{5}} = \dfrac{624}{2{,}598{,}960} = 0.00024$$

f. (1) Imagine constructing a straight (including a straight flush and a royal flush) as a six-step process: step 1 is to choose the lowest denomination of any card of the five (which can be any one of $A, 2, \ldots, 10$), step 2 is to choose a card of that denomination, step 3 is to choose a card of the next higher denomination, and so forth until all five cards have been selected. By the multiplication rule, the number of ways to perform this process is

$$10 \cdot \binom{4}{1}\binom{4}{1}\binom{4}{1}\binom{4}{1}\binom{4}{1} = 10 \cdot 4^5 = 10{,}240.$$

By parts (a) and (b), 40 of these numbers represent royal or straight flushes, so there are $10{,}240 - 40 = 10{,}200$ straights in all.

$$(2) \quad \frac{10{,}200}{\binom{52}{5}} = \frac{10{,}200}{2{,}598{,}960} \cong 0.0039$$

12. The sum of two integers is even if, and only if, either both integers are even or both are odd [*see Example 4.2.3*]. Because $2 = 2 \cdot 1$ and $100 = 2 \cdot 50$, there are 50 even integers and thus 51 odd integers from 1 to 101 inclusive. Hence the number of distinct pairs is the number of ways to choose two even integers from the 50 plus the number of ways to choose two odd integers from the 51:

$$\binom{50}{2} + \binom{51}{2} = 1225 + 1275 = 2500.$$

13. **a.** $2^{10} = 1024$

   d.
   $$\begin{bmatrix} \text{the number of} \\ \text{outcomes with} \\ \text{at least one head} \end{bmatrix} = \begin{bmatrix} \text{the total} \\ \text{number of} \\ \text{outcomes} \end{bmatrix} - \begin{bmatrix} \text{the number of} \\ \text{outcomes with} \\ \text{no heads} \end{bmatrix} = 1{,}024 - 1 = 1{,}023$$

15. **a.** 50   **b.** 50

   c. To get an even sum, both numbers must be even or both must be odd. Hence

   $$\begin{bmatrix} \text{the number of subsets of two} \\ \text{integers chosen from 1 through} \\ \text{100 whose sum is even} \end{bmatrix}$$
   $$= \begin{bmatrix} \text{the number of subsets of two} \\ \text{even integers chosen from} \\ \text{the 50 even integers} \end{bmatrix} + \begin{bmatrix} \text{the number of subsets of two} \\ \text{even integers chosen from} \\ \text{the 50 odd integers} \end{bmatrix}$$
   $$= \binom{50}{2} + \binom{50}{2} = 2{,}450.$$

   d. To obtain an odd sum, one of the numbers must be even and the other odd. Hence the answer is $\binom{50}{1} \cdot \binom{50}{1} = 2{,}500$. Alternatively, note that the answer equals the total number of subsets of two integers chosen from 1 through 100 minus the number of such subsets for which the sum of the numbers is even. This approach gives the answer $\binom{100}{2} - 2{,}450 = 2{,}500$.

17. **a.** Two points determine a line. Hence
   $$\begin{bmatrix} \text{the number of straight lines} \\ \text{determined by the ten points} \end{bmatrix} + \begin{bmatrix} \text{the number of subsets of} \\ \text{two points chosen from ten} \end{bmatrix} = \binom{10}{2} = 45.$$

18. An ordering for the letters in *MISSISSIPPI* can be created as follows:

   Step 1: Choose a subset of one position for the M

   Step 2: Choose a subset of four positions for the I's

   Step 3: Choose a subset of four positions for the S's

   Step 4: Choose a subset of two positions for the P's

   Thus the total number of distinguishable orderings is

   $$\binom{11}{1}\binom{10}{4}\binom{6}{4}\binom{2}{2} = \frac{11!}{1! \cdot 10!} \cdot \frac{10!}{4! \cdot 6!} \cdot \frac{6!}{4! \cdot 2!} \cdot \frac{2!}{2! \cdot 0!} = \frac{11!}{1! \cdot 4! \cdot 4! \cdot 2!} = 34{,}650,$$

   which agrees with the result in Example 9.5.10.

19. **a.** $\dfrac{10!}{2!1!1!1!3!2!1!} = 151,200$ since there are 2 A's, 1 B, 1 H, 3 L's, 2 O's, and 1 U

b. $\dfrac{8!}{2!1!1!1!2!2!} = 5,040$

21. The number of symbols that can be represented in the Morse code using n dots and dashes is 2^n. Therefore, the number of symbols that can be represented in the Morse code using most seven dots and dashes is

$$2 + 2^2 + 2^3 + 2^4 + 2^5 + 2^6 + 2^7 = 2(1 + 2 + 2^2 + 2^3 + 2^4 + 2^5 + 2^6) = 2\left(\frac{2^7-1}{2-1}\right) = 254.$$

23. Let R stand for "right" and U stand for "up." Then the rook must move seven squares to the right and seven squares up. Hence,

$$\begin{bmatrix} \text{the number of} \\ \text{paths the rook} \\ \text{can take} \end{bmatrix} = \begin{bmatrix} \text{the number of} \\ \text{orderings of seven} \\ \text{R's and seven U's} \end{bmatrix} = \frac{14!}{7!7!} = 3,432.$$

24. **a.** Because $210 = 2 \cdot 3 \cdot 5 \cdot 7$, the distinct factorizations of 210 are $1 \cdot 210$, $2 \cdot 105$, $3 \cdot 70$, $5 \cdot 42$, $7 \cdot 30$, $6 \cdot 35$, $10 \cdot 21$, and $14 \cdot 15$. So there are 8 distinct factorizations of 210.

b. *Solution 1:* One factor can be 1, and the other factor can be the product of all the primes. (This gives 1 factorization.) One factor can be one of the primes, and the other factor can be the product of the other three. (This gives $\binom{4}{1} = 4$ factorizations.) One factor can be a product of two of the primes, and the other factor can be a product of the two other primes. The number $\binom{4}{2} = 6$ counts all possible sets of two primes chosen from the four primes, and each set of two primes corresponds to a factorization. Note, however, that the set $\{p_1, p_2\}$ corresponds to the same factorization as the set $\{p_3, p_4\}$, namely, $p_1 p_2 p_3 p_4$ (just written in a different order). In general, each choice of two primes corresponds to the same factorization as one other choice of two primes. Thus the number of factorizations in which each factor is a product of two primes is $\frac{\binom{4}{2}}{2} = 3$.

Solution 1: One factor can be 1, and the other factor can be the product of all the primes. (This gives 1 factorization.) One factor can be one of the primes, and the other factor can be the product of the other three. (This gives $\binom{4}{1} = 4$ factorizations.) One factor can be a product of two of the primes, and the other factor can be a product of the two other primes. The number $\binom{4}{2} = 6$ counts all possible sets of two primes chosen from the four primes, and each set of two primes corresponds to a factorization. Note, however, that the set $\{p_1, p_2\}$ corresponds to the same factorization as the set $\{p_3, p_4\}$, namely, $p_1 p_2 p_3 p_4$ (just written in a different order). In general, each choice of two primes corresponds to the same factorization as one other choice of two primes. Thus the number of factorizations in which each factor is a product of two primes is $\frac{\binom{4}{2}}{2} = 3$.

c. As in the answer to part (b), there are two different ways to look at the solution to this problem.

Solution 1: Separate the factorizations into categories: one category consists only of the factorization in which one factor is 1 and the other factor is the product of all five prime factors *[there is $1 = \binom{5}{0}$ such factorization]*, a second category consists of those factorizations in which one factor is a single prime and the other factor is the product of the four other primes *[there are $\binom{5}{1}$ such factorizations]*, and the third category contains those factorizations in which one factor is a product of two of the primes and the other factor is the product of the other three primes *[there are $\binom{5}{2}$ such factorizations]*. All possible factorizations are included among these categories, and so, by the addition rule, the answer is $\binom{5}{0} + \binom{5}{1} + \binom{5}{2}$ $= 1 + 5 + 10 = 16.$

Solution 2: Let $S = \{p_1, p_2, p_3, p_4, p_5\}$, let $p_1 p_2 p_3 p_4 p_5 = P$, and let $f_1 f_2$ be any factorization of P. The product of the numbers in any subset $A \subseteq S$ can be used for f_1, with the product of the numbers in A^c being f_2. Thus there are as many ways to write $f_1 f_2$ as there are subsets of S, namely $2^5 = 32$ (by Theorem 6.3.1). But given any factors f_1 and f_2, we have that $f_1 f_2 = f_2 f_1$. Thus counting the number of ways to write $f_1 f_2$ counts each factorization twice. So the answer is $\frac{32}{2} = 16$.

Note: In Section 9.7 we will show that $\binom{n}{r} = \binom{n}{n-r}$ whenever $n \geq r \geq 0$. Thus, for example, the answer can be written as

$$\binom{5}{0} + \binom{5}{1} + \binom{5}{2} = \frac{1}{2}\left(\binom{5}{0} + \binom{5}{1} + \binom{5}{2} + \binom{5}{3} + \binom{5}{4} + \binom{5}{5}\right).$$

In Section 9.7 we will also show that for every integer $n \geq 0$,

$$\binom{n}{0} + \binom{n}{1} + \binom{n}{2} + \cdots + \binom{n}{n-2} + \binom{n}{n-1} + \binom{n}{n} = 2^n,$$

and so, in particular,

$$\frac{1}{2}\left[\binom{5}{0} + \binom{5}{1} + \binom{5}{2} + \binom{5}{3} + \binom{5}{4} + \binom{5}{5}\right] = \frac{1}{2} \cdot 2^5 = \frac{32}{2} = 16.$$

These facts illustrate the relationship between the two solutions to part (c) of this exercise.

d. Because the second solution given in parts (b) and (c) is the simplest, we give a general version of it as the answer to this part of the exercise. Let $S = \{p_1, p_2, p_3, \ldots, p_n\}$, let $p_1 p_2 p_3 \cdots p_n = P$, and let $f_1 f_2$ be any factorization of P. The product of the numbers in any subset $A \subseteq S$ can be used for f_1, with the product of the numbers in A^c being f_2. Thus there are as many ways to write $f_1 f_2$ as there are subsets of S, namely 2^n (by Theorem 6.3.1). But given any factors f_1 and f_2, we have that $f_1 f_2 = f_2 f_1$, and so counting the number of ways to write $f_1 f_2$ counts each factorization twice. Hence the answer is $\frac{1}{2^n} = 2^{n-1}$.

25. **a.** There are four choices for where to send the first element of the domain (any element of the co-domain may be chosen), three choices for where to send the second (since the function is one-to-one, the second element of the domain must go to a different element of the co-domain from the one to which the first element went), and two choices for where to send the third element (again since the function is one-to-one). Thus the answer is $4 \cdot 3 \cdot 2 = 24$.

b. none

e. *Hint*: The answer is $n(n-1)(n-2)\cdots(n-m+1)$.

d. Consider functions from a set with four elements to a set with two elements. Denote the set of four elements by $X = \{a, b, c, d\}$ and the set of two elements by $Y = \{u, v\}$. Divide the set of all onto functions from X to Y into two categories. The first category consists of all those that send the three elements in $\{a, b, c\}$ onto $\{u, v\}$ and that send d to either u or v. The functions in this category can be defined by the following two-step process:

Step 1:   Construct an onto function from $\{a, b, c\}$ to $\{u, v\}$.
Step 2:   Choose whether to send d to u or to v.

By part (a), there are six ways to perform step 1, and, because there are two choices for where to send d, there are two ways to perform step 2. Thus, by the multiplication rule, there are $6 \cdot 2 = 12$ ways to define the functions in the first category.

The second category consists of all the other onto functions from X to Y: those that send all three elements in $\{a, b, c\}$ to either u or v and that send d to whichever of u or v is not the image of a, b, and c. Because there are only two choices for where to send the elements in

$\{a, b, c\}$, and because d is simply sent to wherever a, b, and c do not go, there are just two functions in the second category.

Every onto function from X to Y either sends at least two elements of X to the image of d or it does not. If it does, then it is in the first category. If it does not, then it is in the second category. Therefore, all onto functions from X to Y are in one of the two categories and no function is in both categories. So the total number of onto functions is $12 + 2 = 14$.

26. **a.** Let the elements of the domain be called a, b, and c and the elements of the co-domain be called u and v. In order for a function from $\{a, b, c\}$ to $\{u, v\}$ to be onto, two elements of the domain must be sent to u and one to v, or two elements must be sent to v and one to u. There are as many ways to send two elements of the domain to u and one to v as there are ways to choose which elements of $\{a, b, c\}$ to send to u, namely, $\binom{3}{2} = 3$. Similarly, there are $\binom{3}{2} = 3$ ways to send two elements of the domain to v and one to u. Therefore, there are $3 + 3 = 6$ onto functions from a set with three elements to a set with two elements.

b. None.

c. *Hint:* The answer is 6.

27. **a.** A relation on A is any subset of $A \times A$, and $A \times A$ has $8^2 = 64$ elements. So there are 2^{64} relations on A.

b. A reflexive relation must contain (a,a) for all eight elements a in A. Any subset of the remaining 56 elements of $A \times A$ (which has a total of 64 elements) can be combined with these eight to produce a reflexive relation. Therefore, there are as many reflexive relations as there are subsets of a set of 56 elements, namely 2^{56}.

c. Form a relation that is both reflexive and symmetric by a two-step process:

Step 1: Pick a set of elements of the form (a, a) (there are eight such elements, so 2^8 sets).

Step 2: Pick a set of pairs of elements of the form (a, b) and (b, a) where $a \neq b$ (there are $(64 - 8)/2 = 28$ such pairs, so 2^{28} such sets).

The answer is therefore $2^8 \times 2^{28} = 2^{36}$.

d. Form a relation that is both reflexive and symmetric by a two-step process:

Step 1: Pick all eight elements of the form (x, x) where $x \in A$.

Step 2: Pick a set of (distinct) pairs of elements of the form (a, b) and (b, a).

There is just one way to perform step 1, and, as explained in the answer to part (c), there are 2^{28} ways to perform step 2. Therefore, there are 2^{28} relations on A that are reflexive and symmetric.

28. *Hint:* Use the difference rule and the generalization of the inclusion/exclusion rule for 4 sets. (See exercise 48 in Section 9.3.)

30. The error is that the "solution" overcounts the number of poker hands with two pairs. In fact, it counts every such hand twice. For instance, consider the poker hand $\{4\clubsuit, 4\diamondsuit, J\heartsuit, J\spadesuit, 9\clubsuit\}$. If the steps outlined in the false solution in the exercise statement are followed, this hand is first counted when the denomination 4 is chosen in step one, the cards $4\clubsuit$ and $4\diamondsuit$ are chosen in step two, the denomination J is chosen in step three, the cards J $\heartsuit$ and J$\spadesuit$ are chosen in step four, and $9\clubsuit$ is chosen in step five. The hand is counted a second time when the denomination J is chosen in step one, the cards J$\heartsuit$ and J $\spadesuit$ are chosen in step two, the denomination 4 is chosen in step three, the cards $4\clubsuit$ and $4\diamondsuit$ are chosen in step four, and $9\clubsuit$ is chosen in step five.

Section 9.6

1. **a.** $\dbinom{5+3-1}{5} = \dbinom{7}{5} = \dfrac{7 \cdot 6}{2} = 21$

    b. The three elements of the set are 1, 2, and 3. The 5-combinations are [1, 1, 1, 1, 1], [1, 1, 1, 1, 2], [1, 1,1, 1, 3], [1, 1, 1, 2, 2], [1, 1, 1, 2, 3], [1, 1, 1, 3, 3], [1, 1, 2, 2, 2], [1, 1, 2, 2, 3], [1, 1, 3, 3, 3], [1, 2, 2, 2, 2], [1, 2, 2, 2, 3], [1, 2, 2, 3, 3], [1, 2, 3, 3, 3], [1, 3, 3, 3, 3], [2, 2, 2, 2, 2], [2, 2, 2, 2, 3], [2, 2, 2, 3, 3], [2, 2, 3, 3, 3] [2, 3, 3, 3, 3], and [3, 3, 3, 3, 3].

2. **a.** $\dbinom{4+3-1}{4} = \dbinom{6}{4} = \dfrac{6 \cdot 5}{2} = 15$

3. **a.** $\dbinom{20+6-1}{20} = \dbinom{25}{20} = 53,130$

    b. If at least three are éclairs, then 17 additional pastries are selected from six kinds. The number of selections is $\dbinom{17+6-1}{17} = \dbinom{22}{17} = 26,334.$

    Note: In parts (a) and (b), it is assumed that the selections being counted are unordered.

    c. Let T be the set of selections of pastry that may be any one of the six kinds, let $E_{\geq 3}$ be the set of selections containing three or more éclairs, and let $E_{\geq 2}$ be the set of selections containing two or fewer éclairs. Then

    $$N(E_{\leq 2}) = N(T) - N(E_{\geq 3}) \quad \text{because } T = E_{\leq 2} \cup E_{\geq 3} \text{ and } E_{\leq 2} \cap E_{\geq 3} = \emptyset$$
    $$= 53,130 - 26,334 \quad \text{by parts (a) and (b)}$$
    $$= 26,796.$$

    Thus there are 26,796 selections of pastry containing at most two éclairs.

5. The answer equals the number of 4-combinations with repetition allowed that can be formed from a set of n elements. It is

    $$\dbinom{4+n-1}{4} = \dbinom{n+3}{4}$$
    $$= \frac{(n+3)(n+2)(n+1)n(n-1)!}{4!(n-1)!}$$
    $$= \frac{n(n+1)(n+2)(n+3)}{24}.$$

6. By the same reasoning as in Example 9.6.3, the answer is

    $$\dbinom{5+n-1}{5} = \dbinom{n+4}{5} = \frac{(n+4)(n+3)(n+2)(n+1)n}{120}.$$

8. As in Example 9.6.4, the answer is the same as the number of quadruples of integers (i, j, k, m) for which $1 \leq i \leq j \leq k \leq m \leq n$. By exercise 5, this number is As in Example 9.6.4, the answer is the same as the number of quadruples of integers (i, j, k, m) for which $1 \leq i \leq j \leq k \leq m \leq n$. By exercise 5, this number is

    $$\dbinom{n+3}{4} = \frac{n(n+1)(n+2)(n+3)}{24}.$$

9. The number of iterations of the inner loop is the same as the number of integer triples (i, j, k) where $1 \leq k \leq j \leq i \leq n$. As in Example 9.6.3, such triples can be represented as a string of $n-1$ vertical bars and three crosses indicating which three integers from 1 to n are included

in the triple. Thus the number of such triples is the same as the number of strings of $(n-1)$ $|$'s and 3 $\times$'s, which is

$$\binom{n+2}{3} = \frac{n(n+1)(n+2)}{6}.$$

10. Think of the number 20 as divided into 20 individual units and the variables x_1, x_2, and x_3 as three categories into which these units are placed. The number of units in category x_i indicates the value of x_i, in a solution of the equation. By Theorem 9.6.1, the number of ways to select 20 objects from the three categories is $\binom{20+3-1}{20} = \binom{22}{20} = \frac{22 \cdot 21}{2} = 231$, so there are 231 non-negative integer solutions to the equation.

11. The analysis for this exercise is the same as for exercise 10 except that since each $x_i \geq 1$, we can imagine taking 3 of the 20 units, placing one in each category x_1, x_2, and x_3, and then distributing the remaining 17 units among the three categories. The number of ways to do this is

$$\binom{17+3-1}{17} = \binom{19}{17} = \frac{19 \cdot 18}{2} = 171.$$

So there are 171 positive integer solutions to the equation.

12. Think of the number 30 as divided into 30 individual units and the variables (y_1, y_2, y_3, y_4) as four categories into which these units are placed. The number of units in category y_i indicates the value of y_i in a solution of the equation. By Theorem 9.6.1, the number of ways to place 30 objects into four categories is

$$\binom{30+4-1}{30} = \binom{33}{30} = 5456.$$

So there are 5,456 nonnegative integral solutions of the equation.

15. Any number from 1 through 99,999 can be written using 5 digits by adding leading zeroes if necessary. For instance, $385 = 00385$. To count how many such numbers have digits that add up to 10, represent each by a sequence of 4 vertical lines and 10 crosses, where the 4 lines separate the 10 crosses into 5 categories. For instance, $\times\times \mid \mid \times \times \times \times \times \times \mid \times \mid \times\times$ represents the number 20612, and $\mid\mid\mid \times\times\times\times\times \mid \times\times\times\times\times$ represents 00055. By Theorem 9.6.1, there are

$$\binom{10+5-1}{10} = \binom{14}{10} = 1,001$$

ways to place the crosses into the categories, and so there are 1,001 integers from 1 through 99,999 whose digits add up to 10.

16. **a.** Let $L_{\geq 7}$ be the set of selections that include at least seven cans of lemonade. In this case an additional eight cans can be selected from the five types of soft drinks, and so

$$N(L_{\geq 7}) = \binom{8+5-1}{8} = \binom{12}{8} = 495.$$

Let T be the set of selections of cans in which the soft drink may be any one of the five types assuming that there are at least 15 cans of each type assuming that there are at least 15 cans of each type and let $L_{\leq 6}$ be the set of selections that contain at most six cans of lemonade. Then

$$N(L_{\leq 6}) = N(T) - N(L_{\geq 7}) \quad \text{because } T = L_{\leq 6} \cup L \geq 7 \text{and } L_{\leq 6} \cap L_{\geq 7} = \emptyset$$
$$= 3,876 - 495 \qquad \text{by the above and part (a) of Example 9.6.2}$$
$$= 3,381.$$

Thus there are 3,381 selections of fifteen cans of soft drinks that contain at most six cans of lemonade.

b. Let $R_{\leq 5}$ be the set of selections containing at most five cans of root beer, and let $L_{\leq 6}$ be the set of selections containing at most six cans of lemonade. The answer to the question can be represented as $N(R_{\leq 5} \cap L_{\leq 6})$. As in part (a), let T be the set of all the selections of fifteen cans in which the soft drink may be any one of the five types assuming that there are at least 15 cans of each type. If you remove all the selections from T that contain at least six cans of root beer or at least seven cans of lemonade, then you are left with all the selections that contain at most five cans of root beer and at most six cans of lemonade. Thus, in the notation of part (a) and Example 9.6.2,

$$N(R_{\leq 5} \cap L_{\leq 6}) = N(T) - N(R_{\geq 6} \cup L_{\geq 7}). \;(*)$$

Use the inclusion/exclusion rule as follows to compute $N(R_{\geq 6} \cup L_{\geq 7})$:

$$N(R_{\geq 6} \cup L_{\geq 7}) = N(R_{\geq 6}) + N(L_{\geq 7}) - N(R_{\geq 6} \cap L_{\geq 7})$$

To find $N(R_{\geq 6} \cap L_{\geq 7})$, observe that if at least 6 cans of root beer and at least 7 cans of lemonade are selected, then at most 2 additional cans of soft drink can be chosen from the other three types to make up the total of 15 cans. A selection of two such cans can be represented by a string of 2 ×'s and 3 |'s, and a selection of one such can can be represented by a string of 1 × and 3 |'s. Hence

$$N(R_{\geq 6} \cap L_{\geq 7}) = \binom{2+3-1}{2} + \binom{1+3-1}{1} = \binom{4}{2} + \binom{3}{1} 6 + 3 = 9.$$

It follows that

$$N(R_{\geq 6} \cup L_{\geq 7}) = N(R_{\geq 6} + N(L_{\geq 7}) - N(R_{\geq 6} \cap L_{\geq 7}) \quad \text{by the inclusion/exclusion rule}$$

$$= 715 + 495 - 9 \qquad\qquad \text{by part (a) and the computation above, and by part (b) of Example 9.6.2}$$

$$= 1,201.$$

Putting this result together with equation (*) and the value of $N(T)$ from Example 9.6.2(a) gives that

$$N(R_{\leq 5} \cap L_{\leq 6}) = N(T) - N(R_{\geq 6} \cup L_{\geq 7})$$

$$= 3,876 - 1,201 = 2,675.$$

Thus there are 2,675 selections of fifteen soft drinks that contain at most five cans of root beer and at most six cans of lemonade.

17. *Hints:* **a.** The answer is 10,295,472. **b.** See the solution to part (c) of Example 9.6.2. The answer is 9,949,368. **c.** The answer is 9,111,432. **d.** Let T denote the set of all the selections of thirty balloons, assuming that there are at least 30 of each color. Let $R_{\leq 12}$ denote the set of selections from T that contain at most twelve red balloons, let $B_{\leq 8}$ denote the set of selections from T that contain at most eight blue balloons, let $R_{\geq 13}$ denote the set of selections that contain at least thirteen red balloons, and let $B_{\geq 9}$ denote the set of selections that contain at least nine blue balloons. Then the answer to the question can be represented as $N(R_{\leq 12} \cap B_{\leq 8})$. Out of the total of all the balloon selections, if you remove the selections containing at least thirteen red or at least nine blue balloons, then you are left with the selections containing at most twelve red and at most eight blue balloons. Thus $N(R_{\leq 12} \cap B_{\leq 8}) = N(T) - N(R_{\geq 13} \cup B_{\geq 9})$. Compute $N(R_{\geq 13} \cap B_{\geq 9})$, and use the inclusion/exclusion rule to find $N(R_{\geq 13} \cup B_{\geq 9})$.

18. **a.** Think of the 4 kinds of coins as the n categories and the 30 coins to be chosen as the r objects. Each choice of 30 coins is represented by a string of $4-1=3$ vertical bars (to separate the categories) and 30 crosses (to represent the chosen coins). The total number of choices of 30 coins of the 4 different kinds is the number of strings of 33 symbols (3 vertical bars and 30 crosses), namely, $\binom{30+4-1}{30} = \binom{33}{30} = 5,456.$

b. Let T be the set of selections of 30 coins assuming that there are at least 30 coins of each type. Let $Q_{\leq 15}$ be the set of selections containing at most 15 quarters, and let $Q_{\geq 16}$ be the set of selections containing at least 16 quarters. Then

$$T = Q_{\leq 15} \cup Q_{\geq 16} \quad \text{and} \quad Q_{\leq 15} \cap Q_{\geq 16} = \emptyset \quad \text{and so} \quad N(T) = N(Q_{\leq 15}) + N(Q_{\geq 16}).$$

To compute $N(Q_{\geq 16})$, we reason as follows: If at least 16 quarters are included, we can choose the 30 coins by first selecting 16 quarters and then choosing the remaining 14 coins from the four different types. The number of ways to do this is

$$N(Q_{\geq 16}) = \binom{14 + 4 - 1}{14} = \binom{17}{14} = 680.$$

Then $N(T) = 5,456$ *[by part (a)]* and $N(Q_{\geq 16}) = 680$. Therefore, the number of selections containing at most 15 quarters is

$$N(Q_{\leq 15}) = N(T) - N(Q_{\geq 16}) = 5,456 - 680 = 4,776.$$

c. Let T be the set of selections of 30 coins assuming that there are at least 30 coins of each type. Let $D_{\leq 20}$ be he set of selections from T that contain at most 20 dimes, and let $D_{\geq 21}$ be the set of selections from T that contain at least 21 dimes. Then

$$T = D_{\leq 20} \cup D_{\geq 21} \quad \text{and} \quad D_{\leq 20} \cap D_{\geq 21} = \emptyset \quad \text{and so} \quad N(T) = N(D_{\leq 20}) + N(D_{\geq 21}).$$

To compute $N(D_{\geq 21})$, we reason as follows: If at least 21 dimes are included, we can choose the 30 coins by first selecting 21 dimes and then choosing the remaining nine coins from the four different types. The number of ways to do this is

$$N(D_{\geq 21}) = \binom{9 + 4 - 1}{9} = \binom{12}{9} = 220.$$

Then $N(T) = 5,456$ *[by part (a)]* and $N(D_{\geq 21}) = 220$. Therefore, the number of selections containing at most 20 dimes is

$$N(D_{\leq 20}) = N(T) - N(D_{\geq 21}) = 5,456 - 220 = 5,236.$$

d. As in parts (b) and (c), let T be the set of selections of 30 coins assuming that there are at least 30 coins of each type. Let $Q_{\geq 16}$ be the set of selections from T that contain at least 16 quarters, $Q_{\leq 15}$ the set of selections from T that contain at most 15 quarters, $D_{>21}$ the set of selections from T that contain at least 21 dimes, and $D_{\leq 20}$ the set of selections from T that contain at most 20 dimes. If the pile has at most 15 quarters and at most 20 dimes, then the number of combinations of coins that can be chosen is $N(Q_{\leq 15} \cap D_{\leq 20})$, and, by the difference rule,

$$N(Q_{\leq 15} \cap D_{\leq 20}) = N(T) - N(Q_{\geq 16} \cup D_{\geq 21}).$$

In order to find $N(Q_{\geq 16} \cup D_{\geq 21})$, we first compute $N(Q_{\geq 16} \cap D_{\geq 21})$, which is the number of selections of coins containing at least 16 quarters and at least 21 dimes. However, 16 quarters plus 21 dimes would give a total of more than 30 coins. So there are no selections of this type. Thus

$$N(Q_{\leq 15} \cap D_{\leq 20}) = N(T) - N(Q_{\geq 16} \cup D_{\geq 21}) = 5,456 - 0 = 5,456.$$

Then, by the inclusion/exclusion rule,

$$N(Q_{\geq 16} \cup D_{\geq 21}) = N(Q_{\geq 16}) + N(D_{\geq 21}) - N(Q_{\geq 16} \cap D_{\geq 21}) = 680 + 220 - 0 = 900.$$

Therefore the answer to the question is

$$N(Q_{\leq 15} \cap D_{\leq 20}) = N(T) - N(Q_{\geq 16} \cup D_{\geq 21}) = 5,456 - 900 = 4,556.$$

19. *Hints:* The answers are **a.** 51,128 **b.** 46,761

21. Consider those columns of a trace table corresponding to an arbitrary value of k. The values of j go from 1 to k, and for each value of j, the values of i go from 1 to j.

k	k													
j	1	2		3			. . . .	k						
i	1	1	2	1	2	3	. . .	1	2	3		. . .		k

So for each value of k, there are $1 + 2 + 3 + \cdots + k$ columns of the table. Since k goes from 1 to n, the total number of columns in the table is

$$1 + (1+2) + (1+2+3) + \cdots + (1+2+3+\cdots+n)$$

$$= \sum_{k=1}^{1} k + \sum_{k=1}^{2} k + \cdots + \sum_{k=1}^{n-1} k + \sum_{k=1}^{n} k$$

$$= \frac{1 \cdot 2}{2} + \frac{2 \cdot 3}{2} + \cdots + \frac{(n-1) \cdot n}{2} + \frac{n \cdot (n+1)}{2}$$

$$= \frac{1}{2}[1 \cdot 2 + 2 \cdot 3 + \cdots + (n-1) \cdot n + n \cdot (n+1)]$$

$$= \frac{1}{2}\left(\frac{n(n+1)(n+2)}{3}\right) \qquad \text{by exercise 13 of Section 5.2}$$

$$= \frac{n(n+1)(n+2)}{6},$$

which agrees with the result of Example 9.6.4.

Section 9.7

1. $\dbinom{n}{0} = \dfrac{n!}{0!(n-0)!} = \dfrac{n!}{1 \cdot n!} = 1$

3. $\dbinom{n}{2} = \dfrac{n!}{(n-2)! \cdot 2!} = \dfrac{n \cdot (n-1) \cdot (n-2)!}{(n-2)! \cdot 2!} = \dfrac{n(n-1)}{2}$

5. <u>Proof:</u> Suppose n and r are nonnegative integers and $r \leq n$. Then

$$\dbinom{n}{r} = \frac{n!}{r!(n-r)!} \qquad \text{by Theorem 9.5.1}$$

$$= \frac{n!}{(n-(n-r))!(n-r)!} \qquad \text{since } (n-(n-r)) = n-n+r = r$$

$$= \frac{n!}{(n-r)!(n-(n-r))!} \qquad \text{by interchanging the factors in the denominator}$$

$$= \dbinom{n}{n-r} \qquad \text{by Theorem 9.5.1}$$

6. *Solution 1:* Apply formula (9.7.2) with $m+k$ in place of n. This is legal because $m+k \geq 1$.

Solution 2:

$$\dbinom{m+k}{m+k+1} = \frac{(m+k)!}{(m+k-1)![(m+k)-(m+k-1)]!} \qquad \text{by Theorem 9.5.1}$$

$$= \frac{(m+k) \cdot (m+k-1)!}{(m+k-1)!(m+k-m-k+1)!} \qquad \begin{array}{l}\text{since } (m+k)-(m+k-1) \\ = m+k-m-k+1\end{array}$$

$$= \frac{(m+k) \cdot (m+k-1)!}{(m+k-1)! \cdot 1!} \qquad \text{since } m+k-m-k+1 = 1$$

$$= m+k \qquad \begin{array}{l}\text{by canceling } (m+k-1)! \text{ from} \\ \text{numerator and denominator.}\end{array}$$

9. $\dbinom{2(n+1)}{2n} = \dbinom{2n+2}{2n}$

$= \dfrac{(2n+2)!}{(2n)!((2n+2)-2n)!}$

$= \dfrac{(2n+2)(2n+1)(2n)!}{(2n)!2!}$

$= \dfrac{(2n+2)(2n+1)}{2}$

$= \dfrac{2(n+1)(2n+1)}{2}$

$= (n+1)(2n+1)$

10. **a.** From Example 9.7.1, $\dbinom{6}{0} = \dbinom{6}{6} = 1$ and $\dbinom{6}{1} = \dbinom{6}{5} = 6$. We use these numbers, Pascal's formula, and the values for $n = 5$ in Pascal's triangle to compute the following:

$$\binom{6}{2} = 5 + 10 = 15, \quad \binom{6}{3} = 10 + 10 = 20, \quad \binom{6}{4} = 10 + 5 = 15.$$

b. From Example 9.7.1, $\dbinom{7}{0} = \dbinom{7}{7} = 1$. Using the information from part (a), we can compute and $\dbinom{7}{1} = \dbinom{7}{6}$ directly from Pascal's formula along with the rest of the values for $n = 7$.

$$\binom{7}{1} = 1 + 6 = 7, \quad \binom{7}{2} = 6 + 15 = 21, \quad \binom{7}{3} = 15 + 20 = 35,$$

$$\binom{7}{4} = 20 + 15 = 35, \quad \binom{7}{5} = 15 + 6 = 21, \quad \binom{7}{6} = 6 + 1 = 7$$

c. Row for $n = 7$:   1   7   21   35   35   21   7   1

12. $\dbinom{n+3}{r} = \dbinom{n+2}{r-1} + \dbinom{n+2}{r}$

$= \left(\dbinom{n+1}{r-2} + \dbinom{n+1}{r-1} \right) + \left(\dbinom{n+1}{r-1} + \dbinom{n+1}{r} \right)$

$= \dbinom{n+1}{r-2} + 2 \cdot \dbinom{n+1}{r-1} + \dbinom{n+1}{r}$

$= \left(\dbinom{n}{r-3} + \dbinom{n}{r-2} \right) + 2 \cdot \left(\dbinom{n}{r-2} + \dbinom{n}{r-1} \right) + \left(\dbinom{n}{r-1} + \dbinom{n}{r} \right)$

$= \dbinom{n}{r-3} + 3 \cdot \dbinom{n}{r-2} + 3 \cdot \dbinom{n}{r-1} + \dbinom{n}{r}$

13. <u>Proof by mathematical induction:</u> Let the property $P(n)$ be the formula

$$\sum_{i=2}^{n+1} \binom{i}{2} = \binom{n+2}{3}. \quad \leftarrow P(n).$$

Show that $P(1)$ is true:

To prove $P(1)$ we must show that

$$\sum_{i=2}^{1+1} \binom{i}{2} = \binom{1+2}{3}. \quad \leftarrow P(1).$$

Now

$$\sum_{i=2}^{1+1}\binom{i}{2} = \sum_{i=2}^{2}\binom{i}{2} = \binom{2}{2} = 1 = \binom{3}{3} = \binom{1+2}{3},$$

so $P(1)$ is true.

Show that for every integer $k \geq 1$, if $P(k)$ is true, then $P(k+1)$ is true:

Let k be any integer with $k \geq 1$, and suppose that

$$\sum_{i=2}^{k+1}\binom{i}{2} = \binom{k+2}{3}. \quad \leftarrow \begin{array}{l} P(k) \\ \text{inductive hypothesis} \end{array}$$

We must show that

$$\sum_{i=2}^{(k+1)+1}\binom{i}{2} = \binom{(k+1)+2}{3},$$

or, equivalently,

$$\sum_{i=2}^{k+2}\binom{i}{2} = \binom{k+3}{3}. \quad \leftarrow P(k+1)$$

Now the left-hand side of $P(k+1)$ is

$$\begin{aligned}
\sum_{i=2}^{k+2}\binom{i}{2} &= \sum_{i=2}^{k+1}\binom{i}{2} + \binom{k+2}{2} &&\text{by writing the last term separately} \\
&= \binom{k+2}{3} + \binom{k+2}{2} &&\text{by inductive hypothesis} \\
&= \binom{(k+2)+1}{3} &&\text{by Pascal's formula} \\
&= \binom{k+3}{3},
\end{aligned}$$

which is the right-hand side of $P(k+1)$ *[as was to be shown]. [Since we have proved the basis step and the inductive step, we conclude that $P(n)$ is true for every $n \geq 1$.]*

14. *Hint:* Use the results of exercises 3 and 13.

15. <u>Proof (by mathematical induction)</u>: Let r be a fixed nonnegative integer, and let the property $P(n)$ be the formula

$$\sum_{i=r}^{n}\binom{i}{r} = \binom{n+1}{r+1}. \quad \leftarrow P(n)$$

Show that $P(r)$ is true: To prove that $P(r)$ is true, we must show that

$$\sum_{i=r}^{r}\binom{i}{r} = \binom{r+1}{r+1}.$$

Now the left-hand side of this equation is $\binom{r}{r} = 1$, and the right-hand side is $\binom{r+1}{r+1}$, which also equals 1. So $P(r)$ is true.

Show that for every integer $k \geq r$, if $P(k)$ is true then $P(k+1)$ is true: Let k be any integer with $k \geq r$ and suppose that

$$\sum_{i=r}^{k}\binom{i}{r} = \binom{k+1}{r+1}. \quad \leftarrow \begin{array}{l} P(k) \\ \text{inductive hypothesis} \end{array}$$

We must show that

$$\sum_{i=r}^{k+1} \binom{i}{r} = \binom{(k+1)+1}{r+1} \quad \leftarrow P(k+1)$$

The left-hand side of $P(k+1)$ is

$$\sum_{i=r}^{k+1} \binom{i}{r} = \sum_{i=r}^{k} \binom{i}{r} + \binom{k+1}{r} \quad \text{by writing the last term separately}$$

$$= \binom{k+1}{r+1} + \binom{k+1}{r} \quad \text{by inductive hypothesis}$$

$$= \binom{(k+1)+1}{r+1} \quad \text{by Pascal's formula,}$$

and this is the right-hand side of $P(k+1)$ *[as was to be shown].*

17. *Hint:* This follows by letting $m = n = r$ in exercise 16 and using the result of Example 9.7.2.

18. <u>Proof (by mathematical induction)</u>: Let the property $P(n)$ be the equation

$$\binom{m}{0} + \binom{m+1}{1} + \binom{m+2}{2} + \cdots + \binom{m+n}{n} = \binom{m+n+1}{n}. \quad \leftarrow P(n)$$

We will show by mathematical induction that the property is true for every integer $n \geq 0$.

Show that $P(0)$ is true: $P(0)$ is the equation $\binom{m}{0} = \binom{m+0+1}{0} = \binom{m+1}{0}$, is true because by exercise 1 both sides equal 1.

Show that for every integer $k \geq 0$, if $P(k)$ is true then $P(k+1)$ is true: Let k be any integer with $k \geq 0$ and suppose that

$$\binom{m}{0} + \binom{m+1}{1} + \binom{m+2}{2} + \cdots + \binom{m+k}{k} = \binom{m+k+1}{k}. \quad \leftarrow \begin{array}{l} P(k) \\ \text{inductive hypothesis} \end{array}$$

We must show that

$$\binom{m}{0} + \binom{m+1}{1} + \binom{m+2}{2} + \cdots + \binom{m+(k+1)}{k+1} = \binom{m+(k+1)+1}{(k+1)}$$

or, equivalently,

$$\binom{m}{0} + \binom{m+1}{1} + \binom{m+2}{2} + \cdots + \binom{m+k+1}{k+1} = \binom{m+k+2}{k+1}. \quad \leftarrow P(k+1)$$

The left-hand side of $P(k+1)$ is

$$\binom{m}{0} + \binom{m+1}{1} + \binom{m+2}{2} + \cdots + \binom{m+k+1}{k+1}$$

$$= \binom{m+k+1}{k} + \binom{m+k+1}{k+1} \quad \text{by inductive hypothesis}$$

$$= \binom{m+k+2}{k+1} \quad \begin{array}{l} \text{by Pascal's formula with } m+k+1 \\ \text{in place of } n \text{ and } k+1 \text{ in place of } r. \end{array}$$

[This is what was to be shown.]

19. $(1+x)^7$

$$= \binom{7}{0} x^0 \cdot 1^7 + \binom{7}{1} x^1 \cdot 1^6 + \binom{7}{2} x^2 \cdot 1^5 + \binom{7}{3} x^3 \cdot 1^4 + \binom{7}{4} x^4 \cdot 1^3 + \binom{7}{5} x^5 \cdot 1^2 + \binom{7}{6} x^6 \cdot 1^1 + \binom{7}{7} x^7 \cdot 1^0$$

$$= 1 + 7x + 21x^2 + 35x^3 + 35x^4 + 21x^5 + 7x^6 + x^7$$

21. *Solution 1:* $(1 - x)^6$

$= \binom{6}{0}(-x)^0 \cdot 1^6 + \binom{6}{1}(-x)^1 \cdot 1^5 + \binom{6}{2}(-x)^2 \cdot 1^4 + \binom{6}{3}(-x)^3 \cdot 1^3 + \binom{6}{4}(-x)^4 \cdot 1^2 + \binom{6}{5}(-x)^5 \cdot 1^1 + \binom{6}{6}(-x)^6 \cdot 1^0$

$= 1 - 6x + 15x^2 - 20x^3 + 15x^4 - 6x^5 + x^6$

Solution 2: An alternative solution is to first expand and simplify the expression $(a + b)^6$ and then substitute 1 in place of a and $(-x)$ in place of b and further simplify the result. For this approach, first apply the binomial theorem with $n = 6$ to obtain

$(a + b)^6 = \binom{6}{0}a^0 b^6 + \binom{6}{1}a^1 b^5 + \binom{6}{2}a^2 b^4 + \binom{6}{3}a^3 b^3 + \binom{6}{4}a^4 b^2 + \binom{6}{5}a^5 b^1 + \binom{6}{6}a^6 b^0$

$\qquad = b^6 + 6ab^5 + 15a^2 b^4 + 20a^3 b^3 + 15a^4 b^2 + 6a^5 b + a^6.$

Then substitute 1 in place of a and $(-x)$ in place of b to give

$(1-x)^6 = (1)^6 (-x)^0 + 6(1)^5 (-x)^1 + 15(1)^4 (-x)^2 + 20(1)^3 (-x)^3 + 15(1)^2 (-x)^4 + 6(1)^1 (-x)^5 + (-x)^6$

$\qquad = 1 - 6x + 15x^2 - 20x^3 + 15x^4 - 6x^5 + x^6.$

23. $(p - 2q)^4 = \binom{4}{0}p^4 (-2q)^0 + \binom{4}{1}p^3 (-2q)^1 + \binom{4}{2}p^2 (-2q)^2 + \binom{4}{3}p^1 (-2q)^3 + \binom{4}{4}p^0 (-2q)^4$

$\qquad = p^4 - 8p^3 q + 24p^2 q^2 - 32pq^3 + 16q^4$

24. *Solution 1:* $(u^2 - 3v)^4$

$= \binom{4}{0}(u^2)^4 (-3v)^0 + \binom{4}{1}(u^2)^3 (-3v)^1 + \binom{4}{2}(u^2)^2 (-3v)^2 + \binom{4}{3}(u^2)^1 (-3v)^3 + \binom{4}{4}(u^2)^0 (-3v)^4$

$= u^8 - 12u^6 v + 54u^4 v^2 - 108u^2 v^3 + 81v^4$

Solution 2: An alternative solution is to first expand and simplify the expression $(a + b)^4$ and then substitute u^2 in place of a and $(-3v)$ in place of b and further simplify the result. For this approach, first apply the binomial theorem with $n = 4$ to obtain

$(a + b)^4 = \binom{4}{0}u^4 b^0 + \binom{4}{1}a^3 b^1 + \binom{4}{2}a^2 b^2 + \binom{4}{3}a^1 b^3 + \binom{4}{4}b^4$

$\qquad = a^4 + 4a^3 b + 6a^2 b^2 + 4ab^3 + b^4.$

Then substitute u^2 in place of a and $(-3v)$ in place of b to give

$(u^2 - 3v)^4 = (u^2 + (-3v))^4 = (u^2)^4 + 4(u^2)^3 (-3v) + 6(u^2)^2 (-3v)^2 + 4(u^2)(-3v)^3 + (-3v)^4$

$\qquad = u^8 - 12u^6 v + 54u^4 v^2 - 108u^2 v^3 + 81v^4.$

25. $\left(x - \dfrac{1}{x}\right)^5 = \sum_{k=0}^{5} \binom{5}{k} x^{5-k} \left(\dfrac{1}{x}\right)^k$

$= \binom{5}{0}x^5 \left(\dfrac{1}{x}\right)^0 + \binom{5}{1}x^4 \left(\dfrac{1}{x}\right)^1 + \binom{5}{2}x^3 \left(\dfrac{1}{x}\right)^2 + \binom{5}{3}x^2 \left(\dfrac{1}{x}\right)^3 + \binom{5}{4}x^1 \left(\dfrac{1}{x}\right)^4 + \binom{5}{5}x^0 \left(\dfrac{1}{x}\right)^5$

$= x^5 + 5x^3 + 10x + \dfrac{10}{x} + \dfrac{5}{x^3} + \dfrac{1}{x^5}$

27. $\left(x^2 - \dfrac{1}{x}\right)^5$

$= (x^2)^5 + \binom{5}{1}(x^2)^4 \left(-\dfrac{1}{x}\right)^1 + \binom{5}{2}(x^2)^3 \left(-\dfrac{1}{x}\right)^2 + \binom{5}{3}(x^2)^2 \left(-\dfrac{1}{x}\right)^3 + \binom{5}{4}(x^2)^1 \left(-\dfrac{1}{x}\right)^4 + \left(-\dfrac{1}{x}\right)^5$

$= x^{10} - 5x^7 + 10x^4 - 10x + \dfrac{5}{x^2} - \dfrac{1}{x^5}$

29. The term is $\binom{9}{3}x^6 y^3 = 84x^6 y^3$, so the coefficient is 84.

30. The term is $\binom{10}{3}(2x)^7 3^3$, so the coefficient is $\dfrac{10!}{3! \cdot 7!} \cdot 2^7 \cdot 3^3 = 120 \cdot 128 \cdot 27 = 414{,}720.$

31. The term is $\binom{12}{7}a^5 (-2b)^7 = 792a^5 (-128) b^7 = -101{,}376a^5 b^7$, so the coefficient is $-101{,}376.$

33. The term is $\binom{15}{8}(3p^2)^8(-2q)^7 = \binom{15}{8}3^8(-2)^7p^{16}q^7$, so the coefficient is

$\binom{15}{8}3^8(-2)^7 = -5,404,164,480.$

36. Proof: Let $a = 1$, let $b = -1$, and let n be a positive integer. Substitute into the binomial theorem to obtain

$$(1 + (-1))^n = \sum_{k=0}^{n} \binom{n}{k} \cdot 1^{n-k} \cdot (-1)^k$$

$$= \sum_{k=0}^{n} \binom{n}{k}(-1)^k \qquad \text{since } 1^{n-k} = 1.$$

On the other hand, $(1 + (-1))^n = 0^n = 0$, so

$$0 = \sum_{k=0}^{n} \binom{n}{k}(-1)^k$$

$$= \binom{n}{0} - \binom{n}{1} + \binom{n}{2} - \binom{n}{3} + \cdots + (-1)^n \binom{n}{n}.$$

37. *Hint:* $3 = 2 + 1$

38. Proof: Let m be any integer with $m \geq 0$, and apply the binomial theorem with $a = 2$ and $b = -1$. The result is

$$1 = 1^m = (2 + (-1))^m = \sum_{i=0}^{m} \binom{m}{i} 2^{m-i}(-1)^i = = \sum_{i=0}^{m}(-1)^i \binom{m}{i} 2^{m-i}$$

because $1^m = 1$ for every integer m.

39. Proof: Let n be an integer with $n \geq 0$. Apply the binomial theorem with $a = 3$ and $b = -1$ to obtain

$$2^n = (3 + (-1))^n$$

$$= \binom{n}{0}3^n(-1)^0 + \binom{n}{1}3^{n-1}(-1)^1 + \cdots + \binom{n}{i}3^{n-i}(-1)^i + \cdots + \binom{n}{n}3^{n-n}(-1)^n$$

$$= \sum_{i=0}^{n}(-1)^i \binom{n}{i}3^{n-i}.$$

41. *Hint:* Apply the binomial theorem with $a = 1$ and $b = -\frac{1}{2}$, and analyze the resulting equation when n is even and when n is odd.

42. Proof (by mathematical induction): Let the property $P(n)$ be the sentence

> For any set S with n elements, S has 2^{n-1} subsets with an even number of elements and 2^{n-1} subsets with an odd number of elements. $\leftarrow P(n)$

We will prove by mathematical induction that the property is true for every integer $n \geq 1$.

Show that $P(1)$ is true: $P(1)$ is true because any set S with just 1 element, say x, has two subsets: $\emptyset$, which has 0 elements, and $\{x\}$, which has 1 element. Since 0 is even and 1 is odd, the number of subsets of S with an even number of elements equals the number of subsets of S with an odd number of elements, namely, 1, and $1 = 2^0 = 2^{1-1}$.

Show that for every integer $k \geq 1$, if $P(k)$ is true then $P(k+1)$ is true: Let k be any integer with $k \geq 1$ and suppose that

> For any set S with k elements, S has 2^{k-1} subsets with an even number of elements and 2^{k-1} subsets with an odd number of elements.

$P(k)$
$\leftarrow$ inductive hypothesis

We must show that

> For any set S with $k+1$ elements, S has $2^{(k+1)-1}$ subsets with an even number of elements and $2^{(k+1)-1}$ subsets with an odd number of elements.

or, equivalently,

> For any set S with $k+1$ elements, S has 2^k subsets with an even number of elements and 2^k subsets with an odd number of elements.

$\leftarrow P(k+1)$

Call the elements of $S = \{x_1, x_2, \ldots, x_k, x_{k+1}\}$. By inductive hypothesis, $\{x_1, x_2, \ldots, x_k\}$ has 2^{k-1} subsets with an even number of elements and 2^{k-1} subsets with an odd number of elements. Now every subset of $\{x_1, x_2, \ldots, x_k\}$ is also a subset of S, and the only other subsets of S are obtained by taking the union of a subset of $\{x_1, x_2, \ldots, x_k\}$ with $\{x_{k+1}\}$. Moreover, if a subset of $\{x_1, x_2, \ldots, x_k\}$ has an even number of elements, then the union of that subset with $\{x_{k+1}\}$ has an odd number of elements. So 2^{k-1} of the subsets of S that are obtained by taking the union of a subset of $\{x_1, x_2, \ldots, x_k\}$ with $\{x_{k+1}\}$ have an even number of elements and 2^{k-1} have an odd number of elements. Thus the total number of subsets of S with an even number of elements is

$$2^{k-1} + 2^{k-1} - 2 \cdot 2^{k-1} = 2^{1+(k-1)} = 2^k.$$

Similarly, the total number of subsets of S with an odd number of elements is also

$$2^{k-1} + 2^{k-1} = 2^k$$

[as was to be shown].

Alternative justification for the identity in exercise 36: Let n be any positive integer, let E be the largest even integer less than or equal to n, and let O be the largest odd integer less than or equal to n. Let S be any set with n elements. Then the number of subsets of S with an even number of elements is $\binom{n}{0} + \binom{n}{2} + \binom{n}{4} + \cdots + \binom{n}{E}$, and the number of subsets of S with an odd number of elements is $\binom{n}{1} + \binom{n}{3} + \binom{n}{5} + \cdots + \binom{n}{O}$. But there are as many subsets with an even number of elements as there are subsets with an odd number of elements, so if we subtract the second of these quantities from the first we obtain 0:

$$
\begin{aligned}
0 &= \left[\binom{n}{0} + \binom{n}{2} + \binom{n}{4} + \cdots + \binom{n}{E}\right] - \left[\binom{n}{1} + \binom{n}{3} + \binom{n}{5} + \cdots + \binom{n}{O}\right] \\
&= \binom{n}{0} - \binom{n}{1} + \binom{n}{2} - \binom{n}{3} + \cdots + (-1)^n \binom{n}{n}.
\end{aligned}
$$

43. Let n be an integer with $n \geq 0$. Then

$$\sum_{k=0}^{n} \binom{n}{k} 5^k = \sum_{k=0}^{n} \binom{n}{k} 1^{n-k} 5^k = (1+5)^m = 6^n$$

by the binomial theorem with $a = 1$ and $b = 5$ and because 1 raised to any power is 1.

45. Let n be an integer with $n \geq 0$. Then

$$\sum_{i=0}^{n}\binom{n}{i}x^i = \sum_{i=0}^{n}\binom{n}{i}1^{n-i}x^i = (1+x)^n$$

by the binomial theorem with $a = 1$ and $b = x$, because 1 raised to any power is 1, and because $1 \cdot x = x$.

47. Let n be an integer with $n \geq 0$. Then

$$\sum_{j=0}^{2n}(-1)^j\binom{2n}{j}x^j = \sum_{j=0}^{2n}\binom{2n}{j}1^{2n-j}(-x)^j = (1+x^2)^n$$

because $(-1)^j x^j = (-x)^j$ and $1^{2n-j} = 1$ by the laws of exponents and by the binomial theorem with $a = 1$ and $b = x^2$.

48. Let n be an integer with $n \geq 0$. Then

$$\sum_{r=0}^{n}\binom{n}{r}x^{2r} = \sum_{r=0}^{n}\binom{n}{r}1^{n-r}(x^2)^r = (1+x^2)^n$$

because $x^{2r} = (x^2)^r$ and $1^{n-r} = 1$ by the laws of exponents and by the binomial theorem with $a = 1$ and $b = x^2$.

51. Let m be an integer with $m \geq 0$. Then

$$\sum_{i=0}^{m}(-1)^i\binom{m}{i}\frac{1}{2^i} = \sum_{i=0}^{m}\binom{m}{i}1^{m-i}\left(-\frac{1}{2}\right)^i = \left(1-\frac{1}{2}\right)^m = \frac{1}{2^m}$$

because $(-1)^i\frac{1}{2^i} = \left(-\frac{1}{2}\right)^i$ by the laws of exponents and the binomial theorem with $a = 1$ and $b = -\frac{1}{2}$.

53. Let n be an integer with $n \geq 0$. Then

$$\sum_{i=0}^{n}(-1)^i\binom{n}{i}5^{n-i}2^i = \sum_{i=0}^{n}\binom{n}{i}5^{n-i}(-2)^i = (5-2)^n = 3^n$$

because $(-1)^i 2^i = (-2)^i$ by the laws of exponents and by the binomial theorem with $a = 5$ and $b = -2$.

54. Let n be an integer with $n \geq 0$. Then

$$\sum_{k=0}^{n}(-1)^k\binom{n}{k}3^{2n-2k}2^{2k} = \sum_{k=0}^{n}(-1)^k\binom{n}{k}(3^2)^{n-k}(2^2)^k = \sum_{k=0}^{n}\binom{n}{k}9^{n-k}(-4)^k = (9-4)^n = 5^n$$

because $3^{2n-2k} = (3^2)^{n-k}$ and $(-1)^k 2^{2k} = (-1)^k(2^2)^k = (-4)^k$ by the laws of exponents and by the binomial theorem with $a = 9$ and $b = -4$.

55. **b.** $n(1+x)^{n-1} = \sum_{k=1}^{n}\binom{n}{k}kx^{k-1}$

[*The term corresponding to $k = 0$ is zero because $\frac{d}{dx}(x^0) = 0$.*]

c. (i) Substitute $x = 1$ in part (b) above to obtain

$$n(1+1)^{n-1} = \sum_{k=1}^{n}\binom{n}{k}k\cdot 1^{k-1} = \sum_{k=1}^{n}\binom{n}{k}k\cdot 1^{k-1}$$

$$= \sum_{k=1}^{n}\binom{n}{k}k = \binom{n}{1}\cdot 1 + \binom{n}{2}\cdot 2 + \binom{n}{3}\cdot 3 + \cdots + \binom{n}{n}\cdot n.$$

Dividing both sides by n and simplifying gives

$$2^{n-1} = \frac{1}{n}\left[\binom{n}{1} + 2\binom{n}{2} + 3\binom{n}{3} + \cdots + n\binom{n}{n}\right].$$

(ii) Let n be an integer with $n \geq 1$. Apply the formula from part (b) with $x = -1$ to obtain

$$0 = n(1 + (-1))^{n-1} = \sum_{k=1}^{n}\binom{n}{k}k(-1)^{k-1}.$$

Section 9.8

1. By probability axiom 2, $P(\emptyset) = 0$.

2. **a.** By probability axiom 3, $P(A \cup B) = P(A) + P(B) = 0.3 + 0.5 = 0.8$.

   b. Because $A \cup B \cup C = S$ and because A, B, and C are mutually exclusive events, $C = S - (A \cup B)$. Thus, by the formula for the probability of the complement of an event,

$$P(C) = P((A \cup B)^c) = 1 - P(A \cup B) = 1 - 0.8 = 0.2.$$

3. **a.** $P(A \cup B) = 0.4 + 0.2 = 0.6$

   b. By the formula for the probability of a general union and because $S = A \cup B \cup C$,

$$P(S) = ((A \cup B) \cup C) = P(A \cup B) + P(C) - P((A \cup B) \cap C).$$

   Suppose $P(C) = 0.2$. Then, since $P(S) = 1$,

$$1 = 0.6 + 0.2 - P((A \cup B) \cap C) = 0.8 - P((A \cup B) \cap C).$$

   Solving for $P((A \cup B) \cap C)$ gives $P((A \cup B) \cap C) = -0.2$, which is impossible. Hence $P(C) \neq 0.2$.

4. By the formula for the probability of a general union of two events, $P(A \cup B) = P(A) + P(B) - P(A \cap B) = 0.8 + 0.7 - 0.6 = 0.9$.

6. First note that we can apply the formula for the probability of the complement of an event to obtain $0.3 = P(U^c) = 1 - P(U)$. Solving for $P(U)$ gives $P(U) = 0.7$. Second, observe that by De Morgan's law $U^c \cup V^c = (U \cap V)^c$. Thus

$$0.4 = P(U^c \cup V^c) = P((U \cap V)^c) = 1 - P(U \cap V).$$

   Solving for $P(U \cap V)$ gives $P(U \cap V) = 0.6$. So, by the formula for the union of two events,

$$P(U \cup V) = P(U) + P(V) - P(U \cap V) = 0.7 + 0.6 - 0.6 = 0.7.$$

7. **a.** Because $A \cap B = \emptyset$, $P(A \cup B) = P(A) + P(B) = 0.4 + 0.3 = 0.7$.

   b. Note that $C = (A \cup B)^c$. Thus $P(C) = P((A \cup B)^c) = 1 - P(A \cup B) = 1 - 0.7 = 0.3$.

   c. Because $A \cap C = \emptyset$, $P(A \cup C) = P(A) + P(C) = 0.4 + 0.3 = 0.7$.

   d. $P(A^c) = 1 - P(A) = 1 - 0.4 = 0.6$

e. By De Morgan's law $A^c \cap B^c = (A \cup B)^c$. Thus, by part (a) and the formula for the probability of the complement of an event,

$$P(A^c \cap B^c) = P((A \cup B)^c) = 1 - P(A \cup B) = 1 - 0.7 = 0.3$$

f. By De Morgan's law $A^c \cup B^c = (A \cap B)^c$. Thus by the formula for the probability of the complement of an event and because $A \cap B = \emptyset$ and $P(\emptyset) = 0$,

$$P(A^c \cup B^c) = P((A \cap B)^c) = P(\emptyset^c) = P(S) = 1.$$

9. **a.** By the formula for the probability of a general union of two events,

$$P(A \cup B) = P(A) + P(B) - P(A \cap B) = 0.4 + 0.5 - 0.2 = 0.7.$$

b. By part (a), $P(A \cup B) = 0.7$. So, since $C = (A \cup B)^c$, by the formula for the probability of the complement of an event,

$$P(C) = 1 - P(A \cup B) = 1 - 0.7 = 0.3.$$

c. By the formula for the probability of the complement of an event,

$$P(A^c) = 1 - P(A) = 1 - 0.4 = 0.6.$$

d. By De Morgan's law $(A \cup B)^c = A^c \cap B^c$. Thus, by the formula for the probability of the complement of an event,

$$P(A^c \cap B^c) = P((A \cup B)^c) = 1 - P(A \cup B) = 1 - 0.7 = 0.3.$$

e. By De Morgan's law $A^c \cup B^c = (A \cap B)^c$. Thus, by the formula for the probability of the complement of an event,

$$P(A^c \cup B^c) = P((A \cap B)^c) = 1 - P(A \cap B) = 1 - 0.2 = 0.8.$$

f. *Solution 1*: Because $C = S - (A \cup B)$, we have that $C = (A \cup B)^c$. Then

$$
\begin{array}{llll}
B^c \cap C & = & B^c \cap (A \cup B)^c & \text{by substitution} \\
 & = & B^c \cap (A^c \cap B^c) & \text{by De Morgan's law} \\
 & = & (B^c \cap A^c) \cap B^c & \text{by the associative law for } \cap \\
 & = & (A^c \cap B^c) \cap B^c & \text{by the commutative law for } \cap \\
 & = & A^c \cap (B^c \cap B^c) & \text{by the associative law for } \cap \\
 & = & A^c \cap B^c & \text{by the idempotent law for } \cap \\
 & = & (A \cup B)^c & \text{by De Morgan's law} \\
 & = & C & \text{by substitution.}
\end{array}
$$

Hence, by part (b), $P(B^c \cap C) = P(C) = 0.3$.

Solution 2: Because $C = S - (A \cup B)$, we have that $C = (A \cup B)^c$. Thus by De Morgan's law, $C = A^c \cap B^c$. Now $A^c \cap B^c \subseteq B^c$ *[by Theorem 9.2.1(1)b]* and hence $B^c \cap C = C$ *[by Theorem 9.2.3a]*. Therefore $P(B^c \cap C) = P(C) = 0.3$.

11. *Hint:* Since $U \subseteq V$, $V = U \cup (V - U)$

12. <u>Proof 1</u>: Suppose S is any sample space and U and V are any events in S. First note that by the set difference, distributive, universal bound, and identity laws,

$$(V \cap U) \cup (V - U) = (V \cap U) \cup (V \cap U^c) = V \cap (U \cup U^c) = V \cap S = V.$$

Next, observe that if $x \in (V \cap U) \cap (V - U)$, then, by definition of intersection, $x \in (V \cap U)$ and $x \in (V - U)$, and so, by definition of intersection and set difference, $x \in V$, $x \in U$, $x \in V$, and $x \notin U$, and hence, in particular, $x \in U$ and $x \notin U$, which is impossible. It follows that $(V \cap U) \cap (V - U) = \emptyset$. Thus, by substitution and by probability axiom 3 (the formula for the probability of mutually disjoint events),

$$P(V) = P((V \cap U) \cup (V - U)) = P(V \cap U) + P(V - U).$$

Solving for $P(V - U)$ gives

$$P(V - U) = P(V) - P(U \cap V).$$

<u>Proof 2</u>: Suppose S is any sample space and U and V are any events in S. First note that by the set difference, distributive, universal bound, and identity laws,

$$U \cup (V - U) = U \cup (V \cap U^c) = (U \cup V) \cap (U \cup U^c) = (U \cup V) \cap S = U \cup V.$$

Also by the set difference law, and the associative, commutative, and universal bound laws for $\cap$,

$$U \cap (V - U) = U \cap (V \cap U^c) = U \cap (U^c \cap V) = (U \cap U^c) \cap V = \emptyset \cap V = \emptyset.$$

Thus, by probability axiom 3 (the formula for the probability of mutually disjoint events),

$$P(U \cup V) = P(U \cup (V - U)) = P(U) + P(V - U).$$

But also by the formula for the probability of a general union,

$$P(U \cup V) = P(U) + P(V) - P(U \cap V).$$

Equating the two expressions for $P(U \cup V)$ gives

$$P(U) + P(V - U) = P(U) + P(V) - P(U \cap V).$$

Subtracting $P(U)$ from both sides gives

$$P(V - U) = P(V) - P(U \cap V).$$

13. *Hint:* $(A_1 \cup A_2 \cup \ldots \cup A_k) \cap A_{k+1} = \emptyset$ and $A_1 \cup A_2 \cup \ldots \cup A_k \cup A_{k+1} = (A_1 \cup A_2 \cup \ldots \cup A_k) \cup A_{k+1}$.

14. *Solution 1:* The net gain of the grand prize winner is $\$2,000,000 - \$2 = \$1,999,998$. Each of the 10,000 second prize winners has a net gain of $\$20 - \$2 = \$18$, and each of the 50,000 third prize winners has a net gain of $\$4 - \$2 = \$2$. The number of people who do not win anything is $1,500,000 - 1 - 10,000 - 50,000 = 1,439,999$, and each of these people has a net loss of $\$2$. Because all of the 1,500,000 tickets have an equal chance of winning a prize, the expected gain or loss of a ticket is

$$\frac{1}{1500000}(\$1,999,998 \cdot 1 + \$18 \cdot 10000 + \$2 \cdot 50000 + (-\$2) \cdot 1,439,999) = -\$0.40.$$

Solution 2: The total income to the lottery organizer is $\$2$ (per ticket)$\cdot$ 1,500,000 (tickets) $= \$3,000,000$. The payout the lottery organizer must make is $\$2,000,000 + (\$20)(10,000) + (\$4)(50,000) = \$2,400,000$, so the net gain to the lottery organizer is $\$600,000$, which amounts to $\frac{\$600,000}{1,500,000} = \$0.40$ per ticket. Thus the expected net loss to a purchaser of a ticket is $\$0.40$.

15. *Solution 1*: The net gain for the first prize ticket is $40,000 -$20 = \$39,980$, that for the second prize ticket is $\$1,000 -\$20 = \$980$, and that for the third prize ticket is $\$500 - \$20 = \$480$. Each of the other 2,997 raffle tickets has a net loss of $20. Because all of the 3,000 tickets have an equal chance of winning the prizes, the expected gain or loss per ticket is

$$\$39980 \cdot \frac{1}{3000} + \$980 \cdot \frac{1}{3000} + \$480 \cdot \frac{1}{3000} - \$20 \cdot \frac{2997}{3000} \cong -\$6.17,$$

or an expected loss of about $6.17 per ticket.

Solution 2: The total amount spent for the 3,000 tickets is $3,000 \cdot \$20 = \$60,000$. The total amount of prize money awarded is $\$40,000 +\$1,000+ \$500 = \$41,500$. Thus the difference between the value of the prizes and the amount spent on the tickets is $\$60,000 - \$41,500 = \$18,500$, and so the expected loss per ticket is $18500/3000 \cong -\$6.17$.

16. Let 2_1 and 2_2 denote the two balls with the number 2, and let 5 and 6 denote the other two balls. There are $\binom{6}{2}$ subsets of 2 balls that can be chosen from the urn. The following table shows the sums of the numbers on the balls in each set and the corresponding probabilities:

Subset	Sum s	Probability that the sum $= s$
$\{2_1, 2_2\}$	4	1/6
$\{2_1, 5\}, \{2_2, 5\}$	7	2/6
$\{2_1, 6\}\{2_2, 6\}$	8	2/6
$\{5, 6\}$	11	1/6

So the expected value is

$$4 \cdot \frac{1}{6} + 7 \cdot \frac{2}{6} + 8 \cdot \frac{2}{6} + 11 \cdot \frac{1}{6} = 7.5.$$

18. Let 2_1 and 2_2 denote the two balls with the number 2, let 8_1 and 8_2 denote the two balls with the number 8, and let 1 denote the other ball. There are $\binom{5}{3} = 10$ subsets of 3 balls that can be chosen from the urn. The following table shows the sums of the numbers on the balls in each set and the corresponding probabilities:

Subset	Sum s	Probability of s
$\{1, 2_1, 2_2\}$	5	1/10
$\{1, 2_1, 8_1\}, \{1, 2_2, 8_1\}, \{1, 2_1, 8_2\}, \{1, 2_2, 8_2\}$	11	4/10
$\{2_1, 2_2, 8_1\}, \{2_1, 2_2, 8_2\}$	12	2/10
$\{1, 8_1, 8_2\}$	17	1/10
$\{2_1, 8_1, 8_2\}, \{2_2, 8_1, 8_2\}$	18	2/10

Thus the expected value is $5 \cdot \dfrac{1}{10} + 11 \cdot \dfrac{4}{10} + 12 \cdot \dfrac{2}{10} + 17 \cdot \dfrac{1}{10} + 18 \cdot \dfrac{2}{10} = \dfrac{126}{10} = 12.6$.

19. The following table displays the sum of the numbers showing face up on the dice:

	1	2	3	4	5	6
1	2	3	4	5	6	7
2	3	4	5	6	7	8
3	4	5	6	7	8	9
4	5	6	7	8	9	10
5	6	7	8	9	10	11
6	7	8	9	10	11	12

Each cell in the table represents an outcome whose probability is $\frac{1}{36}$. Thus the expected value of the sum is

$$2 \cdot \frac{1}{36} + 3 \cdot \frac{2}{36} + 4 \cdot \frac{3}{36} + 5 \cdot \frac{4}{36} + 6 \cdot \frac{5}{36} + 7 \cdot \frac{6}{36} + 8 \cdot \frac{5}{36} + 9 \cdot \frac{4}{36} + 10 \cdot \frac{3}{36} + 11 \cdot \frac{2}{36} + 12 \cdot \frac{1}{36}$$

$$= \frac{252}{36} = 7.$$

20. *Hint:* The answer is about 7.7 cents.

21. When a coin is tossed 4 times, there are $2^4 = 16$ possible outcomes and there are $\binom{4}{h}$ ways to obtain exactly h heads (as shown by the technique illustrated in Example 9.5.9). The following table shows the possible outcomes of the tosses, the amount gained or lost for each outcome, the number of ways the outcomes can occur, and the probabilities of the outcomes.

Number of Heads	Net Gain (or Loss)	Number of Ways	Probability
0	−$3	$\binom{4}{0} = 1$	1/16
1	−$2	$\binom{4}{1} = 4$	4/16
2	−$1	$\binom{4}{2} = 6$	6/16
3	$2	$\binom{4}{3} = 4$	4/16
4	$3	$\binom{4}{4} = 1$	1/16

Thus the expected value is

$$(-\$3) \cdot \frac{1}{16} + (-\$2) \cdot \frac{4}{16} + (-\$1) \cdot \frac{6}{16} + \$2 \cdot \frac{4}{16} + \$3 \cdot \frac{1}{16} = -\$\frac{6}{16} = -\$0.375.$$

So this game has an expected loss of 37.5 cents.

22. *Hint:* The answer is 1.875.

23. *Hint:* To derive P_{20}, use the distinct roots theorem from Section 5.8. The answer is $P_{20} = \frac{5^{300} - 5^{20}}{5^{300} - 1} \cong 1$.

Section 9.9

1. $P(B) = \dfrac{P(A \cap B)}{P(A|B)} = \dfrac{1/6}{1/2} = \dfrac{1}{3}$

3. Of the students who received A's on the first test, the percent who also received A's on the second test is

$$\frac{\text{the percent of students who received } A\text{'s on both tests}}{\text{the percent of students who received } A\text{'s on the first test}} = \frac{15\%}{25\%} = 0.6 = 60\%.$$

4. **a.** Proof: Suppose S is any sample space and A and B are any events in S such that $P(B) \neq 0$. Note that
   (1) $A \cup A^c = S$ by the complement law for $\cup$.
   (2) $B \cap S = B$ by the identity law for $\cap$.
   (3) $B \cap (A \cup A^c) = (A \cap B) \cup (A^c \cap B)$ by the distributive law and commutative laws for sets.
   (4) $(A \cap B) \cap (A^c \cap B) = \emptyset$ by the complement law for $\cap$ and the commutative and associative laws for sets.
   Thus $B = (A \cap B) \cup (A^c \cap B)$, and, by probability axiom 3, $P(B) = P(A \cap B) + P(A^c \cap B)$. Therefore, $P(A^c \cap B) = P(B) - P(A \cap B)$. By definition of conditional probability, it follows that

   $$P(A^c \mid B) = \frac{P(A^c \cap B)}{P(B)} = \frac{P(B) - P(A \cap B)}{P(B)} = 1 - \frac{P(A \cap B)}{P(B)} = 1 - P(A \mid B).$$

5. *Hints:* (1) $A = (A \cap B) \cup (A \cap B^c)$

   (2) The answer is $P(A \mid B^c) = \dfrac{P(A) - P(A \mid B)P(B)}{1 - P(B)}$

6. **a.**   Let R_1 be the probability that the first ball is red, and let R_2 be the probability that the second ball is red. Then R_1^c is the probability that the first ball is not red, and R_2^c is the probability that the second ball is not red. The tree diagram shows the various relations among the probabilities.

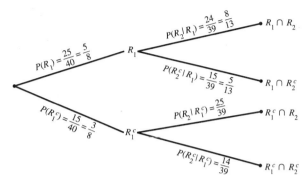

Then

$$P(R_1 \cap R_2) = P(R_2 \mid R_1) \cdot P(R_1) = \frac{8}{13} \cdot \frac{5}{8} = \frac{5}{13} \cong 38.5\%$$

$$P(R_1 \cap R_2^c) = P(R_2^c \mid R_1) \cdot P(R_1) = \frac{5}{13} \cdot \frac{5}{8} = \frac{25}{104} \cong 24\%$$

$$P(R_1^c \cap R_2) = P(R_2 \mid R_1^c) \cdot P(R_1^c) = \frac{25}{39} \cdot \frac{3}{8} = \frac{25}{104} \cong 24\%$$

$$P(R_1^c \cap R_2^c) = P(R_2^c \mid R_1^c) \cdot P(R_1^c) = \frac{14}{39} \cdot \frac{3}{8} = \frac{14}{104} \cong 13.5\%.$$

So the probability that both balls are red is 5/13, the probability that the first ball is red and the second is not is 25/104, the probability that the first ball is not red and the second ball is red is 25/104, and the probability that neither ball is red is 14/104.

b. Note that

$$R_2 = (R_2 \cap R_1) \cup (R_2 \cap R_1^c) \quad \text{and} \quad (R_2 \cap R_1) \cup (R_2 \cap R_1^c) = \emptyset.$$

Thus the probability that the second ball is red is

$$P(R_2) = P(R_2 \cap R_1) + P(R_2 \cap R_1^c) = \frac{5}{13} + \frac{25}{104} = \frac{65}{104} \cong 62.5\%.$$

c. If exactly one ball is red, then either the first ball is red and the second is not or the first ball is not red and the second is red, and these possibilities are mutually exclusive. Thus

$$P(\text{exactly one ball is red}) = P(R_1 \cap R_2^c) + P(R_1^c \cap R_2) = \frac{25}{104} + \frac{25}{104} = \frac{50}{104} = \frac{25}{52} \cong 48.1\%.$$

The probability that both balls are red is $P(R_1 \cap R_2) = \frac{5}{13} \cong 38.5\%$. Then

$$P(\text{at least one ball is red}) = P(\text{exactly one ball is red}) + P(\text{both balls are red})$$
$$= \frac{25}{52} + \frac{5}{13} = \frac{45}{52} \cong 86.5\%.$$

8. **a.** Let W_1 be the event that a woman is chosen on the first draw,

W_2 be the event that a woman is chosen on the second draw,

M_1 be the event that a man is chosen on the first draw,

M_2 be the event that a man is chosen on the second draw.

Then $P(W_1) = \frac{3}{10}$ and $P(W_2 \mid W_1) = \frac{2}{9}$, and thus

$$P(W_1 \cap W_2) = P(W_2 \mid W_1)P(W_1) = \frac{2}{9} \cdot \frac{3}{10} = \frac{1}{15} = 6\frac{2}{3}\%.$$

c. *Hint:* The answer is $\frac{7}{15} = 46\frac{2}{3}\%$.

9. <u>Proof:</u> Suppose that a sample space S is a union of two disjoint events B_1 and B_2, that A is an event in S with $P(A) \neq 0$, and that $P(B_k) \neq 0$ for $k = 1$ and $k = 2$. Because B_1 and B_2 are disjoint, the same reasoning as in Example 9.9.5 establishes that

$$A = (A \cap B_1) \cup (A \cap B_2) \quad \text{and} \quad (A \cap B_1) \cap (A \cap B_2) = \emptyset.$$

Thus

$$P(A) = P(A \cap B_1) + P(A \cap B_2).$$

Moreover, for each $k = 1$ or 2, by definition of conditional probability, we have both

$$P(B_k \mid A) = \frac{P(B_k \cap A)}{P(A)} = \frac{P(A \cap B_k)}{P(A)} \quad \text{and} \quad P(A \cap B_k) = P(A \mid B_k)P(B_k).$$

Putting these results together gives that for each $k = 1$ or 2,

$$P(B_k \mid A) = \frac{P(A \cap B_k)}{P(A)} = \frac{P(A \mid B_k)P(B_k)}{P(A \cap B_1) + P(A \cap B_2)} = \frac{P(A \mid B_k)P(B_k)}{P(A \mid B_1)P(B_1) + P(A \mid B_2)P(B_2)},$$

which is Bayes' theorem for $n = 2$.

11. **a.** Let U_1 be the event that the first urn is chosen, U_2 the event that the second urn is chosen, and B the event that the chosen ball is blue. Then

$$P(B \mid U_1) = \frac{12}{19} \quad \text{and} \quad P(B \mid U_2) = \frac{8}{27}.$$

$$P(B \cup U_1) = P(B \mid U_1)P(U_1) = \frac{12}{19} \cdot \frac{1}{2} = \frac{12}{38}.$$

Also

$$P(A \cap U_2) = P(B \mid U_2)P(U_2) = \frac{8}{27} \cdot \frac{1}{2} = \frac{8}{54}.$$

Now B is the disjoint union of $B \cap U_1$ and $B \cap U_2$. So

$$P(B)P(U_1 \mid B) = P(B \cap U_1) + P(B \cap U_2)$$
$$= \frac{12}{38} + \frac{8}{54} \cong 46.4\%.$$

Thus the probability that the chosen ball is blue is approximately 46.4%.

b. Given that the chosen ball is blue, the probability that it came from the first urn is $P(U_1 \mid B)$. By Bayes' theorem and the computations in part (a),

$$P(U_1 \mid B) = \frac{P(B \mid U_1)P(U_1)}{P(B \mid U_1)P(U_1) + P(B \mid U_2)P(U_2)}$$

$$= \frac{(12/19)(0.5)}{(12/19)(0.5) + (8/27)(0.5)} \cong 68.1\%.$$

12. **a.** Let B_1 be the event that the first urn is chosen, B_2 the event that the second urn is chosen, and A the event that the chosen ball is blue. Then

$$P(A \mid B_1) = \frac{4}{20} \quad \text{and} \quad P(A \mid B_2) = \frac{10}{19}.$$

$$P(A \cap B_1) = P(A \mid B_1)P(B_1) = \frac{4}{20} \cdot \frac{1}{2} = \frac{1}{10}.$$

Also

$$P(A \cap B_2) = P(A \mid B_2)P(B_2) = \frac{10}{19} \cdot \frac{1}{2} = \frac{5}{19}.$$

Now A is the disjoint union of $A \cap B_1$ and $A \cap B_2$. So

$$P(A) = P(A \cap B_1) + P(A \cap B_2) = \frac{1}{10} + \frac{5}{19} = \frac{69}{190} \cong 36.3\%.$$

Thus the probability that the chosen ball is blue is approximately 36.3%.

b. *Solution 1 (using Bayes' theorem)*: Given that the chosen ball is blue, the probability that it came from the first urn is $P(B_1 \mid A)$. By Bayes' theorem and the computations in part (a),

$$P(B_1 \mid A) = \frac{P(A \mid B_1)P(B_1)}{P(A \mid B_1)P(B_1) + P(A \mid B_2)P(B_2)} = \frac{\frac{1}{10}}{\frac{1}{10} + \frac{5}{19}} = \frac{19}{69} \cong 27.5$$

Solution 2 (without explicit use of Bayes' theorem): Given that the chosen ball is blue, the probability that it came from the first urn is $P(B_1 \mid A)$. By the results of part (a),

$$P(B_1 \mid A) = \frac{P(A \cap B_1)}{P(A)} = \frac{\frac{1}{10}}{\frac{69}{190}} = \frac{19}{69} \cong 27.5$$

13. *Hint:* The answers to parts (a) and (b) are approximately 52.9% and 54.0%, respectively.

14. Let A be the event that a randomly chosen person tests positive for drugs, let B_1 be the event that a randomly chosen person uses drugs, and let B_2 be the event that a randomly chosen person does not use drugs. Then A^c is the event that a randomly chosen person does not test positive for drugs, and $P(B_1) = 0.04$, $P(B_2) = 0.96$, $P(A \mid B_2) = 0.03$, and $P(A^c \mid B_1) = 0.02$. Hence $P(A \mid B_1) = 0.98$ and $P(A^c \mid B_2) = 0.97$.

a. $P(B_1 \mid A) = \dfrac{P(A \mid B_1)P(B_1)}{P(A \mid B_1)P(B_1) + P(A \mid B_2)P(B_2)} = \dfrac{(0.97)(0.04)}{(0.97)(0.04) + (0.03)(0.96)} \cong 57.6\%$

b. $P(B_2 \mid A^c) = \dfrac{P(A^c \mid B_2)P(B_2)}{P(A^c \mid B_1)P(B_1) + P(A^c \mid B_2)P(B_2)} = \dfrac{(0.98)(0.96)}{(0.02)(0.04) + (0.98)(0.96)} \cong 99.9\%$ 16.

15. Let B_1 be the event that the part came from the first factory, B_2 the event that the part came from the second factory, and A the event that a part chosen at random from the 180 is defective.

a. The probability that a part chosen at random from the 180 is from the first factory is $P(B_1) = \frac{100}{180}$.

b. The probability that a part chosen at random from the 180 is from the second factory is $P(B_2) = \frac{80}{100}$.

c. The probability that a part chosen at random from the 180 is defective is $P(A)$. Because 2% of the parts from the first factory and 5% of the parts from the second factory are defective, $P(A \mid B_1) = \frac{2}{100}$ and $P(A \mid B_2) = \frac{5}{100}$. By definition of conditional probability,

$$P(A \cap B_1) = P(A \mid B_1)P(B_1) = \frac{2}{100} \cdot \frac{100}{180} = \frac{1}{90}$$

$$P(A \cap B_2) = P(A \mid B_2)P(B_2) = \frac{5}{100} \cdot \frac{80}{180} = \frac{2}{90}.$$

Now because B_1 and B_2 are disjoint and because their union is the entire sample space, A is the disjoint union of $A \cap B_1$ and $A \cap B_2$. Thus the probability that

$$P(A) = P(A \cap B_1) + P(A \cap B_2) = \frac{1}{90} + \frac{2}{90} = \frac{3}{90} \cong 3.3\%.$$

d. *Solution 1 (using Bayes' theorem):* Given that the chosen part is defective, the probability that it came from the first factory is $P(B_1 \mid A)$. By Bayes' theorem and the computations in part (a),

$$P(B_1 \mid A) = \frac{P(A \mid B_1)P(B_1)}{P(A \mid B_1)P(B_1) + P(A \mid B_2)P(B_2)} = \frac{\frac{1}{90}}{\frac{1}{90} + \frac{2}{90}} = \frac{1}{3} \cong 33.3\%.$$

Solution 2 (without explicit use of Bayes' theorem): Given that the chosen ball is green, the probability that it came from the first urn is $P(B_1 \mid A)$. By the results of part (a),

$$P(B_1 \mid A) = \frac{P(A \cap B_1)}{P(A)} = \frac{\frac{1}{90}}{\frac{3}{90}} = \frac{1}{3} \cong 33.3\%.$$

16. *Hint:* The answers to parts (a) and (b) are 11.25% and $21\frac{1}{3}\%$, respectively.

17. Proof: Suppose A and B are events in a sample space S, and $P(A \mid B) = P(A) \neq 0$. Then

$$P(B \mid A) = \frac{P(B \cap A)}{P(A)} = \frac{P(A \mid B)P(B)}{P(A)} = \frac{P(A)P(B)}{P(A)} = P(B).$$

18. Proof: Suppose A and B are events in a sample space S, and $P(A \cap B) = P(A)P(B)$, $P(A) \neq 0$, and $P(B) \neq 0$. Applying the hypothesis to the definition of conditional probability gives

$$P(A \mid B) = \frac{P(A \cap B)}{P(B)} = \frac{P(A)P(B)}{P(B)} = P(A)$$

and

$$P(B \mid A) = \frac{P(A \cap B)}{P(A)} = \frac{P(A)P(B)}{P(A)} = P(B).$$

19. As in Example 6.9.1, the sample space is the set of all 36 outcomes obtained from rolling the two dice and noting the numbers showing face up on each. Let A be the event that the number on the blue die is 2 and B the event that the number on the gray die is 4 or 5. Then

$$A = \{21, 22, 23, 24, 25, 26\},$$

$$B = \{14, 24, 34, 44, 54, 64, 15, 25, 35, 45, 55, 65\}, \quad \text{and} \quad A \cap B = \{24, 25\}.$$

Since the dice are fair (so all outcomes are equally likely), $P(A) = \frac{6}{36}$, $P(B) = \frac{12}{36}$, and $P(A \cap B) = \frac{2}{36}$. By definition of conditional probability,

$$P(A \mid B) = \frac{P(A \cap B)}{P(B)} = \frac{\frac{2}{36}}{\frac{12}{36}} = \frac{1}{6} \quad \text{and} \quad P(B \mid A) = \frac{P(A \cap B)}{P(A)} = \frac{\frac{2}{36}}{\frac{6}{36}} = \frac{1}{3}.$$

Now $P(A) = \frac{6}{36} = \frac{1}{6}$ and $P(B) = \frac{12}{36} = \frac{1}{3}$, and hence $P(A \mid B) = P(A)$ and $P(B \mid A) = P(B)$.

21. If A and B are events in a sample space and $A \cap B = \emptyset$ and A and B are independent, then (by definition of independence) $P(A \cap B) = P(A)P(B)$, and (because $A \cap B = \emptyset$) $P(A \cap B) = 0$. Hence $P(A)P(B) = 0$, and so (by the zero product property) either $P(A) = 0$ or $P(B) = 0$.

23. Let A be the event that the student answers the first question correctly, and let B be the event that the student answers the second question correctly. Because two choices can be eliminated on the first question, $P(A) = \frac{1}{3}$, and because no choices can be eliminated on the second question, $P(B) = \frac{1}{5}$. Thus $P(A^c) = \frac{2}{3}$ and $P(B^c) = \frac{4}{5}$.

a. Because the student's answer choices on each question are assumed to be independent, the probability that the student answers both questions correctly is

$$P(A \cap B) = P(A)P(B) = \frac{1}{3} \cdot \frac{1}{5} = \frac{1}{15} = 6\frac{2}{3}\%.$$

b. The probability that the student answers exactly one question correctly is

$$P((A \cap B^c) \cup (A^c \cap B)) \ = \ P(A \cap B^c) + P(A^c \cap B) \ = \ P(A)P(B^c) + P(A^c)P(B)$$

$$= \ \frac{1}{3} \cdot \frac{4}{5} + \frac{2}{3} \cdot \frac{1}{5} \ = \ \frac{6}{15} \ = \ \frac{2}{5} \ = \ 40\%.$$

c. One solution is to say that the probability that the student answers both questions incorrectly is $P(A^c \cap B^c)$, and $P(A^c \cap B^c) = P(A^c)P(B^c)$ by the result of exercise 22. Thus the answer is

$$P(A^c)\,P(B^c) \ = \ \frac{2}{3} \cdot \frac{4}{5} \ = \ \frac{8}{15} \ = \ 53\frac{1}{3}\%.$$

Another solution uses the fact that the event that the student answers both questions incorrectly is the complement of the event that the student answers at least one question correctly. Thus, by the results of parts (a) and (b), the answer is $1 - (\frac{1}{15} + \frac{2}{5}) = \frac{8}{15} = 53\frac{1}{3}\%$.

24. Let A be the event that a randomly chosen error is missed by proofreader X, and let B be the event that the error is missed by proofreader Y. Then $P(A) = 0.12$ and $P(B) = 0.15$.

a. Because the proofreaders work independently, $P(A \cap B) = P(A)\,P(B)$. Hence the probability that the error is missed by both proofreaders is

$$P(A \cap B) \ = \ P(A)\,P(B) \ = \ (0.12)(0.15) \ = \ 0.018 \ = \ 1.8\%.$$

b. Assuming that the manuscript contains 1000 errors, the expected number of missed errors is $1000 \cdot 0.018\% = 18$.

25. Let H_i be the event that the result of toss i is heads, and let T_i be the event that the result of toss i is tails. Then $P(H_i) = 0.7$ and $P(T_i) = 0.3$ for $i = 1, 2$.

a. Because the results of the tosses are independent, the probability of obtaining exactly two heads is

$$P(H_1 \cap H_2) = P(H_1)P(H_2) = 0.7 \cdot 0.7 = 0.49 = 49\%.$$

b. The probability of obtaining exactly one head is

$$P((H_1 \cap T_2) \cup (T_1 \cap H_2)) \ = \ P(H_1 \cap T_2) + P(T_1 \cap H_2) \ = \ P(H_1)\,P(T_2) + P(T_1)\,P(H_2)$$

$$= \ (0.7)\,(0.3) + (0.3)\,(0.7) \ = \ 42\%.$$

27. *Solution*: The family could have two boys, two girls, or one boy and one girl.

Let the subscript 1 denote the firstborn child (understanding that in the case of twins this might be by only a few moments), and let the subscript 2 denote the secondborn child.

Then we can let $(B_1 G_2, B_1)$ denote the outcome that the firstborn child is a boy, the secondborn is a girl, and the child you meet is the boy.

Similarly, we can let $(B_1 B_2, B_2)$ denote the outcome that both the firstborn and the secondborn are boys and the child you meet is the secondborn boy.

When this notational scheme is used for the entire set of possible outcomes for the genders of the children and the gender of the child you meet, all outcomes are equally likely and the sample space is denoted by

$$\{(B_1 B_2, B_1), (B_1 B_2, B_2), (B_1 G_2, B_1), (B_1 G_2, G_2), (G_1 B_2, G_1), (G_1 B_2, B_2), (G_1 G_2, G_1), (G_1 G_2, G_2)\}.$$

The event that you meet one of the children and it is a boy is

$$\{(B_1 B_2, B_1), (B_1 B_2, B_2), (B_1 G_2, B_1), (G_1 B_2, B_2)\}.$$

The probability of this event is $4/8 = 1/2$.

Discussion: An intuitive way to see this conclusion is to realize that the fact that you happen to meet one of the children and see that it is a boy gives you no information about the gender of the other child. Because each of the children is equally likely to be a boy, the probability that the other child is a boy is 1/2.

Consider the following situation in which the probabilities are identical to the situation described in the exercise. A person tosses two fair coins and immediately covers them so that you cannot see which faces are up. The person then reveals one of the coins, and you see that it is heads. This action on the person's part has given you no information about the other coin; the probability that the other coin has also landed heads up is 1/2.

28. **a.** $P \text{(seven heads)} = \begin{bmatrix} \text{the number of different} \\ \text{ways seven heads can} \\ \text{be obtained in ten tosses} \end{bmatrix} (0.7)^7 (0.3)^3$

$= 120(0.7)^7(0.3)^3 \cong 0.267 = 26.7\%$

29. **a.** $P \text{(none is defective)} = \begin{bmatrix} \text{the number of ways of} \\ \text{having 0 defective items} \\ \text{in a sample of 10 items} \end{bmatrix} P(\text{defective})^0 P(\text{not defective})^{10}$

$= \binom{10}{0} 0.03^0 \cdot 0.97^{10} = 1 \cdot (0.3)^0 (0.97)^{10} \cong 0.737 = 73.7\%$

30. For each $k = 1, 2, \ldots, 10$, let A_k be the event of obtaining a false positive result in year k. By assumption, $A_1, A_2, A_3, \ldots, A_{10}$ are mutually independent events, and so the method of Example 9.9.9 can be used to compute probabilities. The probability of a false positive in any one year is 0.04.

a. $P(0 \text{ false positives}) = \begin{bmatrix} \text{the number of ways} \\ \text{0 false positives can} \\ \text{be obtained over a} \\ \text{ten-year period} \end{bmatrix} \left(P \begin{pmatrix} \text{false} \\ \text{positive} \end{pmatrix} \right)^0 \left(P \begin{pmatrix} \text{not a false} \\ \text{positive} \end{pmatrix} \right)^{10}$

$= \binom{10}{0} 0.96^{10} = 1 \cdot 0.96^{10} \cong 0.665 = 66.5\%$

b. The probability that a woman will have at least one false positive result over a period of ten years is $1 - (0.96)^{10} \cong 1 - 66.5\% = 33.5\%$.

c. $P(2 \text{ false positives}) = \begin{bmatrix} \text{the number of ways} \\ \text{2 false positives can} \\ \text{be obtained over a} \\ \text{ten-year period} \end{bmatrix} \left(P \begin{pmatrix} \text{false} \\ \text{positive} \end{pmatrix} \right)^2 \left(P \begin{pmatrix} \text{not a false} \\ \text{positive} \end{pmatrix} \right)^8$

$= \binom{10}{2} 0.04^2 \cdot 0.96^8 = 45 \cdot 0.04^2 \cdot 0.96^8 = 0.05194 \cong 5.2\%$

d. Let T be the event that a woman's test result is positive one year, and let C be the event that the woman has breast cancer.

(i) By Bayes' formula, the probability of C given T is

$$P(C \mid T) = \frac{P(T \mid C)P(C)}{P(T \mid C)P(C) + P(T \mid C^c)P(C^c)}$$

$$= \frac{(0.98)(0.0002)}{(0.98)(0.0002) + (0.04)(0.9998)}$$

$$\cong 0.00488 = 4.88\%.$$

(ii) The event that a woman's test result is negative one year is T^c. By Bayes formula, the probability of C given T^c is

$$P(C \mid T^c) = \frac{P(T^c \mid C)P(C)}{P(T^c \mid C)P(C) + P(T^c \mid C^c)P(C^c)}$$

$$= \frac{(0.02)(0.0002)}{(0.02)(0.0002) + (0.96)(0.9998)}$$

$$\cong 0.000004 = 0.0004\%.$$

31. **a.** $P(\text{none is male}) \cong 1.3\%$

   b. $P(\text{at least one is male}) = 1 - P(\text{none is male}) \cong 1 - 0.013 = 98.7\%$

33. Suppose a gambler starts with $\$k$. Rolling a fair die leads to one of two disjoint outcomes: winning $\$1$ or losing $\$1$. Let A_k be the event that the gambler is ruined when he has $\$k$. Then A_k is the disjoint union of the following two events: C_k and D_k, where

   C_k is the event that the gambler has $\$k$, wins the next roll, and eventually gets ruined
   and D_k is the event that the gambler has $\$k$, loses the next roll, and eventually gets ruined.

   Now P_k is the probability that the gambler eventually gets ruined when he has $\$k$. By probability axiom 3,
$$P_k = P(C_k) + P(D_k).$$
   Let W be the event that the gambler wins on any given roll. Then
$$P(W) = \frac{1}{6} \quad \text{and} \quad P(W^c) = \frac{5}{6}.$$
   For each integer k with $1 \le k \le 300$, the definition of conditional probability can be used to find $P(C_k)$ and $P(D_k)$:

$$\begin{aligned}
P(C_k) &= P(A_k \cap W) \\
&\qquad \text{by definition of } C_k, A_k, \text{ and } W \\
&= P(A_k \mid W)P(W) \\
&\qquad \text{by definition of conditional probability} \\
&= P(A_{k+1}) \cdot \frac{1}{6} \\
&\qquad \text{because if the gambler wins on a roll when he has } \$k \\
&\qquad \text{then on the next roll he has } \$(k+1) \\
&= P_{k+1} \cdot \frac{1}{6}.
\end{aligned}$$

Similarly,

$$\begin{aligned}
P(D_k) &= P(A_k \cap W^c) \\
&\qquad \text{by definition of } C_k, A_k, \text{ and } W \\
&= P(A_k \mid W^c)P(W^c) \\
&\qquad \text{by definition of conditional probability} \\
&= P(A_{k-1}) \cdot \frac{5}{6} \\
&\qquad \text{because if the gambler loses on a roll when he has } \$k \\
&\qquad \text{then on the next roll he has } \$(k-1) \\
&= P_{k-1} \cdot \frac{5}{6}.
\end{aligned}$$

Thus,

$$P_k = P(C_k) + P(D_k) = P_{k+1} \cdot \frac{1}{6} + P_{k-1} \cdot \frac{5}{6}.$$

34. *Hint:* $P(Y) = P(Y \cap X) + P(Y \cap X^c)$

Review Guide: Chapter 9

Probability

- What is the sample space of an experiment?
- What is an event in the sample space?
- What is the probability of an event when all the outcomes are equally likely?

Counting

- If m and n are integers with $m \leq n$, how many integers are there from m to n inclusive?
- How do you construct a possibility tree?
- What are the multiplication rule, the addition rule, and the difference rule?
- What is the inclusion/exclusion rule?
- What is a permutation? an r-permutation?
- What is $P(n, r)$?
- How does the multiplication rule give rise to $P(n, r)$?
- When should you use the multiplication rule and when should you use the addition rule?
- What are some situations where both the multiplication and the addition or difference rule must be used?
- What is the formula for the probability of the complement of an event?
- How are IP addresses created?
- How is the inclusion/exclusion rule used?
- What is an r-combination?
- What is an unordered selection of elements from a set? *(p. 566)*
- What is complete enumeration? *(p. 567)*
- What formulas are used to compute $\binom{n}{r}$ by hand?
- Describe a situation where both r-combinations and the addition or difference rule must be used?
- Describe a situation where r-combinations, the multiplication rule, and the addition rule are all needed?
- How can r-combinations be used to count the number of permutations of a set with repeated elements?
- What is an r-combination with repetition allowed (or a multiset of size r)?
- How many r-combinations with repetition allowed can be selected from a set of n elements?

The Pigeonhole Principle

- What is the pigeonhole principle?
- How is the pigeonhole principle used to show that rational numbers have terminating or repeating decimal expansions?
- What is the generalized pigeonhole principle?
- What is the relation between one-to-one and onto for a function defined from one finite set to another of the same size?

Pascal's Formula and the Binomial Theorem

- What is Pascal's formula? Can you apply it in various situations?
- What is the algebraic proof of Pascal's formula?

- What is the combinatorial proof of Pascal's formula?
- What is the binomial theorem? Can you apply it in various situations?
- What is the algebraic proof of the binomial theorem?
- What is the combinatorial proof of the binomial theorem?

Probability Axioms and Expected Value

- What is the range of values for the probability of an event?
- What is the probability of an entire sample space?
- What is the probability of the empty set?
- If A and B are disjoint events in a sample space S, what is $P(A \cup B)$?
- If A is an event in a sample space S, what is $P(A^c)$?
- If A and B are any events in a sample space S, what is $P(A \cup B)$?
- How do you compute the expected value of a random experiment or process, if the possible outcomes are all real numbers and you know the probability of each outcome?
- What is the conditional probability of one event given another event?
- What is Bayes' theorem?
- What does it mean for two events to be independent?
- What is the probability of an intersection of two independent events?
- What does it mean for events to be mutually independent?
- What is the probability of an intersection of mutually independent events?

Fill-in-the-Blank Review Questions

Section 9.1

1. A sample space of a random process or experiment is ____.
2. An event in a sample space is ____.
3. To compute the probability of an event using the equally likely probability formula, you take the ratio of the ____ to the ____.
4. If $m \leq n$, the number of integers from m to n inclusive is ____.

Section 9.2

1. The multiplication rule says that if an operation can be performed in k steps and, for each i with $1 \leq i \leq k$, the ith step can be performed in n_i ways (regardless of how previous steps were performed), then the operation as a whole can be performed in ____.
2. A permutation of a set of elements is ____.
3. The number of permutations of a set of n elements equals ____.
4. An r-permutation of a set of n elements is ____.
5. The number of r-permutations of a set of n elements is denoted ____.
6. One formula for the number of r-permutations of a set of n elements is ____ and another formula is ____.

Section 9.3

1. The addition rule says that if a finite set A equals the union of k distinct mutually disjoint subsets $A_1, A_2, \ldots, A_k$, then ____.
2. The difference rule says that if A is a finite set and B is a subset of A, then ____.

3. If S is a finite sample space and A is an event in S, then the probability of A^c equals _____.

4. The inclusion/exclusion rule for two sets says that if A and B are any finite sets, then _____.

5. The inclusion/exclusion rule for three sets says that if A, B, and C are any finite sets, then _____.

Section 9.4

1. The pigeonhole principle states that _____.

2. The generalized pigeonhole principle states that _____.

3. If X and Y are finite sets and f is a function from X to Y then f is one-to-one if, and only if, _____.

Section 9.5

1. The number of subsets of size r that can be formed from a set with n elements is denoted _____, which is read as _____.

2. The number of r-combinations of a set of n elements is _____.

3. Two unordered selections are said to be the same if the elements chosen are the same, regardless of _____.

4. A formula relating $\binom{n}{r}$ and $P(n, r)$ is _____.

5. The phrase "at least n" means _____, and the phrase "at most n" means _____.

Section 9.6

1. Given a set $X = \{x_1, x_2, \ldots x_n\}$, an r-combination with repetition allowed, or a multiset of size r, chosen from X is _____, which is denoted _____.

2. If $X = \{x_1, x_2, \ldots x_n\}$, the number of r-combinations with repetition allowed (or multisets of size r) chosen from X is _____.

3. When choosing k elements from a set of n elements, order may or may not matter and repetition may or may not be allowed.

   - The number of ways to choose the k elements when repetition is allowed and order matters is _____.
   - The number of ways to choose the k elements when repetition is not allowed and order matters is _____.
   - The number of ways to choose the k elements when repetition is not allowed and order does not matter is _____.
   - The number of ways to choose the k elements when repetition is allowed and order does not matter is _____.

Section 9.7

1. If n and r are nonnegative integers with $r \leq n$, then the relation between $\binom{n}{r}$ and $\binom{n}{n-r}$ is _____.

2. Pascal's formula says that if n and r are positive integers with $r \leq n$, then _____.

3. The crux of the algebraic proof of Pascal's formula is that to add two fractions you need to express both of them with a _____.

4. The crux of the combinatorial proof of Pascal's formula is that the set of subsets of size r of a set $\{x_1, x_2, \ldots, x_n\}$ can be partitioned into the set of subsets of size r that contain _____ and those that _____.

5. The binomial theorem says that given any real numbers a and b and any nonnegative integer n, ____.

6. The crux of the algebraic proof of the binomial theorem is that, after making a change of variable so that two summations have the same lower and upper limits, you use the fact that $\binom{m}{k} + \binom{m}{k-1} =$ ____.

7. The crux of the combinatorial proof of the binomial theorem is that the number of ways to arrange k b's and $(n-k)$ a's in order is ____.

Section 9.8

1. If A is an event in a sample space S, $P(A)$ can take values between ____ and ____. Moreover, $P(S) =$ ____, and $P(\emptyset) =$ ____.
2. If A and B are disjoint events in a sample space S, $P(A \cup B) =$ ____.
3. If A is an event in a sample space S, $P(A^c) =$ ____.
4. If A and B are any events in a sample space S, $P(A \cup B) =$ ____.
5. If the possible outcomes of a random process or experiment are real numbers $a_1, a_2, \ldots, a_n$, which occur with probabilities $p_1, p_2, \ldots, p_n$, then the expected value of the process is ____.

Section 9.9

1. If A and B are any events in a sample space S and $P(A) \neq 0$, then the conditional probability of B given A is $P(B|A) =$ ____.
2. Bayes' theorem says that if a sample space S is a union of mutually disjoint events $B_1, B_2, \ldots, B_n$ with nonzero probabilities, if A is an event in S with $P(A) \neq 0$, and if k is an integer with $1 \leq k \leq n$, then ____.
3. Events A and B in a sample space S are independent if, and only if, ____.
4. Events A, B, and C in a sample space S are mutually independent if, and only if, ____, ____, ____, and ____.

Answers for Fill-in-the-Blank Review Questions

Section 9.1

1. the set of all outcomes of the random process or experiment
2. a subset of the sample space
3. number of outcomes in the event; total number of outcomes
4. $n - m + 1$

Section 9.2

1. $n_1 n_2 \cdots n_k$ ways
2. an ordering of the elements of the set in a row
3. $n!$
4. an ordered selection of r of the elements of the set
5. $P(n, r)$
6. $n(n-1)(n-2) \cdots (n-r+1)$; $\dfrac{n!}{(n-r)!}$

Section 9.3

1. the number of elements in A equals $N(A_1) + N(A_2) + \cdots + N(A_k)$
2. the number of elements in $A - B$ is the difference between the number of elements in A minus the number of elements in B
3. $1 - P(A)$
4. $N(A \cup B) = N(A) + N(B) - N(A \cap B)$
5. $N(A \cup B \cup C) = N(A) + N(B) + N(C) - N(A \cap B) - N(A \cap C) - N(B \cap C) + N(A \cap B \cap C)$

Section 9.4

1. if n pigeons fly into m pigeonholes and $n > m$, then at least two pigeons fly into the same pigeonhole (*Or:* a function from one finite set to a smaller finite set cannot be one-to-one)
2. if n pigeons fly into m pigeonholes and, for some positive number k, $km < n$, then at least onee pigeonhole contains $k + 1$ or more pigeons (Or: for any function f from a finite set X with n elements to a finite set Y with m elements and for any positive integer k, if $km < n$, then there is some $y \in Y$ such that y is the image of at least $k + 1$ distinct elements of Y
3. f is onto

Section 9.5

1. $\binom{n}{r}$; n choose r
2. $\binom{n}{r}$ (*Or:* n choose r)
3. the order in which they are chosen
4. $\binom{n}{r} = \dfrac{P(n,r)}{r!}$
5. n or more; n or fewer

Section 9.6

1. an unordered selection of elements taken from X with repetition allowed; $[x_1, x_2, \ldots x_{i_r}]$ where each x_{i_j} is in X and some of the x_{i_j} may equal each other
2. $\binom{r + n - 1}{r}$
3. n^k; $n(n-1)(n-2) \cdots (n - k + 1)$; $\binom{n}{k}$; $\binom{k + n - 1}{k}$

Section 9.7

1. $\binom{n}{r} = \binom{n}{n - r}$
2. $\binom{n + 1}{r} = \binom{n}{r - 1} + \binom{n}{r}$
3. common denominator
4. x_{n+1}; do not contain x_{n+1}
5. $(a + b)^n = \displaystyle\sum_{k=0}^{n} \binom{n}{k} a^{n-k} b^k$

6. $\dbinom{m+1}{k}$

7. $\dbinom{n}{k}$

Section 9.8

1. $0; 1; 1; 0$

2. $P(A) + P(B)$

3. $1 - P(A)$

4. $P(A) + P(B) - P(A \cap B)$

5. $a_1 p_1 + a_2 p_2 + \cdots + a_n p_n$

Section 9.9

1. $\dfrac{P(A \cap B)}{P(A)}$

2. $P(B_k \mid A) = \dfrac{P(A \mid B_k) P(B_k)}{P(A \mid B_1) P(B_1) + P(A \mid B_2) P(B_2) + \cdots + P(A \mid B_n) P(B_n)}$

3. $P(A \cap B) = P(A) \cdot P(B)$

4. $P(A \cap B) = P(A) \cdot P(B); \; P(A \cap C) = P(A) \cdot P(C); \; P(B \cap C) = P(B) \cdot P(C); \; P(A \cap B \cap C) = P(A) \cdot P(B) \cdot P(C)$

Chapter 10: Graphs and Trees

The first section of this chapter introduces the terminology of trails, paths, and circuits and discusses the notion of connectedness and Euler and Hamiltonian circuits. In the chapter as a whole, an attempt is made to balance the presentation of theory and with applications of it. Some exercises are designed to develop facility with terminology and the use of theorems, while others provide opportunities to engage in the kind of reasoning that lies behind the theorems.

Section 10.2 introduces the concept of the adjacency matrix of a graph. The main theorem of the section states that the ijth entry of the kth power of the adjacency matrix equals the number of walks of length k from the ith to the jth vertices in the graph. Matrix multiplication is defined and explored in this section in a way that is intended to be adequate even if you have never seen the definition before.

The concept of graph isomorphism is discussed in Section 10.3. In this section the main theorem gives a list of isomorphic invariants that can be used to show either that two graphs might be isomorphic or that it is impossible for them to be so.

The last three sections of the chapter deal with the subject of trees. Section 10.4 focuses on basic definitions, examples, and theorems giving necessary and sufficient conditions for graphs to be trees, and Section 10.5 contains the definition of rooted tree, binary tree, binary search tree, and the theorems that relate the number of internal to the number of terminal vertices of a full binary tree and the maximum height of a binary tree to the number of its terminal vertices. Section 10.6 on spanning trees and shortest paths contains Kruskal's, Prim's, and Dijkstra's algorithms and proofs of their correctness, as well as applications of minimum spanning trees and shortest paths.

Section 10.1

1. **a.** trail (no repeated edge), not a path (has a repeated vertex, v_1), not a circuit

   b. walk, not a trail (has a repeated edge, e_9), not a circuit

   c. closed walk (starts and ends at the same vertex), trail (no repeated edge since no edge), not a path or a circuit (since no edge)

   d. circuit, not a simple circuit (repeated vertex, v_4)

   e. closed walk (starts and ends at the same vertex but has repeated edges, $\{v_2, v_3\}$ and $\{v_3, v_4\}$)

   f. path

3. **a.** No. The notation $v_1 v_2 v_1$ could equally well refer to $v_1 e_1 v_2 e_2 v_1$ or to $v_1 e_2 v_2 e_1 v_1$, which are different walks.

   b. No, because $e_1 e_2$ could refer either to $v_1 e_1 v_2 e_2 v_1$ or to $v_2 e_1 v_1 e_2 v_2$.

4. **a.** Three. (There are three ways to choose the middle edge.)

   b. $3! + 3 = 9$ (The three paths from part (a) are also trails, and there are an additional $3!$ trails with vertices $v_1, v_2, v_3, v_2, v_3, v_4$. The reason is that from v_2 there are 3 choices of an edge to go to v_3, then 2 choices of a different edge to go back to v_2, and then 1 choice of a different edge to return to v_3.

   c. Infinitely many. (Since a walk may have repeated edges, a walk from v_1 to v_4 may contain an arbitrarily large number of repetitions of edges joining a pair of vertices along the way.)

6. **a.** $\{v_1, v_3\}$, $\{v_2, v_3\}$, $\{v_4, v_3\}$, and $\{v_5, v_3\}$ are all the bridges.

   b. $\{v_7, v_8\}$, $\{v_1, v_2\}$, $\{v_3, v_4\}$

   c. $\{v_2, v_3\}$, $\{v_6, v_7\}$, $\{v_7, v_8\}$, $\{v_9, v_{10}\}$

8. **a.** Three connected components:

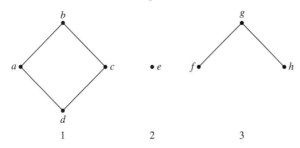

9. **a.** No. This graph has two vertices of odd degree, whereas all vertices of a graph with an Euler circuit have even degree.

b. Yes, by Theorem 10.1.3 since G is connected and every vertex has even degree.

c. Not necessarily. It is not specified that G is connected. For instance, the following graph satisfies the given conditions but does not have an Euler circuit:

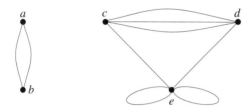

12. One Euler circuit is $e_4e_5e_6e_3e_2e_7e_8e_1$.

14. One Euler circuit is *iabihbchgcdgfdefi*.

15. One Euler circuit is the following: *stuvwxyzrsuwyuzs*.

18. Yes. One Euler circuit is *ABDEACDA*.

19. There is an Euler trail since $\deg(u)$ and $\deg(w)$ are odd, all other vertices have positive even degree, and thegraph is connected. One Euler trail is $uv_1v_0v_7uv_2v_3v_4v_2v_6v_4wv_5v_6w$.

21. One Euler trail from u to w is $uv_1v_2v_3uv_0v_7v_6v_3v_4v_6wv_5v_4w$.

23. **a.** The nonempty subgraphs are as follows:

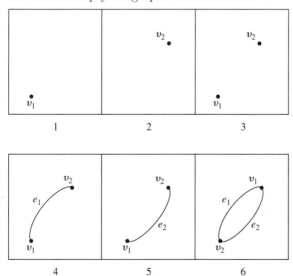

24. **a.**

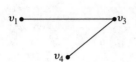

b.

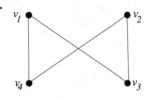

26. **b.**

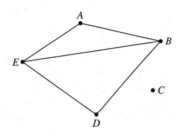

27. When all the vertices and edges of G are combined with all the vertices and edges of G', the result is K_n, the complete graph on n vertices. By Example 4.9.9, K_n has $\dfrac{n(n-1)}{2}$ edges. Since G and G' have no edges in common,

$$\text{the number of edges of } G + \text{ the number of edges of } G' = \frac{n(n-1)}{2}.$$

29. $v_0 v_7 v_1 v_2 v_3 v_5 v_6 v_0$

30. One Hamiltonian circuit is $b\,a\,l\,k\,j\,e\,d\,c\,f\,i\,h\,g\,b$. The only other one traverses this circuit in the opposite direction.

31. *Hint:* See the solution to Example 10.1.9.

32. Here is one sequence of reasoning you could use: Call the given graph G, and suppose G has a Hamiltonian circuit. Then G has a subgraph H that satisfies conditions(1)–(4) of Proposition 10.1.6. Since the degree of b in G is 4 and every vertex in H has degree 2, two edges incident on b must be removed from G to create H. Edge $\{a, b\}$ cannot be removed because doing so would result in vertex d having degree less than 2 in H. Similar reasoning shows that edge $\{b, c\}$ cannot be removed either. So edges $\{b, i\}$ and $\{b, e\}$ must be removed from G to create H. Because vertex e must have degree 2 in H and because edge $\{b, e\}$ is not in H, both edges $\{e, d\}$ and $\{e, f\}$ must be in H. Similarly, since both vertices c and g must have degree 2 in H, edges $\{c, d\}$ and $\{g, d\}$ must also be in H. But then three edges incident on d, namely, $\{e, d\}$, $\{c, d\}$, and $\{g, d\}$, must all be in H, which contradicts the fact that vertex d must have degree 2 in H.

33. Call the given graph G and suppose G has a Hamiltonian circuit. Then G has a subgraph H that satisfies conditions $(1) - (4)$ of Proposition 10.1.6. Since the degree of B in G is five and every vertex in H has degree two, three edges incident on B must be removed from G to create H. Edge $\{B, C\}$ cannot be removed because doing so would result in vertex C having degree less than two in H. Similar reasoning shows that edges $\{B, E\}$, $\{B, F\}$, and $\{B, A\}$ cannot be removed either. It follows that the degree of B in H must be at least four, which contradicts the condition that every vertex in H has degree two in H. Hence no such subgraph H can exist, and so G does not have a Hamiltonian circuit.

34. *Hint:* This graph does not have a Hamiltonian circuit.

36. One Hamiltonian circuit is $v_0v_1v_5v_4v_7v_6v_2v_3v_0$.

38. *Partial answer:*

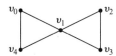

This graph has an Euler circuit $v_0v_1v_2v_3v_1v_4v_0$ but no Hamiltonian circuit.

39. *One answer:*

This graph has a Hamiltonian circuit $v_0v_1v_2v_0$ but no Euler circuit.

Some other examples of graphs with Hamiltonian circuits but not Euler circuits are shown in the statements of exercises 17, 21, 23b, 29, and 30.

40. *Partial answer:*

The walk $v_0v_1v_2v_0$ is both an Euler circuit and a Hamiltonian circuit for this graph.

41. *Partial answer:*

The graph has the Euler circuit $e_1e_2e_3e_4e_5e_6$ and the Hamiltonian circuit $v_0v_1v_2v_3v_0$. These are not the same.

42. It is clear from the map that only a few routes have a chance of minimizing the distance. For instance, one must go to either Düsseldorf or Luxembourg just after leaving Brussels or just before returning to Brussels, and one must either travel from Berlin directly to Munich or the reverse. The possible minimizing routes are those shown below plus the same routes traveled in the reverse direction.

Route	Total Distance (in km)
Bru-Lux-Düss-Ber-Mun-Par-Bru	$219 + 224 + 564 + 585 + 832 + 308 = 2732$
Bru-Düss-Ber-Mun-Par-Lux-Bru	$223 + 564 + 585 + 832 + 375 + 219 = 2798$
Bru-Düss-Lux-Ber-Mun-Par-Bru	$223 + 224 + 764 + 585 + 832 + 308 = 2936$
Bru-Düss-Ber-Mun-Lux-Par-Bru	$223 + 564 + 585 + 517 + 375 + 308 = 2572$

The routes that minimize distance, therefore, are the bottom route shown in the table and that same route traveled in the reverse direction.

43. **a.** <u>Proof</u>: Suppose G is a graph and W is a walk in G that contains a repeated edge e. Let v and w be the endpoints of e. In case $v=w$, then v is a repeated vertex of W. In case $v \neq w$, then one of the following must occur: (1) W contains two copies of vew or of wev (for instance, W might contain a section of the form $vewe'vew$, as illustrated below); (2) W contains separate sections of the form vew and wev (for instance, W might contain a section of the form $vewe'wev$, as illustrated below); or (3) W contains a section of the form $vewev$ or of the form $wevew$ (as illustrated below). In cases (1) and (2), both vertices v and w are repeated, and in case (3), one of v or w is repeated. In all cases, there is at least one vertex in W that is repeated.

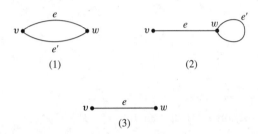

44. <u>Proof</u>: Suppose G is a connected graph and v and w are any particular but arbitrarily chosen vertices of G. *[We must show that u and v can be connected by a path.]* Since G is connected, there is a walk from v to w. If the walk contains a repeated vertex, then delete the portion of the walk from the first occurrence of the vertex to its next occurrence. (For example, in the walk $ve_1v_2e_5v_7e_6v_2e_3w$, the vertex v_2 occurs twice. Deleting the portion of the walk from one occurrence to the next gives $ve_1v_2e_3w$.) If the resulting walk still contains a repeated vertex, do the above deletion process another time. Then check again for a repeated vertex. Continue in this way until all repeated vertices have been deleted. (This must occur eventually, since the total number of vertices is finite.) The resulting walk connects v to w but has no repeated vertex. By exercise 43(b), it has no repeated edge either. Hence it is a path from v to w.

45. <u>Proof</u>: Suppose vertices v and w are part of a circuit in a graph G and one edge e is removed from the circuit. Without loss of generality, we may assume the v occurs before the w in the circuit, and we may denote the circuit by $v_0e_1v_1e_2\ldots e_{n-1}v_{n-1}e_nv_0$ with $v_i = v$, $v_j = w$, $i < j$, and $e_k = e$.

In case either $k \le i$ or $k > j$, then $v = v_ie_{i+1}v_{i+1}\ldots v_{j-1}e_jv_j = w$ is a trail in G from v to w that does not include e.

In case $i < k \le j$, then $v = v_ie_iv_{i-1}e_{i-1}\ldots v_1e_1v_0e_nv_{n-1}\ldots e_{j+1}v_j = w$ is a trail in G from v to w that does not include e.

These possibilities are illustrated by examples (1) and (2) in the diagram below. In both cases there is a trail in G from v to w that does not include e.

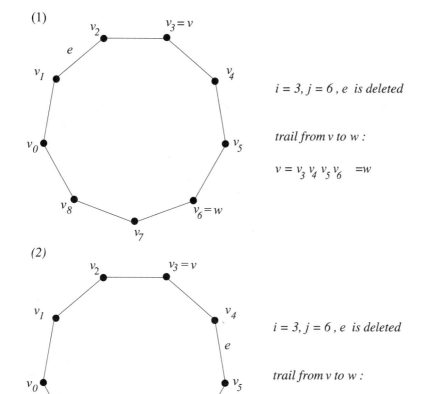

(1)

$i = 3, j = 6$, e is deleted

trail from v to w :

$v = v_3\ v_4\ v_5\ v_6\quad =w$

(2)

$i = 3, j = 6$, e is deleted

trail from v to w :

$v = v_3\ v_2\ v_1\ v_0\ v_8\ \cancel{v_7}\ \cancel{v_6}\quad =w$

46. The graph below contains a circuit, any edge of which can be removed without disconnecting the graph. For instance, if edge e is removed, then the following walk can be used to go from v_1 to v_2 : $v_1 v_5 v_3 v_2$.

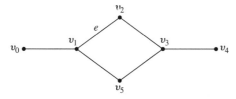

48. <u>Proof</u>: Suppose C is a circuit in a graph G that starts and ends at a vertex v, and suppose w is another vertex in the circuit.

   By definition the circuit has the form $v e_1 v_1 e_2 v_2 \ldots e_{n-1} v_{n-1} e_n v$ where $v_1, v_2, \ldots, v_{n-1}$ are vertices of G, $e_1, e_2, \ldots, e_n$ are distinct edges of G, and $v_i = w$ for some i with $1 \leq i \leq n-1$.

   Then $w = v_i e_{i+1} v_{i+1} \ldots e_n v e_1 v_1 e_2 v_2 \ldots v_i = w$ is a circuit that starts and ends at w.

50. <u>Proof</u>: Let G be a connected graph and let C be a circuit in G. Let G' be the subgraph obtained by removing all the edges of C from G and also any vertices that become isolated when the edges of C are removed. *[We must show that there exists a vertex v such that v is in both C and G'.]* Pick any vertex v of C and any vertex w of G'. Since G is connected, there is a path from v to w (by Lemma 10.1.1(a)). Let i be the largest subscript such that v_i is in

C.

$$
v = v_0 e_1 v_1 e_2 v_2 \ldots v_{i-1} e_i \quad v_i \quad e_{i+1} \quad v_{i+1} \quad \ldots v_{n-1} e_n v_n = \quad w
$$

$$
\uparrow \qquad\qquad\qquad\qquad\qquad \uparrow \qquad\quad \uparrow \qquad\qquad\qquad\qquad\qquad \uparrow
$$

$$
\text{in } C \qquad\qquad\qquad\qquad\qquad \text{in } C \quad \text{not in } C \qquad\qquad\qquad\qquad \text{in } G'
$$

If $i = n$, then $v_n = w$ is in C and also in G', and we are done. If $i < n$, then v_i is in C and v_{i+1} is not in C. This implies that e_{i+1} is not in C (for if it were, both endpoints would be in C by definition of circuit). Hence when G' is formed by removing the edges and resulting isolated vertices from G, then e_{i+1} is not removed. That means that v_i does not become an isolated vertex, so v_i is not removed either. Hence v_i is in G'. Consequently, v_i is in both C and G' *[as was to be shown].*

51. Proof: Suppose G is a graph with an Euler circuit. If G has only one vertex, then G is automatically connected. If v and w are any two vertices of G, then v and w each appear at least once in the Euler circuit (since an Euler circuit contains every vertex of the graph). The section of the circuit between the first occurrence of one of v or w and the first occurrence of the other is a walk from one of the two vertices to the other. Since the choice of v and w was arbitrary, given any two vertices in G there is a walk from one to the other. So, by definition, G is connected.

54. **a.** Let m and n be positive integers and let $K_{m,n}$ be a complete bipartite graph on (m, n) vertices. Since $K_{m,n}$ is connected, by Theorem 10.1.4 it has an Euler circuit if, and only if, every vertex has even degree. But $K_{m,n}$ has m vertices of degree n and n vertices of degree m. So $K_{m,n}$ has an Euler circuit if, and only if, both m and n are even.

    b. Let m and n be positive integers, let $K_{m,n}$ be a complete bipartite graph on (m, n) vertices, and suppose $V_1 = \{v_1, v_2, \ldots, v_m\}$ and $V_2 = \{w_1, w_2, \ldots, w_n\}$ are the disjoint sets of vertices such that each vertex in V_1 is joined by an edge to each vertex in V_2 and no vertex within V_1 or V_2 is joined by an edge to any other vertex within the same set. If $m = n \geq 2$, then $K_{m,n}$ has the following Hamiltonian circuit: $v_1 w_1 v_2 w_2 \ldots v_m w_m v_1$. If $K_{m,n}$ has a Hamiltonian circuit, then $m = n$ because the vertices in any Hamiltonian circuit must alternate between V_1 and V_2 (since no edges connect vertices within either set) and because no vertex, except the first and last, appears twice in a Hamiltonian circuit. If $m = n = 1$, then $K_{m,n}$ does not have a Hamiltonian circuit because $K_{1,1}$ contains just one edge joining two vertices. Therefore, $K_{m,n}$ has a Hamiltonian circuit if, and only if, $m = n \geq 2$.

56. **b.** *Hint:* Divide the proof into three parts. (1) Show that if G is any graph containing a closed walk with an odd number of edges, then G contains a circuit with an odd number of edges. (2) Show that if G is any connected graph that does not have a circuit with an odd number of edges, then G is bipartite.(3) Show that if G is any graph with at least two vertices and is such that G does not have a circuit with an odd number of edges, then G is bipartite.

57. Suppose G is a connected graph in which every every vertex has even degree. Let C be the path $v_1 e_1 v_2 e_2 v_3 \ldots e_n v_{n+1}$ and suppose that C has maximum length in G. That is, C has at least as many vertices and edges as any other circuit in G.We first show that $v_1 = v_{n+1}$. Once C leaves v_1 each subsequent entrance and exit in C contributes 2 to the degree of v_1. So if $v_{n+1} \neq v_1$, then v_1 has odd degree, which contradictions the assumption that every vertex in G has even degree. Next let H be the subgraph of G that contains all the vertices and edges in C, and suppose that $H \neq G$. We first show that H contains every vertex of G. Suppose there is a vertex u of G that is not in H. Since G is connected there is a path P from u to v_1. Let v_i be the first vertex in C that is not in P and let w be the vertex in P that connects to v_i by an edge e. Form a path P' as follows:

$$
P': we v_i e_i v_{i+1} \ldots e_n v_{n+1} = v_1 e_1 v_2 \ldots e_{i-1} v_i.
$$

Then P' has one more vertex than C, which contradicts the assumption that C is a path with at least as many vertices as any other path in G. Next we show that H contains every edge of

G. Suppose there is an edge e in G that is not in H. Because H contains every vertex in G, there are vertices v_i and v_j in G that are incident on e. Create a path C' by adding e between v_i and v_j as follows:

$$C': \ v_i e v_j e_{j+1} v_{j+1} \ldots e_n v_{n+1} = v_1 e_1 v_2 \ldots e_{j-1} v_j.$$

Then C' is a path with one more edge than C, which contradictions the assumption that C is a path with at least as many edges as any other path in G.

Section 10.2

1. **a.** Equate corresponding entries to find that
$$a + b = 1 \qquad a - c = 0 \qquad c = -1 \qquad b - a = 3.$$
Since $c = -1$, then $a - (-1) = a + 1 = 0$, and so $a = -1$.

Consequently, $a + b = (-1) + b = 1$, and hence $b = 2$.

To check these calculations, substitute $a = -1$ and $b = 2$ into $b - a$.

The result is $b - a = 2 - (-1) = 3$, which agrees with the given value.

Thus $a = -1$, $b = 2$, and $c = -1$.

2. **a.**

	v_1	v_2	v_3	v_4
v_1	0	0	1	1
v_2	0	0	2	0
v_3	1	2	0	0
v_4	1	0	0	1

3. **a.**

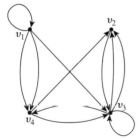

Any labels may be applied to the edges because the adjacency matrix does not determine edge labels.

b.

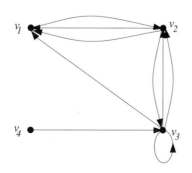

Any labels may be applied to the edges because the adjacency matrix does not determine edge labels.

4. **a.**

	v_1	v_2	v_3	v_4
v_1	0	0	1	1
v_2	0	0	2	0
v_3	1	2	0	0
v_4	1	0	0	1

c.

$$\begin{array}{c} \\ v_1 \\ v_2 \\ v_3 \\ v_4 \end{array} \begin{array}{cccc} v_1 & v_2 & v_3 & v_4 \end{array} \\ \begin{bmatrix} 0 & 1 & 1 & 1 \\ 1 & 0 & 1 & 1 \\ 1 & 1 & 0 & 1 \\ 1 & 1 & 1 & 0 \end{bmatrix}$$

5. a.

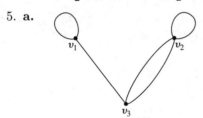

Any labels may be applied to the edges because the adjacency matrix does not determine edge labels.

6. a. The graph is connected.

b. The graph is not connected; the matrix shows that there are no edges joining the vertices from the set $\{v_1, v_2\}$ to those in the set $\{v_3, v_4\}$.

8. a. $2 \cdot 1 + (-1) \cdot 3 = -1$

9. a. $\begin{bmatrix} 3 & -3 & 12 \\ 1 & -5 & 2 \end{bmatrix}$ **b.** $\begin{bmatrix} 0 & 8 \\ -5 & 4 \end{bmatrix}$ **c.** $\begin{bmatrix} -2 & -3 \\ 4 & 6 \end{bmatrix}$ **d.** $\begin{bmatrix} 7 & 0 \\ 0 & 7 \end{bmatrix}$

10. a. No product. (**A** has three columns, and **B** has two rows.)

b. BA $= \begin{bmatrix} -2 & -2 & 2 \\ 1 & -5 & 2 \end{bmatrix}$ **f. B**$^2 = \begin{bmatrix} 4 & 0 \\ 1 & 9 \end{bmatrix}$ **i. AC** $= \begin{bmatrix} 2 & -1 \\ -5 & -2 \end{bmatrix}$

12. One among many possible examples is $\mathbf{A} = \mathbf{B} = \begin{bmatrix} 0 & 1 \\ 0 & 0 \end{bmatrix}$.

14. *Hint:* If the entries of the $m \times m$ identity matrix are denoted by δ_{ik}, then $\delta_{ik} = \begin{cases} 0 & \text{if } i \neq k \\ 1 & \text{if } i = k. \end{cases}$

If $\mathbf{A} = (a_{ij})$, then the ijth entry of $\mathbf{IA}$ is $\sum\limits_{k=1}^{m} \delta_{ik} a_{kj}$.

15. <u>Proof</u>: Suppose $\mathbf{A}$ is an $m \times m$ symmetric matrix. Then for all integers i and j with $1 \leq i, j \leq m$,

$$(\mathbf{A}^2)_{ij} = \sum_{k=1}^{m} \mathbf{A}_{ik} \mathbf{A}_{kj} \quad \text{and} \quad (\mathbf{A}^2)_{ji} = \sum_{k=1}^{m} \mathbf{A}_{jk} \mathbf{A}_{ki}.$$

Now since $\mathbf{A}$ is symmetric, $\mathbf{A}_{ik} = \mathbf{A}_{ki}$ and $\mathbf{A}_{kj} = \mathbf{A}_{jk}$ for all i, j, and k, and thus $\mathbf{A}_{ik} \mathbf{A}_{kj} = \mathbf{A}_{jk} \mathbf{A}_{ki}$ *[by the commutative law for multiplication of real numbers]*. Hence $(\mathbf{A}^2)_{ij} = (\mathbf{A}^2)_{ji}$ for all integers i and j with $1 \leq i, j \leq m$.

17. <u>Proof (by mathematical induction)</u>: Let $\mathbf{A}$ be any $m \times m$ matrix and let the property $P(n)$ be the equation

$$\mathbf{A}^n \mathbf{A} = \mathbf{A} \mathbf{A}^n. \qquad \leftarrow P(n)$$

We will prove that $P(n)$ is true for every integer $n \geq 1$.

Show that $P(1)$ is true: We must show that $\mathbf{A}^1 \mathbf{A} = \mathbf{A} \mathbf{A}^1$. This is true because $\mathbf{A}^1 = \mathbf{A}$ and $\mathbf{A} \mathbf{A} = \mathbf{A} \mathbf{A}$.

Show that for every integer $k \geq 1$, if $P(k)$ is true then $P(k+1)$ is true: Let k be any integer such that $k \geq 1$, and suppose that

$$\mathbf{A}^k \mathbf{A} = \mathbf{A} \mathbf{A}^k. \qquad \leftarrow \begin{array}{l} P(k) \\ \text{inductive hypothesis} \end{array}$$

We must show that
$$\mathbf{A}^{k+1}\mathbf{A} = \mathbf{A}\mathbf{A}^{k+1}. \quad \leftarrow P(k+1)$$

The left-hand side of $P(k+1)$ is

$$\begin{aligned}
\mathbf{A}^{k+1}\mathbf{A} &= \left(\mathbf{A}\mathbf{A}^k\right)\mathbf{A} & \text{by definition of matrix power}\\
&= \mathbf{A}\left(\mathbf{A}^k\mathbf{A}\right) & \text{by exercise 16}\\
&= \mathbf{A}\left(\mathbf{A}\mathbf{A}^k\right) & \text{by inductive hypothesis}\\
&= \mathbf{A}\mathbf{A}^{k+1} & \text{by definition of matrix power,}
\end{aligned}$$

and this is the right-hand side of $P(k+1)$.

Since the choice of $\mathbf{A}$ was arbitrary, this proof shows that $\mathbf{A}^n\mathbf{A} = \mathbf{A}\mathbf{A}^n$ for every integer $n \geq 1$ and every $m \times m$ matrix $\mathbf{A}$.

18. Proof (by mathematical induction: Let $\mathbf{A}$ be any $m \times m$ symmetric matrix and let the property $P(n)$ be the sentence

$$\mathbf{A}^n \text{ is symmetric.} \quad \leftarrow P(n)$$

We will prove that $P(n)$ is true for every integer $n \geq 1$.

Show that $P(1)$ is true: $P(1)$ is true because by assumption $\mathbf{A}$ is a symmetric matrix.

Show that for every integer $k \geq 1$, if $P(k)$ is true then $P(k+1)$ is true: Let k be any integer with $k \geq 1$, and suppose

$$\mathbf{A}^k \text{ is symmetric.} \quad \leftarrow \begin{array}{c} P(k) \\ \text{inductive hypothesis} \end{array}$$

We must show that

$$\mathbf{A}^{k+1} \text{ is symmetric.} \quad \leftarrow P(k+1)$$

Let $\mathbf{A}^k = (b_{ij})$. Then for all $i, j = 1, 2, \ldots, m$,

$$\begin{aligned}
\text{the } ij\text{th entry of } \mathbf{A}^{k+1} &= \text{the } ij\text{th entry of } \mathbf{A}\mathbf{A}^k & \text{by definition of matrix power}\\
&= \sum_{r=1}^{m} a_{ir}b_{rj} & \text{by definition of matrix multiplication}\\
&= \sum_{r=1}^{m} a_{ri}b_{jr} & \begin{array}{l}\text{because } A \text{ is symmetric by hypothesis and}\\ A^k \text{ is symmetric by inductive hypothesis}\end{array}\\
&= \sum_{r=1}^{m} b_{jr}a_{ri} & \begin{array}{l}\text{because multiplication of real numbers}\\ \text{is commutative}\end{array}\\
&= \text{the } ji\text{th entry of } \mathbf{A}^k\mathbf{A} & \text{by definition of matrix multiplication}\\
&= \text{the } ji\text{th entry of } \mathbf{A}\mathbf{A}^k & \text{by exercise 17}\\
&= \text{the } ji\text{th entry of } \mathbf{A}^{k+1} & \text{by definition of matrix power.}
\end{aligned}$$

Thus $\mathbf{A}^{k+1}$ is symmetric *[as was to be shown]*.

Since the choice of $\mathbf{A}$ was arbitrary, this proof shows that $\mathbf{A}^n$ is symmetric for every symmetric $m \times m$ matrix $\mathbf{A}$ and every integer $n \geq 1$.

19. **a.**
$$\mathbf{A}^2 = \begin{bmatrix} 1 & 1 & 2 \\ 1 & 0 & 1 \\ 2 & 1 & 0 \end{bmatrix} \begin{bmatrix} 1 & 1 & 2 \\ 1 & 0 & 1 \\ 2 & 1 & 0 \end{bmatrix} = \begin{bmatrix} 6 & 3 & 3 \\ 3 & 2 & 2 \\ 3 & 2 & 5 \end{bmatrix}$$

$$\mathbf{A}^3 = \begin{bmatrix} 1 & 1 & 2 \\ 1 & 0 & 1 \\ 2 & 1 & 0 \end{bmatrix} \begin{bmatrix} 6 & 3 & 3 \\ 3 & 2 & 2 \\ 3 & 2 & 5 \end{bmatrix} = \begin{bmatrix} 15 & 9 & 15 \\ 9 & 5 & 8 \\ 15 & 8 & 8 \end{bmatrix}$$

20. **a.** 2 since $\left(\mathbf{A}^2\right)_{23} = 2$ **b.** 3 since $\left(\mathbf{A}^2\right)_{34} = 3$ **c.** 6 since $\left(\mathbf{A}^3\right)_{14} = 6$ **d.** 17 since $\left(\mathbf{A}^3\right)_{23} = 17$

21. Proof (by mathematical induction: Let $\mathbf{A}$ be the adjacency matrix for K_3, the complete graph on three vertices, and let property $P(n)$ be the sentence

>All the entries along the main diagonal of $\mathbf{A}^n$ are equal to each other $\leftarrow P(n)$
>and all the entries off the main diagonal are also equal to each other.

We will prove that $P(n)$ is true for every integer $n \geq 1$.

Show that $P(1)$ is true: $P(1)$ is true because

$$\mathbf{A}^1 = \mathbf{A} = \begin{bmatrix} 0 & 1 & 1 \\ 1 & 0 & 1 \\ 1 & 1 & 0 \end{bmatrix},$$

which is the adjacency matrix for K_3, and all the entries along the main diagonal of $\mathbf{A}$ are 0 *[because K_3 has no loops]* and all the entries off the main diagonal are 1 *[because each pair of vertices is connected by exactly one edge]*.

Show that for every integer $m \geq 1$, if $P(m)$ is true then $P(m+1)$ is true: Let m be any integer with $m \geq 1$, and suppose

>All the entries along the main diagonal of $\mathbf{A}^m$ are equal to each other $P(m)$
>and all the entries off the main diagonal are also equal to each other. $\leftarrow$ inductive hypothesis

We must show that

>All the entries along the main diagonal of $\mathbf{A}^{m+1}$ are equal to each other $\leftarrow P(m+1)$
>and all the entries off the main diagonal are also equal to each other.

By inductive hypothesis,

$$\mathbf{A}^m = \begin{bmatrix} b & c & c \\ c & b & c \\ c & c & b \end{bmatrix} \quad \text{for some integers } b \text{ and } c.$$

It follows that

$$\mathbf{A}^{m+1} = \mathbf{A}\mathbf{A}^m = \begin{bmatrix} 0 & 1 & 1 \\ 1 & 0 & 1 \\ 1 & 1 & 0 \end{bmatrix} \begin{bmatrix} b & c & c \\ c & b & c \\ c & c & b \end{bmatrix} = \begin{bmatrix} 2c & b+c & b+c \\ b+c & 2c & b+c \\ b+c & b+c & 2c \end{bmatrix}$$

As can be seen, all the entries of $\mathbf{A}^{m+1}$ along the main diagonal are equal to each other and all the entries off the main diagonal are equal to each other. So the property is true for $n = m+1$.

22. **b.**   *Hint:* If G is bipartite, then its vertices can be partitioned into two sets V_1 and V_2 so that no vertices in V_1 are connected to each other by an edge and no vertices in V_2 are connected to each other by an edge. Label the vertices in V_1 as $v_1, v_2, \ldots, v_k$ and label the vertices in V_2 as $v_{k+1}, v_{k+2}, \ldots, v_n$. Now look at the matrix of G formed according to the given vertex labeling.

23. **b.** *Hint:* Consider the ijth entry of

$$\mathbf{A} + \mathbf{A}^2 + \mathbf{A}^3 + \cdots + \mathbf{A}^n.$$

If G is connected, then given the vertices v_i and v_j, there is a walk connecting v_i and v_j. If this walk has length k, then by Theorem 10.2.2, the ijth entry of $\mathbf{A}^k$ is not equal to 0. Use the facts that all entries of each power of $\mathbf{A}$ are nonnegative and that a sum of nonnegative numbers is positive provided that at least one of the numbers is positive.

Section 10.3

1. The graphs are isomorphic. One way to define the isomorphism is as follows:

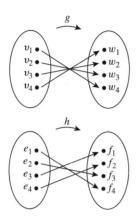

2. The graphs are not isomorphic. G has five vertices and G' has six.

3. The graphs are isomorphic. One way to define to isomorphism is as follows.

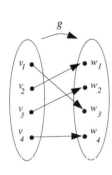 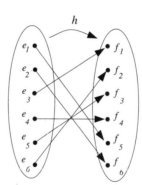

6. The graphs are isomorphic. One isomorphism is the following:

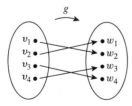

8. The graphs are not isomorphic. G has a simple circuit of length 3; G' does not.

9. The graphs are isomorphic. One way to define the isomorphism is as follows.

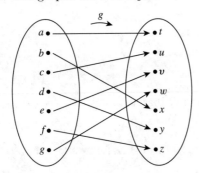

10. The graphs are isomorphic. One way to define the isomorphism is as follows:

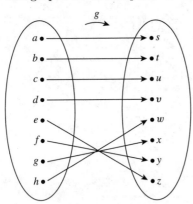

12. The graphs are isomorphic. One isomorphism is the following:

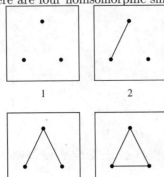

14. There are four nonisomorphic simple graphs with three vertices.

15. Of all nonisomorphic simple graphs with four vertices, there is one with 0 edges, one with 1 edge, two with 2 edges, three with 3 edges, two with 4 edges, one with 5 edges, and one with 6 edges. These eleven graphs are shown below.

16.

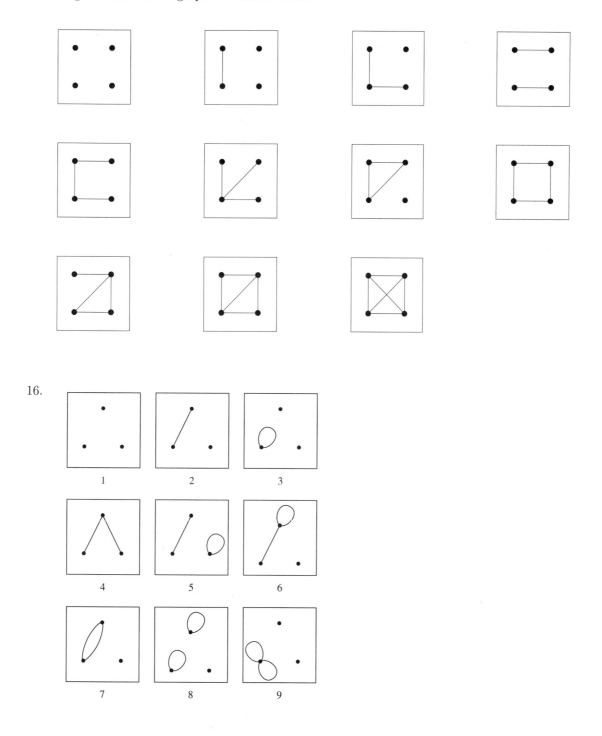

18. There are three nonisomorphic graphs with four vertices and three edges in which all 3 edges are loops, five in which 2 edges are loops and 1 is not a loop, six in which 1 edge is a loop and 2 edges are not loops, and six in which none of the 3 edges is a loop. These twenty graphs are shown as follows.

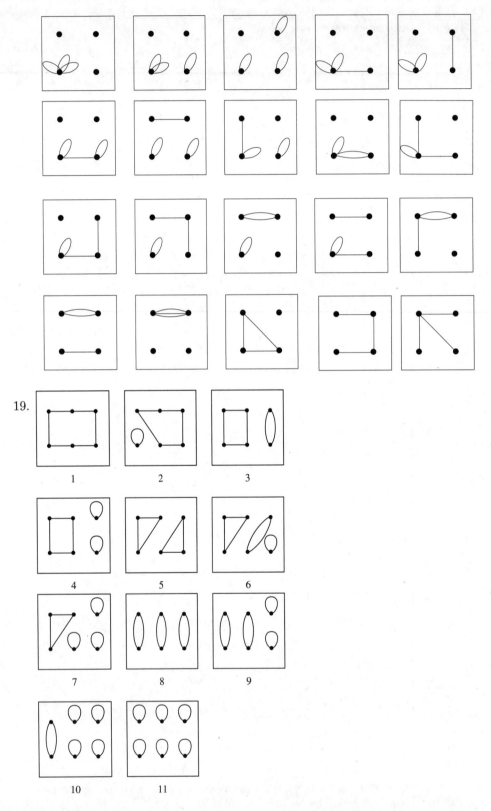

21. <u>Proof</u>: Suppose G and G' are isomorphic graphs and G has n vertices, where n is a nonnegative integer. *[We must show that G' has n vertices.]* By definition of graph isomorphism, there is

a one-to-one correspondence $g\colon V(G) \to V(G')$ sending vertices of G to vertices of G'. Since $V(G)$ is a finite set and g is a one-to-one correspondence, the number of vertices in $V(G')$ equals the number of vertices in $V(G)$. Hence G' has n vertices *[as was to be shown]*.

23. <u>Proof</u>: Suppose G and G' are isomorphic graphs and suppose G has a circuit C of length k, where k is a nonnegative integer. Let

$$C \text{ be } v_0 e_1 v_1 e_2 \ldots e_k v_k (=v_0).$$

By definition of graph isomorphism, there are one-to-one correspondences $g\colon V(G) \to V(G')$ and $h\colon E(G) \to E(G')$ that preserve the edge-endpoint functions in the sense that for each v in $V(G)$ and each e in $E(G)$, v is an endpoint of $e \Leftrightarrow g(v)$ is an endpoint of $h(e)$. Let

$$C' \text{ be } g(v_0)h(e_1)g(v_1)h(e_2)\ldots.h(e_k)g(v_k)(=g(v_0)).$$

Then C' is a circuit of length k in G'. The reasons are that

(1) because g and h preserve the edge-endpoint functions, both $g(v_i)$ and $g(v_{i+1})$ are incident on $h(e_{i+1})$ for each $i = 0, 1, \ldots, k-1$, and so C' is a walk from $g(v_0)$ to $g(v_0)$, and

(2) since C is a circuit, then $e_1, e_2, \ldots, e_k$ are distinct, and since h is a one-to-one correspondence, $h(e_1), h(e_2), \ldots, h(e_k)$ are also distinct, which implies that C' has k distinct edges. Therefore, G' has a circuit C of length k.

24. <u>Proof</u>: Suppose G and G' are isomorphic graphs and suppose G has a simple circuit C of length k, where k is a nonnegative integer. By definition of graph isomorphism, there are one-to-one correspondences $g\colon V(G) \to V(G')$ and $h\colon E(G) \to E(G')$ that preserve the edge-endpoint functions in the sense that for all v in $V(G)$ and e in $E(G)$, v is an endpoint of $e \Leftrightarrow g(v)$ is an endpoint of $h(e)$.

Let C be $v_0 e_1 v_1 e_2 \ldots e_k v_k (= v_0)$, and let C' be $g(v_0)h(e_1)g(v_1)h(e_2)\ldots h(e_k)g(v_k)(= g(v_0))$. By the same reasoning as in the solution to exercise 23 in Appendix B, C' is a circuit of length k in G'.

Suppose C' is not a simple circuit. Then C' has a repeated vertex, say $g(v_i) = g(v_j)$ for some $i, j = 0, 1, 2, \ldots, k-1$ with $i \neq j$. But since g is a one-to-one correspondence this implies that $v_i = v_j$, which is impossible because C is a simple circuit. Hence the supposition is false, and so we conclude that C' is a simple circuit. Therefore G' has a simple circuit of length k.

25. *Hint:* Suppose G and G' are isomorphic and G has m vertices of degree k; call them $v_1, v_2, \ldots, v_m$. Since G and G' are isomorphic, there are one-to-one correspondences $g\colon V(G) \to V(G')$ and $h\colon E(G) \to E(G')$. Show that $g(v_1), g(v_2), \ldots, g(v_m)$ are m distinct vertices of G', each of which has degree k.

27. <u>Proof</u>: Suppose G and G' are isomorphic graphs and suppose G is connected. By definition of graph isomorphism, there are one-to-one correspondences $g\colon V(G) \to V(G')$ and $h\colon E(G) \to E(G')$ that preserve the edge-endpoint functions in the sense that for all v in $V(G)$ and e in $E(G)$, v is an endpoint of $e \Leftrightarrow g(v)$ is an endpoint of $h(e)$. Suppose w and x are any two vertices of G'. Then $u = g^{-1}(w)$ and $v = g^{-1}(x)$ are distinct vertices in G (because g is a one-to-one correspondence). Since G is connected, there is a walk in G connecting u and v. Say this walk is $u e_1 v_1 e_2 v_2 \ldots e_n v$. Because g and h preserve the edge-endpoint functions, $w = g(u)h(e_1)g(v_1)h(e_2)g(v_2)\ldots h(e_n)g(v) = x$ is a walk in G' connecting w and x.

30. Suppose that G and G' are isomorphic via one-to-one correspondences $g\colon V(G) \to V(G')$ and $h\colon E(G) \to E(G')$, where g and h preserve the edge-endpoint functions. Now w_6 has degree one in G', and so by the argument given in Example 10.4.4, w_6 must correspond to one of the vertices of degree one in G: either $g(v_1) = w_6$ or $g(v_6) = w_6$. Similarly, since w_5 has degree three in G', w_5 must correspond to one of the vertices of degree three in G: either $g(v_3) = w_5$ or $g(v_4) = w_5$. Because g and h preserve the edge-endpoint functions, edge f_6 with endpoints

w_5 and w_6 must correspond to an edge in G with endpoints v_1 and v_3, or v_1 and v_4, or v_6 and v_3, or v_6 and v_4. But this contradicts the fact that none of these pairs of vertices are connected by edges in G. Hence the supposition is false, and G and G' are not isomorphic.

Section 10.4

1. **a.** Math 110

2. **a.**

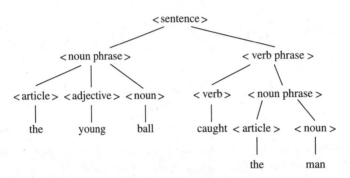

3. By Theorem 10.4.2, a tree with n vertices (where $n \geq 1$) has $n - 1$ edges, and so by the handshake theorem (Theorem 4.9.1), its total degree is twice the number of edges, or $2(n-1) = 2n - 2$.

4. **a.**

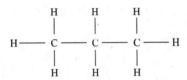

   d. *Hint:* Each carbon atom in G is bonded to four other atoms in G, because otherwise an additional hydrogen atom could be bonded to it, and this would contradict the assumption that G has the maximum number of hydrogen atoms for its number of carbon atoms. Also each hydrogen atom is bonded to exactly one carbon atom in G, because otherwise G would not be connected.

5. *Hint:* Revise the algorithm given in the proof of Lemma 10.4.1 to keep track of which vertex and edge were chosen in step 1 (by, say, labeling them v_0 and e_0). Then after one vertex of degree 1 is found, return to v_0 and search for another vertex of degree 1 by moving along a path outward from v_0 starting with another edge incident on v_0. Such an edge exists because v_0 has degree at least 2.

6. Define an infinite graph G as follows: $V(G) = \{v_i \mid i \in \mathbf{Z}\} = \{\ldots, v_{-2}, v_{-1}, v_0, v_1, v_2, \ldots\}$, $E(G) = \{e_i \mid i \in \mathbf{Z}\} = \{\ldots, e_{-2}, e_{-1}, e_0, e_1, e_2, \ldots\}$, and the edge-endpoint function is defined by the rule $f(e_i) = \{v_{i-1}, v_i\}$ for all $i \in \mathbf{Z}$. Then G is circuit-free, but each vertex has degree two. G is illustrated as follows.

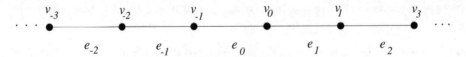

7. **a.** Leaves (or terminal vertices): v_1, v_5, v_7 Internal (or branch) vertices: v_2, v_3, v_4, v_6

8. Any tree with nine vertices has eight edges, not nine. Thus there is no tree with nine vertices and nine edges.

9. One such graph is

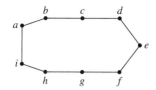

10. One such graph is

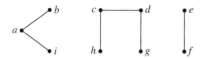

11. There is no tree with six vertices and a total degree of 14. Any tree with six vertices has five edges and hence, by the handshake theorem (Theorem 4.9.1) it has a total degree of 10, not 14.

12. One such tree is shown.

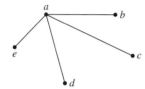

13. No such graph exists. By Theorem 10.4.4, a connected graph with six vertices and five edges is a tree. Hence such a graph cannot have a nontrivial circuit.

14.

15. One circuit-free graph with seven vertices and four edges is shown below.

18. Any tree with five vertices has four edges. By the handshake theorem (Theorem 4.9.1), the total degree of such a graph is eight, not ten. Hence there is no tree with five vertices and total degree ten.

21. Any tree with ten vertices has nine edges. By the handshake theorem (Theorem 4.9.1), the total degree of such a tree is 18, not 24. Hence there is no such graph.

22. Yes. Since it is connected and has 12 vertices and 11 edges, by Theorem 10.4.4 it is a tree. It follows from Lemma 10.5.1 that it has vertex of degree 1.

24. Yes, G' is connected. To see why, suppose u and w are any two vertices of G'. Then u and w are vertices of G and neither is equal to v. Since G is connected, there is a walk in G from u to w, and so by Lemma 10.1.1, there is a path in G from u to w. This path does not include edge e or vertex v because a path does not have a repeated edge, and e is the unique edge incident on v. *[If a path from u to w leads into v, then it must do so via e. But then it cannot emerge from v to continue on to w because no edge other than e is incident on v.]* Thus this

path is a path in G'. It follows that any two vertices of G' are connected by a walk in G', and so G' is connected.

25. No. Suppose there were a connected graph with eight vertices and six edges. Either the graph itself would be a tree or edges could be eliminated from its circuits to obtain a tree. In either case, there would be a tree with eight vertices and six or fewer edges. But by Theorem 10.4.2, a tree with eight vertices has seven edges, not six or fewer. This contradiction shows that the supposition is false, so there is no connected graph with eight vertices and six edges.

26. *Hint:* See the answer to exercise 25.

27. Yes, G is connected. To see why, suppose G is a circuit-free graph with ten vertices and nine edges. Let $G_1, G_2, \ldots, G_k$ be the connected components of G. *[To show that G is connected, we will show that $k = 1$.]* Each G_i is a tree since each G_i is connected and circuit-free. For each $i = 1, 2, \ldots, k$, let G_i have n_i vertices. Note that since G has ten vertices in all,

$$n_1 + n_2 + \cdots n_{k=} = 10.$$

By Theorem 10.4.2,

$$G_1 \text{ has } n_1 - 1 \text{ edges,}$$
$$G_2 \text{ has } n_2 - 1 \text{ edges,}$$
$$\vdots$$
$$G_k \text{ has } n_k - 1 \text{ edges.}$$

So the number of edges of G equals

$$(n_1 - 1) + (n_2 - 1) + \cdots + (n_k - 1) = (n_1 + n_2 + \cdots n_k) - \underbrace{(1 + 1 + \cdots + 1)}_{k \text{ 1's}} = 10 - k.$$

But we are given that G has nine edges. Hence $10 - k = 9$, and so $k = 1$. Thus G has just one connected component, G_1, and so G is connected.

28. *Hint:* See the answer to exercise 27 and the proof of Corollary 10.4.5.

30. A tree with five vertices must have four edges and, therefore, a total degree of 8. Since at least two vertices have degree 1 and no vertex has degree greater than 4, the possible degrees of the five vertices are as follows: 1,1,1,1,4; 1,1,1,2,3; and 1,1,2,2,2. The corresponding trees are shown below.

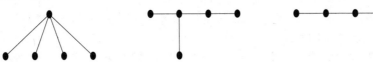

31. **b.** *Hint:* There are six.

Section 10.5

1. **a.** 3   **b.** 0   **c.** 5   **d.** u, v **e.** d **f.** k, l **g.** m, s, t, x, y **h.** 12

3. **a.**

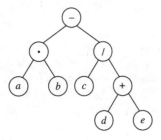

b.

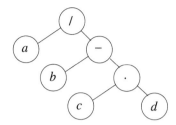

4.

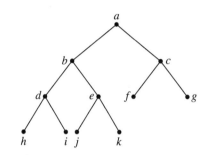

5. There is no full binary tree with the given properties because any full binary tree with five internal vertices has six leaves, not seven.

6. Any full binary tree with four internal vertices has five leaves for a total of nine–not seven– vertices in all. Thus there is no full binary tree with the given properties.

7. There is no full binary tree with 12 vertices because any full binary tree has $2k + 1$ vertices, where k is the number of internal vertices. But $2k + 1$ is always odd, and 12 is even.

8.

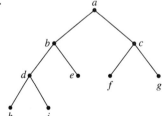

9.

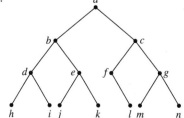

10.

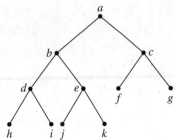

11. There is no binary tree that has height 3 and nine leaves because any binary tree of height 3 has at most $2^3 = 8$ leaves.

12. There is no tree with the given properties because any full binary tree with eight internal vertices has nine terminal vertices, not seven.

15. There is no tree with the given properties because a full binary tree with five internal vertices has $2 \cdot 5 + 1$ or eleven vertices in all, not nine.

18. There is no full binary tree with sixteen vertices because a full binary tree has $2k + 1$ vertices, where k is the number of internal vertices, and $16 \neq 2k + 1$ for any integer k.

20. **a.** The height of the tree is $\geq \log_2 25 \cong 4.6$. So since the height of any tree is an integer, the height of this tree must be at least 5.

21. **a.**

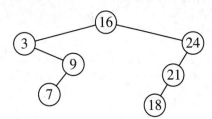

b.

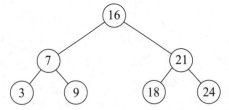

22. **a.**

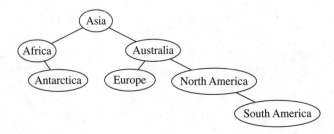

23.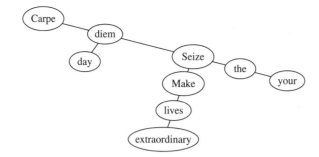

24.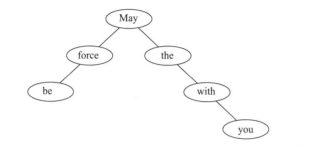

Section 10.6

1.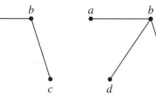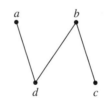

3. One of many spanning trees is as follows:

5. Minimum spanning tree:

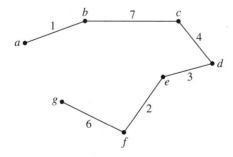

Order of adding the edges:
$\{a, b\}, \{e, f\}, \{e, d\}, \{d, c\}, \{g, f\}, \{b, c\}$

6. Minimum spanning tree:

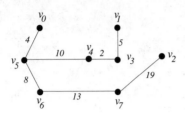

Order of adding the edges: $\{v_3, v_4\}, \{v_0, v_5\}, \{v_1, v_3\}, \{v_5, v_6\}, \{v_4, v_5\}, \{v_6, v_7\}, \{v_2, v_7\}$

7. The minimum spanning tree is the same as in exercise 5:

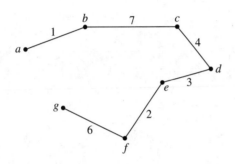

Order of adding the edges: $\{a, b\}, \{b, c\}, \{c, d\}, \{d, e\}, \{e, f\}, \{f, g\}$

9. There are four minimum spanning trees:

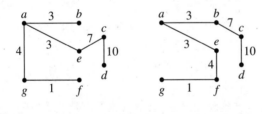

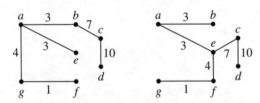

When Prim's algorithm is used, edges are added in any of the orders obtained by following one of the eight paths from left to right across the diagram below.

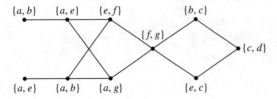

When Kruskal's algorithm is used, edges are added in any of the orders obtained by following one of the eight paths from left to right across the diagram below.

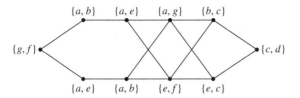

12. Let N = Nashville, S = St. Louis, Lv = Louisville, Ch = Chicago, Cn = Cincinnati, D = Detroit, Mw = Milwaukee, and Mn = Minneapolis. The following tables show the results of following the steps of Dijkstra's algorithm.

Step	V(T)	E(T)	F
0	$\{N\}$	$\emptyset$	$\{N\}$
1	$\{N\}$	$\emptyset$	$\{Lv, Mn\}$
2	$\{N, Lv\}$	$\{\{N, Lv\}\}$	$\{Mn, S, Cn, Ch, D, Mw\}$
3	$\{N, Lv, Cn\}$	$\{\{N, Lv\}, \{Lv, Cn\}\}$	$\{Mn, S, Ch, D, Mw\}$
4	$\{N, Lv, Cn, S\}$	$\{\{N, Lv\}, \{Lv, Cn\},$ $\{Lv, S\}\}$	$\{Mn, Ch, D, Mw\}$
5	$\{N, Lv, Cn,$ $S, Ch\}$	$\{\{N, Lv\}, \{Lv, Cn\},$ $\{Lv, S\}, \{Lv, Ch\}\}$	$\{Mn, D, Mw\}$
6	$\{N, Lv, Cn,$ $S, Ch, D\}$	$\{\{N, Lv\}, \{Lv, Cn\}, \{Lv, S\},$ $\{Lv, Ch\}\{Lv, D\}\}$	$\{Mn, Mw\}$
7	$\{N, Lv, Cn,$ $S, Ch, D, Mw\}$	$\{\{N, Lv\}, \{Lv, Cn\}, \{Lv, S\},$ $\{Lv, Ch\}\{Lv, D\}, \{Ch, Mw\}\}$	$\{Mn\}$
8	$\{N, Lv, Cn,$ $S, Ch, Mw,$ $Mn\}$	$\{\{N, Lv\}, \{Lv, Cn\}, \{Lv, S\},$ $\{Lv, Ch\}, \{Lv, D\}, \{Ch, Mw\},$ $\{N, Mn\}\}$	

Step	L(N)	L(S)	L(Lv)	L(Cn)	L(Ch)	L(D)	L(Mw)	L(Mn)
0	**0**	∞	∞	∞	∞	∞	∞	∞
1	0	∞	**151**	∞	∞	∞	∞	695
2	0	393	151	**234**	420	457	499	695
3	0	**393**	151	234	420	457	499	695
4	0	393	151	234	**420**	457	499	695
5	0	393	151	234	420	**457**	494	695
6	0	393	151	234	420	457	**494**	695
7	0	393	151	234	420	457	494	**695**

Thus the shortest path from Nashville to Minneapolis has length $L(Mn)$ = 695 miles.

In step 2 $D(Lv) = N$, in step 3 $D(Cn) = Lv$, in step 4 $D(S) = Cn$, in step 5 $D(Ch) = Lv$, in step 6 $D(D) = Lv$, in step 7 $D(Mw) = Ch$, and in step 8 $D(Mn) = N$. Tracing backwards from Mn gives $D(Mn) = N$, which is the starting point. So the shortest path is the direct route from Nashville to Minneapolis, without any intermediary stops.

13. The table below follows the steps of Dijkstra's algorithm to show that the shortest path from a to z has length 8.

Step	V(T)	E(T)	F	L(a)	L(b)	L(c)	L(d)	L(e)	L(z)
0	$\{a\}$	$\emptyset$	$\{a\}$	**0**	∞	∞	∞	∞	∞
1	$\{a\}$	$\emptyset$	$\{b, d\}$	0	2	∞	**1**	∞	∞
2	$\{a, d\}$	$\{\{a, d\}\}$	$\{b, c, e\}$	0	**2**	6	1	11	∞
3	$\{a, b, d\}$	$\{\{a, d\}, \{a, b\}\}$	$\{c, e\}$	0	2	**5**	1	6	∞
4	$\{a, b,$ $c, d\}$	$\{\{a, d\}, \{a, b\},$ $\{b, c\}\}$	$\{e, z\}$	0	2	5	1	**6**	13
5	$\{a, b, c,$ $d, e\}$	$\{\{a, d\}, \{a, b\},$ $\{b, c\}, \{c, e\}\}$	$\{z\}$	0	2	5	1	6	**8**
6	$\{a, b, c,$ $d, e, z\}$	$\{\{a, d\}, \{a, b\},$ $\{b, c\}, \{c, e\},$ $\{e, z\}\}$							

Thus the shortest path from a to z has length $L(z) = 8$. In step 2 $D(d) = a$, in step 3 $D(b) = b$, in step 4 $D(c) = b$, in step 5 $D(e) = c$, and in step 6 $D(z) = e$. Tracing backwards from z gives $D(z) = e$, $D(e) = c$, $D(c) = b$, and $D(b) = a$. So the shortest path from a to z is $abcez$.

15. The table below follows the steps of Dijkstra's algorithm to show that the shortest path from a to z has length 5.

Step	V(T)	E(T)	F	L(a)	L(b)	L(c)	L(d)	L(e)	L(f)	L(g)
0	$\{a\}$	$\emptyset$	$\{a\}$	0	∞	∞	∞	∞	∞	∞
1	$\{a\}$	$\emptyset$	$\{b, e, g\}$	0	**3***	∞	∞	3	∞	4
2	$\{a, b\}$	$\{\{a, b\}\}$	$\{c, e, g\}$	0	3	10	∞	**3**	∞	4
3	$\{a, b, e\}$	$\{\{a, b\}, \{a, e\}\}$	$\{c, d, f, g\}$	0	3	10	14	3	7	**4**
4	$\{a, b, e, g\}$	$\{\{a, b\}, \{a, e\}, \{a, g\}\}$	$\{c, d, f\}$	0	3	10	14	3	**5**	4
5	$\{a, b, e,$ $g, f\}$	$\{\{a, b\}, \{a, e\}, \{a, g\}, \{g, f\}\}$								

*At this point, vertex e could have been chosen instead of vertex b.

In step 2, $D(b) = a$, in step 3 $D(e) = a$, in step 4 $D(g) = a$, and in step 5 $D(f) = g$. Tracing backwards from f gives $D(f) = g$ and $D(g) = a$. So the shortest path is a, g, f.

18. **a.** If there were two distinct paths from one vertex of a tree to another, they (or pieces of them) could be patched together to obtain a circuit. Since a tree cannot have a circuit, it is impossible to have two distinct paths from one vertex of a tree to another.

b. Proof by contradiction: Suppose not. Suppose that some tree T has distinct vertices u and v and that P_1 and P_2 are two distinct paths joining u and v. *[We must deduce a contradiction. To do so, we will show that T would have to contain a circuit.]* Let P_1 be denoted $u = v_0, v_1, v_2, \ldots, v_m = v$, and let P_2 be denoted $u = w_0, w_1, w_2, \ldots, w_n = v$. Because P_1 and P_2 are distinct, and T has no parallel edges, the sequence of vertices in P_1 must diverge from the sequence of vertices in P_2 at some point. Let i be the least integer such that $v_i \neq w_i$. Then $v_{i-1} = w_{i-1}$. Let j and k be the least integers greater than i so that $v_j = w_k$. *[There must be such integers because $v_m = w_n = v$.]* Then

$$v_{i-1}v_i v_{i+1} \cdots v_j (=w_k) w_{k-1} \ldots w_i w_{i-1} (= v_{i-1})$$

is a circuit in T. The existence of such a circuit contradicts the fact that T is a tree. Hence the supposition must be false. That is, given any tree with vertices u and v, there is a unique path joining u and w.

20. <u>Proof</u>: Let G be a connected graph and let T be a circuit-free subgraph of G. Assume that if any edge e of G not in T is added to T, the resulting graph contains a circuit. We will prove by contraction that T is a spanning tree for G. Suppose not. That is, suppose T is not a spanning tree for G. *[We will derive a contradiction.]*

Case 1 (T is not connected): In this case, there are vertices u and v in T such that there is no walk in T from u to v. Now, since G is connected, there is a walk in G from u to v, and hence, by Lemma 10.2.1, there is a path in G from u to v. Let $e_1, e_2, \ldots, e_k$ be the edges of this path that are not in T. When these edges is added to T, the result is a graph T' in which u and v are connected by a path. In addition, by hypothesis, each of the edges e_i creates a circuit when added to T. Now remove these edges one by one from T'. By the same argument used in the proof of Lemma 10.5.3, each such removal leaves u and v connected since, by hypothesis, each e_i is an edge of a circuit when added to T. Hence, after all the e_i have been removed, u and v remain connected. But this contradicts the fact that there is no walk in T from u to v.

Case 2 (T is connected): In this case, since T is not a spanning tree and T is circuit-free, there is a vertex v in G such that v is not in T. *[For if T were connected, circuit- free, and contained every vertex in G, then T would be a spanning tree for G.]* Since G is connected, v is not isolated. Thus there is an edge e in G with v as an endpoint. Let T' be the graph obtained from T by adding e and v. *[Note that e is not already in T because if it were, its endpoint v would also be in T and it is not.]* Then T' contains a circuit because, by hypothesis, addition of any edge to T creates a circuit. Also T' is connected because T is and because when e is added to T, e becomes part of a circuit in T'. Now deletion of an edge from a circuit does not disconnect a graph, so if e is deleted from T' the result is a connected graph. But the resulting graph contains v, which means that there is an edge in T connecting v to another vertex of T. This implies that v is in T *[because both endpoints of any edge in a graph must be part of the vertex set of the graph]*, which contradicts the fact that v is not in T.

Thus, in either case, the supposition that T is not a spanning tree leads to a contradiction. Hence the supposition is false, and T is a spanning tree for G.

21. **a.** No.

<u>Counterexample</u>: Let G be the following graph.

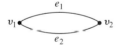

The two spanning trees for G shown below do not have an edge in common. So it is not the case that any two spanning trees for a graph must have an edge in common.

b. <u>Counterexample</u>: Let G be the following simple graph.

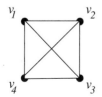

The two spanning trees for G shown below do not have an edge in common. So it is not the case that any two spanning trees for a simple graph must have an edge in common.

22. *Hint*: Suppose e is contained in every spanning tree of G and the graph obtained by removing e from G is connected. Let G' be the subgraph of G obtained by removing e, and let T' be a spanning tree for G'. How is T' related to G?

24. <u>Proof</u> : Suppose that T is a minimum spanning tree for a connected, weighted graph G and that G contains an edge e (not a loop) that is not in T. Let v and w be the endpoints of e. We know from exercise 18 that there is a unique path in T from v to w. Let e' be any edge of this path. We will prove by contradiction that $w(e') \leq w(e)$.

So suppose $w(e') > w(e)$. *[We will show that this supposition leads to a contradiction.]*

Form a new graph T' by adding e to T and deleting e'. By exercise 20, adding an edge to a spanning tree creates a circuit, and by Lemma 10.5.3, deleting an edge from a circuit does not disconnect a graph. Consequently, T' is also a spanning tree for G. Furthermore, $w(T') < w(T)$ because
$$w(T') = w(T) - w(e') + w(e) = w(T) - (w(e') - w(e)) < w(T)$$
[since $w(e') > w(e)$, which implies that $w(e') - w(e) > 0$]. But this contradicts the fact that T is a minimum spanning tree for G. Hence the supposition is false, and so $w(e') \leq w(e)$.

25. *Hint:* Suppose e is an edge that has smaller weight than any other edge of G, and suppose T is a minimum spanning tree for G that does not contain e. Create a new spanning tree T' by adding e to T and removing another edge of T (which one?). Then $w(T') < w(T)$.

26. Yes. <u>Proof by contradiction</u>: Suppose G is a weighted graph in which all the weights of all the edges are distinct, and suppose G has two distinct minimum spanning trees T_1 and T_2. *[We will show that this supposition leads to a contradiction.]*

Let e be the edge of least weight that is in one of the trees but not the other. Without loss of generality, we may say that e is in T_1. Add e to T_2 to obtain a graph G'. By exercise 19, G' contains a nontrivial circuit. At least one other edge f of this circuit is not in T_1 because otherwise T_1 would contain the whole circuit, which would contradict the fact that T_1 is a tree.

Now f has weight greater than e because all edges have distinct weights, f is in T_2 and not in T_1, and e is the edge of least weight that is in one of the trees and not the other. Remove f from G' to obtain a tree T_3.

Then $w(T_3) < w(T_2)$ because T_3 is the same as T_2 except that it contains e rather than f and $w(e) < w(f)$. Consequently, T_3 is a spanning tree for G of smaller weight than T_2. This contradicts the supposition that T_2 is a minimum spanning tree for G. Thus G cannot have more than one minimum spanning tree, and so the supposition is false and G does not have two distinct minimum spanning trees.

27. <u>Proof by contradiction</u>: Suppose not. That is, suppose there exists a connected, weighted graph G with n vertices and an edge e that (1) has larger weight than any other edge of G, (2) is in a circuit C of G, and (3) is in a minimum spanning tree T for G. *[We will show that this supposition leads to a contradiction.]*

Let the endpoints of e be vertices v and w, and let H be the graph obtained from T by removing e. In other words, $V(H) = V(T)$ and $E(H) = E(T) - \{e\}$. Then H is a circuit-free subgraph of T that contains all the vertices of G but only $n - 2$ edges, too few to be a tree.

Now H consists of two components, one, say H_v, contains v and the other, say H_w, contains w. Let e' be a "bridge" from H_w to H_v. That is, as shown in the solution to exercise 45 in Section 10.1, there is a trail in G from v to w that does not include e, and so we may let e' be the edge in the trail that immediately precedes the first vertex in the trail that is in H_v. Let T' be the graph obtained from H by adding e'. More precisely,

$$V(T') = V(H) \, and \, E(T') = E(H) \cup \{e\}.$$

Then T' is connected, contains every vertex of G (as does T), and has $n-1$ edges (the same as T). Hence, by Theorem 10.4.4, T' is a spanning tree for G. Now

$$w(T') = w(T) - w(e) + w(e') = w(T) - (w(e) - w(e')) < w(T)$$

because $w(e) > w(e')$. Thus T' is a spanning tree of smaller weight than a minimum spanning tree for G, which is a contradiction. Hence the supposition is false, and the given statement is true.

28. The output will be a "minimum spanning forest" for the graph. It will contain a minimum spanning tree for each connected component of the input graph.

30. At the end of step 4 of Algorithm 10.6.3, for each vertex x in T, $D(x)$ is the vertex at the other end of the edge that was used to connect x to T. Modify the algorithm by adding the following after part 4:

5. Initialize the path from a to z to be $\emptyset$.

[The path is developed by starting with vertex z and adding the edge $D(z)$ that joined z to T. Then the edge that joined $D(z)$ to T is added, and so forth, until vertex a is reached. Because the graph is finite and circuit-free, vertex a must eventually be reached.]

  Let x have the initial value z.

  while $(D(x) \neq a)$

      a. Add the edge $\{D(x), x\}$ to the path from a to z

      b. Let $x := D(x)$

  end while

Review Guide: Chapter 10

Definitions: How are the following terms defined?

- graph, edge-endpoint function
- loop in a graph, parallel edges, adjacent edges, isolated vertex, edge incident on an endpoint
- directed graph
- simple graph
- complete graph on n vertices
- complete bipartite graph on (m, n) vertices
- subgraph
- degree of a vertex in a graph, total degree of a graph
- walk, trail, path, closed walk, circuit, simple circuit
- connected vertices, connected graph
- connected component of a graph
- Euler circuit in a graph
- Euler trail in a graph
- Hamiltonian circuit in a graph
- adjacency matrix of a directed (or undirected) graph
- symmetric matrix
- $n \times n$ identity matrix
- powers of a matrix
- isomorphic graphs
- isomorphic invariant for graphs
- circuit-free graph
- tree, forest, trivial tree
- parse tree, syntactic derivation tree
- terminal vertex (or leaf), internal vertex (or branch vertex)
- rooted tree, level of a vertex in a rooted tree, height of a rooted tree
- parents, children, siblings, descendants, and ancestors in a rooted tree
- binary tree, full binary tree, subtree
- spanning tree
- weighted graph, minimum spanning tree

Graphs

- How can you use a graph as a model to help solve a problem?
- What does the handshake theorem say? In other words, how is the total degree of a graph related to the number of edges of the graph?
- How can you use the handshake theorem to determine whether graphs with specified properties exist?
- If an edge is removed from a circuit in a graph, does the graph remain connected?
- A graph has an Euler circuit if, and only if, it satisfies what two conditions?
- A graph has a Hamiltonian circuit if, and only if, it satisfies what four conditions?
- What is the traveling salesman problem?
- How do you find the adjacency matrix of a directed (or undirected) graph? How do you find the graph that corresponds to a given adjacency matrix?
- How can you determine the connected components of a graph by examining the adjacency matrix of the graph?
- How do you multiply two matrices?

- How do you use matrix multiplication to compute the number of walks from one vertex to another in a graph?
- How do you show that two graphs are isomorphic?
- What are some invariants for graph isomorphisms?
- How do you establish that two simple graphs are isomorphic?

Trees

- How do you show that a saturated carbon molecule with k carbon atoms has $2k + 2$ hydrogen atoms?
- If a tree has more than one vertex, how many vertices of degree 1 does it have? Why?
- If a tree has n vertices, how many edges does it have? Why?
- If a connected graph has n vertices, what additional property guarantees that it will be a tree? Why?
- How can you represent an algebraic expression using a binary tree?
- Given a full binary tree, what is the relation among the number of its internal vertices, terminal vertices, and total number of vertices?
- Given a binary tree, what is the relation between the number of its terminal vertices and its height?
- What property characterizes a binary search tree?
- How do you build a binary search tree?
- What is the relation between the number of edges in two different spanning trees for a graph?
- How does Kruskal's algorithm work?
- How do you know that Kruskal's algorithm produces a minimum spanning tree?
- How does Prim's algorithm work?
- How do you know that Prim's algorithm produces a minimum spanning tree?
- How does Dijkstra's shortest path algorithm work?
- How do you know that Dijkstra's shortest path algorithm produces a shortest path?

Fill-in-the-Blank Review Questions

Section 10.1

1. Let G be a graph and let v and w be vertices in G.

    (a) A walk from v to w is _____.
    (b) A trail from v to w is _____.
    (c) A path from v to w is _____.
    (d) A closed walk is _____.
    (e) A circuit is _____.
    (f) A simple circuit is _____.
    (g) A trivial circuit is _____.
    (h) Vertices v and w are connected if, and only if, _____.

2. A graph is connected if, and only if, _____.

3. Removing an edge from a circuit in a graph does not _____.

4. An Euler circuit in a graph is _____.

5. A graph has an Euler circuit if, and only if, _____.

6. Given vertices v and w in a graph, there is an Euler trail from v to w if, and only if, _____.

7. A Hamiltonian circuit in a graph is _____.

8. If a graph G has a Hamiltonian circuit, then G has a subgraph H with the following properties: ____, ____, ____, and ____.

9. A traveling salesman problem involves finding a ____ that minimizes the total distance traveled for a graph in which each edge is marked with a distance.

Section 10.2

1. In an adjacency matrix for a directed graph, the entry in the ith row and jth column is ____.

2. In an adjacency matrix for an undirected graph, the entry in the ith row and jth column is ____.

3. An $n \times n$ square matrix is called symmetric if, and only if, for all integers i and j from 1 to n, the entry in row ____ and column ____ equals the entry in row ____ and column ____.

4. The ijth entry in the product of two matrices $\mathbf{A}$ and $\mathbf{B}$ is obtained by multiplying row ____ of $\mathbf{A}$ by row ____ of $\mathbf{B}$.

5. In an $n \times n$ identity matrix the entries along the diagonal are all ____ and the off-diagonal entries are all ____.

6. If G is a graph with vertices $v_1, v_2, \ldots, v_m$ and $\mathbf{A}$ is the adjacency matrix of G, for each positive integer n and for all integers i and j with $i, j = 1, 2, \ldots, m$, the ijth entry of $\mathbf{A}^n =$ ____.

Section 10.3

1. If G and G' are graphs, then G is isomorphic to G' if, and only if, there exist a one-to-one correspondence g from the vertex set of G to the vertex set of G' and a one-to-one correspondence h from the edge set of G to the edge set of G' such that for all vertices v and edges e in G, v is an endpoint of e if, and only if, ____.

2. A property P is an isomorphic invariant for graphs if, and only if, given any graphs G and G', if G has property P and G' is isomorphic to G then ____.

3. Some invariant properties for graph isomorphisms are ____, ____, ____, ____, ____, ____, ____, ____, ____, and ____.

Section 10.4

1. A circuit-free graph is a graph with ____.

2. A forest is a graph that is ____, and a tree is a graph that is ____.

3. A trivial tree is a graph that consists of ____.

4. Any tree with at least two vertices has at least one vertex of degree ____.

5. If a tree T has at least two vertices, then a terminal vertex (or leaf) in T is a vertex of degree ____ and an internal vertex (or branch vertex) in T is a vertex of degree ____.

6. For any positive integer n, any tree with n vertices has ____.

7. For any positive integer n, if G is a connected graph with n vertices and $n - 1$ edges then ____.

Section 10.5

1. A rooted tree is a tree in which ____. The level of a vertex in a rooted tree is ____. The height of a rooted tree is ____.

2. A binary tree is a rooted tree in which ____.

3. A full binary tree is a rooted tree in which ____.

4. If k is a positive integer and T is a full binary tree with k internal vertices, then T has a total of ____ vertices and has ____ terminal vertices.

5. If T is a binary tree that has t terminal vertices and height h, then t and h are related by the inequality ____.

Section 10.6

1. A spanning tree for a graph G is ____.

2. A weighted graph is a graph for which ____, and the total weight of the graph is ____.

3. A minimum spanning tree for a connected weighted graph is ____.

4. In Kruskal's algorithm, the edges of a connected, weighted graph are examined one by one in order of ____.

5. In Prim's algorithm, a minimum spanning tree is built by expanding outward ____.

6. In Dijkstra's algorithm, a vertex is in the fringe if it is adjacent to ____ vertex in the tree that is being built up.

7. At each stage of Dijkstra's algorithm, the vertex that is added to the tree is a vertex in the fringe whose label is a ____.

Answers for Fill-in-the-Blank Review Questions

Section 10.1

1.
   (a) a finite alternating sequence of adjacent vertices and edges of G
   (b) a walk that does not contain a repeated edge
   (c) a trail that does not contain a repeated vertex
   (d) a walk that starts and ends at the same vertex
   (e) a closed walk that contains at least one edge and does not contain a repeated edge
   (f) a circuit that does not have any repeated vertex other than the first and the last
   (g) a walk consisting of a single vertex and no edge
   (h) there is a walk from v to w

2. given any two vertices in the graph there is a walk from one to the other

3. disconnect the graph

4. a circuit that contains every vertex and every edge of the graph

5. the graph is connected and every vertex has positive, even degree

6. the graph is connected, v and w have odd degree, and all other vertices have positive, even degree

7. a simple circuit that includes every vertex of the graph

8. H contains every vertex of G; H is connected; H has the same number of edges as vertices; every vertex of H has degree 2

9. Hamiltonian circuit

Section 10.2

1. the number of arrows from v_i (the ith vertex) to v_j (the jth vertex)

2. the number of edges connecting v_i (the ith vertex) and v_j (the jth vertex)

3. i; j; j; i

4. i; j

5. 1; 0

6. the number of walks of length n from v_i to v_j

Section 10.3

1. $g(v)$ is an endpoint of $h(e)$
2. G' has property P
3. has n vertices; has m edges; has a vertex of degree k; has m vertices of degree k; has a circuit of length k; has a simple circuit of length k; has m simple circuits of length k; is connected; has an Euler circuit; has a Hamiltonian circuit

Section 10.4

1. no circuits
2. circuit-free and not connected; connected and circuit-free
3. a single vertex (and no edges)
4. 1
5. 1; greater than 1 (*Or:* at least two)
6. $n-1$ edges
7. G is a tree

Section 10.5

1. one vertex is distinguished from the others and is called the root
   the number of edges along the unique path between it and the root
   the maximum level of any vertex of the tree
2. every parent has at most two children
3. every parent has exactly two children
4. $2k+1$; $k+1$
5. $t \le 2^h$, or, equivalently, $\log_2 t \le h$

Section 10.6

1. a subgraph of G that contains every vertex of G and is a tree
2. each edge has an associated real number weight; the sum of the weights of all the edges of the graph
3. a spanning tree that has the least possible total weight compared to all other spanning trees for the graph
4. weight; an edge of least weight
5. initial vertex; adjacent vertices and edges
6. adjacent to a
7. minimum among all those in the fringe

Chapter 11: Analysis of Algorithm Efficiency

The focus of Chapter 11 is the analysis of algorithm efficiency in Sections 11.3 and 11.5. The chapter opens with a brief review of the properties of function graphs that are especially important for understanding O-, Ω-, and Θ-notations, which are introduced in Section 11.2. For simplicity, the examples in Section 11.2 emphasize polynomial functions. Section 11.3 introduces the analysis of algorithm efficiency with examples that include sequential search, insertion sort, selection sort (in the exercises), and polynomial evaluation (in the exercises). Section 11.4 discusses the properties of logarithms that are particularly important in the analysis of algorithms and other areas of computer science, and Section 11.5 applies the properties to analyze algorithms whose orders involve logarithmic functions. Examples in Section 11.5 include binary search and merge sort.

Section 11.1

1. **a.** $f(0)$ is positive.

   b. $f(x) = 0$ when $x = -2$ and $x = 3$ (approximately)

   c. $x_1 = -1$ and $x_2 = 2$ (approximately)

   d. $x = 1$ or $x = -\frac{1}{2}$ (approximately)

   e. increase

   f. decrease

3.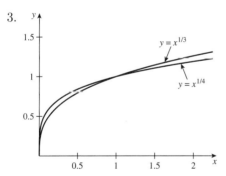

When $0 < x < 1, x^{1/3} < x^{1/4}$. When $x > 1, x^{1/3} > x^{1/4}$.

5.

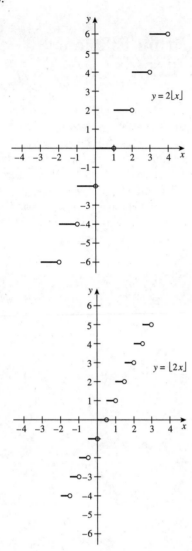

The graphs show that $2 \lfloor x \rfloor \neq \lfloor 2x \rfloor$ for many values of x.

6.

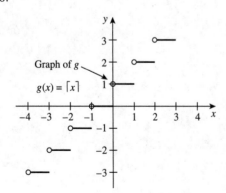

8.

x	$F(x) = \lfloor x^{1/2} \rfloor$
0	0
1/2	0
1	1
2	1
3	1
4	2

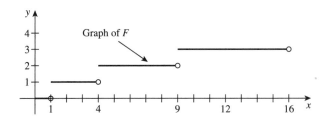

9.

x	$G(x)$
0	0
1/2	1/2
3/4	3/4
1	0
1 1/2	1/2
1 3/4	3/4
2	0

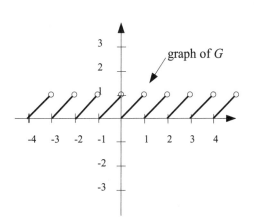

10.

n	$f(n) = |n|$
0	0
1	1
2	2
3	3
−1	1
−2	2
−3	3

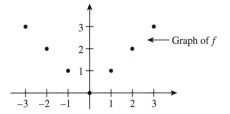

12.

n	$h(n) = \lfloor n/2 \rfloor$
0	0
1	0
2	1
3	1
4	2
5	2
6	3
7	3
8	4
9	4

Graph of h

14. f is increasing on the intervals $\{x \in \mathbf{R} \mid -3 < x < -2\}$ and $\{x \in \mathbf{R} \mid 0 < x < -2 < x < -2\}$, and f is decreasing on $\{x \in \mathbf{R} \mid -2 < x < 0\}$ and $\{x \in \mathbf{R} \mid 2.5 < x < 4\}$ (approximately).

15. <u>Proof</u>: Suppose that x_1 and x_2 are particular but arbitrarily chosen real numbers such that $x_1 < x_2$. *[We must show that $f(x_1) < f(x_2)$.]* Since

$$x_1 < x_2 \quad \text{then} \quad 2x_1 < 2x_2 \quad \text{and so} \quad 2x_1 - 3 < 2x_2 - 3$$

by basic properties of inequalities. Thus, by definition of f, $f(x_1) < f(x_2)$ *[as was to be shown]*. Hence f is increasing on the set of all real numbers.

17. **a.** <u>Proof</u>: Suppose x_1 and x_2 are real numbers with $x_1 < x_2 < 0$. *[We must show that $h(x_1) > h(x_2)$.]* Multiply both sides of $x_1 < x_2$ by x_1 to obtain $(x_1)^2 > x_1 x_2$ *[by T23 of Appendix A since $x_1 < 0$]*, and multiply both sides of $x_1 < x_2$ by x_2 to obtain $x_1 x_2 > (x_2)^2$ *[by T23 of Appendix A since $x_2 < 0$]*. By transitivity of order *[Appendix A, T18]* $(x_2)^2 < (x_1)^2$, and so, by definition of h, $h(x_2) < h(x_1)$.

18. **a.** *Preliminaries:* If both x_1 and x_2 are positive, then by the rules for working with inequalities (see Appendix A),

$$\frac{x_1 - 1}{x_1} < \frac{x_2 - 1}{x_2} \Rightarrow x_2(x_1 - 1) < x_1(x_2 - 1) \qquad \text{by multiplying both sides by } x_1 x_2 \text{ (which is positive)}$$

$$\Rightarrow x_1 x_2 - x_2 < x_1 x_2 - x_1 \qquad \text{by multiplying out}$$

$$\Rightarrow -x_2 < -x_1 \qquad \text{by subtracting } x_1 x_2 \text{ from both sides}$$

$$\Rightarrow x_2 > x_1 \qquad \text{by multiplying by } -1.$$

Are these steps reversible? Yes!

Proof: Suppose that x_1 and x_2 are positive real numbers and $x_1 < x_2$. *[We must show that $k(x_1) < k(x_2)$.]* Then

$$x_1 < x_2 \Rightarrow -x_2 < -x_1 \qquad \text{by multiplying by } -1$$

$$\Rightarrow x_1 x_2 - x_2 < x_1 x_2 - x_1 \qquad \text{by adding } x_1 x_2 \text{ to both sides}$$

$$\Rightarrow x_2(x_1 - 1) < x_1(x_2 - 1) \qquad \text{by factoring both sides}$$

$$\Rightarrow \frac{x_1 - 1}{x_1} < \frac{x_2 - 1}{x_2} \qquad \begin{array}{l}\text{by dividing both sides by the}\\ \text{positive number } x_1 x_2\end{array}$$

$$\Rightarrow k(x_1) < k(x_2) \qquad \text{by definition of } k.$$

[This is what was to be shown.]

b. When $x < 0$, k is increasing.

Proof: Suppose x_1 and x_2 are any real numbers such that $x_1 < x_2 < 0$. Multiplying both sides of this inequality by -1 gives $-x_1 > -x_2$, and adding $x_1 x_2$ to both sides gives $x_1 x_2 - x_1 > x_1 x_2 - x_2$. Now, since x_1 and x_2 are both negative, $x_1 x_2$ is positive, and hence

$$\frac{x_1 x_2 - x_1}{x_1 x_2} > \frac{x_1 x_2 - x_2}{x_1 x_2}.$$

Simplifying the two fractions gives

$$\frac{x_2 - 1}{x_2} > \frac{x_1 - 1}{x_1}$$

and so $k(x_1) < k(x_2)$ by definition of k.

19. Proof: Suppose $f: \mathbf{R} \to \mathbf{R}$ is increasing. *[We must show that f is one-to-one. In other words, we must show that for all real numbers x_1 and x_2, if $x_1 \neq x_2$ then $f(x_1) \neq f(x_2)$.]* Suppose x_1 and x_2 are real numbers and $x_1 \neq x_2$. By the trichotomy law *[Appendix A, T17]* $x_1 < x_2$, or $x_1 > x_2$. In case $x_1 < x_2$, then since f is increasing, $f(x_1) < f(x_2)$ and so $f(x_1) \neq f(x_2)$. Similarly, in case $x_1 > x_2$, then $f(x_1) > f(x_2)$ and so $f(x_1) \neq f(x_2)$. Thus in either case, $f(x_1) \neq f(x_2)$ *[as was to be shown]*.

21. **a.** Proof: Suppose u and v are nonnegative real numbers with $u < v$. *[We must show that $f(u) < f(v)$.]* Note that $v = u + h$ for some positive real number h. By substitution and the binomial theorem,

$$v^m = (u + h)^m = u^m + \left[\binom{m}{1} u^{m-1} h + \binom{m}{2} u^{m-2} h^2 + \cdots + \binom{m}{m-1} u h^{m-1} + h^m \right].$$

The bracketed sum is positive because $u \geq 0$ and $h > 0$, and a sum of nonnegative terms that includes at least one positive term is positive. Hence

$$v^m = u^m + \text{a positive number},$$

and so $f(u) = u^m < v^m = f(v)$ *[as was to be shown]*.

b. Proof by contradiction:

Suppose that g is not increasing. Then there exist real numbers x_1 and x_2 such that $0 < x_1 < x_2$ and $g(x_1) \geq g(x_2)$. By definition of g,

$$x_1^{\frac{m}{n}} \geq x_2^{\frac{m}{n}}.$$

Applying part (a) to this inequality gives

$$\left(x_1^{\frac{m}{n}} \right)^n \geq \left(x_2^{\frac{m}{n}} \right)^n.$$

By the laws of exponents, $x_1^{\frac{m}{n}\cdot n} = x_1^{m}$ and $x_2^{\frac{m}{n}\cdot n} = x_2^{m}$, and so

$$x_1^{m} \geq x_2^{m}.$$

But, by part (a), $x_1^{m} < x_2^{m}$, and so we have reached a contradiction. Hence the supposition is false, and thus g is increasing.

22.

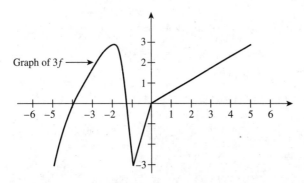

24. <u>Proof</u>: Suppose that f is a real-valued function of a real variable, f is decreasing on a set S, and M is any positive real number. *[We must show that Mf is decreasing on S. In other words, we must show that for all x_1 and x_2 in the set S, if $x_1 < x_2$ then $(Mf)(x_1) > (Mf)(x_2)$.]*

Suppose x_1 and x_2 are in S and $x_1 < x_2$. Since f is decreasing on S, $f(x_1) > f(x_2)$, and since M is positive, $Mf(x_1) > Mf(x_2)$ *[because when both sides of an inequality are multiplied by a positive number, the direction of the inequality is unchanged]*. It follows by definition of Mf that $(Mf)(x_1) > (Mf)(x_2)$ *[as was to be shown]*.

27. To find the answer algebraically, solve the equation $2x^2 = x^2 + 10x + 11$ for x. Subtracting x^2 from both sides gives $x^2 - 10x - 11 = 0$, and either using the quadratic formula or factoring $x^2 - 10x - 11 = (x - 11)(x + 1)$ gives $x = 11$ (since $x > 0$). To find an approximate answer with a graphing calculator, plot both $f(x) = x^2 + 10x + 11$ and $2g(x) = 2x^2$ for $x > 0$, as shown in the figure, and find that $2g(x) > f(x)$ when $x > 11$ (approximately). You can obtain only an approximate answer from a graphing calculator because the calculator computes values only to an accuracy of a finite number of decimal places.

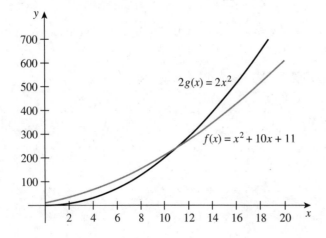

Section 11.2

1. **a.** *Formal version of negation*: $f(n)$ is not $\Omega(g(n))$ if, and only if, $\forall$ positive real numbers a and A, $\exists$ an integer $n \geq a$ such that $Ag(n) > f(n)$.

   b. *Informal version of negation*: $f(n)$ is not $\Omega(g(n))$ if, and only if, no matter what positive real numbers a and A might be chosen, it is possible to find an integer n greater than or equal to a with the property that $Ag(n) > f(n)$.

3. **a.** *Formal version of negation*: $f(n)$ is not $\Theta(g(n))$ if, and only if, $\forall$ positive real numbers k, A, and B, $\exists$ an integer $n \geq k$ such that either $Ag(n) > f(n)$ or $f(n) > Bg(n)$.

   b. *Informal version of negation*: $f(n)$ is not $\Theta(g(n))$ if, and only if, no matter what positive real numbers k, A, and B might be chosen, it is possible to find an integer n greater than or equal to k with the property that either $Ag(n) > f(n)$ or $f(n) > Bg(n)$.

4. $n - \left\lfloor \dfrac{n}{2} \right\rfloor + 1$ is $\Omega(n)$

5. $n - \left\lfloor \dfrac{n}{2} \right\rfloor + 1$ is $O(n)$

6. $3n(n-2)$ is $\Theta(n^2)$

9. $n^2 \left(\left\lceil \dfrac{n}{3} \right\rceil - 1 \right)$ is $\Theta(n^3)$

10. **a.** For each integer $n \geq 1$, $0 \leq 2n^2 + 15n + 4$ because all terms in $2n^2 + 15n + 4$ are positive. Moreover,

$$
\begin{aligned}
2n^2 + 15n + 4 &\leq 2n^2 + 15n^2 + 4n^2 \quad \text{because when n} \geq 1, 15n \leq 15n^2 \text{and } 4 \leq 4n^2 \\
&= 21n^2 \quad \text{by combining like terms.}
\end{aligned}
$$

Therefore, by transitivity of equality and order,

$$0 \leq 2n^2 + 15n + 4 \leq 21n^2 \quad \text{for each integer } n \geq 1.$$

b. For each integer $n \geq 1$, $2n^2 \leq 2n^2 + 15n + 4$ because $15n + 4 > 0$ since n is positive.

c. Sketch of graph

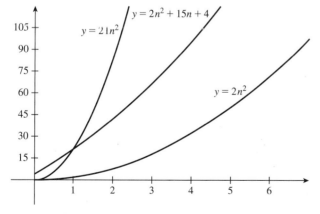

d. Let $A = 2$ and $a = 1$. Then, by substitution from the result of part (b),

$$An^2 < 2n^2 + 15n + 4 \quad \text{for each integer } n \geq a,$$

and hence, by definition of Ω-notation, $2n^2 + 15n + 4$ is $\Omega(n^2)$. Let $B = 21$ and $b = 1$. Then, by substitution from the result of part (a),

$$0 < 2n^2 + 15n + 4 \leq Bn^2 \quad \text{for each integer } n \geq b,$$

and hence by definition of O-notation, $2n^2 + 15n + 4$ is $O(n^2)$.

e. *Solution 1:* Let $A = 2$, $B = 21$, and $k = 1$. By the results of parts (a) and (b),

$$An^2 \leq 2n^2 + 15n + 4 \leq Bn^2 \quad \text{for each integer } n \geq k,$$

and hence, by definition of Θ-notation, $2n^2 + 15n + 4$ is $\Theta(n^2)$.

Solution 2: By part (d), $2n^2 + 15n + 4$ is both $\Omega(n^2)$ and $O(n^2)$. Hence, by Theorem 11.2.1,

$$2n^2 + 15n + 4 \text{ is } \Theta(n^2).$$

12. **a.** For every integer $n \geq 1$,
$$0 \leq 7n^3 + 10n^2 + 3$$

because all terms in $7n^3 + 10n^2 + 3$ are positive. Moreover,

$$7n^3 + 10n^2 + 3 \quad \leq \quad 7n^3 + 10n^3 + n^3 \quad \text{because when } n \geq 1, \ 10n^2 \leq 10n^3 \text{ and } 3 \leq 3n^3$$
$$= \quad 20n^3 \qquad\qquad \text{by combining like terms.}$$

Therefore, by transitivity of equality and order,

$$0 \leq 7n^3 + 10n^2 + 3 \leq 20n^3 \qquad \text{for every integer } n \geq 1.$$

b. For every integer $n \geq 1$,
$$7n^3 \leq 7n^3 + 10n^2 + 3$$

because $10n^2 + 3 > 0$ since n is positive.

c. Sketch of graph

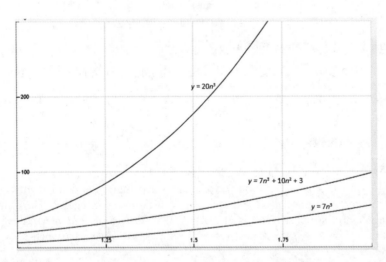

d. Let $A = 7$ and $a = 1$. Then, by substitution from the result of part (b),

$$An^3 \leq 7n^3 + 10n^2 + 3 \qquad \text{for every integer } n \geq a,$$

and hence, by definition of Ω-notation, $7n^3 + 10n^2 + 3$ is $\Omega(n^3)$. Let $B = 20$ and $b = 1$. Then, by substitution from the result of part (a),

$$7n^3 + 10n^2 + 3 \leq Bn^3 \qquad \text{for every integer } n \geq b,$$

and hence by definition of O-notation, $7n^3 + 10n^2 + 3$ is $O(n^3)$.

e. *Solution 1:* Let $A = 7$, $B = 20$, and $k = 1$. By the results of parts (a) and (b),

$$An^3 \leq 7n^3 + 10n^2 + 3 \leq Bn^3 \qquad \text{for every integer } n \geq k,$$

and hence, by definition of Θ-notation, $7n^3 + 10n^2 + 3$ is $\Theta(n^3)$.

Solution 2: By part (d), $7n^3 + 10n^2 + 3$ is both $\Omega(n^3)$ and $O(n^3)$. Hence, by Theorem 11.2.1,

$$7n^3 + 10n^2 + 3 \text{ is } \Theta(n^3).$$

13. For each integer $n \geq 1$,

$$5n^3 \leq 5n^3 + 65n + 30$$

because $65n + 30 > 0$ since n is positive. Moreover,

$$\begin{aligned} 5n^3 + 65n + 30 \quad &\leq \quad 5n^3 + 65n^3 + 30n^3 \quad \text{because when } n \geq 1, \ 65n^2 \leq 65n^3 \text{ and } 30 \leq 30n^3 \\ &= \quad 100n^3 \qquad\qquad\qquad \text{by combining like terms.} \end{aligned}$$

Therefore, by transitivity of order and equality,

$$5n^3 \leq 5n^3 + 65n + 30 \leq 100n^3.$$

Thus, let $A = 5, B = 100$, and $k = 1$. Then

$$An^3 \leq 5n^3 + 65n + 30 \leq Bn^3 \quad \text{for each integer } n \geq k,$$

and hence, by definition of Θ-notation, $5n^3 + 65n + 30$ is $\Theta(n^3)$.

15. For each integer $n \geq 1$,

$$n \leq n + \frac{1}{2} < n + 1,$$

and so $\lfloor n + \frac{1}{2} \rfloor = n$, by definition of floor, and $\lfloor n + \frac{1}{2} \rfloor$ is nonnegative. In addition, when $n \geq 1$, then $n + 1 \leq n + n = 2n$, and thus, by transitivity of equality and order,

$$n \leq \left\lfloor n + \frac{1}{2} \right\rfloor \leq 2n.$$

Let $A = 1, B = 2$, and $k = 1$. Then

$$An \leq \left\lfloor n + \frac{1}{2} \right\rfloor \leq Bn \text{ for every integer } n \geq k,$$

and hence, by definition of Θ-notation, $\lfloor n + \frac{1}{2} \rfloor$ is $\Theta(n)$.

18. Proof of Theorem 11.2.7(b): Suppose f and g are real-valued functions defined on the same set of nonnegative integers, suppose $f(n) \geq 0$ and $g(n) \geq 0$ for every integer $n \geq r$, where r is a positive real number, and suppose $f(n)$ is $\Theta(g(n))$. *[We must show that $g(n)$ is $\Theta(f(n))$.]* By definition of Θ-notation, there exist positive real numbers A, B, and k with $k \geq r$ such that for each integer $n \geq k$,

$$Ag(n) \leq f(n) \leq Bg(n).$$

Dividing the left-hand inequality by A and the right-hand inequality by B gives that

$$g(n) \leq \frac{1}{A} f(n) \quad \text{and} \quad \frac{1}{B} f(n) \leq g(n),$$

and combining the resulting inequalities produces

$$\frac{1}{B} f(n) \leq g(n) \leq \frac{1}{A} f(n) \quad \text{for each integer } n \geq k.$$

Now both $f(n) \geq 0$ and $g(n) \geq 0$ for each integer $n \geq k$. Also, since both A and B are positive real numbers, so are $1/A$ and $1/B$. Thus, by definition of Θ-notation, $g(n)$ is $\Theta(f(n))$.

20. Proof (by contradiction): Suppose not. That is, suppose n^5 is $O(n^5)$. *[We must show that this supposition leads to a contradiction.]* By definition of O-notation, there exist positive real numbers B and b such that

$$0 \leq n^5 \leq Bn^2 \quad \text{for each integer } n \geq b.$$

Dividing the inequalities by n^5 and taking the cube root of both sides gives

$$0 \leq n \leq \sqrt[3]{B} \quad \text{for each integer } n \geq b.$$

These two conditions are contradictory because on the one hand n can be any integer greater than or equal to b, but when n is greater than b, then n is less than $\sqrt[3]{B}$, which is a fixed integer. Thus the supposition leads to a contradiction, and hence the supposition is false.

21. Proof of Theorem 11.2.4 (by contradiction): Suppose there exists a real-valued function f defined on a set of nonnegative integers such that $f(n)$ is $\Omega(n^m)$, where m is a positive integer, and suppose that $f(n)$ is also $O(n^p)$ for some positive real number $p < m$. By definition of O-notation, there exist real numbers B and b such that

$$0 \leq f(n) \leq Bn^p \quad \text{for every integer } n \geq b.$$

But also, by definition of Ω-notation, there exist positive real numbers A and a such that

$$An^m \leq f(n) \quad \text{for every integer } n \geq a.$$

Let t be the larger of b and a Then by transitivity of order,

$$An^m \leq Bn^p \quad \text{for every integer } n \geq t.$$

Dividing by An^p and taking the $(m-p)$th root of both sides gives that

$$n \leq \sqrt[m-p]{B/A} \quad \text{for every integer } n \geq t.$$

Now$\equiv$ n can be any integer greater than or equal to b, and so, in particular, it can be greater than $\sqrt[m-p]{B/A}$, , which contradicts its being less than or equal to $\sqrt[m-p]{B/A}$. Thus the supposition leads to a contradiction, and hence the supposition is false.

22. **a.** *Solution 1 (using ad hoc calculations):* Let $\Leftrightarrow$ stand for the words "if, and only if," and observe that

(*) $$\frac{1}{2}n^4 \leq 2n^4 - 90n^3 + 3$$

$\Leftrightarrow$ $\quad n^4 \leq 4n^4 - 180n^3 + 6 \quad$ because dividing or multiplying both sides of an inequality by 2, which is positive, preserves the direction of the inequality

$\Leftrightarrow \quad 180n^3 - 6 \leq 3n^4 \quad$ because adding or subtracting $180n^3 - 6$ to both sides of an inequality preserves the direction of the inequality

(**) $\Leftrightarrow \quad 60 - \dfrac{2}{n^3} \leq n \quad$ because dividing or multiplying both sides of an inequality by $3n^3$, which is positive, preserves the direction of the inequality.

Because all the inequalities are equivalent (that is, each inequality is true if, and only if, all the others are true), any value of n that makes inequality (**) true makes inequality (*) true also. Now

$$\text{if} \quad n \geq 60, \quad \text{then} \quad n \geq 60 - \frac{2}{n^3},$$

which is inequality (**). Therefore, inequality (*) is also true for every integer $n \geq 60$:

$$\frac{1}{2}n^4 \leq 2n^4 - 90n^3 + 3.$$

Let $A = \frac{1}{2}$ and $a = 60$. Then for every integer $n \geq a$,

$$An^4 \leq 2n^4 - 90n^3 + 3.$$

and so, by definition of Ω-notation, $2n^4 - 90n^3 + 3$ is $\Omega(n^4)$.

Solution 2 (using the general procedure):

To use the general procedure from Example 11.2.4 for showing that $2n^4 - 90n^3 + 3$ is $\Omega(n^4)$, let

$$A = \frac{1}{2} \cdot 2 = 1 \quad \text{and} \quad a = \frac{2}{2}(|-90| + |3|) = 93$$

and note that $a \geq 1$. We will show that $n^4 \leq 2n^4 - 90n^3 + 3$ for every integer $n \geq a$.

Now $n \geq a$ implies that

$$n \geq 90 + 3.$$

Multiplying both sides by n^3 gives

$$n^4 \geq 90n^3 + 3n^3$$

and subtracting first $3n^3$ and then 3 from the right-hand side gives that

$$n^4 \geq 90n^3 \geq 90n^3 - 3 \quad \text{for every integer } n \geq a.$$

Subtracting the right-hand side from the left-hand side and adding n^4 to both sides gives

$$2n^4 - 90n^3 + 3 \geq n^4 \quad \text{for every integer } n \geq a.$$

Thus since $A = 1$,

$$2n^4 - 90n^3 + 3 \geq An^4 \text{ for every integer } n \geq a,$$

and so, by definition of Ω-notation, $2n^4 - 90n^3 + 3$ is $\Omega(n^4)$.

b. To show that $2n^4 - 90n^3 + 3$ is $O(n^4)$, observe that for every integer $n \geq 1$,

$$2n^4 - 90n^3 + 3 \quad \leq \quad 2n^4 + 90n^3 + 3 \qquad \text{because when } n \geq 1, 90n^3 \text{ is positive}$$

$$\leq \quad 2n^4 + 90n^4 + 3n^4 \qquad \begin{array}{l}\text{by Theorem 11.2.2 since n} \geq 1, \ n^3 \leq n^4 \text{ and} \\ 1 \leq n^4, \text{ and so } 90n^3 \leq 90n^4 \text{ and } 3 \leq 3n^4\end{array}$$

$$= \quad 95n^4 \qquad \text{because } 2 + 90 + 3 = 95.$$

Thus, by transitivity of order and equality, for every integer $n \geq 1$,

$$2n^4 - 90n^3 + 3 \leq 95n^4.$$

In addition, by part (a), for every integer $n \geq 60$,

$$\frac{1}{2}n^4 \leq 2n^4 - 90n^3 + 3$$

so since $0 \leq \frac{1}{2}n^4$, transitivity of order gives that for every integer $n \geq 60$,

$$0 \leq 2n^4 - 90n^3 + 3 \leq 95n^4.$$

Let $B = 14$ and $b = 60$. Then, for every integer $n \geq b$,

$$0 \leq 2n^4 - 90n^3 + 3 \leq Bn^4$$

and hence, by definition of O-notation,

$$2n^4 - 90n^3 + 3 \text{ is } O(n^4).$$

c. *Solution 1:* Let $A = \frac{1}{2}$, $B = 95$, and $k = 60$. By the results of parts (a) and (b), for every integer $n \geq k$,

$$An^4 \leq 2n^4 - 90n^3 + 3 \leq Bn^4$$

and hence, by definition of Θ-notation,

$$2n^4 - 90n^3 + 3 \text{ is } \Theta(n^4).$$

Solution 2: By parts (a) and (b), $2n^4 - 90n^3 + 3$ is both $\Omega(n^4)$ and $O(n^4)$. Hence, by Theorem 11.2.1,

$$2n^4 - 90n^3 + 3 \text{ is } \Theta(n^4).$$

24. **a.** *Solution 1 (Using ad hoc calculations):* Let $\Leftrightarrow$ stand for the words "if, and only if," and observe that

(*) $$\frac{1}{8}n^5 \leq \frac{1}{4}n^5 - 50n^3 + 3n + 12$$

$\Leftrightarrow$ $50n^3 - 3n - 12 \leq \frac{1}{8}n^5$ because adding or subtracting $50n^3 - 3n - 12$ to both sides of an inequality preserves the direction of the inequality

(**) $\Leftrightarrow$ $\dfrac{400}{n} - \dfrac{24}{n^3} - \dfrac{96}{n^4} \leq n$ because dividing or multiplying both sides of an inequality by $8/n^4$, which is positive, preserves the direction of the inequality.

Because all the inequalities are equivalent (that is, each inequality is true if, and only, if all the others are true), any value of n that makes inequality (**) true makes inequality (*) true also. Observe that if $n \geq 20$, then

$$\frac{1}{n} \leq \frac{1}{20} \quad \text{and so} \quad \frac{400}{n} \leq \frac{400}{20} = 20.$$

In addition,

$$\frac{400}{n} - \frac{24}{n^3} - \frac{96}{n^4} \leq \frac{400}{n},$$

and so, by transitivity of order and equality, if $n \geq 20$, then

$$\frac{400}{n} - \frac{24}{n^3} - \frac{96}{n^4} \leq n.$$

It follows that when $n \geq 20$, inequality (**) is true and thus inequality (*) is also true. Hence,

$$\frac{1}{8}n^5 \leq \frac{1}{4}n^5 - 50n^3 + 3n + 12 \quad \text{for every integer } n \geq 20.$$

Let $A = \dfrac{1}{8}$ and $a = 20$. Then

$$An^5 \leq \frac{1}{4}n^5 - 50n^3 + 3n + 12 \quad \text{for every integer } n \geq a$$

and so, by definition of Ω-notation, $\frac{1}{4}n^5 - 50n^3 + 3n + 12$ is $\Omega(n^5)$.

Solution 2 (Using the general procedure):

To use the general procedure to show that $\frac{1}{4}n^5 - 50n^3 + 3n + 12$ is $\Omega(n^5)$, take

$$A = \frac{1}{2} \cdot \frac{1}{4} = \frac{1}{8} \quad \text{and} \quad a = \left(\frac{2}{\frac{1}{4}}\right)(|-50| + |3| + |12|) = 8 \cdot 65 = 520$$

and note that $a \geq 1$. Then $n \geq a$ implies that

$$n \geq 520 = 8(50 + 3 + 12)$$

Multiplying both sides by $\frac{1}{8}n^4$ gives

$$\frac{1}{8}n^5 \geq 50n^4 + 3n^4 + 12n^4$$

$$\geq 50n^3 - (3n + 12) \quad \text{because } 3n + 12 \text{ is positive}$$

$$= 50n^3 - 3n - 12 \quad \text{by algebra.}$$

Thus, by transitivity of order and equality,

$$\frac{1}{8}n^5 \geq 50n^3 - 3n - 12 \quad \text{for every integer } n \geq a.$$

Subtracting the right-hand side from the left-hand side and adding $\frac{1}{8}n^5$ to both sides gives

$$\frac{1}{4}n^5 - 50n^3 + 3n + 12 \geq \frac{1}{8}n^5 \quad \text{for every integer } n \geq a.$$

Thus since $A = \frac{1}{8}$,

$$\frac{1}{4}n^5 - 50n^3 + 3n + 12 \geq An^5 \quad \text{for every integer } n \geq a,$$

and so, by definition of Ω-notation, $\frac{1}{4}n^5 - 50n^3 + 3n + 12$ is $\Omega(n^5)$.

b. To show that $\frac{1}{4}n^5 - 50n^3 + 3n + 12$ is $O(n^5)$, observe that for every integer $n \geq 1$,

$$\frac{1}{4}n^5 - 50n^3 + 3n + 12 \leq \frac{1}{4}n^5 + 50n^3 + 3n + 12$$

because when $n \geq 1$, $50n^3$ is positive

$$\leq \frac{1}{4}n^5 + 50n^5 + 3n^5 + 12n^5$$

by Theorem 11.2.2 since $n \geq 1$, $n^3 \leq n^5$ and $1 \leq n^5$, and so $50n^3 \leq 50n^5$, $3n \leq 3n^5$ and $12 \leq 12n^5$

$$= \frac{261}{4}n^5$$

because $2 + 90 + 3 = 95$.

Thus, by transitivity of order and equality,

$$\frac{1}{4}n^5 - 50n^3 + 3n + 12 \;\le\; \frac{261}{4}n^5 \quad \text{for every integer } n \ge 1.$$

In addition, by part (a)

$$\frac{1}{8}n^5 \;\le\; \frac{1}{4}n^5 - 50n^3 + 3n + 12 \quad \text{for every integer } n \ge 20,$$

so since $0 \le \frac{1}{8}n^5$, transitivity of order gives that

$$0 \;\le\; \frac{1}{4}n^5 - 50n^3 + 3n + 12 \;\le\; \frac{261}{4}n^5 \quad \text{for every integer } n \ge 20.$$

Let $B = \dfrac{261}{4}$ and $b = 20$. Then

$$0 \;\le\; \frac{1}{4}n^5 - 50n^3 + 3n + 12 \;\le\; Bn^5 \quad \text{for every integer } n \ge b,$$

and hence, by definition of O-notation, $\frac{1}{4}n^5 - 50n^3 + 3n + 12$ is $O(n^5)$.

c. *Solution 1*: Let $A = \dfrac{1}{8}$, $B = \dfrac{261}{4}$, and $k = 20$. By the results of parts (a) and (b),

$$An^5 \;\le\; \frac{1}{4}n^5 - 50n^3 + 3n + 12 \;\le\; Bn^5 \quad \text{for every integer } n \ge k,$$

and hence, by definition of Θ-notation, $\frac{1}{4}n^5 - 50n^3 + 3n + 12$ is $\Theta(n^5)$.

Solution 2: By part (d), $\frac{1}{4}n^5 - 50n^3 + 3n + 12$ is both $\Omega(n^5)$ and $O(n^5)$. Hence, by Theorem 11.2.1,

$$\frac{1}{4}n^5 - 50n^3 + 3n + 12 \text{ is } \Theta(n^5).$$

25. <u>Proof:</u> Suppose $P(n) = a_m n^m + a_{m-1}n^{m-1} + a_{m-2}n^{m-2} + \cdots + a_1 n + a_0$ where all the coefficients $a_0, a_1, \ldots, a_m$ are real numbers and $a_m > 0$.

a. ***Proof that $P(n)$ is $\Omega(n^m)$*):** According to the general procedure described in Example 11.2.4, we let

$$A = \frac{1}{2}a_m, \quad d = 2\left(\frac{|a_{m-1}| + |a_{m-2}| + \cdots + |a_2| + |a_1| + |a_0|}{a_m}\right), \text{ and } a = \max(d, 1).$$

Then $n \ge a$ implies that

$$n \ge 2\left(\frac{|a_{m-1}| + |a_{m-2}| + \cdots + |a_2| + |a_1| + |a_0|}{a_m}\right).$$

Multiplying both sides by $\frac{1}{2}a_m n^{m-1}$ gives

$$
\begin{aligned}
\frac{1}{2}a_m n^m \;\ge\;\; & (|a_{m-1}| + |a_{m-2}| + \cdots + |a_2| + |a_1| + |a_0|)n^{m-1} \\[4pt]
=\;\; & |a_{m-1}|n^{m-1} + |a_{m-2}|n^{m-1} + \cdots + |a_2|n^{m-1} + |a_1|n^{m-1} + |a_0|n^{m-1} \\[4pt]
\ge\;\; & |a_{m-1}|n^{m-1} + |a_{m-2}|n^{m-2} + \cdots + |a_2|n^2 + |a_1|n + |a_0| \\[2pt]
& \text{because } n^{m-1} \ge n^r \text{ for every } r \le m-1 \text{ since } n \ge 1 \\[4pt]
\ge\;\; & a_{m-1}n^{m-1} + a_{m-2}n^{m-2} + \cdots + a_2 n^2 + a_1 n + a_0.
\end{aligned}
$$

Thus, by transitivity of order and equality,

$$\left(\frac{a_m}{2}\right) n^m \geq |a_{m-1}| n^{m-1} + |a_{m-2}| n^{m-1} + \cdots + |a_1| n + |a_0| \quad \text{for every integer } n \geq a.$$

Subtracting the right side from the left gives

$$\left(\frac{a_m}{2}\right) n^m - |a_{m-1}| n^{m-1} - |a_{m-2}| n^{m-1} - \cdots - |a_1| n - |a_0| \geq 0 \quad \text{for every integer } n \geq a.$$

It follows that

$$\left(\frac{a_m}{2}\right) n^m + |a_{m-1}| n^{m-1} + |a_{m-2}| n^{m-2} + \cdots + |a_1| n + |a_0| \geq 0 \quad \text{for every integer } n \geq a$$

because, by definition of absolute value, each $-|a_i| \leq a_i$. Adding $\left(\frac{a_m}{2}\right) n^m$ to both sides gives that

$$a_m n^m + |a_{m-1}| n^{m-1} + |a_{m-2}| n^{m-2} + \cdots + |a_1| n + |a_0| \geq \left(\frac{a_m}{2}\right) n^m \quad \text{for every integer } n \geq a.$$

Therefore, with $A = \dfrac{a_m}{2}$ and $a = \max(d, 1)$, we have that

$$a_m n^m + |a_{m-1}| n^{m-1} + |a_{m-2}| n^{m-2} + \cdots + |a_1| n + |a_0| \geq A n^m \quad \text{for every integer } n \geq a,$$

and so, by definition of Ω-notation, $a_m n^m + a_{m-1} n^{m-1} + a_{m-2} n^{m-2} + \cdots + a_1 n + a_0$ is $\Omega(n^m)$, which finishes a proof for the theorem on polynomial orders.

b. *Proof that $P(n)$ is $O(n^m)$*: Observe that for every integer $n \geq 1$,

$$a_m n^m + a_{m-1} n^{m-1} + a_{m-2} n^{m-2} + \cdots + a_1 n + a_0$$

$$\leq \quad |a_m| n^m + |a_{m-1}| n^{m-1} + |a_{m-2}| n^{m-2} + \cdots + |a_2| n^2 + |a_1| n + |a_0|$$

because by definition of absolute value each $a_i \leq |a_i|$

$$\leq \quad |a_m| n^m + |a_{m-1}| n^m + |a_{m-2}| n^m + \cdots + |a_2| n^m + |a_1| n^m + |a_0| n^m$$

by Theorem 11.2.2 since $n \geq 1$, $n^{m-i} \leq n^m$ for every integer i from 0 through m

$$= \quad (|a_m| + |a_{m-1}| + |a_{m-2}| + \cdots + |a_2| + |a_1| + |a_0|) n^m$$

Let $B = |a_m| + |a_{m=1}| + |a_{m=2}| + \cdots + |u_2| + |u_1| + |u_0|$. Then, by transitivity of order and equality,

$$a_m n^m + a_{m-1} n^{m-1} + a_{m-2} n^{m-2} + \cdots + a_1 n + a_0 \leq B n^m \quad \text{for every integer } n \geq 1.$$

In addition, by part (a), there exists a positive real number a such that

$$\frac{a_m}{2} n^m \leq a_m n^m + a_{m-1} n^{m-1} + a_{m-2} n^{m-2} + \cdots + a_1 n + a_0 \quad \text{for every integer } n \geq a.$$

Now $\dfrac{a_m}{2} n^m > 0$ because $a_m > 0$, and thus, transitivity of order gives that

$$0 \leq a_m n^m + a_{m-1} n^{m-1} + a_{m-2} n^{m-2} + \cdots + a_1 n + a_0 \quad \text{for every integer } n \geq a.$$

Let $b = \max(1, a)$. Then

$$0 \leq a_m n^m + a_{m-1} n^{m-1} + a_{m-2} n^{m-2} + \cdots + a_1 n + a_0 \leq B n^m \quad \text{for every integer } n \geq b,$$

and hence, by definition of O-notation, $a_m n^m + a_{m-1} n^{m-1} + a_{m-2} n^{m-2} + \cdots + a_1 n + a_0$ is $O(n^m)$.

    c. *Proof that $P(n)$ is $\Theta(n^m)$:* By parts (a) and (b),

$$a_m n^m + a_{m-1} n^{m-1} + a_{m-2} n^{m-2} + \cdots + a_1 n + a_0$$

is both $\Omega(n^m)$ and $O(n^m)$. Hence, by Theorem 11.2.1,

$$a_m n^m + a_{m-1} n^{m-1} + a_{m-2} n^{m-2} + \cdots + a_1 n + a_0 \quad \text{is} \quad \Theta(n^m).$$

26. $\dfrac{(n+1)(n-2)}{4} = \dfrac{1}{4}(n^2 + n - 2n - 2) = \dfrac{1}{4}(n^2 - n - 2) = \dfrac{1}{4}n^2 - \dfrac{1}{4}n - \dfrac{1}{2}$, which is $\Theta(n^2)$ by the theorem on polynomial orders.

27. $\dfrac{n}{3}\left(4n^2 - 1\right) = \dfrac{4}{3}n^3 - \dfrac{1}{3}n$, which is $\Theta(n^3)$ by the theorem on polynomial orders.

29. $\dfrac{n(n+1)(2n+1)}{6} = \dfrac{1}{6}[n(n+1)(2n+1)] = \dfrac{1}{6}[(n^2 + n)(2n+1)] = \dfrac{1}{6}(2n^3 + 3n^2 + n)$

         $= \dfrac{1}{3}n^3 + \dfrac{1}{2}n^2 + \dfrac{1}{6}n$, which is $\Theta(n^3)$ by the theorem on polynomial orders.

30. $\left(\dfrac{n(n+1)}{2}\right)^2 = \dfrac{n^2(n^2 + 2n + 1)}{4} = \dfrac{1}{4}n^4 + \dfrac{1}{2}n^3 + \dfrac{1}{4}n^2$, which is $\Theta(n^4)$ by the theorem on polynomial orders.

32. By exercise 10 of Section 5.2,

$$1^2 + 2^2 + 3^2 + \cdots + n^2 = \frac{n(n+1)(2n+1)}{6},$$

which is $\Theta(n^3)$ by exercise 29 above. Hence $1^2 + 2^2 + 3^2 + \cdots + n^2$ is $\Theta(n^3)$.

33. By exercise 11 of Section 5.2, $1^3 + 2^3 + 3^3 + \cdots + n^3 = \left(\dfrac{n(n+1)}{2}\right)^2$, and by exercise 30 above this is $\Theta(n^4)$. Hence $1^3 + 2^3 + 3^3 + \cdots + n^3$ is $O(n^4)$.

34. Note that

$$
\begin{aligned}
2 + 4 + 6 + \cdots + 2n \;&=\; 2(1 + 2 + 3 + \cdots + n) && \text{by factoring out a 2} \\[4pt]
&=\; 2\left(\frac{n(n+1)}{2}\right) && \text{by Theorem 5.2.1} \\[4pt]
&=\; n^2 + n && \text{by algebra,}
\end{aligned}
$$

and so, by the theorem on polynomial orders,

$$2 + 4 + 6 + \cdots + 2n \text{ is } \Theta(n^2).$$

36. Note that

$$
\begin{aligned}
\sum_{i=1}^{n}(4i - 9) \;&=\; 4\sum_{i=1}^{n} i - \sum_{i=1}^{n} 9 && \text{by Theorem 5.1.1} \\[4pt]
&=\; 4\left(\frac{n(n+1)}{2}\right) - \underbrace{(9 + 9 + \cdots + 9)}_{n \text{ terms}} && \text{by Theorem 5.2.1} \\[4pt]
&=\; 2n^2 + 2n - 9n && \text{by definition of multiplication} \\[4pt]
&=\; 2n^2 - 7n && \text{by algebra,}
\end{aligned}
$$

38. *Hint:* Use the result of exercise 13 from Section 5.2.

39. Note that

$$\sum_{k=1}^{n}(k^2 - 2k) = \sum_{k=1}^{n}k^2 - 2\sum_{k=1}^{n}k \qquad \text{by Theorem 5.1.1}$$

$$= \frac{n(n+1)(2n+1)}{6} - 2\left(\frac{n(n+1)}{2}\right) \qquad \text{by exercise 10, Section 5.2, \& Theorem 5.2.1}$$

$$= \frac{2n^3 + 3n^2 + n}{6} + n^2 + n \qquad \text{by definition of multiplication}$$

$$= \frac{1}{3}n^3 + \frac{3}{2}n^2 + \frac{7}{6}n \qquad \text{by algebra,}$$

and so, by the theorem on polynomial orders,

$$\sum_{k=1}^{n}(k^2 - 2k) \text{ is } \Theta(n^3).$$

40. **a.** <u>Proof:</u> Suppose c is a positive real number and f is a real-valued function defined on a set of nonnegative integers with $f(n) \geq 0$ for each integer n greater than or equal to a positive real number k. Now if we let $A = B = c$, we have that for each integer $n \geq k$,

$$Af(n) \leq cf(n) \leq Bf(n)$$

and so, by definition of Θ-notation, $cf(n)$ is $\Theta(f(n))$.

b. Let $c = 3$ and $f(n) = n$. Then f is a real-valued function and $f(n) \geq 0$ for each integer $n \geq 0$. So by part (a), $cf(n)$ is $\Theta(f(n))$, or, by substitution, $3n$ is $\Theta(n)$.

42. If a function f has the property that $f(n)$ is $\Theta(1)$, then, by definition of Θ-notation, there exist positive real numbers A, B, and k such that

$$A \cdot 1 \leq f(n) \leq B \cdot 1 \quad \text{for every integer } n \geq k.$$

In other words for $n \geq k$ all the numbers $f(n)$ are located between the two real numbers A and B. Geometrically speaking, for every value of n to the right of k, the graph of f lies between two horizontal lines, the lower at height A and the upper at height B.

43. By exercise 15, $\left\lfloor \frac{n+1}{2} \right\rfloor$ is $\Theta(n)$, and by exercise 40(b), $3n$ is also $\Theta(n)$. Thus $\left\lfloor \frac{n+1}{2} \right\rfloor + 3n$ is $\Theta(n)$ by Theorem 11.2.9(a).

44. By exercise 28, $\frac{n(n-1)}{2}$ is $\Theta(n^2)$, by exercise 17, $\left\lfloor \frac{n}{2} \right\rfloor$ is $\Theta(n)$, and by exercise 41 (with $f(n) = 1$), 1 is $\Theta(1)$. Now $n \leq n^2$ and $1 \leq n^2$ for each integer $n \geq 1$. Thus $\frac{n(n-1)}{2} + \left\lfloor \frac{n}{2} \right\rfloor + 1$ is $\Theta(n^2)$ by Theorem 11.2.9(c).

45. By exercise 17, $\left\lfloor \frac{n}{2} \right\rfloor$ is $\Theta(n)$, by exercise 40(b), $4n$ is $\Theta(n)$, and by exercise 41 (with $f(n) = 3$), 3 is $\Theta(1)$. Now $n \leq n$ and $1 \leq n$ for every integer $n \geq 1$. Thus $\left\lfloor \frac{n}{2} \right\rfloor + 4n + 3$ is $\Theta(n)$.

46. **a.** <u>Proof (by mathematical induction):</u> Let the property $P(m)$ be the sentence

$$\text{If } n \text{ is any integer with } n > 1, \text{ then } n^m > 1. \qquad \leftarrow P(m)$$

Show that $P(1)$ is true: We must show that if n is any integer with $n > 1$, then $n^1 > 1$. But this is true because $n^1 = n$ and $n > 1$. So $P(1)$ is true.

Show that for every integer $k \geq 1$, if $P(k)$ is true then $P(k+1)$ is true: Let k be any particular but arbitrarily chosen integer with $k \geq 1$, and suppose that

$$\text{If } n \text{ is any integer with } n > 1, \text{ then } n^k > 1. \qquad \leftarrow \quad \begin{array}{c} P(k) \\ \text{inductive hypothesis} \end{array}$$

We must show that

$$\text{If } n \text{ is any integer with } n > 1, \text{ then } n^{k+1} > 1. \qquad \leftarrow P(k+1)$$

So suppose n is any integer with $n > 1$. By inductive hypothesis, $n^k > 1$, and multiplying both sides by the positive number n gives $n \cdot n^k > n \cdot 1$, or, equivalently, $n^{k+1} > n$. Thus $n^{k+1} > n$ and $n > 1$, and so, by transitivity of order, $n^{k+1} > 1$ *[as was to be shown]*.

b. Proof: Suppose n is any integer with $n > 1$ and r and s are integers with $r < s$. Then $s - r$ is an integer with $s - r \geq 1$, and so, by part (a), $n^{s-r} > 1$. Multiplying both sides by n^r gives $n^r \cdot n^{s-r} > n^r \cdot 1$, and so, by the laws of exponents, $n^s > n^r$ *[as was to be shown]*.

47. **a.** The following proof is contained in Appendix B. An alternative proof that does not use mathematical induction is shown below it.

Proof (by mathematical induction): Let the property $P(m)$ be the sentence

$$\text{If } 0 < x \leq 1, \text{ then } x^m \leq 1. \qquad \leftarrow P(m)$$

Show that $P(1)$ is true: We must show that if $0 < x \leq 1$, then $x^1 \leq 1$. But $x \leq 1$ by assumption and $x^1 = x$. So $P(1)$ is true.

Show that for every integer $k \geq 1$, if $P(k)$ is true then $P(k+1)$ is true: Let k be any integer with $k \geq 1$, and suppose that

$$\text{If } 0 < x \leq 1, \text{ then } x^k \leq 1. \qquad \leftarrow \quad \begin{array}{c} P(k) \\ \text{inductive hypothesis} \end{array}$$

We must show that

$$\text{If } 0 < x \leq 1, \text{ then } x^{k+1} \leq 1. \qquad \leftarrow P(k+1)$$

So let x be any number with $0 < x \leq 1$. By inductive hypothesis, $x^k \leq 1$, and multiplying both sides of this inequality by the nonnegative number x gives $x \cdot x^k \leq x \cdot 1$. Thus, by the laws of exponents, $x^{k+1} \leq x$.

Then $x^{k+1} \leq x$ and $x \leq 1$, and hence, by the transitive property of order (T18 in Appendix A), $x^{k+1} \leq 1$.

Proof (without mathematical induction):

By exercise 21(a) of Section 11.1, for any positive integer n, the function f defined by $f(x) = x^m$ is increasing on the set of nonnegative real numbers. Hence if $0 < x \leq 1$, then $x^m \leq 1^m = 1$.

b. *Hint:* What is the contrapositive of the statement in part (a)?

48. Proof of Theorem 11.2.6(b): Let f and g be real-valued functions defined on the same set of nonnegative integers, and suppose there is a positive real number r such that $f(n) \geq 0$ and $g(n) \geq 0$ for each integer $n \geq r$. Suppose also that $g(n)$ is $O(f(n))$. We will show that $f(n)$ is $\Omega(g(n))$. By definition of O-notation, there are positive real numbers B and b such that $b \geq r$, and, for each integer $n \geq b$,

$$0 \leq g(n) \leq Bf(n).$$

Divide the right-hand inequality by B to obtain

$$\frac{1}{B}g(n) \leq f(n),$$

for each integer $n \geq b$. Let $A = 1/B$ and $a = b$. Then for each integer $n \geq a$,

$$Ag(n) \leq f(n)$$

and so $f(n)$ is $\Omega(g(n))$ by definition of Ω-notation.

50. **a.** <u>Proof of Theorem 11.2.8(a)</u>: Let f and g be real-valued functions defined on the same set of nonnegative integers, and suppose there is a positive real number r such that $f(n) \geq 0$ and $g(n) \geq 0$ for each $n \geq r$. Suppose also that $f(n)$ is $\Omega(g(n))$ and c is any positive real number. *[We will show that $cf(n)$ is $\Omega(g(n))$.]* By definition of Ω-notation, there are positive real numbers A and a such that $a \geq r$, and, for each integer $n \geq a$,

$$Ag(n) \leq f(n).$$

Multiply both sides of the inequality by c to obtain

$$cAg(n) \leq cf(n),$$

and let $A' = Ac$. Then A' is a positive real number because both A and c are positive real numbers.

Hence there are positive real numbers A' and a such that $a \geq r$, and, for each integer $n \geq a$,

$$A'g(n) \leq cf(n).$$

Thus $cf(n)$ is $\Omega(g(n))$ by definition of Ω-notation.

51. **a.** <u>Proof of Theorem 11.2.9(a)</u>: Let f_1, f_2, and g be real-valued functions defined on the same set of nonnegative integers, and suppose there is a positive real number r such that $f_1(n) \geq 0$, $f_2(n) \geq 0$, and $g(n) \geq 0$ for every integer $n \geq r$. Suppose also that $f_1(n)$ is $\Theta(g(n))$ and $f_2(n)$ is $\Theta(g(n))$. *[We will show that $(f_1(n) + f_2(n))$ is $\Theta(g(n))$.]* By definition of Θ-notation, there are positive real numbers A, B, A', B', k, and k' such that $k \geq r$, $k' \geq r$ and

$$Ag(n) \leq f_1(n) \leq Bg(n) \quad \text{for every integer } n \geq k$$

and

$$A'g(n) \leq f_2(n) \leq B'g(n) \quad \text{for every integer } n \geq k'.$$

Let $k'' = \max(k, k')$. Adding the two inequalities gives that

$$Ag(n) + A'g(n) \leq f_1(n) + f_2(n) \leq Bg(n) + B'g(n) \quad \text{for every integer } n \geq k'',$$

or, equivalently,

$$(A + A')g(n) \leq f_1(n) + f_2(n) \leq (B + B')g(n) \quad \text{for every integer } n \geq k''.$$

Let $A'' = A + A'$ and let $B'' = B + B'$. Then both A'' and B'' are positive real numbers and

$$A''g(n) \leq f_1(n) + f_2(n) \leq B''g(n) \quad \text{for every integer } n \geq k''.$$

Thus, $(f_1(n) + f_2(n))$ *is* $\Theta(g(n))$ by definition of Θ-notation.

b. <u>Proof of Theorem 11.2.8(b)</u>: Let f_1, f_2, g_1 and g_2 be real-valued functions defined on the same set of nonnegative integers, and suppose there is a positive real number r such that $f_1(n) \geq 0$, $f_2(n) \geq 0$, $g_1(n) \geq 0$, and $g_2(n) \geq 0$ for every integer $n \geq r$. Suppose that $f_1(n)$ is $\Theta(g_1(n))$ and $f_2(n)$ is $\Theta(g_2(n))$. *[We will show that $(f_1(n)f_2(n))$ is $\Theta(g_1(n)g_2(n))$.]* By definition of Θ-notation, there are positive real numbers A, B, A', B', k, and k' such that $k \geq r$, $k' \geq r$ and

$$Ag_1(n) \leq f_1(n) \leq Bg_1(n) \quad \text{for every integer } n \geq k$$

and

$$A'g_2(n) \leq f_2(n) \leq B'g_2(n) \quad \text{for every integer } n \geq k'.$$

Let $k'' = \max(k, k')$. Since all quantities are positive, we can multiply the two inequalities to obtain

$$Ag_1(n) \cdot A'g_2(n) \leq f_1(n) \cdot f_2(n) \leq Bg_1(n) \cdot B'g_2(n) \quad \text{for every integer } n \geq k'',$$

or, equivalently,

$$(AA')g_1(n)g_2(n) \leq f_1(n)f_2(n) \leq (BB')g_1(n)g_2(n) \quad \text{for every integer } n \geq k''.$$

Let $A'' = AA'$ and let $B'' = BB'$. Then both A'' and B'' are positive real numbers and

$$A''g_1(n)g_2(n) \leq f_1(n)f_2(n) \leq B''g_1(n)g_2(n) \quad \text{for every integer } n \geq k''.$$

Thus, $(f_1(n)f_2(n))$ *is* $\Theta(g_1(n)g_2(n))$ by definition of Θ-notation.

c. Proof of Theorem 11.2.9(c): Let f_1, f_2, g_1 and g_2 be real-valued functions defined on the same set of nonnegative integers, and suppose there is a positive real number r such that $f_1(n) \geq 0$, $f_2(n) \geq 0$, $g_1(n) \geq 0$, and $g_2(n) \geq 0$ for every integer $n \geq r$. If $f_1(n)$ is $\Theta(g_1(n))$ and $f_2(n)$ is $\Theta(g_2(n))$ and if there is a real number s so that $g_1(n) \leq g_2(n)$ for every integer $n \geq s$. *[We will show that $(f_1(n) + f_2(n))$ is $\Theta(g_2(n))$.]* By definition of Θ-notation, there are positive real numbers A, B, A', B', k, and k' such that $k \geq r$, $k' \geq r$ and

$$Ag_1(n) \leq f_1(n) \leq Bg_1(n) \quad \text{for every integer } n \geq k$$

and

$$A'g_2(n) \leq f_2(n) \leq B'g_2(n) \quad \text{for every integer } n \geq k'.$$

Let $k'' = \max(k, k', s)$. Adding the two inequalities and using the fact that all quantities are positive and that $g_1(n) \leq g_2(n)$ for every integer $n \geq s$ gives that for every integer $n \geq k''$

$$A'g_2(n) \leq Ag_1(n) + A'g_2(n) \leq f_1(n) + f_2(n) \leq Bg_1(n) + B'g_2(n) \leq (B + B')g_2(n).$$

Let $B'' = (B + B')$. Then

$$A'g_2(n) \leq f_1(n) + f_2(n) \leq B''g_2(n) \quad \text{for every integer } n \geq k'',$$

and so, $(f_1(n) + f_2(n))$ is $\Theta(g_2(n))$ by definition of Θ-notation.

Section 11.3

1. **a.** $\log_2(200) = \dfrac{\ln 200}{\ln 2} \cong 7.6$ nanoseconds $= 0.0000000076$ second

   d. $200^2 = 40,000$ nanoseconds $= 0.00004$ second

   e. $200^8 = 2.56 \times 10^{18}$ nanoseconds $\cong \dfrac{2.56 \times 10^{18}}{10^9 \cdot 60 \cdot 60 \cdot 24 \cdot (365.25)}$ years $\cong 81.1215$ years *[because there are 10^9 nanoseconds in a second, 60 seconds in a minute, 60 minutes in an hour, 24 hours in a day, and approximately 365.25 days in a year on average].*

2. **a.** When the input size is increased from m to $2m$, the number of operations increases from cm^2 to $c(2m)^2 = 4cm^2$.

   b. By part (a), the number of operations increases by a factor of $\dfrac{4cm^2}{cm^2} = 4$.

   c. When the input size is increased by a factor of 10 (from m to $10m$), the number of operations increases by a factor of $\dfrac{c(10m)^2}{cm^2} = \dfrac{100cm^2}{cm^2} = 100$.

3. **a.** When the input size is increased from m to $2m$, the number of operations increases from cm^3 to $c(2m)^3 = 8cm^3$.

   b. By part (a), the number of operations increases by a factor of $\dfrac{8cm^3}{cm^3} = 8$.

   c. When the input size is increased by a factor of 10 (from m to $10m$), the number of operations increases by a factor of $\dfrac{c(10m)^3}{cm^3} = \dfrac{1000cm^3}{cm^3} = 1000$.

6. **a.** There are two multiplications, one addition, and one subtraction for each iteration of the loop, so there are four times as many operations as there are iterations of the loop. The loop is iterated $(n-1) - 3 + 1 = n - 3$ times (since the number of iterations equals the top minus the bottom index plus 1). Thus the total number of operations is $4(n-3) = 4n - 12$.

   b. By the theorem on polynomial orders, $4n - 12$ is $\Theta(n)$, so the algorithm segment has order n.

8. **a.** There is one addition for each iteration of the loop, and there are $\lfloor n/2 \rfloor$ iterations of the loop.

   b. Because

$$\lfloor n/2 \rfloor = \begin{cases} n/2 & \text{if } n \text{ is even} \\ (n-1)/2 & \text{if } n \text{ is odd,} \end{cases} = \begin{cases} \frac{1}{2}n & \text{if } n \text{ is even} \\ \frac{1}{2}n - \frac{1}{2} & \text{if } n \text{ is odd,} \end{cases}$$

then $\lfloor n/2 \rfloor$ is $\Theta(n)$ by theorem on polynomial orders. So the algorithm segment has order n.

9. **a.** There are $2n$ iterations of the inner loop for each iteration of the outer loop, and there are n iterations of the outer loop, as illustrated in the following table:

k	1				2				$\cdots$	n			
i	1	2	$\cdots$	$2n$	1	2	$\cdots$	$2n$	$\cdots$	1	2	$\cdots$	$2n$

$$\underbrace{\qquad}_{2n} \qquad \underbrace{\qquad}_{2n} \qquad \underbrace{\qquad}_{2n}$$

Therefore, the number of iterations of the inner loop is $2n \cdot n = 2n^2$. Now for each iteration of the inner loop, there are two operations: one multiplication and one addition. Thus, the total number of elementary operations that must be performed when the algorithm is executed is $2 \cdot 2n^2 = 4n^2$.

   b. Since $4n^2$ is $\Theta(n^2)$ (by the theorem on polynomial orders), the algorithm segment has order n^2.

11. **a.** There is one addition for each iteration of the inner loop. The number of iterations in the inner loop equals the number of columns in the table below, which shows the values of k and j for which the inner loop is executed.

k	1		2			3				$\cdots$	$n-1$			
j	1	2	1	2	3	1	2	3	4	$\cdots$	1	2	3	$\cdots$	n

$$\underbrace{\qquad}_{2} \quad \underbrace{\qquad}_{3} \quad \underbrace{\qquad}_{4} \qquad \underbrace{\qquad}_{n}$$

So, by Theorem 5.2.1, the total number of iterations of the inner loop is

$$2 + 3 + \cdots + n = (1 + 2 + 3 + \cdots + n) - 1 = \frac{n(n+1)}{2} - 1 = \frac{n^2 + n}{2} - 1 = \frac{1}{2}n^2 + \frac{1}{2}n - 1.$$

Because one operation is performed for each iteration of the inner loop, the total number of operations is $\frac{1}{2}n^2 + \frac{1}{2}n - 1$.

   b. By the theorem on polynomial orders, $\frac{1}{2}n^2 + \frac{1}{2}n - 1$ is $\Theta(n^2)$, and so the algorithm segment has order n^2.

12. **a.** For each iteration of the inner loop there is one comparison. The number of iterations of the inner loop can be deduced from the following table, which shows the values of k and i for which the inner loop is executed.

k	1				2				$\cdots$	$n-2$		$n-1$
i	2	3	$\cdots$	n	3	4	$\cdots$	n	$\cdots$	$n-1$	n	n

$$\underbrace{\qquad\qquad}_{n-1} \quad \underbrace{\qquad\qquad}_{n-2} \quad \underbrace{\qquad}_{2} \quad \underbrace{\ }_{1}$$

Therefore, by Theorem 5.2.1, the number of iterations of the inner loop is

$$(n-1) + (n-2) + \cdots + 2 + 1 = \frac{n(n-1)}{2}.$$

The total number of elementary operations that must be performed when the algorithm is executed is the number performed during each iteration of the inner loop times the number of iterations of the inner loop:

$$1 \cdot \left(\frac{n(n-1)}{2}\right) = \frac{1}{2}n^2 - \frac{1}{2}n.$$

b. By the theorem on polynomial orders, $\frac{1}{2}n^2 - \frac{1}{2}n$ is $\Theta(n^2)$, and so the algorithm segment has order n^2.

14. **a.** There is one addition for each iteration of the inner loop, and there is one additional addition and one multiplication for each iteration of the outer loop. The number of iterations in the inner loop equals the number of columns in the following table, which shows the values of i and j for which the inner loop is executed.

i	1	2		3			$\cdots$	n			
j	1	1	2	1	2	3	$\cdots$	1	2	3	$\cdots$	n

$$\underbrace{\ }_{1} \quad \underbrace{\ }_{2} \quad \underbrace{\qquad}_{3} \quad \underbrace{\qquad\qquad}_{n}$$

So, by Theorem 5.2.1, the total number of iterations of the inner loop is

$$1 + 2 + 3 + \cdots + n = (1 + 2 + 3 + \cdots + n) = \frac{n(n+1)}{2} = \frac{n^2 + n}{2} = \frac{1}{2}n^2 + \frac{1}{2}n.$$

Because one addition is performed for each iteration of the inner loop, the number of operations performed when the inner loop is executed is $\frac{1}{2}n^2 + \frac{1}{2}$. Now an additional two operations are performed each time the outer loop is executed, and because the outer loop is executed n times, this gives an additional $2n$ operations. Therefore, the total number of operations is

$$\frac{1}{2}n^2 + \frac{1}{2}n + 2n = \frac{1}{2}n^2 + \frac{5}{2}n.$$

b. By the theorem on polynomial orders, $\frac{1}{2}n^2 + \frac{5}{2}n$ is $\Theta(n^2)$, and so the algorithm segment has order n^2.

15. **a.** There are three multiplications for each iteration of the inner loop, and there is one additional addition for each iteration of the outer loop. The number of iterations of the inner loop can be deduced from the following table, which shows the values of i and j for which the inner loop is executed.

i	1				2				$\cdots$	$n-2$		$n-1$
j	2	3	$\cdots$	n	3	4	$\cdots$	n	$\cdots$	$n-1$	n	n

$$\underbrace{\qquad\qquad}_{n-1} \quad \underbrace{\qquad\qquad}_{n-2} \quad \underbrace{\qquad}_{2} \quad \underbrace{\ }_{1}$$

Hence, by Theorem 5.2.1, the total number of iterations of the inner loop is

$$(n-1) + (n-2) + \cdots + 2 + 1 = \frac{n(n-1)}{2}.$$

Because three multiplications are performed for each iteration of the inner loop, the number of operations that are performed when the inner loop is executed is

$$3 \cdot \frac{n(n-1)}{2} = \frac{3}{2}(n^2 - n) = \frac{3}{2}n^2 - \frac{3}{2}n.$$

Now an additional operation is performed each time the outer loop is executed, and because the outer loop is executed n times, this gives an additional n operations. Therefore, the total number of operations is

$$\left(\frac{3}{2}n^2 - \frac{3}{2}n\right) + n = \frac{3}{2}n^2 - \frac{1}{2}n.$$

b. By the theorem on polynomial orders, $\frac{3}{2}n^2 - \frac{1}{2}n$ is $\Theta(n^2)$, and so the algorithm segment has order n^2.

17. **a.** There are two subtractions and one multiplication for each iteration of the inner loop. If n is odd, the number of iterations of the inner loop equals the number of columns in the following table, which shows the values of i and j for which the inner loop is executed.

i	1	2	3		4		5			6			...	$n-1$		...			n			...	
$\frac{i+1}{2}$	1	1	2		2		3			3			...	$\frac{n-1}{2}$		...			$\frac{n+1}{2}$			...	
j	1	1	1	2	1	2	1	2	3	1	2	3	...	1	2	...	$\frac{n-1}{2}$		1	2	...		$\frac{n+1}{2}$

$$\underbrace{\qquad}_{1}\quad\underbrace{\qquad}_{1}\quad\underbrace{\qquad}_{2}\quad\underbrace{\qquad}_{2}\quad\underbrace{\qquad}_{3}\quad\underbrace{\qquad}_{3}\qquad\underbrace{\qquad}_{\frac{n-1}{2}}\qquad\underbrace{\qquad}_{\frac{n+1}{2}}$$

Thus the number of iterations of the inner loop is

$$1 + 1 + 2 + 2 + \cdots + \frac{n-1}{2} + \frac{n-1}{2} + \frac{n+1}{2}$$

$$= 2 \cdot \left(1 + 2 + 3 + \cdots + \frac{n-1}{2}\right) + \frac{n+1}{2}$$

$$= 2 \cdot \frac{\frac{n-1}{2}\left(\frac{n-1}{2}+1\right)}{2} + \frac{n+1}{2} \qquad \text{by Theorem 5.2.1}$$

$$= \frac{n^2 - 2n + 1}{4} + \frac{n-1}{2} + \frac{n-1}{2}$$

$$= \frac{1}{4}n^2 + \frac{1}{2}n + \frac{1}{4}.$$

By similar reasoning, if n is even, then the number of iterations of the inner loop is

$$1 + 1 + 2 + 2 + 3 + 3 + \cdots + \frac{n}{2} + \frac{n}{2} = 2 \cdot \left(1 + 2 + 3 + \cdots + \frac{n}{2}\right)$$

$$= 2 \cdot \left(\frac{\frac{n}{2}\left(\frac{n}{2}+1\right)}{2}\right) \qquad \text{by Theorem 5.2.1}$$

$$= \frac{n^2}{4} + \frac{n}{2}.$$

Because three operations are performed for each iteration of the inner loop, the answer is $3\left(\frac{n^2}{4} + \frac{n}{2}\right)$ when n is even and $3\left(\frac{1}{4}n^2 + \frac{1}{2}n + \frac{1}{4}\right)$ when n is odd.

b. Since $3\left(\frac{n^2}{4} + \frac{n}{2}\right)$ is $\Theta(n^2)$ and $3\left(\frac{1}{4}n^2 + \frac{1}{2}n + \frac{1}{4}\right)$ is also $\Theta(n^2)$ (by the theorem on polynomial orders), this algorithm segment has order n^2.

18. **a.** There is one multiplication for each iteration of the inner loop. If n is odd, the number of iterations of the inner loop can be deduced from the following table, which shows the values of

i and j for which the inner loop is executed. (In calculating values, note that when $i = n - 2$, $\left\lfloor \frac{i+1}{2} \right\rfloor = \left\lfloor \frac{(n-2)+1}{2} \right\rfloor = \left\lfloor \frac{n-1}{2} \right\rfloor = \frac{n-1}{2}$, when $i = n - 1$, $\left\lfloor \frac{i+1}{2} \right\rfloor = \left\lfloor \frac{(n-1)+1}{2} \right\rfloor = \left\lfloor \frac{n}{2} \right\rfloor = \frac{n-1}{2}$, and when $i = n$, $\left\lfloor \frac{i+1}{2} \right\rfloor = \left\lfloor \frac{n+1}{2} \right\rfloor = \frac{n+1}{2}$.)

i	1			2			$\cdots$	$n-2$				$n-1$				n			
$\left\lfloor \frac{i+1}{2} \right\rfloor$	1			1			$\cdots$	$\frac{n-1}{2}$				$\frac{n-1}{2}$				$\frac{n+1}{2}$			
j	1	2	$\cdots$ n	1	2	$\cdots$ n	$\cdots$	$\frac{n-1}{2}$	$\frac{n-1}{2}+1$	$\cdots$	n	$\frac{n-1}{2}$	$\frac{n-1}{2}+1$	$\cdots$	n	$\frac{n+1}{2}$	$\frac{n+1}{2}+1$	$\cdots$	n

Braces: n ; n ; $n - \frac{n-1}{2} + 1 = \frac{n+3}{2}$; $n - \frac{n-1}{2} + 1 = \frac{n+3}{2}$; $n - \frac{n+1}{2} + 1 = \frac{n+1}{2}$

Thus, if n is odd, the number of iterations of the inner loop is

$$n + n + (n - 1) + (n - 1) + \cdots + \frac{n+3}{2} + \frac{n+3}{2} + \frac{n+1}{2}$$

$$= 2\left(n + (n-1) + \cdots + \frac{n+3}{2}\right) + \frac{n+1}{2}$$

$$= 2 \cdot \left(\sum_{k=1}^{n} k - \sum_{k=1}^{(n+1)/2} k\right) + \frac{n+1}{2} \qquad \text{because } \frac{n+3}{2} - 1 = \frac{n+1}{2}$$

$$= n(n+1) - \frac{n+1}{2}\left(\frac{n+1}{2} + 1\right) + \frac{n+1}{2} \qquad \text{by Theorem 5.2.1}$$

$$= \frac{4n(n+1)}{4} - \frac{(n+1)^2}{4}$$

$$= \frac{4n^2 + 4n - n^2 - 2n - 1}{4}$$

$$= \frac{3n^2 + 2n - 1}{4}$$

$$= \frac{3}{4}n^2 + \frac{1}{2}n - \frac{1}{4}.$$

Similarly, if n is even, the number of iterations of the inner loop can be deduced from the following table, which shows the values of i and j for which the inner loop is executed. (In calculating values, note that when $i = n - 2$, $\left\lfloor \frac{i+1}{2} \right\rfloor = \left\lfloor \frac{(n-2)+1}{2} \right\rfloor = \left\lfloor \frac{n-1}{2} \right\rfloor = \frac{n-2}{2}$, and when $i = n - 1$, $\left\lfloor \frac{i+1}{2} \right\rfloor = \left\lfloor \frac{(n-1)+1}{2} \right\rfloor = \left\lfloor \frac{n}{2} \right\rfloor = \frac{n}{2}$.)

i	1			2			$\cdots$	$n-2$				$n-1$				n			
$\left\lfloor \frac{i+1}{2} \right\rfloor$	1			1			$\cdots$	$\frac{n-2}{2}$				$\frac{n}{2}$				$\frac{n}{2}$			
j	1	2	$\cdots$ n	1	2	$\cdots$ n	$\cdots$	$\frac{n-1}{2}$	$\frac{n-1}{2}+1$	$\cdots$	n	$\frac{n}{2}$	$\frac{n-1}{2}+1$	$\cdots$	n	$\frac{n}{2}$	$\frac{n}{2}+1$	$\cdots$	n

Braces: n ; n ; $n - \frac{n-1}{2} + 1 = \frac{n+3}{2}$; $n - \frac{n}{2} + 1 = \frac{n+2}{2}$; $n - \frac{n}{2} + 1 = \frac{n+2}{2}$

Thus, if n is even, the number of iterations of the inner loop is

$$n + n + (n - 1) + (n - 1) + \cdots + \frac{n+2}{2} + \frac{n+2}{2}$$

$$= 2\left(n + (n-1) + \cdots + \frac{n+2}{2}\right)$$

$$= 2 \cdot \left(\sum_{k=1}^{n} k - \sum_{k=1}^{n/2} k\right) \qquad \text{because } \frac{n+2}{2} - 1 = \frac{n}{2}$$

$$= n(n+1) - \frac{n}{2}\left(\frac{n}{2} + 1\right) \qquad \text{by Theorem 5.2.1}$$

$$= \frac{4n(n+1)}{4} - \frac{n^2}{4} - \frac{2n}{4}$$

$$= \frac{4n^2 + 4n - n^2 - 2n}{4}$$

$$= \frac{3n^2 + 2n}{4}$$

$$= \frac{3}{4}n^2 + \frac{1}{2}n.$$

Because one operation is performed for each iteration of the inner loop, the answer is that, when n is odd,

$$1 \cdot \left(\frac{3}{4}n^2 + n - \frac{3}{4} \right) = \frac{3}{4}n^2 + n - \frac{3}{4}$$

elementary operations are performed, and, when n is even,

$$1 \cdot \left(\frac{3}{4}n^2 + n - 1 \right) = \frac{3}{4}n^2 + n - \frac{1}{2}n$$

elementary operations are performed.

b. By the theorem on polynomial orders, $\frac{3}{4}n^2 + n - \frac{3}{4}$ is $\Theta(n^2)$ and $\frac{3}{4}n^2 + \frac{1}{2}n$ is also $\Theta(n^2)$ and so this algorithm segment has order n^2.

19. *Hint:* See Section 9.6 for a discussion of how to count the number of iterations of the innermost loop.

20.

	a[1]	a[2]	a[3]	a[4]	a[5]
Initial order	6	2	1	8	4
Result of step 1	2	6	1	8	4
Result of step 2	1	2	6	8	4
Result of step 3	1	2	6	8	4
Final order	1	2	4	6	8

21.

	a[1]	a[2]	a[3]	a[4]	a[5]
initial order	7	3	6	9	5
result of step k = 2	3	7	6	9	5
result of step k = 3	3	6	7	9	5
result of step k = 4	3	6	7	9	5
result of step k = 5	3	5	6	7	9

22.

n	5										
a[1]	6	2			1						
a[2]	2	6		1	2						
a[3]	1		6					4			
a[4]	8						4	6			
a[5]	4					8					
k	2		3		4		5				
x	2		1		8		4				
j	1	0	2	1	0	3	0	4	3	2	0

24. There are seven comparisons between values of x and values of $a[j]$: one $k = 2$, two when $k = 3$, one when $k = 4$, and three when $k = 5$.

27. *Hint:* $E_n = \frac{1}{2}[3 + 4 + \cdots + (n+1)]$, which equals $\frac{1}{2}[(1 + 2 + 3 + \cdots + (n+1)) - (1 + 2)]$.

28. The top row of the table shows the initial values of the array, and the bottom row shows the final values. The results for executing each step in the for-next loop are shown in separate rows.

k	a[1]	a[2]	a[3]	a[4]	a[5]
Initial	7	3	8	4	2
1	2	3	8	4	7
2	2	3	8	4	7
3	2	3	4	8	7
4	2	3	4	7	8
5	2	3	4	7	8

30.

n	5												
a[1]	7	3			2								
a[2]	3	7											
a[3]	8									4			
a[4]	4									8		7	
a[5]	2						7					8	
k	1				2			3		4			
IndexOfMin	1	2		5		2		3	4	4	5		
i	2		3	4	5		3	4	5	4	5		5
temp						7				8			7

32. There is one comparison for each combination of values of k and i: namely, $4 + 3 + 2 + 1 = 10$.

33. As i goes from $k + 1$ to 5 through $5 - (k + 1) + 1 = 5 - k$ values (where k goes from 1 to 4), the number of comparisons is

$$(5 - 1) + (5 - 2) + (5 - 3) + (5 - 4) = 4 + 3 + 2 + 1 = 10.$$

35. **b.** $n - 3 + 1 = n - 2$ **d.** *Hint:* The answer is n^2.

36.

n	3								
a[0]	2								
a[1]	1								
a[2]	−1								
a[3]	3								
x	2								
polyval	2	4				0			24
i	1	2				3			
term	1	2	−1	−2	−4	3	6	12	24
j	1		1	2		1	2	3	

38. Number of multiplications = number of iterations of the inner loop

$$= 1 + 2 + 3 + \cdots + n = \frac{n(n+1)}{2} \quad \text{by Theorem 5.2.1}$$

Number of additions = number of iterations of the outer loop $= n$

Hence the total number of multiplications and additions is $\dfrac{n(n+1)}{2} + n = \dfrac{1}{2}n^2 + \dfrac{3}{2}n.$

39. By the result of exercise 38, $s_n = \dfrac{1}{2}n^2 + \dfrac{3}{2}n$, which is $\Theta(n^2)$ by the theorem on polynomial orders.

40.

n	3			
$a[0]$	2			
$a[1]$	1			
$a[2]$	-1			
$a[3]$	3			
x	2			
$polyval$	3	5	11	24
i	1	2	3	

42. There are two operations (one addition and one multiplication) per iteration of the loop, and there are n iterations of the loop. Therefore, $t_n = 2n.$

Section 11.4

1.

x	$f(x) = 3^x$
0	$3^0 = 1$
1	$3^1 = 3$
2	$3^2 = 9$
-1	$3^{-1} = 1/3$
-2	$3^{-2} = 1/9$
$1/2$	$3^{1/2} \cong 1.7$
$-(1/2)$	$3^{-(1/2)} \cong 0.6$

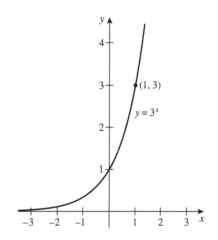

3.

x	$h(x) = \log_{10} x$
1	0
10	1
100	2
1/10	-1
1/100	-2

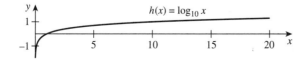

5.

x	$\lfloor \log_2 x \rfloor$
$1 \leq x < 2$	0
$2 \leq x < 4$	1
$4 \leq x < 8$	2
$8 \leq x < 16$	3
$1/2 \leq x < 1$	-1
$1/4 \leq x < 1/2$	-2

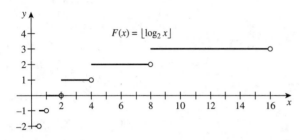

6.

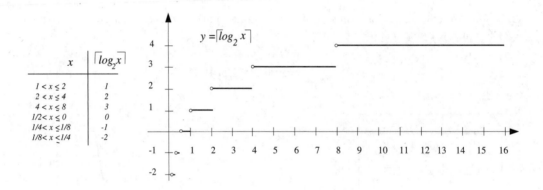

x	$\lceil \log_2 x \rceil$
$1 < x \leq 2$	1
$2 < x \leq 4$	2
$4 < x \leq 8$	3
$1/2 < x \leq 0$	0
$1/4 < x \leq 1/8$	-1
$1/8 < x < 1/4$	-2

7.

x	$x \log_2 x$
1	$1. 0 = 0$
2	$2 \cdot 1 = 2$
4	$4 \cdot 2 = 8$
8	$8 \cdot 3 = 24$
1/8	$(1/8) \cdot (-3) = -3/8$
1/4	$(1/4) \cdot (-2) = -1/2$
3/8	$(3/8) \cdot (\log_2(3/8)) \cong -0.53$

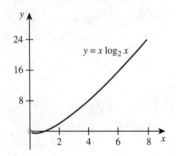

9. The distance above the axis is

$$(2^{64} \text{ units}) \cdot \left(\frac{1 \text{ inch}}{4 \text{ unit}}\right) = \frac{2^{64}}{4} \text{ inches} = \frac{2^{64}}{4 \cdot 12 \cdot 5280} \text{ miles} \cong 72{,}785{,}448{,}520{,}000 \text{ miles}.$$

The ratio of the height of the point to the average distance of the earth to the sun is approximately $72785448520000/93000000 \cong 782{,}639$. (If you perform the computation using metric units and the approximation 0.635 cm $\cong 1/4$ inch, the ratio comes out to be approximately $780{,}912$.)

10. **b.** By definition of logarithm, $\log_b x$ is the exponent to which b must be raised to obtain x. Thus when b is actually raised to this exponent, x is obtained. That is, $b^{\log_b(x)} = x$.

11. **b.**

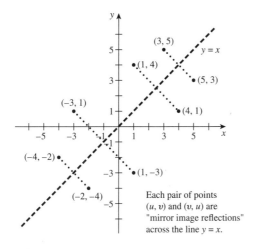

Each pair of points (u, v) and (v, u) are "mirror image reflections" across the line $y = x$.

12. When $\frac{1}{2} < x < 1$, then $-1 < \log_2 x < 0$.

When $\frac{1}{4} < x < \frac{1}{2}$, then $-2 < \log_2 x < -1$.

When $\frac{1}{8} < x < \frac{1}{4}$, then $-3 < \log_2 x < -2$.

And so forth.

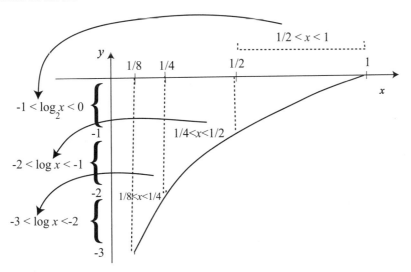

13. *Hints:* (1) $\lfloor \log_{10} x \rfloor = m$, (2) See Example 11.4.1.

15. No. Counterexample: Let $n = 2$. Then $\lceil \log_2(n-1) \rceil = \lceil \log_2 1 \rceil = \lceil 0 \rceil = 0$, whereas $\lceil \log_2 n \rceil = \lceil \log_2 2 \rceil = \lceil 11 \rceil = 1$.

16. *Hint:* The statement is true.

17.

18. $\lfloor \log_2 148206 \rfloor + 1 = 18$

21. **a.**

$$a_1 \;=\; 1$$

$$\left.\begin{aligned}
a_2 &= a_{\lfloor 2/2 \rfloor} + 2 = a_1 + 2 = 1 + 2 \\
a_3 &= a_{\lfloor 3/2 \rfloor} + 2 = a_1 + 2 = 1 + 2
\end{aligned}\right\}$$

$$\left.\begin{aligned}
a_4 &= a_{\lfloor 4/2 \rfloor} + 2 = a_2 + 2 = (1+2) + 2 = 1 + 2 \cdot 2 \\
a_5 &= a_{\lfloor 5/2 \rfloor} + 2 = a_2 + 2 = (1+2) + 2 = 1 + 2 \cdot 2 \\
a_6 &= a_{\lfloor 6/2 \rfloor} + 2 = a_3 + 2 = (1+2) + 2 = 1 + 2 \cdot 2 \\
a_7 &= a_{\lfloor 7/2 \rfloor} + 2 = a_3 + 2 = (1+2) + 2 = 1 + 2 \cdot 2
\end{aligned}\right\}$$

$$\left.\begin{aligned}
a_8 &= a_{\lfloor 8/2 \rfloor} + 2 = a_4 + 2 = (1 + 2 \cdot 2) + 2 = 1 + 3 \cdot 2 \\
a_9 &= a_{\lfloor 9/2 \rfloor} + 2 = a_4 + 2 = (1 + 2 \cdot 2) + 2 = 1 + 3 \cdot 2 \\
&\;\;\vdots \\
a_{15} &= a_{\lfloor 15/2 \rfloor} + 2 = a_7 + 2 = (1 + 2 \cdot 2) + 2 = 1 + 3 \cdot 2
\end{aligned}\right\}$$

$$\left.\begin{aligned}
a_{16} &= a_{\lfloor 16/2 \rfloor} + 2 = a_8 + 2 = (1 + 3 \cdot 2) + 2 = 1 + 4 \cdot 2 \\
a_{17} &= a_{\lfloor 17/2 \rfloor} + 2 = a_8 + 2 = (1 + 3 \cdot 2) + 2 = 1 + 4 \cdot 2 \\
&\;\;\vdots
\end{aligned}\right\}$$

Guess: $a_n \;=\; 1 + 2 \lfloor \log_2 n \rfloor$

b. Proof (by strong mathematical induction): Suppose the sequence $a_1, a_2, a_3, \ldots$ is defined recursively as follows: ~~be a sequence that satisfies the recurrence relation~~

$$a_k = a_{\lfloor k/2 \rfloor} + 2 \text{ for every integer } k \geq 2, \text{ with initial condition } a_1 = 1,$$

and let the property $P(n)$ be the equation

$$a_n = 1 + 2 \lfloor \log_2 n \rfloor. \qquad \leftarrow P(n)$$

We will show by strong mathematical induction that $P(n)$, is true for each integer $n \geq 1$.

Show that $P(1)$ is true: $P(1)$ is the equation $1 + 2_{\lfloor \log_2 1 \rfloor} = 1 + 2 \cdot 0 = 1$, and 1 is the value of a_1.

Show that for any integer $k \geq 1$, if $P(i)$ is true for every integer i from 1 through k, then $P(k + 1)$ is true: Let k be any integer with $k \geq 1$, and suppose that

$$a_i = 1 + 2_{\lfloor \log_2 i \rfloor} \text{ for every integer } i \text{ with } 1 \leq i \leq k. \quad \leftarrow \begin{array}{l} \text{inductive} \\ \text{hypothesis} \end{array}$$

We must show that

$$a_{k+1} = 1 + 2 \lfloor \log_2(k+1) \rfloor$$

Case 1 (k is odd): In this case $k + 1$ is even, and

$$\begin{aligned}
a_{k+1} &= a_{\lfloor (k+1)/2 \rfloor} + 2 && \text{by the recursive definition of } a_1, a_2, a_3, \ldots \\
&= a_{(k+1)/2} + 2 && \text{by Theorem 4.6.2 because } k + 1 \text{ is even} \\
&= 1 + 2 \lfloor \log_2((k+1)/2) \rfloor + 2 && \text{by inductive hypothesis} \\
&= 3 + 2 \lfloor \log_2(k+1) - \log_2 2 \rfloor && \text{by Theorem 7.2.1(b)} \\
&= 3 + 2 \lfloor \log_2(k+1) - 1) \rfloor && \text{because } \log_2 2 = 1 \\
&= 3 + 2 \lfloor (\log_2(k+1) \rfloor - 1) && \begin{array}{l} \text{because for every real number x, } \lfloor x - 1 \rfloor = \lfloor x \rfloor - 1 \\ \text{by exercise 15, Section 4.6} \end{array} \\
&= 1 + 2 \lfloor \log_2(k+1) \rfloor && \text{by algebra.}
\end{aligned}$$

Case 2 (k is even): In this case $k + 1$ is odd, and

$$
\begin{aligned}
a_{k+1} &= a_{\lfloor (k+1)/2 \rfloor} + 2 && \text{by the recursive definition of } a_1, a_2, a_3, \ldots \\
&= a_{k/2} + 2 && \text{by Theorem 4.6.2 because } k + 1 \text{ is odd} \\
&= 1 + 2 \lfloor \log_2 (k/2) \rfloor + 2 && \text{by inductive hypothesis} \\
&= 3 + 2 \lfloor \log_2 k - \log_2 2 \rfloor && \text{by Theorem 7.2.1(b)} \\
&= 3 + 2 \lfloor \log_2 k - 1 \rfloor && \text{because } \log_2 2 = 1 \\
&= 3 + 2(\lfloor \log_2 k \rfloor - 1) && \text{because for every real number x, } \lfloor x - 1 \rfloor = \lfloor x \rfloor - 1 \\
& && \text{by exercise 15, Section 4.6} \\
&= 1 + 2 \lfloor \log_2 k \rfloor && \text{by algebra} \\
&= 1 + 2 \lfloor \log_2 (k + 1) \rfloor && \text{by property 11.4.3}
\end{aligned}
$$

Thus in either case, $a_{k+1} = 1 + 2 \lfloor \log_2(k + 1) \rfloor$ *[as was to be shown].*

23. *Hint:* When $k \geq 2$, then $k^2 \geq 2k$, and so $k \leq \frac{k^2}{2}$. Hence $\frac{k^2}{2} + k \leq \frac{k^2}{2} + \frac{k^2}{2} = k^2$. Also, when $k \geq 2$ then $k^2 > 1$, and so $\frac{1}{2} < \frac{k^2}{2}$. Consequently, $\frac{k^2}{2} + \frac{1}{2} < \frac{k^2}{2} + \frac{k^2}{2} = k^2$.

24. <u>Proof (by strong mathematical induction):</u> Let $c_1, c_2, c_3, \ldots .$ be a sequence that satisfies the recurrence relation

$$
c_k = 2c_{\lfloor k/2 \rfloor} + k \text{ for every integer } k \geq 2, \text{ with initial condition } c_1 = 0,
$$

and let the property $P(n)$ be the inequality

$$
c_n \leq n \log_2 n. \qquad \leftarrow P(n)
$$

Show that P(1) is true: For $n = 1$ the inequality states that $c_1 \leq 1 \cdot \log_2 1 = 1 \cdot 0 = 0$, which is true because $c_1 = 0$. So $P(1)$ is true.

Show that if $k \geq 1$ and $P(i)$ is true for every integer i from 1 through k, then $P(k + 1)$ is true: Let k be any integer with $k \geq 1$, and suppose that

$$
c_i \leq i \log_2 i \text{ for every integer } i \text{ with } 1 \leq i \leq k. \qquad \leftarrow \text{ inductive hypothesis}
$$

We must show that

$$
c_{k+1} \leq (k + 1) \log_2(k + 1).
$$

First note that because k is greater than 1 and by definition of floor,

$$
1 \leq \left\lfloor \frac{k + 1}{2} \right\rfloor \leq \frac{k + 1}{2}.
$$

Also, because k is an integer with $k \geq 1$, we have

$$
1 \leq k \Rightarrow k + 1 \leq k + k \Rightarrow k + 1 \leq 2k \Rightarrow \frac{k + 1}{2} \leq k.
$$

Thus, by the transitive property of order,

$$
\left\lfloor \frac{k + 1}{2} \right\rfloor \leq k.
$$

Then
$$c_{k+1} = 2c_{\lfloor (k+1)/2 \rfloor} + (k+1)$$

by definition of $c_1, c_2, c_3, \ldots$

$$\leq 2\left(\left\lfloor \frac{k+1}{2} \right\rfloor \log_2 \left\lfloor \frac{k+1}{2} \right\rfloor\right) + (k+1)$$

by inductive hypothesis because $\left\lfloor \frac{k+1}{2} \right\rfloor \leq k$

$$\leq (k+1) \log_2\left(\frac{k+1}{2}\right) + (k+1)$$

since $1 \leq \left\lfloor \frac{k+1}{2} \right\rfloor \leq \frac{k+1}{2}$, we have by property (11.4.1)
that $\log_2 \left\lfloor \frac{k+1}{2} \right\rfloor \leq \log_2 \left(\frac{k+1}{2}\right)$

$$= (k+1)\left[\log_2(k+1) - \log_2 2\right] + (k+1)$$

by Theorem 7.2.1(b)

$$= (k+1)\left[\log_2(k+1) - 1\right] + (k+1)$$

because $\log_2 2 = 1$

$$= (k+1)\log_2(k+1)$$

by algebra,

Therefore, by transitivity of equality and order, $c_{k+1} \leq (k+1)\log_2(k+1)$ *[as was to be shown]*.

25. *Solution 1:* One way to solve this problem is to compare values for $\log_2 x$ and $x^{1/10}$ for conveniently chosen, large values of x. For instance, if powers of 10 are used, the following results are obtained:

$$\log_2(10^{10}) = 10\log_2 10 \cong 33.2 \quad \text{and} \quad (10^{10})^{1/10} = 10^{10 \cdot (1/10)} = 10^1 = 10.$$

Thus, when $x = 10^{10}$, it is not the case that $\log_2(x) < (x)^{1/10}$.

However, since

$$\log_2(10^{20}) = 20\log_2 10 = 66.4 \quad \text{and} \quad (10^{20})^{1/10} = 10^{20 \cdot (1/10)} = 10^2 = 100,$$

and since $66.4 < 100$, it is the case that $\log_2(x) < (x)^{1/10}$ when $x = 10^{20}$.

Solution 2: Another approach is to use a graphing calculator or computer to sketch graphs of $y = \log_2 x$ and $y = x^{1/10}$, taking seriously the hint to "think big" in choosing the interval size for the x's. A few tries and use of the zoom and trace features make it appear that the graph of $y = x^{1/10}$ crosses above the graph of $y = \log_2 x$ at about 4.9155×10^{17}. Thus, for values of x larger than this, $x^{1/10} > \log_2 x$.

27. By Theorem 11.2.7, n is $\Theta(n)$ and $\log_2 n$ is $\Theta(\log_2 n)$, and, by Theorem 11.2.8(c), $2n$ is $\Theta(n)$. In addition, by property 11.4.9, there is a positive real number s such that for each integer $n \geq s$, $\log_2 n \leq n$. Finally, if n is any integer with $n \geq 1$, then $n \geq 0$. Thus it follows from Theorem 11.2.9(c) that $2n + \log_2 n$ is $\Theta(n)$.

29. By Theorem 11.2.7, n^2 is $\Theta(n^2)$ and 2^n is $\Theta(2^n)$. In addition, by property 11.4.10, there is a positive real number s such that for each integer $n \geq s$, $n^2 \leq 2^n$. Finally, if n is any integer, then $2^n \geq 0$. Thus it follows from Theorem 11.2.9(c) that $n^2 + 2^n$ is $\Theta(2^n)$.

30. For every integer $n > 0$,
$$2^n \leq 2^{n+1} \leq 2 \cdot 2^n.$$

Thus, let $A = 1$, $B = 2$, and $k = 0$. Then

$$A \cdot 2^n \leq 2^{n+1} \leq B \cdot 2^n \quad \text{for every integer } n > k,$$

and so, by definition of Θ-notation, 2^{n+1} is $\Theta(2^n)$.

31. *Hint:* Use a proof by contradiction. Start by supposing that 4^n is $O(2^n)$. That is, that there are positive real numbers B and b such that $O \leq 4^n \leq B \cdot 2^n$ for every real number $n > b$ and use the fact that $\frac{4^n}{2^n} = \left(\frac{4}{2}\right)^n = 2^n$ to obtain a contradiction.

32. By Theorem 5.2.2, for each integer $n \geq 0$,

$$1 + 2 + 2^2 + \cdots + 2^n = \frac{2^{n+1} - 1}{2 - 1} = 2^{n+1} - 1.$$

Also,

$$2^n \leq 2^{n+1} - 1 \leq 2^{n+1} = 2 \cdot 2^n.$$

Thus, by transitivity of order,

$$2^n \leq 1 + 2 + 2^2 + \cdots + 2^n \leq 2 \cdot 2^n.$$

Moreover,

$$2^n \geq 0 \text{ for each integer } n.$$

Let $A = 1, B = 2$, and $k = 1$. Then, for each integer $n > k$,

$$A \cdot 2^n \leq 1 + 2 + 2^2 + \cdots + 2^n \leq B \cdot 2^n.$$

Thus, by definition of Θ-notation, $1 + 2 + 2^2 + \cdots + 2^n$ is $\Theta(2^n)$.

33. By factoring out a 4 and using the formula for the sum of a geometric sequence (Theorem 5.2.2), we have that for every integer $n > 1$,

$$
\begin{aligned}
4 + 4^2 + 4^3 + \cdots + 4^n &= 4(1 + 4 + 4^2 + \cdots + 4^{n-1}) \\
&= 4\left(\frac{4^{(n-1)+1} - 1}{4 - 1}\right) \\
&= \frac{4}{3}(4^n - 1) \\
&= \frac{4}{3} \cdot 4^n - \frac{4}{3} \\
&\leq \frac{4}{3} \cdot 4^n.
\end{aligned}
$$

Moreover, because

$$4 + 4^2 + 4^3 + \cdots + 4^{n-1} \geq 0, \quad \text{then} \quad 4^n \leq 4 + 4^2 + 4^3 + \cdots + 4^{n-1} + 4^n.$$

So let $A = 1$, $B = 4/3$, and $k = 1$. Then, because all quantities are positive,

$$A \cdot 4^n \leq 4 + 4^2 + 4^3 + \cdots + 4^n \leq B \cdot 4^n \quad \text{for every integer } n > k,$$

and thus, by definition of Θ-notation, $4 + 4^2 + 4^3 + \cdots + 4^n$ is $\Theta(4^n)$.

36. Given the expression $n + \dfrac{n}{2} + \dfrac{n}{4} + \cdots + \dfrac{n}{2^n}$, factor out n to obtain

$$
\begin{aligned}
n + \frac{n}{2} + \frac{n}{4} + \cdots + \frac{n}{2^n} &= n\left(1 + \frac{1}{2} + \frac{1}{4} + \cdots + \frac{1}{2^n}\right) \\[2mm]
&= n\left(\frac{\left(\frac{1}{2}\right)^{n+1} - 1}{\frac{1}{2} - 1}\right) \qquad \text{by Theorem 5.2.2} \\[2mm]
&= n\left(\frac{1 - 2^{n+1}}{2^n(1 - 2)}\right) \qquad \begin{array}{l}\text{by multiplying numerator.}\\ \text{and denominator by } 2^{n+1}\end{array} \\[2mm]
&= n\left(\frac{2^{n+1} - 1}{2^n}\right) \\[2mm]
&= n\left(2 - \frac{1}{2^n}\right) \qquad \text{by algebra.}
\end{aligned}
$$

Now $1 \le 2 - \dfrac{1}{2^n} \le 2$ when $n > 1$. Thus

$$
1 \cdot n \ \le\ n\left(2 - \frac{1}{2^n}\right) \ \le\ 2 \cdot n,
$$

and so, by substitution,

$$
1 \cdot n \ \le\ n + \frac{n}{2} + \frac{n}{4} + \cdots + \frac{n}{2^n} \ \le\ 2 \cdot n.
$$

Let $A = 1, B = 2$, and $k = 1$. Then, for each integer $n > k$,

$$
A \cdot n \ \le\ n + \frac{n}{2} + \frac{n}{4} + \cdots + \frac{n}{2^n} \ \le\ B \cdot n.
$$

Hence, by definition of Θ-notation,

$$
n + \frac{n}{2} + \frac{n}{4} + \cdots + \frac{n}{2^n} \text{ is } \Theta(n).
$$

39. $1 + \dfrac{1}{2} = \dfrac{3}{2}, \quad 1 + \dfrac{1}{2} + \dfrac{1}{3} = \dfrac{11}{6}, \quad 1 + \dfrac{1}{2} + \dfrac{1}{3} + \dfrac{1}{4} = \dfrac{50}{24} = \dfrac{25}{12}, \quad 1 + \dfrac{1}{2} + \dfrac{1}{3} + \dfrac{1}{4} + \dfrac{1}{5} = \dfrac{137}{60}$

40. If n is any integer with $n \ge 3$, then

$$
n + \frac{n}{2} + \frac{n}{3} + \cdots + \frac{n}{n} = n\left(1 + \frac{1}{2} + \frac{1}{3} + \cdots + \frac{1}{n}\right).
$$

By Example 11.4.7 and by Theorem 11.2.7(a),

$$
1 + \frac{1}{2} + \frac{1}{3} + \cdots + \frac{1}{n} \text{ is } \Theta(\log_2 n) \text{ and } n \text{ is } \Theta(n).
$$

Thus, by Theorem 11.2.9(c),

$$
n + \frac{n}{2} + \frac{n}{3} + \cdots + \frac{n}{n} \text{ is } \Theta(n \log_2 n).
$$

41. <u>Proof</u>: If n is any positive integer, then $\log_2 n$ is defined and by definition of floor,

$$
\lfloor \log_2 n \rfloor \le \log_2 n < \lfloor \log_2 n \rfloor + 1.
$$

If, in addition, n is greater than 2, then since the logarithmic function with base 2 is increasing

$$\log_2 n > \log_2 2 = 1.$$

Thus, by definition of floor,

$$1 \le \lfloor \log_2 n \rfloor.$$

Adding $\lfloor \log_2 n \rfloor$ to both sides of this inequality gives

$$\lfloor \log_2 n \rfloor + 1 \le 2 \lfloor \log_2 n \rfloor.$$

Hence, by the transitive property of order (T18 in Appendix A),

$$\log_2 n \le 2 \lfloor \log_2 n \rfloor,$$

and dividing both sides by 2 gives

$$\frac{1}{2} \log_2 n \le \lfloor \log_2 n \rfloor.$$

Let $A = 1/2, B = 1$, and $k = 2$. Then $A \log_2 n \le \lfloor \log_2 n \rfloor \le B \log_2 n$ for every integer $n \ge k$. Therefore, by definition of Θ-notation, $\lfloor \log_2 n \rfloor$ is $\Theta(\log_2 n)$.

42. <u>Proof</u>: If n is any positive integer, then $\log_2 n$ is defined and by definition of ceiling,

$$\lceil \log_2 n \rceil - 1 < \log_2 n \le \lceil \log_2 n \rceil.$$

Adding 1 to both sides of the left-hand inequality gives

$$\lceil \log_2 n \rceil < \log_2 n + 1.$$

If, in addition, n is greater than 2, then since the logarithmic function with base 2 is increasing

$$\log_2 n > \log_2 2 = 1.$$

Thus,

$$\lceil \log_2 n \rceil < \log_2 n + 1 < \log_2 n + \log_2 n = 2 \log_2 n.$$

Hence, by the transitive property of order (T18 in Appendix A),

$$\log_2 n \le \lfloor \log_2 n \rfloor \le 2 \log_2 n$$

Let $A = 1, B = 2$, and $k = 2$. Then

$$A \log_2 n \le \lceil \log_2 n \rceil \le B \log_2 n \quad \text{for every integer } n \ge k.$$

Therefore, by definition of Θ-notation, $\lceil \log_2 n \rceil$ is $\Theta(\log_2 n)$.

43. <u>Proof (by mathematical induction)</u>: Let the property $P(n)$ be the inequality

$$n \le 10^n. \qquad \leftarrow P(n)$$

Show that $P(1)$ is true: When $n = 1$, the inequality is $1 \le 10$, which is true.

Show that for every integer $k \ge 1$, if $P(k)$ is true then $P(k+1)$ is true: Let k be any integer with $k \ge 1$ and suppose that

$$k \le 10^k. \qquad \leftarrow \begin{array}{l} P(k) \\ \text{inductive hypothesis} \end{array}$$

We must show that

$$k + 1 \le 10^{k+1}. \quad \leftarrow P(k+1)$$

By inductive hypothesis, $k \le 10^k$. Adding 1 to both sides gives $k + 1 \le 10^k + 1$. And when $k \ge 1$,

$$10^k + 1 \le 10^k + 9 \cdot 10^k = 10 \cdot 10^k = 10^{k+1}.$$

Thus, by transitivity of order, $k + 1 \le 10^{k+1}$ *[as was to be shown]*.

44. *Hint:* To prove the inductive step, use the fact that if $k > 1$, then $k + 1 \leq 2k$. Apply the logarithmic function with base 2 to both sides of this inequality, and use properties of logarithms.

45. <u>Proof:</u> Suppose n is a variable that takes positive integer values. Then whenever $n \geq 2$,

$$2^n = \underbrace{2 \cdot 2 \cdot 2 \cdot 2 \cdot 2 \cdots 2}_{n \text{ factors}} \leq \underbrace{2 \cdot 2 \cdot 3 \cdot 4 \cdot 5 \cdots n}_{n \text{ factors}} \leq 2n!.$$

Let $B = 2$ and $b = 2$. Since 2^n and $n!$ are positive for every integer n,

$$2^n \leq Bn! \text{ for every integer } n \geq b.$$

Hence by definition of O-notation 2^n is $O(n!)$.

46. **a.** Example 11.4.6 showed that if n is any integer with $n \geq 1$, then $n! \leq n^n$. So, because the logarithmic function with base 2 is increasing,

$$\log_2(n!) \leq \log_2(n^n) \ (= n \log_2(n^n)).$$

Also, when $n \geq 1$, then $\log_2(n!) \geq \log_2 1 \geq 0$. Thus let $B = 1$ and $b = 1$. Then

$$0 \leq \log_2(n!) \leq Bn \log_2(n^n) \text{ for every integer } n \geq b.$$

So, by definition of O-notation, $\log_2(n!)$ is $O(n \log_2 n)$.

b. *Hint:*
$$(n!)^2 = n! \cdot n! = (1 \cdot 2 \cdot 3 \cdots 3)\,(n \cdot (n-1) \cdots 3 \cdot 2 \cdot 1)$$
$$= \left(\prod_{r=1}^{n} r\right)\left(\prod_{r=1}^{n}(n - r + 1)\right) = \prod_{r=1}^{n} r(n - r + 1).$$

Show that for each integer $r = 1, 2, \ldots, n$, $nr - n^2 + r \geq n$.

47. Let n be a positive integer, and suppose that $x > (2^n)^{2n}$. By properties of logarithms,

$$\log_2 x = (2n)\left(\frac{1}{2n}\right)(\log_2 x)$$
$$= (2n)\log_2\left(x^{\frac{1}{2n}}\right) < 2nx^{\frac{1}{2n}} \qquad (*)$$

(where the last inequality holds by substituting $x^{\frac{1}{2n}}$ in place of u in $\log_2 u < u$). Now raising both sides of $x > (2n)^{2n}$ to the $1/2$ power gives

$$x^{1/2} > \left((2n)^{2n}\right)^{1/2} = (2n)^n.$$

When both sides are multiplied by $x^{1/2}$, the result is

$$x = x^{1/2}x^{1/2} > x^{1/2}(2n)^n = x^{1/2}(2n)^n,$$

or, more compactly,

$$x^{1/2}(2n)^n < x.$$

Then, since the power function defined by $x \to x^{1/n}$ is increasing for every $x > 0$ (see exercise 21 of Section 11.1), we can take the nth root of both sides of the inequality and use the laws of exponents to obtain

$$\left(x^{1/2}(2n)^n\right)^{1/n} < x^{1/n},$$

or, equivalently,

$$2nx^{\frac{1}{2n}} < x^{1/n}. \qquad (**)$$

Finally use transitivity of order (Appendix A, T18) to combine $(*)$ and $(**)$ and conclude that $\log_2 x < x^{1/n}$ *[as was to be shown]*.

48. Let n be any positive integer and x a real number with $x > (2n)^{2n}$. By exercise 47,

$$\log_2 x < x^{1/n}$$

for every positive integer n. Now if $n \geq 2$, then $x^{1/n} < x^{1/2}$ (by property (11.2.1)). So, in particular,

$$\log_2 x < x^{1/2}.$$

But since $x > (2n)^{2n}$, then by properties of inequalities and exercise 21 of Section 11.1,

$$x > n^2 \;\Rightarrow\; \sqrt{x} > n \;\Rightarrow\; \frac{1}{n}\sqrt{x} > 1 \;\Rightarrow\; \frac{1}{n}\sqrt{x}\cdot\sqrt{x} > 1\cdot\sqrt{x} \;\Rightarrow\; \frac{1}{n}x > x^{1/2}.$$

Putting the inequalities $\log_2 x < x^{1/2}$ and $x^{1/2} < \frac{1}{n}x$ together gives

$$\log_2 x < \frac{1}{n}x.$$

Applying the exponential function with base 2 to both sides results in

$$2^{\log_2 x} < 2^{\frac{1}{n}x} \;\Rightarrow\; x < (2^x)^{1/n} \;\Rightarrow\; x^n < 2^x.$$

49. **a.** <u>Proof (by mathematical induction):</u> Let b be any real number with $b > 1$, and let the property $P(n)$ be the equation

$$\lim_{x\to\infty}\left(\frac{x^n}{b^x}\right) = 0 \qquad \leftarrow\ P(n)$$

Show that $P(1)$ is true: By L'Hôpital's rule,

$$\lim_{x\to\infty}\left(\frac{x^1}{b^x}\right) = \lim_{x\to\infty}\left(\frac{1}{b^x(\ln b)}\right) = 0.$$

Thus $P(1)$ is true.

Show that for every integer $k \geq 1$, if $P(k)$ is true then $P(k+1)$ is true: Let k be any integer with $k \geq 1$, and suppose that

$$\lim_{x\to\infty}\left(\frac{x^k}{b^x}\right) = 0. \qquad \langle \quad \begin{array}{l} P(k) \\ \text{inductive hypothesis} \end{array}$$

We must show that

$$\lim_{x\to\infty}\left(\frac{x^{k+1}}{b^x}\right) = 0. \qquad \leftarrow\ P(k+1)$$

Now by L'Hôpital's rule,

$$\begin{aligned}
\lim_{x\to\infty}\left(\frac{x^{k+1}}{b^x}\right) &= \lim_{x\to\infty}\frac{(k+1)x^k}{(\ln b)b^x} \\
&= \frac{(k+1)}{(\ln b)}\left[\lim_{x\to\infty}\left(\frac{x^k}{b^x}\right)\right] \\
&= \frac{(k+1)}{(\ln b)}\cdot 0 \qquad \text{by inductive hypothesis} \\
&= 0
\end{aligned}$$

[This is what was to be shown.]

b. By the result of part (a) and the definition of limit, given any real number $\varepsilon > 0$, there exists an integer N such that $\left|\dfrac{x^n}{b^n} - 0\right| < \varepsilon$ for every $x > N$. In this case take $\varepsilon = 1$. It follows that for every $x > N$, $\dfrac{x^n}{b^x} < 1$ since x and b are positive. Multiply both sides by b^x to obtain $x^n < b^x$. Let $B = 1$. Then $0 < x^n < B \cdot b^x$ for every $x > N$. Hence, by definition of O-notation, x^n is $O(b^x)$.

51. Completion of proof from Example 11.4.4:

   Case 2 (k is odd): In this case $k + 1$ is even, and

a_{k+1}	$= \ 2a_{\lfloor (k+1)/2 \rfloor}$	by definition of $a_1, a_2, a_3, \ldots$
	$= \ 2a_{(k+1)/2}$	because $k+1$ is even $\lfloor (k+1)/2 \rfloor = (k+1)/2$ (Theorem 4.6.2)
	$= \ 2 \cdot 2^{\lfloor \log_2((k+1)/2) \rfloor}$	by inductive hypothesis because, since $k+1$ is even, $k+1 \geq 2$, and so $(k+1)/2 \geq 1$
	$= \ 2^{\lfloor \log_2((k+1)/2) \rfloor + 1}$	by the laws of exponents from algebra (7.2.1)
	$= \ 2^{\lfloor \log_2(k+1) - \log_2 2 \rfloor + 1}$	by Theorem 7.2.1(b)
	$= \ 2^{\lfloor \log_2(k+1) - 1 \rfloor + 1}$	because $\log_2 2 = 1$
	$= \ 2^{\lfloor \log_2(k+1) \rfloor - 1 + 1}$	by substituting $x = \log_2(k+1)$ into the identity $\lfloor x - 1 \rfloor = \lfloor x \rfloor - 1$ derived in exercise 15 of Section 4.6
	$= \ 2^{\lfloor \log_2(k+1) \rfloor}$	by algebra.

Section 11.5

1. $\log_2 1{,}000 = \log_2(10^3) = 3 \log_2 10 \cong 3(3.32) \cong 9.96$

   $\log_2(1{,}000{,}000) = \log_2(10^6) = 6 \log_2 10 \cong 6(3.32) \cong 19.92$

   $\log_2(1{,}000{,}000{,}000{,}000) = \log_2(10^{12}) = 12 \log_2 10 \cong 12(3.32) = 39.84$

2. **a.** If $m = 2^k$, where k is a positive integer, then the algorithm requires $c \lfloor \log_2(2^k) \rfloor = c \lfloor k \rfloor = ck$ operations. If the input size is increased to $m^2 = (2^k)^2 = 2^{2k}$, then the number of operations required is $c \lfloor \log_2(2^{2k}) \rfloor = c \lfloor 2k \rfloor = 2(ck)$. Since $\dfrac{2(ck)}{ck} = 2$, the number of operations doubles.

   b. As in part (a), for an input of size $m = 2^k$, where k is a positive integer, the algorithm requires ck operations. If the input size is increased to $m^{10} = (2^k)^{10} = 2^{10k}$, then the number of operations required is $c \lfloor \log_2(2^{10k}) \rfloor = c \lfloor 10k \rfloor = 10(ck)$. Thus the number of operations increases by a factor of 10.

   c. When the input size is increased from 2^7 to 2^{28}, the factor by which the number of operations increases is $\dfrac{c \lfloor \log_2(2^{28}) \rfloor}{c \lfloor \log_2(2^7) \rfloor} = \dfrac{28c}{7c} = 4$.

3. A little numerical exploration can help find an initial window to use to draw the graphs of $y = x$ and $y = \lfloor 50 \log_2 x \rfloor$. Note that when $x = 2^8 = 256$,

$$\lfloor 50 \log_2 x \rfloor = \lfloor 50 \log_2(2^8) \rfloor = \lfloor 50 \cdot 8 \rfloor = \lfloor 400 \rfloor = 400 > 256 = x.$$

But when $x = 2^9 = 512$,

$$\lfloor 50 \log_2 x \rfloor = \lfloor 50 \log_2(2^9) \rfloor = \lfloor 50 \cdot 9 \rfloor = \lfloor 450 \rfloor = 450 < 512 = x.$$

So a good choice of initial window would be the interval from 256 to 512. Drawing the graphs, zooming if necessary, and using the trace feature reveal that when $n < 438, n < \lfloor 50 \log_2 n \rfloor$.

5. **a.**

index	0			1
bot	1			
top	10	4	1	
mid		5	2	1
x	Chia			

b.

index	0				
bot	1	6		7	
top	10		7		6
mid		5	8	6	7
x	Max				

6. **a.**

index	0			
bot	1			
top	10	4	1	0
mid		5	2	1
x	Amanda			

b.

index	0		8
bot	1	6	
top	10		
mid		5	8
x	Roy		

7. **a.** The array has $top - bot + 1$ elements.

b. <u>Proof</u>: Suppose top and bot are particular but arbitrarily chosen positive integers such that $top - bot + 1$ is an odd number. Then, by definition of odd, there is an integer k such that

$$top - bot + 1 = 2k + 1.$$

Add $2 \cdot bot - 1$ to both sides to obtain

$$bot + top = 2 \cdot bot - 1 + 2k + 1 = 2(bot + k).$$

Since $bot + k$ is an integer, by definition of even, $bot + top$ is even.

8.

n	27	13	6	3	1	0

9. For each positive integer n, $n \ div \ 2 = \lfloor n/2 \rfloor$. Thus when the algorithm segment is run for a particular n, each time the **while** loop iterates one time, the input to the next iteration is $\lfloor n/2 \rfloor$. It follows that the number of iterations of the loop for n is one more than the number of iterations for $\lfloor n/2 \rfloor$. That is, $a_n = 1 + a_{\lfloor n/2 \rfloor}$. Also note that $a_1 = 1$.

10. The recurrence relation and initial condition of $a_1, a_2, a_3, \ldots$ that were derived in exercise 9 are the same as those for the sequence $w_1, w_2, w_3, \ldots$ discussed in the worst-case analysis of the binary search algorithm. Thus the general formulas for the two sequences are the same. That is, $a_n = 1 + \lfloor \log_2 n \rfloor$, for each integer $n \geq 1$.

11. In the analysis of the binary search algorithm, it was shown that $1 + \lfloor \log_2 n \rfloor$ is $\Theta(\log_2 n)$. Thus the given algorithm segment has order $\log_2 n$.

12.

n	424	141	47	15	5	1	0

14. *Hint:* The formula is $b_n = 1 + \lfloor \log_3 n \rfloor$ for every integer $n \geq 1$.

15. If $n \geq 3$, then

$$
\begin{aligned}
b_n &= 1 + \lfloor \log_3 n \rfloor && \text{by the result of exercise 14} \\
\Rightarrow \quad b_n &\leq 1 + \log_3 n && \text{because } \lfloor \log_3 n \rfloor \leq \log_3 n \text{ by definition of floor} \\
\Rightarrow \quad b_n &\leq \log_3 n + \log_3 n && \text{because if } n \geq 3 \text{ then } \log_3 n \geq 1 \\
\Rightarrow \quad b_n &\leq 2 \log_3 n && \text{by algebra.}
\end{aligned}
$$

Furthermore, because $\log_3 n \geq 0$ for $n > 2$, we may write

$$|\log_3 n| < |\lfloor \log_3 n \rfloor + 1| \leq 2 |\log_3 n|.$$

Let $A = 1$, $B = 2$, and $k = 2$. Then all quantities are positive, and so

$$A |\log_3 n| < |\lfloor \log_3 n \rfloor + 1| \leq B |\log_3 n| \quad \text{for every integer } n > k.$$

Hence by definition of Θ-notation, $b_n = 1 + \lfloor \log_3 n \rfloor$ is $\Theta(\log_3 n)$, and thus the algorithm segment has order $\log_3 n$.

18. Suppose an array of length k is input to the **while** loop and the loop is iterated one time. The elements of the array can be matched with the integers from 1 to k with $m = \left\lceil \dfrac{k+1}{2} \right\rceil$, as shown below:

	left subarray				right subarray		
$a[bot]$	$a[bot+1]$	$\ldots$	$a[mid-1]$	$a[mid]$	$a[mid+1]$	$\ldots$	$a[top-1]$	$a[top]$
$\updownarrow$	$\updownarrow$		$\updownarrow$	$\updownarrow$	$\updownarrow$		$\updownarrow$	$\updownarrow$
1	2		$m-1$	m	$m+1$		$k-1$	k

Case 1 (k is even): In this case $m = \left\lceil \dfrac{k+1}{2} \right\rceil = \left\lceil \dfrac{k}{2} + \dfrac{1}{2} \right\rceil = \dfrac{k}{2} + 1$, and so the number of elements in the left subarray equals $m - 1 = (\dfrac{k}{2} + 1) - 1 = \dfrac{k}{2} = \left\lfloor \dfrac{k}{2} \right\rfloor$. The number of elements in the right subarray equals $k - (m+1) - 1 = k - m = k - (\dfrac{k}{2} + 1) = \dfrac{k}{2} - 1 < \left\lfloor \dfrac{k}{2} \right\rfloor$. Hence both subarrays (and thus the new input array) have length at most $\left\lfloor \dfrac{k}{2} \right\rfloor$.

Case 2 (k is odd): In this case $m = \left\lceil \dfrac{k+1}{2} \right\rceil = \dfrac{k+1}{2}$, and so the number of elements in the left subarray equals $m - 1 = \dfrac{k+1}{2} - 1 = \dfrac{k-1}{2} = \left\lfloor \dfrac{k}{2} \right\rfloor$. The number of elements in the right subarray equals $k - m = k - \dfrac{k+1}{2} = \dfrac{k-1}{2} = \left\lfloor \dfrac{k}{2} \right\rfloor$ also. Hence both subarrays (and thus the new input array) have length $\left\lfloor \dfrac{k}{2} \right\rfloor$.

The arguments in cases 1 and 2 show that the length of the new input array to the next iteration of the **while** loop has length at most $\lfloor k/2 \rfloor$.

20.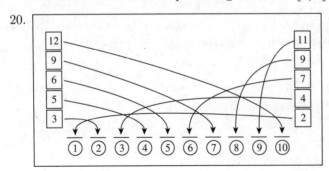

21.

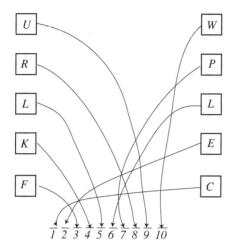

22.

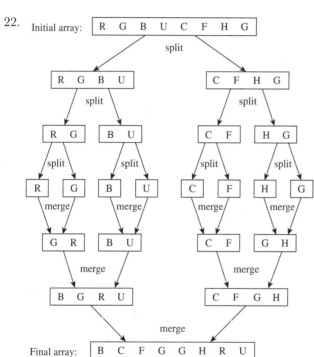

24. **a.** Refer to Figure 11.5.3. Observe that when k is odd, the subarray

$$a[mid + 1], a[mid + 2], \ldots, a[top]$$

has length

$$k - \left(\frac{k+1}{2} + 1\right) + 1 = \frac{2k}{2} - \frac{k+1}{2} = \frac{k-1}{2} = \left\lfloor \frac{k}{2} \right\rfloor$$

and that when k is even, it has length

$$k - \left(\frac{k}{2} + 1\right) + 1 = \frac{2k}{2} - \frac{k}{2} = \frac{k}{2} = \left\lfloor \frac{k}{2} \right\rfloor.$$

b. Refer to Figure 11.5.3 and observe that when k is even, the subarray

$$a[mid], a[mid + 1], \ldots, a[top]$$

has length $k - \left(\frac{k}{2} + 1\right) + 1 = \frac{k}{2} = \left\lceil \frac{k}{2} \right\rceil$.

25. *Hint:* The following are the steps for part (a) in the case where k is odd and $k + 1$ is even:

$$
\begin{aligned}
m_{k+1} &= m_{\lfloor (k+1)/2 \rfloor} + m_{\lceil (k+1)/2 \rceil} + (k+1) - 1 && \text{by definition of } m_1, m_2, m_3, \ldots \\
&= m_{(k+1)/2} + m_{(k+1)/2} + k && \text{by algebra and the definition of floor and ceiling} \\
&= 2m_{(k+1)/2} + k && \text{by algebra} \\
&\geq 2 \cdot \frac{1}{2} \cdot \frac{k+1}{2} \log_2 \left(\frac{k+1}{2} \right) + k && \text{by inductive hypothesis} \\
&= \frac{1}{2}(k+1)(\log_2(k+1) - \log_2 2) + k && \text{by Theorem 7.2.1(a)} \\
&= \frac{1}{2}(k+1)(\log_2(k+1) - 1) + k && \text{because } \log_2 2 = 1 \\
&= \frac{1}{2}(k+1)\log_2(k+1) - \frac{1}{2}(k+1) + k && \text{by algebra} \\
&= \frac{1}{2}(k+1)\log_2(k+1) + \frac{1}{2}(k-1) && \text{by algebra} \\
&= \frac{1}{2}(k+1)\log_2(k+1) && \text{because } k \geq 1.
\end{aligned}
$$

So by transitivity, $m_{k+1} \geq \frac{1}{2}(k+1)\log_2(k+1)$.

Review Guide: Chapter 11

Definitions: How are the following terms defined?

- real-valued function of a real variable
- graph of a real-valued function of a real variable
- power function with exponent a
- floor function
- multiple of a real-valued function of a real variable
- increasing function
- decreasing function
- $f(x)$ is $\Omega(g(x)$, where f and g are real-valued functions of a real variable defined on the same set of nonnegative integers with $g(n) \geq 0$ for every integer $n \geq r$, where r is a positive real number
- $f(x)$ is $O(g(x)$, where f and g are real-valued functions of a real variable defined on the same set of nonnegative integers with $g(n) \geq 0$ for every integer $n \geq r$, where r is a positive real number
- $f(x)$ is $\Theta(g(x)$, where f and g are real-valued functions of a real variable defined on the same set of nonnegative integers with $g(n) \geq 0$ for every integer $n \geq r$, where r is a positive real number
- algorithm A is $\Theta(g(n)$ (*Or: A* has order $g(n)$)
- algorithm A is $\Omega(g(n)$ (*Or: A* has a best case order $g(n)$)
- algorithm A is $O(g(n)$ (*Or: A* has a worst case order $g(n)$)
- polynomial time algorithms, NP class, NP-complete problems, the P vs. NP problem, tractable and intractable problems

Polynomial and Rational Functions and Their Orders

- What is the graph of the floor function?
- What is the difference between the graph of a function defined on an interval of real numbers and the graph of a function defined on a set of integers?
- How do you graph a multiple of a real-valued function of a real variable?
- How do you prove that a function is increasing (decreasing)?
- What are some properties of O-, Ω-, and Θ-notation? Can you prove them?
- If n is an integer with $n \geq 1$, what is the relationship between n^r and n^s, where r and s are positive rational numbers and $r < s$?
- Suppose f is a real-valued functions of a real variable defined on a set of nonnegative integers. How could it be possible for f to be both $O(n^3)$ and $O(n^5)$?
- Given a polynomial of degree m, how do you use the definition of Θ-notation to show that the polynomial has order n^m?
- What is the theorem on polynomial orders?
- What is an order for the sum of the first n integers?

Efficiency of Algorithms

- How do you compute the order of an algorithm segment that contains a loop? a nested loop?
- How do you find the number of times a loop will iterate when an algorithm segment is executed?
- How do you use the theorem on polynomial orders to help find the order of an algorithm segment?
- What is the sequential search algorithm? How do you compute its worst case order? its average case order?
- What is the insertion sort algorithm? How do you compute its best and worst case orders?

Logarithmic and Exponential Orders

- What do the graphs of logarithmic and exponential functions look like?
- What can you say about the base 2 logarithm of a number that is between two consecutive powers of 2?
- How do you compute the number of bits needed to represent a positive integer in binary notation?
- How are logarithms used to solve recurrence relations?
- If $b > 1$, what can you say about the relation among $\log_b x$, x^r, and $x \log_b x$?
- If $b > 1$ and $c > 1$, how are orders of $\log_b n$ and $\log_c n$ related?
- What is an order for a harmonic sum?
- What is a divide-and-conquer algorithm?
- What is the binary search algorithm?
- What is the worst case order for the binary search algorithm, and how do you find it?
- What is the merge sort algorithm?
- What is the worst case order for the merge sort algorithm, and how do you find it?

Fill-in-the-Blank Review Questions

Section 11.1

1. If f is a real-valued function of a real variable, then the domain and co-domain of f are both _____.

2. A point (x, y) lies on the graph of a real-valued function of a real variable f if, and only if, _____.

3. If a is any nonnegative real number, then the power function with exponent a, p_a, is defined by _____.

4. Given a function $f: \mathbf{R} \to \mathbf{R}$ and a real number M, the function Mf is defined by _____.

5. Given a function $f: \mathbf{R} \to \mathbf{R}$, to prove that f is increasing, you suppose that _____ and then you show that _____.

6. Given a function $f: \mathbf{R} \to \mathbf{R}$, to prove that f is decreasing, you suppose that _____ and then you show that _____.

Section 11.2

1. A sentence of the form "$Ag(n) \le f(n)$ for every $n \ge a$," translates into Ω-notation as _____.

2. A sentence of the form "$f(n) \le Bg(n)$ for every $n \ge b$," translates into O-notation as _____.

3. A sentence of the form "$Ag(n) \le f(n) \le Bg(n)$ for every $n \ge k$," translates into Θ-notation as _____.

4. When $n \ge 1$, n_____ n^2 and n^2_____ n^5.

5. According to the theorem on polynomial orders, if $p(x)$ is a polynomial in x, then $p(n)$ is $\Theta(n^m)$, where m is _____.

6. If n is a positive integer, then $1 + 2 + 3 + \cdots + n$ has order _____.

Section 11.3

1. When an algorithm segment contains a nested **for-next** loop, you can find the number of times the loop will iterate by constructing a table in which each column represents _____.
2. In the worst case for an input array of length n, the sequential search algorithm has to look through _____ elements of the input array before it terminates.
3. The worst case order of the insertion sort algorithm is _____, and its average case order is _____.

Section 11.4

1. The domain of the exponential function is _____, and its range is _____.
2. The domain of the logarithmic function is _____, and its range is _____.
3. If k is an integer and $2^k \leq x < 2^{k+1}$, then $\lfloor \log_2 x \rfloor =$ _____.
4. If b is a real number with $b > 1$, then there is a positive real number s with the property that for any real number x that is greater than or equal to s, when the quantities x, x^2, $\log_b x$, and $x \log_b x$ are arranged in order of increasing size, the result is _____.
5. If n is a positive integer, then $1 + \frac{1}{2} + \frac{1}{3} + \cdots + \frac{1}{n}$ has order _____.

Section 11.5

1. To solve a problem using a divide-and-conquer algorithm, you reduce it to a fixed number of smaller problems of the same kind, which _____, and so forth until _____.
2. To search an array using the binary search algorithm in each step, you compare a middle element of the array to _____. If the middle element is less than _____, you _____, and if the middle element is greater than _____, you _____.
3. The worst case order of the binary search algorithm is _____.
4. To sort an array using the merge sort algorithm, in each step until the last one you split the array into approximately two equal sections and sort each section using _____. Then you _____ the two sorted sections.
5. The worst case order of the merge sort algorithm is _____.

Answers for Fill-in-the-Blank Review Questions

Section 11.1

1. sets of real numbers
2. $y = f(x)$
3. $p_a(x) = x^a$ for each real number x
4. $(Mf)(x) = M \cdot f(x)$ for each $x \in \mathbf{R}$
5. x_1 and x_2 are any real numbers such that $x_1 < x_2$
   $f(x_1) < f(x_2)$
6. x_1 and x_2 are any real numbers such that $x_1 < x_2$
   $f(x_1) > f(x_2)$

Section 11.2

1. $f(x)$ is $\Omega(g(n))$
2. $f(x)$ is $O(g(n))$
3. $f(x)$ is $\Theta(g(n))$
4. $\leq, \leq$
5. the degree of $p(n)$

6. n^2

Section 11.3

1. one iteration of the innermost loop
2. n
3. n^2, n^2

Section 11.4

1. the set of all real numbers, the set of all positive real numbers
2. the set of all positive real numbers, the set of all real numbers
3. k
4. $\log_b x < x < x \log_b x < x^2$
5. $\ln x$ (or, equivalently, $\log_2 x$)

Section 11.5

1. a fixed number of smaller problems of the same kind;
   can themselves be reduced to the same finite number of smaller problems of the same kind;
   easily resolved problems are obtained
2. the element you are looking for;
   the element you are looking for;
   apply the binary search algorithm to the lower half of the array;
   the element you are looking for;
   apply the binary search algorithm to the upper half of the array

3. $\log_2 n$, where n is the length of the array
4. merge sort; merge
5. $n \log_2 n$

Chapter 12: Regular Expressions and Finite-State Automata

This chapter opens with some historical background about the connections between computers and formal languages. Section 12.1 focuses on regular expressions and emphasizes their utility for pattern matching, whether for compilers or for general text processing.

Section 12.2 introduces the concept of finite-state automaton. In one sense, it is a natural sequel to the discussions of digital logic circuits in Section 2.4 and Boolean functions in Section 7.1, with the next-state function of an automaton governing the operation of sequential circuit in much the same way that a Boolean function governs the operation of a combinatorial circuit. The section also provides practice in finding a finite-state automaton that corresponds to a regular expression and shows how to write a program to implement a finite-state automaton. Both abilities are useful for computer programming. The section ends with a statement and partial proof of Kleene's theorem, which describes the exact nature of the relationship between finite-state automata and regular languages.

The equivalence and simplification of finite-state automata, discussed in Section 12.3, provides an additional application for the concept of equivalence relation, introduced in Section 8.3. Note the parallel between the simplification of digital logic circuits discussed in Section 2.4 and the simplification of finite-state automata developed in this section. Both kinds of simplification have obvious practical use.

Section 12.1

1. $L_1 = \{\lambda, x, y, xx, yy, xxx, xyx, yxy, yyy, xxxx, xyyx, yxxy, yyyy\}$

3. **a.** $(a + b) \cdot (c + d)$

   b. $L = \{11*, 11/, 12*, 12/, 21*, 21/, 22*, 22/\}$

   $11* = 1 * 1 = 1$, $11/ = 1/1 = 1$, $12* = 1 * 2 = 2$, $12/ = 1/2 = 0.5$, $21* = 2 * 1 = 2$, $21/ = 2/1 = 2$, $22* = 2 * 2 = 4$, $22/ = 2/2 = 1$

4. $L_1 L_2$ is the set of all strings of a's and b's that start with an a and contain an odd number of a's.

   $L_1 \cup L_2$ is the set of all strings of a's and b's that contain an even number of a's or that start with an a and contain only that one a. (Note that because 0 is an even number, both λ and b are in $L_1 \cup L_2$.)

   $(L_1 \cup L_2)^*$ is the set of all strings of a's and b's. The reason is that a and b are both in $L_1 \cup L_2$, and thus every string in a and b is in $(L_1 \cup L_2)^*$.

6. $L_1 L_2$ is the set of strings of 0's and 1's that both start and end with a 0.

   $L_1 \cup L_2$ is the set of strings of 0's and 1's that start with a 0 or end with a 0 (or both).

   $(L_1 \cup L_2)^*$ is the set of strings of 0's and 1's that start with a 0 or end with a 0 (or both) or that contain 00.

7. $(a \,|\, ((b^*)\, b)) \,((a^*) \,|\, (ab))$

9. $(((x \,|\, (y(z^*)))^*)((yx) \,|\, (((yz)^*)z)))$

10. $(ab^* \,|\, cb^*)\, (ac \,|\, bc)$

12. $xy(x^*y)^* \,|\, (yx \,|\, y)y^*$

13. $L(\lambda \,|\, ab) \;=\; L(\lambda) \cup L(ab) \;=\; \{\lambda\} \cup L(a)L(b) \;=\; \{\lambda\} \cup \{xy \,|\, x \in L(a) \text{ and } y \in L(b)\}$
    $$= \{\lambda\} \cup \{xy \,|\, x \in \{a\} \text{ and } y \in \{b\}\} \;=\; \{\lambda\} \cup \{ab\} \;=\; \{\lambda,\, ab\}$$

15. $L((a \,|\, b)c) \;=\; L(a \,|\, b)L(c) \;=\; (L(a) \cup L(b))L(c) \;=\; (\{a\} \cup \{b\})\{c\} \;=\; (\{a, b\})\{c\} \;=\; \{ac, bc\}$

16. Here is a sample of five strings out of infinitely many: 0101, 1, 01, 10000, and 011100.

18. $x, yxxy, xx, xyxxy, xyxxyyxxy, \ldots$

19. The language consists of all strings of a's and b's that contain exactly three a's and end in an a.

21. The language consists of the set of all strings of x's and y's that start with xy or yy followed by any string of x's and y's.

22. *aaaba* is in the language but *baabb* is not because if a string in the language contains a b to the right of the left-most a, then it must contain another a to the right of all the b's.

24. The string 120 does not belong to the language defined by $(01^*2)^*$ because it does not start with 0. However, 01202 does belong to the language because 012 and 02 are both defined by 01^*2 and the language is closed under concatenation.

25. One solution is $0^*10^*(0^*10^*10^*)^*$.

27. $x \mid y^* \mid y^*(xyy^*)(\lambda \mid x)$

28. $$\begin{aligned} L((r \mid s)t) &= L(r \mid s)L(t) = (L(r) \cup L(s))L(t) \\ &= \{xy \mid (x \in L(r) \cup L(s)) \text{ and } y \in L(t)\} \\ &= \{xy \mid (x \in L(r) \text{ or } x \in L(s)) \text{ and } y \in L(t)\} \\ &= \{xy \mid (x \in L(r) \text{ and } y \in L(t)) \text{ or } (x \in L(s) \text{ and } y \in L(t))\} \\ &= \{xy \mid xy \in L(rt) \text{ or } xy \in L(st)\} = L(rt) \cup L(st) = L(rt \mid st) \end{aligned}$$

30. Note that for any regular expression x, $(x^*)^*$ defines the set of all strings obtained by concatenating a finite number of a finite number of concatenations of copies of x. But any such string can equally well be obtained simply by concatenating a finite number of copies of x, and thus $(x^*)^* = x^*$. Hence the given languages are the same: $L((rs)^*) = L(((rs)^*)^*)$.

31. $pre[a-z]^+$

33. $[a-z]\{3\}[a-z]^*ly$

34. $[a-z]^*(a \mid e \mid i \mid o \mid u)[a-z]^*$

36. $[\hat{}\,AEIOU][A-Z]^*[A \mid E \mid I \mid O \mid U]\{2\}[A-Z]^*$

37. $[0-9]\{3\} - [0-9]\{2\} - 3[0-9]\{2\}6$

39. $([+-] \mid \lambda)[0-9]^*(\backslash. \mid \lambda)[0-9]^*$

40. *Hint:* Leap years from 1980 to 2079 are 1980, 1984, 1988, 1992, 1996, 2000, 2004, and so forth. Note that the fourth digit is 0, 4, or 8 for the years whose third digit is even and that the fourth digit is 2 or 6 for the years whose third digit is odd.

Section 12.2

1. **a.** \$1 or more deposited

2. **a.** s_0, s_1, s_2 **b.** 0, 1   **c.** s_0 **d.** s_2 **e.** Annotated next-state table:

	State	Input 0	Input 1
→	s_0	s_1	s_0
	s_1	s_1	s_2
◎	s_2	s_2	s_2

3. **a.** U_0, U_1, U_2, U_3 **b.** a, b **c.** U_0 **d.** U_3 **e.** Annotated next-state table:

		input	
		a	b
$\rightarrow$	U_0	U_2	U_1
	U_1	U_2	U_3
state	U_2	U_2	U_2
◎	U_3	U_3	U_3

5. **a.** A, B, C, D, E, F **b.** x, y **c.** A **d.** D, E **e.** Annotated next-state table:

		Input	
		x	y
$\rightarrow$	A	C	B
	B	F	D
State	C	E	F
◉	D	F	D
◉	E	E	F
	F	F	F

6. **a.** s_0, s_1, s_2, s_3 **b.** $0, 1$ **c.** s_0 **d.** s_0 **e.** Annotated next-state table:

			input	
			0	1
$\rightarrow$	◎	s_0	s_0	s_1
state		s_1	s_1	s_2
	s_2	s_2	s_3
	s_3	s_3	s_0

7. **a.** s_0, s_1, s_2, s_3 **b.** $0, 1$ **c.** s_0 **d.** s_0, s_2 **e.** Annotated next-state table:

			Input	
			0	1
$\rightarrow$	◉	s_0	s_0	s_1
State		s_1	s_1	s_2
◉	s_2	s_2	s_3
	s_3	s_3	s_0

8. **a.** s_0, s_1, s_2 **b.** $0, 1$ **c.** s_0 **d.** s_2 **e.** Transition diagram:

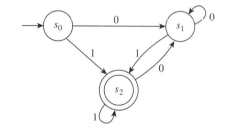

9. **a.** s_0, s_1, s_2, s_3 **b.** $0, 1$ **c.** s_0 **d.** s_1 **e.** Transition diagram:

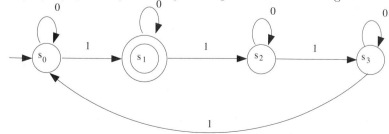

10. **a.** $N(s_1, 1) = s_2$, $N(s_0, 1) = s_3$ **c.** $N^*(s_0, 10011) = s_2$, $N^*(s_1, 01001) = s_2$

11. **a.** $N(s_3, 0) = s_4$, $N(s_2, 1) = s_4$ **c.** $N^*(s_0, 010011) = s_3$, $N^*(s_3, 01101) = s_4$

There are multiple correct answers for part (d) of exercises 12 and 13, part (b) of exercises 14–19, and exercises 20–48.

12. **a.** (i) s_2 (ii) s_2 (iii) s_1

b. The strings in (i) and (ii) − namely 1110001 and 0001000 − send the automaton to an accepting state, but the string in (iii) − namely 11110000 − does not.

c. The language accepted by this automaton is the set of all strings of 0's and 1's that contain at least one 0 followed (not necessarily immediately) by at least one 1.

d. $1^*00^*1\,(0\,|\,1)^*$

14. **a.** The language accepted by this automaton is the set of all strings of 0's and 1's that end in 00.

b. $(0\,|\,1)^*\,00$

15. **a.** The language accepted by this automaton is the set of all strings of x's and y's of length at least two that consist either entirely of x's or entirely of y's.

b. $xxx^*\,|\,yyy^*$

17. **a.** The language accepted by this automaton is the set of all strings of 0's and 1's with the following property: If n is the number of 1's in the string, then $n\ mod\ 4 = 0$ or $n\ mod\ 4 = 2$. This is equivalent to saying that n is even.

b. $0^*\,|\,(0^*10^*10^*)^*$

18. **a.** The language accepted by this automaton is the set of all strings of 0's and 1's that end in 1.

b. $(0\,|\,1)^*\,1$

20. **a.** Call the automaton being constructed A. Acceptance of a string by A depends on the values of three consecutive inputs. Thus A requires at least four states:

s_0: initial state

s_1: state indicating that the last input character was a 1

s_2: state indicating that the last two input characters were 1's

s_3: state indicating that the last three input characters were 1's, the acceptance state

If a 0 is input to A when it is in state s_0, no progress is made toward achieving a string of three consecutive 1's. Hence A should remain in state s_0. If a 1 is input to A when it is in state s_0, it goes to state s_1, which indicates that the last input character of the string is a 1. From state s_1, A goes to state s_2 if a 1 is input. This indicates that the last two characters of the string are 1's. But if a 0 is input, A should return to s_0 because the wait for a string of three consecutive 1's must start over again. When A is in state s_2 and a 1 is input, then a string of three consecutive 1's is achieved, so A should go to state s_3. If a 0 is input when A is in state s_2, then progress toward accumulating a sequence of three consecutive 1's is lost, so A should return to s_0. When A is in a state s_3 and a 1 is input, then the final three symbols of the input string are 1's, and so A should stay in state s_3. If a 0 is input when A is in state s_3, then A should return to state s_0 to await the input of more 1's. Thus the transition diagram is as follows:

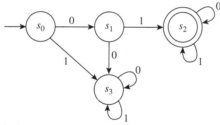

b. $(0\,|\,1)^*111$

21. **a.** Call the automaton being constructed A. Acceptance of a string by A depends on the values of two consecutive inputs. Thus A requires at least three states:

s_0: initial state

s_1: state indicating that the first input character is a 0

s_2: state indicating that the first two input characters are 01. (So the string is accepted.)

If A is in state s_0 and a 1 is input, then whatever additional symbols are input, the resulting string will not start with 01 and hence will not be accepted. Thus an additional state, s_3 is needed from which it is not possible to go to an accepting state. One way to achieve this is to keep A in state s_3 if either a 0 or a 1 is input to it when it is in state s_3.

If A is in state s_0 and a 0 is input, it goes to state s_1, and if a 1 is input to A when it is state s_1, then it goes to state s_2. In that case, its first two symbols are 01, and so whatever additional symbols are input, the resulting string is accepted. So inputting either 0 or 1 to A in state s_2 keeps A in s_2.

If A is in state s_1 and a 0 is input, then the first two input symbols are 00, and so whatever additional symbols are input, A should go to state s_3 because the resulting string will not start with 01 and hence will not be accepted. As a consequence, the transition diagram has the following appearance:

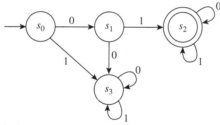

b. $01(0\,|\,1)^*$

22. *Hint:* Use five states: s_0 (the initial state), s_1 (the state indicating that the previous input symbol was an a), s_2 (the state indicating that the previous input symbol was a b), s_3 (the state indicating that the previous two input symbols were a's), and s_4 (the state indicating that the previous two input symbols were b's).

23. **a.**

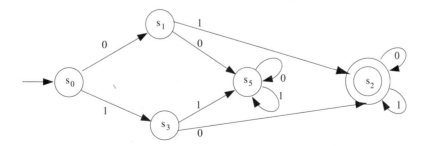

b. $(01\,|\,10)(0\,|\,1)^*$

24. a.

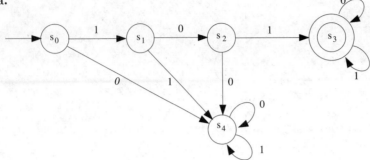

b. $101(0 \mid 1)^*$

25. a.

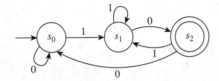

b. $(0 \mid 1)^*10$

26. a.

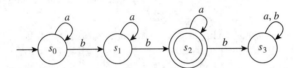

b. $a^*ba^*ba^*$

27. a.

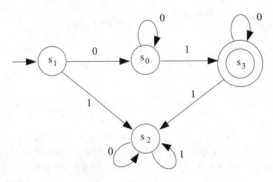

b. 00^*10^* (or using the $^+$ notation: 0^+10^*)

28. a.

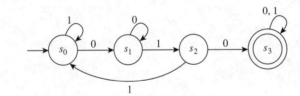

b. $(0 \mid 1)^*010(0 \mid 1)^*$

29.

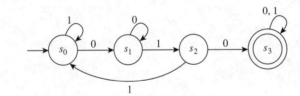

30.

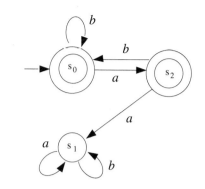

31.

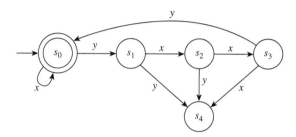

33.

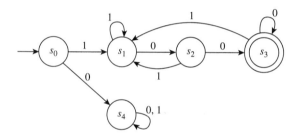

36.

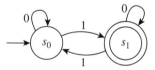

39. Let $\hat{P}$ denote a list of all letters of a lowercase alphabet except p, $\hat{R}$ denote a list of all the letters of a lowercase alphabet except r, and $\hat{E}$ denote a list of all the letters of a lowercase alphabet except e.

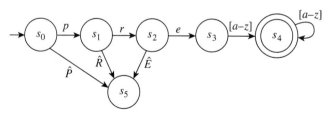

42. Let $\mathscr{S}$ denote a list of all the consonants in a lowercase alphabet.

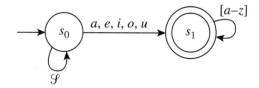

45.

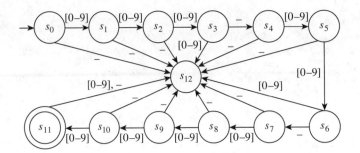

48. Let d represent the character class $[0 - 9]$.

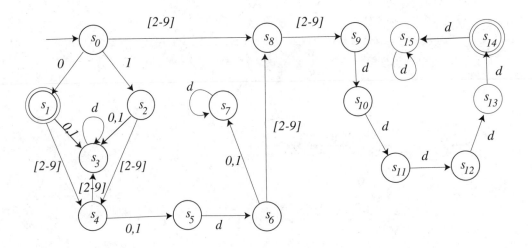

51. Proof (by contradiction): Suppose there were a finite-state automaton A that accepts L. Consider all strings of the form a^i for some integer $i \geq 0$.

Since the set of all such strings is infinite and the number of states of A is finite, by the pigeonhole principle at least two of these strings, say a^p and a^q with $p < q$, must send A to the same state, say s, when input to A starting in its initial state. (The strings of the given form are the pigeons, the states are the pigeonholes, and each string is associated with the state to which A goes when the string is input to A starting in its initial state.)

Because A accepts L, A accepts $a^q b^q$ but A does not accept $a^p b^q$.

Now since $a^q b^q$ is accepted by A, A goes to an accepting state if, starting from the initial state, first a^q is input to it (sending it to state s) and then b^q is input to it. But A also goes to state s after a^p is input to it. Hence, inputting b^q to A after inputting a^p also sends A to an accepting state. In other words, A accepts $a^p b^q$.

Thus $a^p b^q$ is accepted by A and yet it is not accepted by A, which is a contradiction. Hence the supposition is false: there is no finite-state automaton that accepts L.

53. *Hint:* Suppose the automaton A has N states. Choose an integer m such that $(m+1)^2 - m^2 > N$. Consider strings of a's of lengths between m^2 and $(m+1)^2$.

Since there are more strings than states, at least two strings must send A to the same state s_i:

$$(m+1)^2$$

$$\overbrace{\underbrace{aa\dots aaa}_{m^2}\dots aaa\dots aaa\dots a}$$

after both of these
inputs, A is in state s_i

It follows (by removing the a's shown in color) that the automaton must accept a string of the form a^k, where $m^2 < k < (m+1)^2$.

54. **a.** Proof: Suppose A is a finite-state automaton with input alphabet $\sum$, and suppose $L(A)$ is the language accepted by A.

Define a new automaton A' as follows: Both the states and the input symbols of A' are the same as the states and input symbols of A. The only difference between A and A' is that each accepting state of A is a non-accepting state of A', and each non-accepting state of A is an accepting state of A'.

It follows that each string in $\sum^*$ that is accepted by A is not accepted by A', and each string in $\sum^*$ that is not accepted by A is accepted by A'. Thus $L(A') = (L(A))^c$.

b. Proof: Let A_1 and A_2 be finite-state automata, and let $L(A_1)$ and $L(A_2)$ be the languages accepted by A_1 and A_2, respectively.

By part (a), there exist automata A_1' and A_2' such that $L(A_1') = (L(A_1))^c$ and $L(A_2') = (L(A_2))^c$.

Hence, by Kleene's theorem (part 1), there are regular expressions r_1 and r_2 that define $(L(A_1))^c$ and $(L(A_2))^c$, respectively. So we may write $(L(A_1))^c = L(r_1)$ and $(L(A_2))^c = L(r_2)$.

Now by definition of regular expression, $r_1 \mid r_2$ is a regular expression, and, by definition of the language defined by a regular expression, $L(r_1 \mid r_2) = L(r_1) \cup L(r_2)$.

Thus, by substitution and De Morgan's law, $L(r_1 \mid r_2) = (L(A_1))^c \cup (L(A_2))^c = (L(A_1) \cap L(A_2))^c$, and so, by Kleene's theorem (part(2)), there is a finite-state automaton, say A, that accepts $(L(A_1) \cap L(A_2))^c$.

It follows from part (a) that there is a finite-state automaton, A', that accepts $((L(A_1) \cap L(A_2))^c)^c$. But, by the double complement law for sets, $((L(A_1) \cap L(A_2))^c)^c = L(A_1) \cap L(A_2)$.

So there is a finite-state automaton, A', that accepts $L(A_1) \cap L(A_2)$, and hence, by Kleene's theorem and the definition of regular language, $L(A_1) \cap L(A_2)$ is a regular language.

Section 12.3

1. **a.** 0-equivalence classes: $\{s_0, s_1, s_3, s_4\}, \{s_2, s_5\}$

   1-equivalence classes: $\{s_0, s_3\}, \{s_1, s_4\}, \{s_2, s_5\}$

   2-equivalence classes: $\{s_0, s_3\}, \{s_1, s_4\}, \{s_2, s_5\}$

   b.

3. **a.** 0-equivalence classes: $\{s_1, s_3\}, \{s_0, s_2\}$

   1-equivalence classes: $\{s_1, s_3\}, \{s_0, s_2\}$

b. transition diagram for $\bar{A}$:

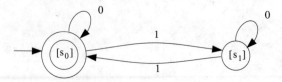

4. **a.** 0-equivalence classes: $\{s_0, s_1, s_2\}, \{s_3, s_4, s_5\}$

1-equivalence classes: $\{s_0, s_1, s_2\}, \{s_3, s_5\}, \{s_4\}$

2-equivalence classes: $\{s_0, s_2\}, \{s_1\}, \{s_3, s_5\}, \{s_4\}$

3-equivalence classes: $\{s_0, s_2\}, \{s_1\}, \{s_3, s_5\}, \{s_4\}$

b.

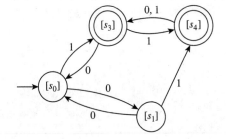

6. **a.** 0-equivalence classes: $\{s_0, s_1, s_3, s_4, s_5\}, \{s_2, s_6\}$

1-equivalence classes: $\{s_0, s_4, s_5\}, \{s_1, s_3\}, \{s_2\}, \{s_6\}$

2-equivalence classes: $\{s_0, s_4\}, \{s_5\}, \{s_1\}, \{s_3\}, \{s_2\}, \{s_6\}$

3-equivalence classes: $\{s_0\}, \{s_4\}, \{s_5\}, \{s_1\}, \{s_3\}, \{s_2\}, \{s_6\}$

b. The transition diagram for $\overline{A}$ is the same as the one given for A except that the states are denoted $[s_0], [s_1], [s_2], [s_3], [s_4], [s_5], [s_6]$.

7. Yes. For A:

0-equivalence classes: $\{s_0, s_2\}, \{s_1, s_3\}$

1-equivalence classes: $\{s_0\}, \{s_2\}, \{s_1, s_3\}$

2-equivalence classes: $\{s_0\}, \{s_2\}, \{s_1, s_3\}$

Transition diagram for $\bar{A}$:

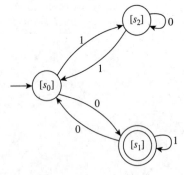

For A':

0-equivalence classes: $\{s_0', s_1', s_2'\}, \{s_3'\}$

1-equivalence classes: $\{s_0', s_2'\}, \{s_1'\}, \{s_3'\}$

2-equivalence classes: $\{s_0', s_2'\}, \{s_1'\}, \{s_3'\}$

Transition diagram for $\bar{A}'$:

Except for the labeling of the states, the transition diagrams for $\bar{A}$ and $\bar{A}'$ are identical. Hence $\bar{A}$ and $\bar{A}'$ accept the same language, and so, by Theorem 12.3.3, A and A' also accept the same language. Thus A and A' are equivalent automata.

9. For A:

0-equivalence classes: $\{s_1, s_2, s_4, s_5\}, \{s_0, s_3\}$

1-equivalence classes: $\{s_1, s_2\}, \{s_4, s_5\}, \{s_0, s_3\}$

2-equivalence classes: $\{s_1\}, \{s_2\}, \{s_4, s_5\}, \{s_0, s_3\}$

3-equivalence classes: $\{s_1\}, \{s_2\}, \{s_4, s_5\}, \{s_0, s_3\}$

Therefore, the states of $\bar{A}$ are the 3-equivalence classes of A.

For A':

0-equivalence classes: $\{s_2', s_3', s_4', s_5'\}, \{s_0', s_1'\}$

1-equivalence classes: $\{s_2', s_3', s_4', s_5'\}, \{s_0', s_1'\}$

Therefore, the states of $\bar{A}'$ are the 1-equivalence classes of A'.

According to the text, two automata are equivalent if, and only if, their quotient automata are isomorphic, provided inaccessible states have first been removed. Now A and A' have no inaccessible states, and $\bar{A}$ has four states, whereas $\bar{A}'$ has only two states. Therefore, A and A' are not equivalent.

This result can also be obtained by noting, for example, that the string 11 is accepted by A' but not by A.

11. *Partial answer:* Suppose A is a finite-state automaton with set of states S and relation R_* of $*$-equivalence of states. *[To show that R_* is an equivalence relation, we must show that R is reflexive, symmetric, and transitive.]*

Proof that R_* is symmetric:

[We must show that for all states s and t, if $s\,R_*t$ then $t\,R_*s$.] Suppose that s and t are any states of A such that $s\,R_*t$. *[We must show that $t\,R_*s$.]* Since $s\,R_*t$, then for every input string w,

$$\left[\begin{array}{c} N^*(s, w) \text{ is an} \\ \text{accepting state} \end{array}\right] \Leftrightarrow \left[\begin{array}{c} N^*(t, w) \text{ is an} \\ \text{accepting state} \end{array}\right],$$

where N^* is the eventual-state function on A. It follows from the symmetry of the $\Leftrightarrow$ relation that for every input string w,

$$\left[\begin{array}{c} N^*(t, w) \text{ is an} \\ \text{accepting state} \end{array}\right] \Leftrightarrow \left[\begin{array}{c} N^*(s, w) \text{ is an} \\ \text{accepting state} \end{array}\right].$$

Hence $t\,R_*s$ *[as was to be shown]*, and so R_* is symmetric.

12. Proof of Property (12.3.2): Suppose A is a finite-state automaton with set of states S and relation R_k of k-equivalence of states. We will show that R_k is reflexive, symmetric, and transitive.

    R_k **is reflexive**: Suppose that s is a state of A. It is certainly true that for every input string w of length less than or equal to k, $N^*(s, w)$ is an accepting state $\Leftrightarrow N^*(s, w)$ is an accepting state. So by definition of R_k, $s\ R_k\ s$.

    R_k **is symmetric**: [We must show that for all states s and t, if $s\ R_k\ t$ then $t\ R_k\ s$.] Suppose that s and t are any states of A such that $s\ R_k\ t$. [We must show that $t\ R_k\ s$.] Since $s\ R_k\ t$, then for every input string w of length less than or equal to k,

$$\begin{bmatrix} N^*(s, w) \text{ is an} \\ \text{accepting state} \end{bmatrix} \Leftrightarrow \begin{bmatrix} N^*(t, w) \text{ is an} \\ \text{accepting state} \end{bmatrix},$$

    where N^* is the eventual-state function on A. It follows from the symmetry of the $\Leftrightarrow$ relation that for every input string w of length less than or equal to k,

$$\begin{bmatrix} N^*(t, w) \text{ is an} \\ \text{accepting state} \end{bmatrix} \Leftrightarrow \begin{bmatrix} N^*(s, w) \text{ is an} \\ \text{accepting state} \end{bmatrix}.$$

    Hence $t\ R_k\ s$.

    R_k **is transitive**: Suppose that s, t, and u are states of A such that $s\ R_k\ t$ and $t\ R_k\ u$. By definition of R_k, for every input string w of length less than or equal to k, $N^*(s, w)$ is an accepting state $\Leftrightarrow N^*(t, w)$ is an accepting state and $N^*(t, w)$ is an accepting state $\Leftrightarrow N^*(u, w)$ is an accepting state. It follows by transitivity of the $\Leftrightarrow$ relation that $N^*(s, w)$ is an accepting state $\Leftrightarrow N^*(u, w)$ is an accepting state. Hence by definition of R_k, $s\ R_k u$.

    Conclusion: Since R_k is reflexive, symmetric, and transitive, R_k is an equivalence relation.

    Note: This proof is identical to the proof in the solution to exercise 11 except that every occurrence of "for each input string w" is replaced by "for each input string w of length less than or equal to k."

13. Proof of Property (12.3.3): By property (12.3.2), for each integer k, k-equivalence is an equivalence relation. Now by Theorem 10.3.4, the distinct equivalence classes of an equivalence relation form a partition of the set on which the relation is defined. In this case, the relation is defined on the set of all states of the automaton. So the k-equivalence classes form a partition of the set of all states of the automaton.

15. Proof of Property (12.3.5): Suppose k is an integer such that $k \geq 1$ and C_k is a k-equivalence class. We must show that there is a $k - 1$ equivalence class, C_{k-1}, such that $C_k \subseteq C_{k-1}$.

    By property (12.3.3), the $(k - 1)$-equivalence classes partition the set of all states of A in to a union of mutually disjoint subsets.

    Let s be any state in C_k. Then s is in *some* $(k - 1)$-equivalence class; call it C_{k-1}.

    Let t be any other state in C_k. [We will show that $t \in C_{k-1}$ also.] Then $t\ R_k\ s$, and so for every input string of length k, $N^*(t, w)$ is an accepting state $\Leftrightarrow N^*(s, w)$ is an accepting state.

    Since $k - 1 < k$, it follows that for every input string of length $k - 1$, $N^*(t, w)$ is an accepting state $\Leftrightarrow N^*(s, w)$ is an accepting state.

    Consequently, $t\ R_{k-1}\ s$, and so t and s are in the same $(k - 1)$-equivalence class.

    But $s \in C_{k-1}$. Hence $t \in C_{k-1}$ also. We, therefore, conclude that $C_k \subseteq C_{k-1}$.

17. *Hint*: If $m < k$, then every input string of length less than or equal to m has length less than or equal to k.

18. <u>Proof of Property (12.3.7)</u>: Suppose A is an automaton and C is a $*$-equivalence class of states of A.

    By Theorem 12.3.2, there is an integer $K \geq 0$ such that C is a K-equivalence class of A. Suppose C contains both an accepting state s and a nonaccepting state t of A.

    Since both s and t are in the same K-equivalence class, s is K-equivalent to t (by exercises 36 and 37 of Section 8.3), and so by exercise 17, s is 0-equivalent to t.

    But this is impossible because there are only two 0-equivalence classes, the set of all accepting states and the set of all nonaccepting states, and these two sets are disjoint.

    Hence the supposition that C contains both an accepting and a nonaccepting state is false: C consists entirely of accepting states or entirely of nonaccepting states.

19. *Hint:* Suppose two states s and t are equivalent. You must show that for any input symbol m, the next-states $N(s, m)$ and $N(t, m)$ are equivalent. To do this, use the definition of equivalence and the fact that for any string w', input symbol m, and state s, $N^*(N(s, m), w') = N^*(s, mw')$.

Review Guide: Chapter 12

Definitions: How are the following terms defined?

- alphabet, string over an alphabet, formal language over an alphabet
- Σ^n, Σ^* (the Kleene closure of Σ), and Σ^+ (the positive closure of Σ), where Σ is an alphabet
- concatenation of x and y, where x and y are strings
- concatenation of L and L', where L and L' are languages
- union of L and L', where L and L' are languages
- Kleene closure of L , where L is a language
- regular expression over an alphabet
- language defined by a regular expression
- character class
- finite-state automaton, next-state function
- language accepted by a finite-state automaton
- eventual-state function for a finite-state automaton
- regular language
- $*$-equivalence of states in a finite-state automaton
- k-equivalence of states in a finite-state automaton
- quotient automaton
- equivalent automata

Regular Expressions

- What is the order of precedence for the operations in a regular expression?
- How do you find the language defined by a regular expression?
- Given a language, how do you find a regular expression that defines the language?
- What are some practical uses of regular expressions?

Finite-State Automata

- How do you construct an annotated next-state table for a finite-state automaton given the transition diagram for the automaton?
- How do you construct a transition diagram for a finite-state automaton given its next-state table?
- How do you find the state to which a finite-state automaton goes if the characters of a string are input to it?
- How do you find the language accepted by a finite-state automaton?
- Given a simple formal language, how do you construct a finite-state automaton to accept the language?
- How can you use software to simulate the action of a finite-state automaton?
- What do the two parts of Kleene's theorem say about the relation between the language accepted by a finite-state automaton and the language defined by a regular expression?
- How can the pigeonhole principle be used to show that a language is not regular?
- How do you find the k-equivalence classes for a finite-state automaton?
- How do you find the $*$-equivalence classes for a finite-state automaton?
- How do you construct the quotient automaton for a finite-state automaton?
- What is the relation between the language accepted by a finite-state automaton and the language accepted by the corresponding quotient automaton?

Fill-in-the-Blank Review Questions

Section 12.1

1. If x and y are strings, the concatenation of x and y is ____.

2. If L and L' are languages, the concatenation of L and L' is ____.

3. If L and L' are languages, the union of L and L' is ____.

4. If L is a language, the Kleene closure of L is ____.

5. The set of regular expressions over a finite alphabet Σ is defined recursively. The base for the definition is the statement that ____. The recursion for the definition specifies that if r and s are any regular expressions in the set, then the following are also regular expressions in the set: ____, ____, and ____.

6. The function that associates a language to each regular expression over an alphabet Σ is defined recursively. The base for the definition is the statement that $L(\emptyset) = $ ____, $L(\lambda) = $ ____, and $L(a) = $ ____ for every $a \in \Sigma$. The recursion for the definition specifies that if $L(r)$ and $L(r')$ are the languages defined by the regular expressions r and r' over Σ, then $L(rr') = $ ____, $L(r \mid r') = $ ____, and $L(r^*) = $ ____.

7. The notation $[A - C]$ is an example of a ____ and denotes the regular expression ____.

8. Use of a single dot in a regular expression stands for ____.

9. The symbol $^\wedge$, placed at the beginning of a character class, indicates ____.

10. If r is a regular expression, the notation $r+$ denotes ____.

11. If r is a regular expression, the notation $r?$ denotes ____.

12. If r is a regular expression, the notation $r\{n\}$ means that ____ and the notation $r\{m, n\}$ means that ____.

Section 12.2

1. The five objects that make up a finite-state automaton are ____, ____, ____, ____, and ____.

2. The next-state table for an automaton shows the values of ____.

3. In the annotated next-state table, the initial state is indicated with an ____ and the accepting states are marked by ____.

4. A string w consisting of input symbols is accepted by a finite-state automaton A if, and only if, ____.

5. The language accepted by a finite-state automaton A is ____.

6. If N is the next-state function for a finite-state automaton A, the eventual-state function N^* is defined as follows: for each state s of A and for each string w that consists of input symbols of A, $N^*(s, w) = $ ____.

7. One part of Kleene's theorem says that given any language that is accepted by a finite-state automaton, there is ____.

8. The second part of Kleene's theorem says that given any language defined by a regular expression, there is ____.

9. A regular language is ____.

10. Given the language consisting of all strings of the form $a^k b^k$, where k is a positive integer, the pigeonhole principle can be used to show that the language is ____.

Section 12.3

1. Given a finite-state automaton A with eventual-state function N^* and given any states s and t in A, we say that s and t are *-equivalent if, and only if, ____.

2. Given a finite-state automaton A with eventual-state function N^* and given any states s and t in A, we say that s and t are k-equivalent if, and only if; _____.

3. Given states s and t in a finite-state automaton A, s is 0-equivalent to t if, and only if, either both s and t are _____ or both are _____. Moreover, for every integer $k \geq 1$, s is k-equivalent to t if, and only if, (1) s and t are $(k-1)$-equivalent and (2) _____.

4. If A is a finite-state automaton, then for some integer $K \geq 0$, the set of K-equivalence classes of states of A equals the set of _____-equivalence classes of A, and for all such K these are both equal to the set of _____.

5. Given a finite-state automaton A, the set of states of the quotient automaton $\bar{A}$ is _____.

Answers for Fill-In-the-Blank Review Questions

Section 12.1

1. the string obtained by writing all the characters of x followed by all the characters of y
2. $\{xy \mid x \in L \text{ and } y \in L'\}$
3. $\{s \mid s \in L \text{ or } s \in L'\}$
4. $\{t \mid t \text{ is a concatenation of any finite number of strings in } L\}$
5. $\emptyset$, λ, and each individual symbol in Σ are regular expressions over Σ; (rs); $(r \mid s)$; (r^*)
6. $\emptyset$; $\{\lambda\}$; $\{a\}$; $L(r)L(r')$; $L(r) \cup L(r')$; $(L(r))^*$
7. character class; $(A \mid B \mid C)$
8. an arbitrary character
9. a character of the same type as those in the range of the class is to occur at that point in the string except for one of the specific characters indicated after the $^\wedge$ sign
10. the concatenation of r with itself any positive finite number of times
11. $(\lambda \mid r)$
12. the concatenation of r with itself exactly n times; the concatenation of r with itself anywhere from m through n times

Section 12.2

1. a finite set of input symbols; a finite set of states; a designated initial state; a designated set of accepting states; a next-state function that associates a "next-state" with each state and input symbol of the automaton
2. the next-state function for each state and input symbol of the automaton
3. arrow; double circles
4. when the symbols in the string are input to the automaton in sequence from left to right, starting from the initial state, the automaton ends up in an accepting state
5. the set of strings that are accepted by A
6. the state to which A goes if it is in state s and the characters of w are input to it in sequence
7. a regular expression that defines the same language
8. a finite-state automaton that accepts the same language
9. a language defined by a regular expression (*Or:* a language accepted by a finite-state automaton)
10. not regular

Section 12.3

1. for every input string w, either $N^*(s,w)$ and $N^*(t,w)$ are both accepting states or both are nonaccepting states

2. for every input string w of length less than or equal to k, either $N^*(s,w)$ and $N^*(t,w)$ are both accepting states or both are nonaccepting states

3. accepting states, nonaccepting states; for any input symbol m, $N(s,m)$ and $N(t,m)$ are also $(k-1)$-equivalent

4. $(K+1)$; $*$-equivalence classes of states of A

5. the set of $*$-equivalence classes of states of A

Tips for Success with Proofs and Disproofs

Make sure your proofs are genuinely convincing. Write in complete sentences. Express yourself carefully and completely – but concisely! If you are confident that both you and all your fellow students would understand the reason for a step, you could omit it or just refer to it in the briefest way. If you are not sure about this, include the reason, but try to find a balance between using more words than necessary and using enough to make sure that everyone understands.

Disproof by Counterexample

- To disprove a universal statement, give a counterexample.
- Write the word "Counterexample" at the beginning of a counterexample.
- Write counterexamples in complete sentences.
- Give values of the variables that you believe show the property is false.
- Include the computations that prove beyond any doubt that these values really do make the property false.

All Proofs

- Write the word "Proof" at the beginning of a proof.
- Write proofs in complete sentences.
- Start each sentence with a capital letter and finish with a period.

Direct Proof

- Begin each direct proof with the word "Suppose."
- In the "Suppose" sentence:
  - Introduce a variable or variables (indicating the general set they belong to - e.g., integers, real numbers etc.), and
  - Include the hypothesis that the variables satisfy.
- Identify the conclusion that you will need to show in order to complete the proof.
- Reason carefully from the "suppose" to the "conclusion to be shown."
- Include the little words (like "Then," "Thus," "So," "It follows that") that make your reasoning clear.
- Give a reason to support each assertion you make in your proof.

Proof by Contradiction

- Begin each proof by contradiction by writing "Suppose not. That is, suppose...," and continue this sentence by carefully writing the negation of the statement to be proved.
- After you have written the "suppose," you need to show that this supposition leads logically to a contradiction.
- Once you have derived a contradiction, you can conclude that the thing you supposed is false. Since you supposed that the given statement was false, you now know that the given statement is true.

Proof by Contraposition

- Look to see if the statement to be proved is a universal conditional statement.
- If so, you can prove it by writing a direct proof of its contrapositive.

Some Mathematical Conventions

1. When introducing a new variable into a discussion, the convention is to place the new variable to the left of the equal sign and the expression that defines it to the right. This convention is identical to the one used in computer programming. For example, in a computer program, if a and b have previously been defined, and you want to assign the value of $a + b$ to a new variable s, you would write something like

$$s := a + b.$$

Similarly, in a mathematical proof, if a and b have previously been introduced into a discussion, and you want to let s be their sum,

instead of writing "Let $a + b = s$," you should write, "Let $s = a + b$."

2. It is considered good mathematical writing to avoid starting a sentence with a variable. That is one reason that mathematical writing frequently uses words and phrases such as Then, Thus, So, Therefore, It follows that, Hence, etc. For example, in a proof that any sum of even integers is even, instead of writing,

By definition of even, $m = 2a$ and $n = 2b$ for some integers a and b.

$$m + n = 2a + 2b \ldots$$

write

By definition of even, $m = 2a$ and $n = 2b$ for some integers a and b.
Then

$$m + n = 2a + 2b \ldots$$

The fact that $m + n = 2a + 2b$ is a consequence of the facts that $m = 2a$ and $n = 2b$. Including the word "Then" in your proof alerts your reader to this reasoning.

3. Standard mathematical writing avoids repeating the left-hand side in a sequence of equations in which the left-hand side remains constant. For example, if $n = 5q + 4$, instead of writing

$$
\begin{aligned}
n^2 &= (5q + 4)^2 \\
n^2 &= 25q^2 + 40q + 16 \\
n^2 &= 25q^2 + 40q + 15 + 1 \\
n^2 &= 5(5q^2 + 8q + 3) + 1
\end{aligned}
$$

all the n^2 except the first are omitted and each subsequent equal sign is read as "which equals," as shown below:

$$
\begin{aligned}
n^2 &= (5q + 4)^2 \\
&= 25q^2 + 40q + 16 \\
&= 25q^2 + 40q + 15 + 1 \\
&= 5(5q^2 + 8q + 3) + 1
\end{aligned}
$$

4. Respecting the equal sign is one of the most important mathematical conventions. An equal sign should only be used between quantities that are equal, not as a substitute for words like "is," or "means that," or "if and only if," or $\Leftrightarrow$, or "is equivalent to." For example, if $a = 4$ and $b = 12$, students occasionally explain that a divides b by writing:

$$a \mid b = 4 \mid 12 \text{ since } 12 = 4 \cdot 3.$$

When read out loud, this becomes,

"a divides b equals 4 divides 12 since 12 equals 4 times 3," which makes no sense. A correct version uses the words "if, and only if," or the symbol $\Leftrightarrow$, as in

$$a \mid b \Leftrightarrow 4 \mid 12, \text{ which is true because } 12 = 4 \cdot 3$$

or it can be expressed as

$$a \mid b \text{ because } 4 \mid 12 \text{ since } 12 = 4 \cdot 3.$$

5. It is unnecessary, and even risky, to place full statements of definitions and theorems inside the bodies of proofs. The reason is that the variables used to express them can become confused with variables that are introduced as steps in the proof. So instead of including the statement of the definition of divisibility, for example, just write, "by definition of divisibility." Similarly, instead of including the statement of, say, Theorem 8.4.3, just write, "by Theorem 8.4.3." For instance, to prove that a sum of any even integer plus any odd integer is odd, someone might write the following:

> Suppose m is any even integer and n is any odd integer.
> For an integer to be even means that it equals $2k$ for some integer k, and
> for an integer to be odd, means that it equals $2k + 1$ for some integer k.
> Thus $m + n = 2k + (2k + 1) = 4k + 1\ldots$

The problem is that although the letter k appears in the statements of the definitions in the text, it refers to a different quantity in each one. However, when the statements are combined together in the proof, the letter k can have only one interpretation. The result is that the argument in the "proof" only applies to an even integer and the next successive odd integer, not to *any* even integer and *any* odd integer.

Find the Mistake

All of the following problems contain a mistake. Identify and correct each one.

1. **Section 2.2**: The negation of "$1 < a < 5$" is "$1 \geq a \geq 5$."

2. **Section 2.2**: "P only if Q" means "if Q then P."

3. **Section 3.2**
   (a) The negation of "For every real number x, if $x > 2$ then $x^2 > 4$" is "For every real number x, if $x > 2$ then $x^2 \leq 4$."
   (b) The negation of "For every real number x, if $x > 2$ then $x^2 > 4$" is "There exist real numbers x such that if $x > 2$ then $x^2 \leq 4$."
   (c) The negation of "For every real number x, if $x > 2$ then $x^2 > 4$" is "There exists a real number x such that $x > 2$ and $x^2 < 4$."

4. **Section 3.2**: The contrapositive of "For every real number x, if $x > 2$ then $x^2 > 4$" is "For every real number x, if $x \leq 2$ then $x^2 \leq 4$."

5. **Section 3.3**: Statement: $\exists$ a real number x such that $\forall$ real numbers y, $x + y = 0$. Proposed negation: $\forall$ real number x, if y is a real number then $x + y \neq 0$.

6. **Section 4.2**: A person is asked to prove that the square of any odd integer is odd. Toward the end of a proof the person writes: "Therefore $n^2 = 2k + 1$, which is the definition of odd."

7. **Section 4.2**: *Prove:* The square of any even integer is even.
   Beginning of proof: Suppose that r is any integer. Then if m is any even integer, $m = 2r$.

8. **Section 4.2**: *Prove directly from the definition of even:* For every even integer n, $(-1)^n = 1$.
   Beginning of proof: Suppose n is any even integer. Then $n = 2r$ for some integer r. By substitution, $(-1)^n = (-1)^{2r} = 1$ because $2r$ is even....

9. **Section 4.2**: *Prove directly from the definition of even:* For every even integer n, $(-1)^n = 1$.
   Beginning of proof: Suppose n is any even integer. Then $n = 2r$ for some integer r. By substitution, $(-1)^{2r} = ((-1)^2)^r$....

10. **Section 4.4**: *Prove:* For all integers a and b, if a and b are divisible by 3 then $a + b$ is divisible by 3.
    Beginning of proof: Suppose that for all integers a and b, if a and b are divisible by 3 then $a + b$ is divisible by 3.

11. **Section 4.4**: *Prove:* For every integer a, if 3 divides a, then 3 divides a^2.
    Beginning of proof: Suppose a is any integer such that 3 divides a. Then $a = 3k$ for any integer k.

12. **Section 4.4**: *Prove:* For every integer a, if $a = 3b + 1$ for some integer b, then $a^2 - 1$ is divisible by 3.
    Beginning of proof:
    (1) Let a be any integer such that $a = 3b + 1$ for some integer b.
    (2) We will prove that $a^2 - 1$ is divisible by 3.
    (3) Saying that $a^2 - 1$ is divisible by 3 means that $a^2 - 1 = 3q$ for some integer q.
    (4) Then $(3b + 1)^2 - 1 = 3q$.
    (5) Since q is an integer, by definition of divisibility, $a^2 - 1$ is divisible by 3.

13. **Section 4.4**: *Prove:* For all integers a, b, and c, if a divides b and b divides c, then a divides c.

    Beginning of proof: Suppose that for all integers a, b, and c, a divides b and b divides c. We must show that a divides c.

14. **Section 4.5**: *Prove:* For every integer a, $a^2 - 2$ is not divisible by 3.

    Beginning of proof: Suppose a is any integer. By the quotient-remainder theorem with divisor $d = 3$, there exist unique integers q and r such that $a = 3q + r$, where $0 < r \leq 3$.

15. **Section 4.7**: *Prove by contradiction:* The product of any irrational number and any rational number is irrational.

    Beginning of proof: Suppose not. That is, suppose the product of any irrational number and any rational number is rational.

16. **Section 4.7**: The negation of "n is not divisible by any prime number greater than 1 and less than or equal to $\sqrt{n}$" is "n is divisible by any prime number greater than 1 and less than or equal to $\sqrt{n}$."

17. **Section 5.2**: The equation $1 + 2 + 3 + \cdots + n = \dfrac{n(n+1)}{2}$ is true for $n = 1$ because $1 + 2 + 3 + \cdots + 1 = \dfrac{1(1+1)}{2}$ is true.

18. **Section 5.2**: The equation $1 + 2 + 3 + \cdots + n = \dfrac{n(n+1)}{2}$ is true for $n = 1$ because

$$1 = \frac{1(1+1)}{2} \Rightarrow 1 = \frac{2}{2} \Rightarrow 1 = 1.$$

19. **Section 5.2**: *Prove by mathematical induction:* For all integers $n \geq 1$,

$$1 + 2 + 3 + \cdots + n = \frac{n(n+1)}{2}.$$

    Beginning of proof: Let the property $P(n)$ be

$$1 + 2 + 3 + \cdots + n = \frac{n(n+1)}{2} \text{ for every integer } n \geq 1.$$

20. **Section 5.2**: *Prove by mathematical induction:* For every integer $n \geq 1$,

$$1 \cdot 2 + 2 \cdot + 3 \cdot 4 + \cdots + (n-1) \cdot n = \frac{n(n-1)(n+1)}{3}.$$

    Proof of the inductive step:

    Suppose k is any integer such that $k \geq 1$ and

$$(k-1) \cdot k = \frac{k(k-1)(k+1)}{3}.$$

    We must show that

$$((k+1)-1) \cdot (k+1) = \frac{(k+1)((k+1)-1)((k+1)+1)}{3}.$$

21. **Section 6.1**: Given sets A and B, to show that A is a subset of B, we must show that there is an element x such that x is in A and x is in B.

22. **Section 6.1**: Given sets A and B, to show that A is a subset of B, we must show that for every x, x is in A and x is in B.

23. **Section 7.2**: To prove that $F\colon A \to B$ is one-to-one, assume that if $F(x_1) = F(x_2)$ then $x_1 = x_2$.

24. **Section 7.2**: To prove that $F\colon A \to B$ is one-to-one, we must show that for all x_1 and x_2 in A, $F(x_1) = F(x_2)$ and $x_1 = x_2$.

25. **Section 8.2**: Let R be the relation defined on the set of all integers by $a\,R\,b$ if, and only if, $ab > 0$. Prove that R is symmetric.
    Beginning of proof: To show that R is symmetric, assume that for all integers a and b, $a\,R\,b$. We will show that $b\,R\,a$.

Find the Mistake: Solutions

All of the following problems contain a mistake. Identify and correct each one.

1. **Section 2.2**: The negation of "$1 < a < 5$" is "$1 \geq a \geq 5$."

   Answer: A statement of the form "$1 < a < 5$" is an *and* statement.
   Thus, by De Morgan's law, its negation is an *or* statement.
   Therefore, the correct negation is "$1 \geq a$ or $a \geq 5$."

2. **Section 2.2**: "P only if Q" means "if Q then P."

   Answer: "P only if Q" means that the only way P can occur is for Q to occur.
   In other words, if Q does not occur, then P cannot occur, or, equivalently, "if P occurs then Q must have occurred."
   So "P only if Q" means "if P then Q."

3. **Section 3.2**
   (a) The negation of "For every real number x, if $x > 2$ then $x^2 > 4$" is "For every real number x, if $x > 2$ then $x^2 \leq 4$."

   (b) The negation of "For every real number x, if $x > 2$ then $x^2 > 4$" is "There exist real numbers x such that if $x > 2$ then $x^2 \leq 4$."

   (c) The negation of "For every real number x, if $x > 2$ then $x^2 > 4$" is "There exists a real number x such that $x > 2$ and $x^2 < 4$."

   Answer to a, b, and c: All three proposed negations are incorrect. The negation of a "For every" statement is a "There exists" statement, the negation of "if p then q" is "p and not q," and the negation of "$x^2 > 4$" is "$x^2 \leq 4$."
   So a correct negation in all three cases is "There exists a real number x such that $x > 2$ and $x^2 \leq 4$."

4. **Section 3.2**: The contrapositive of "For every real number x, if $x > 2$ then $x^2 > 4$" is "For every real number x, if $x \leq 2$ then $x^2 \leq 4$."

   Answer: The contrapositive of "if p then q" is "if not q then not p."
   In this case p is $x > 2$ and q is $x^2 > 4$.
   Thus a correct answer is "For every real number x, if $x^2 \leq 4$ then $x \leq 2$."

5. **Section 3.3**: Statement: $\exists$ a real number x such that $\forall$ real numbers y, $x + y = 0$. Proposed negation: $\forall$ real number x, if y is a real number then $x + y \neq 0$.

   Answer: The proposed negation began correctly with "$\forall$ real number x," but the correct continuation is an existential statement.
   Thus a correct answer is, "$\forall$ real number x, $\exists$ a real number y such that $x + y \neq 0$."

6. **Section 4.2**: A person is asked to prove that the square of any odd integer is odd. Toward the end of a proof the person writes: "Therefore $n^2 = 2k + 1$, which is the definition of odd."

   Answer: For an integer to be odd means that it equals 2 times some integer plus 1.
   So it is not correct to say that "$2k + 1$ *is* the definition of odd."
   The person should have written something like, "Therefore $n^2 = 2k + 1$, where k is an integer, and so n^2 is odd by definition of odd."

7. **Section 4.2**: *Prove:* The square of any even integer is even.

   Beginning of proof: Suppose that r is any integer. Then if m is any even integer, $m = 2r$.

   Answer: To prove that the square of any even integer is even, you must start by supposing you have a *[particular but arbitrarily chosen]* even integer.

By using the definition of even, you can *deduce* what the even integer must look like, namely that it must equal $2 \cdot$ (some integer).

A correct proof would start with an even integer m and deduce the existence of an integer r such that $m = 2r$.

This "proof" has it backwards.

8. **Section 4.2**: *Prove directly from the definition of even: For every even integer n, $(-1)^n = 1$.*

   Beginning of proof: Suppose n is any even integer. Then $n = 2r$ for some integer r. By substitution, $(-1)^n = (-1)^{2r} = 1$ because $2r$ is even....

   Answer: By claiming that $(-1)^{2r} = 1$, this "proof" assumes that (-1) raised to an even power equals 1.

   Thus this proof assumes what is to be proved.

9. **Section 4.2**: *Prove directly from the definition of even: For every even integer n, $(-1)^n = 1$.*

   Beginning of proof: Suppose n is any even integer. Then $n = 2r$ for some integer r. By substitution, $(-1)^{2r} = ((-1)^2)^r$....

   Answer: The reason given, "by substitution," doesn't apply here. When we write "by substitution," we need to show what is being substituted for what. The actual reason for the given equation is the following property of exponents: For every real number b and for all integers m and n, $(b^m)^n = b^{mn}$.

   The following revision would be correct:
   Prove directly from the definition of even: For all even integers n, $(-1)^n = 1$.
   Beginning of proof: Suppose n is any even integer. Then $n = 2r$ for some integer r, and so

   $$\begin{aligned} (-1)^n &= (-1)^{2r} & \text{by substitution} \\ &= ((-1)^2)^r & \text{by a property of exponents...} \end{aligned}$$

10. **Section 4.4**: *Prove: For all integers a and b, if a and b are divisible by 3 then $a+b$ is divisible by 3.*

    Beginning of proof: Suppose that for all integers a and b, if a and b are divisible by 3 then $a + b$ is divisible by 3.

    Answer: This proof begins by assuming exactly what is to be proved.
    If one assumes what is to be proved, there is nothing left to do!

11. **Section 4.4**: *Prove: For every integer a, if 3 divides a, then 3 divides a^2.*

    Beginning of proof: Suppose a is any integer such that 3 divides a. Then $a = 3k$ for any integer k.

    Answer: It is incorrect to write, "$a = 3k$ for any integer k."
    The reason is that in order to complete the proof k is not just "any" integer: k is actually $a/3$. A correct version is, "Then $a = 3k$ for some integer k" or "Then there is an integer k such that $a = 3k$."

12. **Section 4.4**: *Prove: For every integer a, if $a = 3b + 1$ for some integer b, then $a^2 - 1$ is divisible by 3.*

    Beginning of proof:
    (1) Let a be any integer such that $a = 3b + 1$ for some integer b.
    (2) We will prove that $a^2 - 1$ is divisible by 3.
    (3) Saying that $a^2 - 1$ is divisible by 3 means that $a^2 - 1 = 3q$ for some integer q.
    (4) Then $(3b + 1)^2 - 1 = 3q$.
    (5) Since q is an integer, by definition of divisibility, $a^2 - 1$ is divisible by 3.

    Answer: This "proof" assumes something equivalent to what is to be proved.
    After stating "We will prove that $a^2 - 1$ is divisible by 3" it is correct to state that "Saying

that $a^2 - 1$ is divisible by 3 means that $a^2 - 1 = 3q$ for some integer q."
However, Statement (5) in the proof assumes that the integer q has been shown to exist, but that is not the case.

13. **Section 4.4**: *Prove:* For all integers a, b, and c, if a divides b and b divides c, then a divides c.

    Beginning of proof: Suppose that for all integers a, b, and c, a divides b and b divides c. We must show that a divides c.

    Answer: Supposing that "for all integers a, b, and c, a divides b and b divides c " is the same as supposing that no matter what three integers you pick, the first will divide the second and the second will divide the third.
    This is false.
    For instance, if you pick $a = 2$, $b = 3$, and $c = 5$, it is not true that 2 divides 3 and it is also not true that 3 divides 5.
    A correct proof would begin, "Suppose that a, b, and c, are *[particular but arbitrarily chosen]* integers such that a divides b and b divides c. We must show that a divides c.

14. **Section 4.5**: *Prove:* For every integer a, $a^2 - 2$ is not divisible by 3.

    Beginning of proof: Suppose a is any integer. By the quotient-remainder theorem with divisor $d = 3$, there exist unique integers q and r such that $a = 3q + r$, where $0 < r \leq 3$.

    Answer: The inequality is incorrect.
    The correct inequality from the quotient-remainder theorem is $0 \leq r < 3$.

15. **Section 4.7**: *Prove by contradiction:* The product of any irrational number and any rational number is irrational.

    Beginning of proof: Suppose not. That is, suppose the product of any irrational number and any rational number is rational.

    Answer: A proof by contradiction should start with the negation of the statement to be proved.
    In this case, the statement to be proved is universal, and so its negation is existential. However, this proposed proof begins with a universal statement. A correct way to begin the proof is the following:
    Beginning of proof: Suppose not. That is, suppose there exists an irrational number and a rational number whose product is rational.

16. **Section 4.7**: The negation of "n is not divisible by any prime number greater than 1 and less than or equal to $\sqrt{n}$" is "n is divisible by any prime number greater than 1 and less than or equal to $\sqrt{n}$."

    Answer: Consider negating the statement "He *is not accused by any* person."
    The negation is not "He *is accused by any* person."
    The negation is "He *is accused by some* person."
    Similarly, the negation of "n *is not divisible by any* prime number greater than 1 and less than or equal to $\sqrt{n}$" is not "n *is divisible by any* prime number greater than 1 and less than or equal to $\sqrt{n}$."
    It is "n *is divisible by some* prime number greater than 1 and less than or equal to $\sqrt{n}$," or "There exists a prime number greater than 1 and less than or equal to $\sqrt{n}$ that divides n."

17. **Section 5.2**: The equation $1 + 2 + 3 + \cdots + n = \dfrac{n(n+1)}{2}$ is true for $n = 1$ because $1 + 2 + 3 + \cdots + 1 = \dfrac{1(1+1)}{2}$ is true.

    Answer: When $n = 1$, the expression $1 + 2 + 3 + \cdots + n$ equals 1.
    It is confusing to write it as $1 + 2 + 3 + \cdots + 1$.

18. **Section 5.2**: The equation $1 + 2 + 3 + \cdots + n = \dfrac{n(n+1)}{2}$ is true for $n = 1$ because

$$1 = \frac{1(1+1)}{2} \Rightarrow 1 = \frac{2}{2} \Rightarrow 1 = 1.$$

Answer: A false statement can imply a true conclusion.
So being able to deduce a true conclusion from a given statement, is not a valid way to prove that the given statement is true.

19. **Section 5.2**: *Prove by mathematical induction:* For every integer $n \geq 1$,

$$1 + 2 + 3 + \cdots + n = \frac{n(n+1)}{2}.$$

Beginning of proof: Let the property $P(n)$ be

$$1 + 2 + 3 + \cdots + n = \frac{n(n+1)}{2} \text{ for every integer } n \geq 1.$$

Answer: The job of a proof by mathematical induction is to prove that a given property is true for all integers greater than or equal to a given integer.
In this example, the property $P(n)$ is simply the equation

$$1 + 2 + 3 + \cdots + n = \frac{n(n+1)}{2}$$

without the words "for every integer $n \geq 1$."
Adding the words "for every integer $n \geq 1$" as part of $P(n)$ makes $P(n)$ identical with what is to be proved.
So the inductive step in the proof would assume what is to be proved.

20. **Section 5.2**: *Prove by mathematical induction:* For every integer $n \geq 1$,

$$1 \cdot 2 + 2 \cdot 3 + 3 \cdot 4 + \cdots + (n-1) \cdot n = \frac{n(n-1)(n+1)}{3}.$$

Proof of the inductive step:

Suppose k is any integer such that $k \geq 1$ and

$$(k-1) \cdot k = \frac{k(k-1)(k+1)}{3}.$$

We must show that

$$((k+1) - 1) \cdot (k+1) = \frac{(k+1)((k+1) - 1)((k+1) + 1)}{3}.$$

Answer: In this example, the property $P(n)$ is the equation

$$1 \cdot 2 + 2 \cdot 3 + 3 \cdot 4 + \cdots + (n-1) \cdot n = \frac{n(n-1)(n+1)}{3}.$$

In the inductive step $P(k)$ and $P(k+1)$ are obtained by substituting k and $k+1$ in place of n in $P(n)$:

$$P(k) \text{ is } 1 \cdot 2 + 2 \cdot 3 + 3 \cdot 4 + \cdots + (k-1) \cdot k = \frac{k(k-1)(k+1)}{3},$$

and

$$P(k+1) \text{ is } 1 \cdot 2 + 2 \cdot 3 + 3 \cdot 4 + \cdots + ((k+1) - 1) \cdot (k+1) = \frac{(k+1)((k+1) - 1)((k+1) + 1)}{3}.$$

The left-hand sides of both $P(k)$ and $P(k+1)$ are summations, not just the final term of a summation.

21. **Section 6.1**: Given sets A and B, to show that A is a subset of B, we must show that there is an element x such that x is in A and x is in B.

    Answer: This answer implies that for A to be a subset of B, it is enough for there to be a single element that is in both sets.
    But this is false.
    For instance, if $A = \{1, 2\}$ and $B = \{2, 3\}$, then 2 is in both A and B, but A is not a subset of B because 1 is in A and 1 is not in B.
    In fact, for A to be a subset of B means that for every x, *if* x is in A *then* x must be in B.

22. **Section 6.1**: Given sets A and B, to show that A is a subset of B, we must show that for every x, x is in A and x is in B.

    Answer: There are two problems with this answer.
    One is that it implies that A and B are identical sets, whereas for A to be a subset of B it is possible for B to contain elements that are not in A.
    In addition, because no domain is specified for x, it appears to say that everything in the universe is in both A and B, which is not the case for most sets A and B.

23. **Section 7.2**: To prove that $F: A \to B$ is one-to-one, assume that if $F(x_1) = F(x_2)$ then $x_1 = x_2$.

    Answer: Assuming that "if $F(x_1) = F(x_2)$ then $x_1 = x_2$" is essentially the same as assuming that F is one-to-one.
    In other words, it essentially assumes what needs to be proved.

24. **Section 7.2**: To prove that $F: A \to B$ is one-to-one, we must show that for all x_1 and x_2 in A, $F(x_1) = F(x_2)$ and $x_1 = x_2$.

    Answer: This statement implies that for all x_1 and x_2 in A, $x_1 = x_2$.
    In other words, it implies that there is only one element in A, which is very seldom the case.

25. **Section 8.2**: Let R be the relation defined on the set of all integers by $a\,R\,b$ if, and only if, $ab > 0$. Prove that R is symmetric.
    Beginning of proof: To show that R is symmetric, assume that for all integers a and b, $a\,R\,b$. We will show that $b\,R\,a$.

    Answer: The problem with these statements is that saying "assume that for all integers a and b, $a\,R\,b$" is equivalent to saying that every integer is related to every other integer by R. This is not the case.
    For instance, -1 is not related to 1 because $(-1) \cdot 1 = -1$ and $-1 \not> 0$.